Rockets, Missiles, and Men in Space

ALSO BY WILLY LEY

Exotic Zoology

Engineers' Dreams

Watchers of the Skies

The Conquest of Space
with paintings by Chesley Bonestell

Beyond the Solar System
with paintings by Chesley Bonestell

COLLABORATIONS

The Exploration of Mars
with Wernher von Braun

Across the Space Frontier
with Wernher von Braun, *et al.*

Conquest of the Moon
with Wernher von Braun, *et al.*

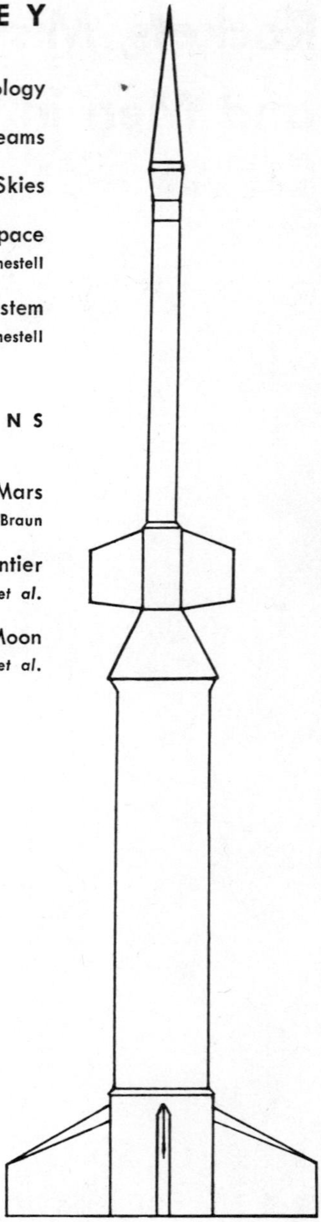

Rockets, Missiles, and Men in Space

BY WILLY LEY

NEW YORK / THE VIKING PRESS

Rockets, Missiles, and Men in Space
First published in 1968 by The Viking Press, Inc.
625 Madison Avenue, New York, N.Y. 10022

Published simultaneously in Canada by
The Macmillan Company of Canada Limited

Library of Congress catalog card number: 67–20676

Printed in U.S.A.

Published originally by The Viking Press, Inc., as *Rockets* (1944), this book has undergone several major revisions issued under the titles *Rockets and Space Travel* (1947) and *Rockets, Missiles, and Space Travel* (1951, rev. ed. 1957, 2nd rev. ed. 1961). Together they had twenty printings before the printing of this newly revised and expanded edition.

The new version of my book
is dedicated to
the space explorers of the next generation
who will want to know
what their fathers thought and did

Foreword, 1967

Once upon a time this book was called *Rockets* and its length was 288 pages. In May 1944, when it was first published, it had ten chapters. The first six were historical; the others predicted rockets for meteorological research, rockets capable of rising beyond the atmosphere, manned spaceships, and a manned artificial satellite.

A second printing appeared in September 1944, the month the V-2s began falling on London. For a while the book could limp along with "notes" tacked on in the appendix. But then new projects started and the book had to be rewritten. The new edition appeared in 1947 with the expanded title *Rockets and Space Travel.* During the next three years wartime secrets, both American and German, became known and another rewriting was necessary; this time the title became *Rockets, Missiles, and Space Travel.*

Then the Space Age dawned and, as a consequence, here is a new version with still another title change. But the interesting and important development is not that the book grew from 304 to 576 pages. In the original, 157 pages were history; the others, less than half of the whole, were prediction. With each new version the percentage changed: more and more was history and less and less prediction. What you are holding now is virtually all history; the amount of prediction that remains is negligible and what is still in the prediction stage is already a "project," with time tables, organization charts, and everything that goes with a proper project.

Some people may feel that this is a sad state of affairs—it was so nice to dream. But there is no reason for regret. When all the current projects have been carried out they will form a firm basis on which to build still another set of dreams.

WILLY LEY

Jackson Heights, N.Y.
October 2, 1967

Contents

Illustrations

PLATES

(Grouped following page 174)

FIGURES

Rockets, Missiles, and Men in Space

1 The Beginnings of an Idea

There are four several ways whereby this flying in the air hath been or may be attempted. Two of them by the strength of other things, and two of them by our own strength: (1) By spirits, or angels. (2) By the help of fowls. (3) By wings fastened immediately to the body. (4) By a flying chariot.

—Bishop John Wilkins, *Mathematicall Magick* (1648)

The story of an idea is, of necessity, the story of many things. Ideas, like large rivers, never have just one source. Just as the water of a river near its mouth is composed largely of the waters of many tributaries, so an idea, in its final form, is composed largely of later additions. And because this is so, it is often difficult to find the source of a river or the beginnings of an idea.

Luckily the idea of space travel, of trips away from the earth, has to have a mainstream which can be recognized without fail. One cannot conceive of a trip away from the earth unless there exists a previous concept of other worlds. And it so happens that the gradual rise of the concept of other worlds in addition to our own earth took place in historical times.

While it is impossible to put a date on the first appearance of that concept, we can pick a number of instances from history and declare that at that time, and in that place, the idea of space travel could not possibly exist because the astronomical foundation was still lacking. We can say, for example, that space travel is a thought that could not have existed in ancient Babylon.

We know that the Babylonians, by means of patient and long observation of the sky in a favorable climate, possessed an extensive knowledge of the apparent motions of the heavenly bodies. A modern man who, looking out of his window, notices a reddish star near the horizon, may wonder whether it might not be Mars. In thinking that, he will remember that Mars is a planet farther away from the sun than the earth, that it is smaller than the earth, and that it takes about two of our years to revolve around the sun. He will wonder whether it might, after all, be "inhabited."

The train of thought of an astronomer-priest of ancient Babylon was different. He would have known at once that the red star *was* Mars, in fact he would have expected it. He would have been able to trace its apparent movement across the sky; he would have known after how many months and days it would reappear and where.

But Mars, to the priest, was not the next planet outside the orbit of the earth, having about two-thirds the diameter of the earth and moving along its orbit at a steadily varying distance. Mars, to the priest, was not even a solid body. To him it was the abode of a certain god and his knowledge of that abode, as well as his *duty,* consisted in knowing the star's movements. Nothing else. There is not the smallest indication that the Babylonians, who possessed a patiently acquired and enormous knowledge of the movements of the sun, the moon, and the planets among the so-called fixed stars, ever attempted to determine their distance or their properties. It may be that they felt that such inquiries should not be made about the abodes of the gods; in any event, they never tried. Their astronomical knowledge was extensive and surprisingly reliable as far as the apparent movements of the lights in the sky went. Aside from that, their astronomical knowledge was nil and their conceptions were childish. The absence of the idea that there were other worlds, somehow comparable in size to the size of the earth, precluded all further knowledge.

Ancient Chinese astronomy exhibits the same limitation. The ancient Chinese astronomers, again by patient observation, learned to recognize certain cycles. They knew what lights would be in the sky at what time and where. They predicted eclipses. They even invented a number of astronomical instruments—not telescopes, but instruments that are a great help in ascertaining the position of a planet or a star. But they believed that the earth was flat and they did not even guess that the lights in the sky which they observed so diligently might be other worlds.

Similar beliefs were held in ancient Greece. In addition to the straight statements of later Greek writers that their ancestors once believed the earth to be flat and surrounded by the ocean, we have an interesting bit of literary evidence. It can be derived from the great epic which everybody at least knows by name, the *Odyssey,* more particularly from books IX to XII, which the Frenchman Victor Bérard has called the "Tales of Odysseus at the Court of King Alkinoös." In that narrative, you remember, "everything" happens to Odysseus of Ithaka. The story is a collection of all the adventures which tradition and mythology made "possible": sirens and goddesses, man-eating giants and sea monsters, storms and whirlpools. Because everything "possible" happened to Odysseus, what did *not* happen to him is, in retrospect, just as important. His ship, for example, did not run into the Mountain of Lodestone; it did not find itself retarded by the "jellied sea"; the men did not kindle a fire on the back of a dozing *Kraken.* Odysseus did not meet the "Flying Dutchman." Naturally not, because none of these legends existed in, say, 800 B.C. Nor was Odysseus's ship caught by a storm and carried to the moon. For the moon was still regarded as a silver shield in the sky, its spots probably being the reflections of the land and water below. If one could only see them more clearly and if one could ascertain one's own position, one might be able to study terrestrial geography by watching the moon.

But Greek astronomy, unlike Babylonian and Chinese, finally overcame the flat-earth concept. It began to go its own way around the year 540 B.C., a century after a Babylonian astronomer had founded a school on the island of Cos. Thales of Miletos, who died in 548 B.C. and who is often called the Father of Greek Astronomy, may have been a pupil of that Babylonian. Of course he may have been a pupil of a pupil, and in any event we do not really know much about what was taught in that school on Cos. One of the things taught was probably the *saros,* but that word did not mean then what it does today. Now it means a cycle of 223 lunations, or 18 years and 11 days. After such a cycle the moon returns to its original position in the sky, as far as the naked eye can tell. But this use of the word goes back only to Dr. Edmond Halley. In the time of Thales a saros meant 3600 years.

Then came Pythagoras of Samos. Mainly because he traveled a great deal in Egypt and in the East, he learned to identify "Hesperos" and "Phosphoros" as the same planet, the one we call Venus. He recognized the obliquity of the ecliptic and grew used to the concept of earth as a sphere freely poised in space. Not quite two centuries later, Heraclides of Pontus, who became a disciple of Plato in 360 B.C., taught that the sun was the center of revolution for the two planets Mercury and Venus. But he still believed that the sun revolved around the earth.

Still, the Greeks, who came close to the invention of the steam engine, also almost succeeded in arriving at a true picture of the solar system. Aristarchus of Samos actually proposed what we now call the Copernican system, with the sun in the center and all the planets revolving around it. The same Aristarchus tried to measure the relative distances of the sun and the moon from the earth. The attempt miscarried, partly because of the lack of magnifying instruments for observation, partly because his method was not quite adequate. That attempt was made about 280 B.C.

A few years later Eratosthenes was born at Cyrene. He became a famous philosopher, so famous that Ptolemy Euergetes called him to Alexandria. He was given the post of Royal Librarian at Alexandria, the center of the cultural and intellectual life of the time. It was there that Eratosthenes decided to measure the earth. He had been told that at Syene (Assuan) a vertical pole would cast no shadow at noon on the day of the summer solstice and that a well was illuminated by direct sunlight all the way to the bottom. That meant that Syene, which in his opinion was due south of Alexandria, was situated on the Tropic of Cancer. Measuring the length of a shadow of a pole at Alexandria at noon on the day of the summer solstice proved to Eratosthenes that the distance from Alexandria to Syene was one-fiftieth of the circumference of the earth. Now Assuan is not precisely under the Tropic of Cancer. Nor is it due south of Alexandria. But the various small mistakes seem to have canceled out, for the figure of 250,000 stadia at which Eratosthenes arrived was surprisingly close to the truth. To all intents and purposes he had established the true size of the earth a little over two centuries before the birth of Christ.

The next great astronomer was Hipparchus, who lived from about 180 B.C. to about 125 B.C. Hipparchus carefully verified a number of Babylonian discoveries that had to do with apparent motions; he determined the length of the year, the obliquity of the ecliptic, and many other things on the same order. His most important work was the cataloguing of 1080 "fixed" stars—he divided them into six magnitudes according to their apparent brightness, a system which is still in use.

These were great and valuable achievements but Hipparchus marred them, at least in our eyes, by making the bad mistake of discarding Aristarchus's correct conception of the sun as the center, or approximate center, of the movements of the planets. Eratosthenes' measurement of the earth may have frightened him—this was too big a globe to set in motion. Thus Hipparchus gave way to the emotionally and geometrically easier system of placing the earth in the center and of letting the planets, including the sun and moon, revolve around the earth.

This is what is usually called the Ptolemaic System, but Claudius Ptolemaeus, after whom it is named, did not invent it. Claudius Ptolemaeus, who came some 250 years after Hipparchus, rounded out the work of the earlier master. The result of this work was a book which preserved "the Hipparchus" for us, all the more important since only one early work by Hipparchus survived in its original form. The book into which Claudius Ptolemaeus had expanded the work of his master Hipparchus was called the Great Collection, in Greek *Megale Syntaxis.* Some 700 years later an unnamed translator was called upon by the astronomically inclined Caliph al-Mámun to translate it into Arabic. The translator not only did that; he provided a new title too. Using the Greek word *megiste* (greatest) with the Arab prefix *al,* he gave to it the title under which it has been known ever since: the *Almagest.*

While astronomical thought developed, philosophic commentary and speculation could not stay silent. Through the whole period from Pythagoras in about 540 B.C. to about 100 B.C. there was a kind of subterranean philosophic rumble, called the debate about the plurality of worlds. Because that debate flared up again during the Middle Ages, and then once more during and after the Renaissance, in different forms, its original meaning is always well misunderstood. Of course it is of very little help that most of the original writings are lost and that our knowledge of what so-and-so is supposed to have said is derived from a chance repetition of somebody else's quotation from what may or may not have been the original. In fact many old opinions are known only through a book written by one Hippolytus in about the second quarter of the third century A.D. This book, called *Philosophumena* and for many centuries wrongly ascribed to Origenes, quoted older ideas about the plurality of worlds for the purpose of refuting them as heresies.

One can't exactly call that an unbiased source, but so much seems to be clear: that the Greek philosophers of the time before Christ did not mean what the philosophers of the fifteenth, sixteenth, and seventeenth centuries meant when they discussed the plurality of worlds. To the latter the term referred to the question of whether the other planets, especially Venus and Mars, should be considered "earths." To the Greek philosophers the term "world" meant a whole Ptolemaic system. It meant the earth with the sun and the moon and the other planets plus an array of fixed stars around them, all wrapped up in an impenetrable outer shell. One might say that it meant a planetary system with the earth in the center and a starry sky boxed for storage in the universe. If somebody held that there were "more worlds than one," he did not mean that there were other planets besides earth, he meant a plurality of such sphere-inclosed systems. And if two philosophers agreed that there were several such systems, there still was plenty of room for disagreement on their number, on their distribution in the universe, and especially on their contents.

Some, even though they were adventurous enough to believe in several worlds, still asserted that they were all identical. Others added nonuniformity to plurality and if Hippolytus can be trusted with correct reporting, Democritus believed that the various "worlds differ in size, that in some is neither sun nor moon, that in others both are greater than with us and yet in others more in number." Furthermore, that "the intervals between the worlds are unequal, here more and here less."

Compared to this the ideas of the Pythagoreans appear simply tame. They merely invented an additional planet, the *Antikhthon* or counter-earth. It was supposed to follow the same course as the earth around the "central fire," of which the sun was only a reflection, and to have, like the earth, an "uninhabitable hemisphere," namely the one turned toward the "central fire." Since this assumed body was a counter-*earth,* it seemed likely that it was a counterpart of the earth in every respect, including inhabitants. But the counter-earth was one of the specialties of the Pythagoreans; nobody else paid any attention to the idea.

The one man who definitely opposed all these speculations was Aristotle. He did not agree with the Pythagoreans, but he also had no use for the quite independently established thought that the moon was a palpable body.[1] He completely rejected any idea involving a "plurality of worlds," without being able to say anything valid in support of his rejection. He just *said* that the unchanging sky prevented the belief in other earths. He simply asserted that, since all matter is contained in one world, there can be no others. He added that since earths are heavy, they *must* come together in the center of the universe. The best answer to the last assertion was given some 1500 years later during the Middle Ages. When this statement was quoted by a theologian during a debate, another theologian replied rather pointedly that the heavy earths would stay where the Lord God had put them.

Aristotle's attitude, in this and in other respects, would have handicapped science much less if he had not been so wholeheartedly accepted by Christian teachers a thousand years later. It literally came to the point where thinkers set out with the notion that all wisdom could be found in the Bible, all astronomy in the *Almagest*, and all science in the writings of Aristotle. Not only was it simply forbidden to teach anything that contradicted or diverged from Aristotle's statements, it was also denied that there was anything that Aristotle had not known.

But this attitude, which made life difficult for Giordano Bruno, Nicholas Copernicus, and Galileo Galilei, came much later. Those who came soon after Aristotle considered him but one philosopher among several and felt that they had learned a great deal since the time of his death.

Metrodorus remarked rather simply that "it seems absurd that in a large field only one stalk should grow and in infinite space only one world exist." And Plutarch poked fun at the notion that the earth should be situated in the middle of the universe, since the universe was infinite and without boundaries, hence also without a center. This same Plutarch, who is still well known in our times as author of the *Lives*, wrote a book *On the Face That Can Be Seen in the Orb of the Moon* (*De Facie in Orbe Lunae*). It was a great summary of all the earlier ideas and thoughts and Plutarch emerged with the conviction that the moon is a second earth. He accepted the statement of Anaxagoras that the moon might be larger than the whole mainland of Greece. Consequently it was a smaller earth, but very much like earth in other respects, with mountains and valleys. And it is inhabited,

[1] "More current than the conception of an inhabited counter-earth was the belief that the moon is similar to our globe. Orpheus describes it as a 'celestial earth,' Thales is said to consider it of an earthly nature, and the Pythagoreans are recorded as believing 'the moon is terraneous, is inhabited as our earth is, and contains animals of a larger size and plants of a rarer beauty than our globe affords. The animals in their virtues and energy are fifteen degrees superior to ours, emit nothing excrementitious, and the days are fifteen times longer.' Philolaus considers the moon 'a body like the earth, with plants and animals,' and Xenophanes thinks of it as 'inhabited, and a land of many cities and mountains.' To Anaxagoras and Democritus it is 'a solid, condensed, and fiery body, in which there are countries, mountains, and valleys.' Heraclitus regards it as 'an earth covered with a bright cloud.' "—Dr. Grant McColley, in *Annals of Science* (1936), Vol. 1, No. 4.

but not by people. Its inhabitants are "demons" and on occasion one of them visits the earth. The "demon" of Socrates was one of them and the Oracle of Delphi probably another one.

The existence of Plutarch's book shows that, in his time, at least the moon was generally accepted as a solid body in the sky. The true nature of those luminous pin points might be debatable, but there was at least one other "earth." The fundamental conception necessary for the growth of the idea of space travel—that there are other worlds besides our own—had been established.

Plutarch died in A.D. 120. Precisely forty years later the first novel of a voyage to the moon was written. Its author was the Greek sophist and satirist, Lucian (properly Lukian) of Samosata.

Lukian called the book *True History*—it is often referred to under its Latin title *Vera Historia*—but warned the reader at the outset: "I write of things which I have neither seen nor suffered nor learned from another, things which are not and never could have been, and therefore my readers should by no means believe them."

The story itself, which was to exert great literary influence some fourteen centuries later, is the "missing adventure" of Odysseus—an adventure that could have been in the *Odyssey* and that I believe would have been in the *Odyssey* if the necessary astronomical knowledge had existed in Homer's time.

The *Vera Historia* begins with a number of adventures which take place on earth, but in that strange and unknown region west of the Pillars of Hercules, west of Gibraltar, where adventures of incredible nature seemed the rule rather than the exception. It seems that the Phoenicians in their time had invented and told gruesome tales of that region in wholesale lots, presumably for the purpose of frightening possible competitors away from the West and especially from the Atlantic Ocean. These tales also enabled them to drive harder bargains for their trade goods; wares that had been acquired with so much personal danger naturally commanded higher prices. In Lukian's time many of these old tales were still lingering, at least to the extent that they could be used for literary purposes, even though they no longer frightened sailors and traders.

After these preliminary adventures, when the reader might expect that the vessel is ready to sail for home, the great adventure strikes. The ship is caught in a terrible whirlwind and lifted out of the sea. The wind carries it high above the ocean and for seven days and seven nights the travelers do not know what is going to happen to them. But on the eighth day, as the first English translation (Oxford, 1634) has it: "wee came in view of a great countrie in the aire, like to a shining Island."

The ship has reached the moon.

The travelers are soon surrounded by warriors riding on three-headed gigantic birds and are brought before the king, who greets them in classic

Greek, since he is Endymion himself. The travelers are sufficiently astonished by this, but they are even more astonished when they learn that the moon is on the brink of a war with the sun. Both sides have prepared extensively for this war and both have acquired allies so that their armies are very impressive. The infantry alone numbers 60,000,000 men. Then there are 80,000 men of the aerial cavalry, riding the three-headed birds, and 20,000 cabbage-bird riders, having mounts of enormous size covered with a dense growth of cabbage deep enough for a man to hide in. There also are giant spiders, the smallest of them larger than an island; there are garlic throwers, flea riders (30,000 of them), and gigantic ants, the latter in the army of the sun.

Lukian's satirical mind must have worked overtime when he invented these armies and their components, and outside the narrow globe of the earth he found real opportunities for enthusiastic satire. He took from the Pythagoreans the idea that the inhabitants of the moon do not eject anything excrementitious and then went even further. His Selenites did not die either, since death, even the most decorous death, cannot help having a somewhat messy aftermath. But neither are they immortal. When their time comes they dissolve into vapor and smoke, a quiet and terribly clean exit from life. Nor do they indulge in sex, as far as one can find out, nor in childbirth of the terrestrial variety. The Selenites being ultra-ultra in every respect, just as the old philosophers had demanded, their young emerge with exquisite cleanliness.

While the travelers of the *Vera Historia* make their trip to the moon accidentally, the hero of Lukian's other moon-travel story, *Icaromenippus,* plans his trip carefully. Observing the night sky for a long long time, and wondering about the matter that composes the radiant sun, but being especially attracted by the moon, Lukian's second hero finally decides to go to the moon and to the stars. He provides himself with two wings, one of a vulture and one of an eagle, and begins to practice. He does learn how to fly and when he feels sure of himself in the new element he takes off for the moon from the summit of Mt. Olympus.

He does reach the moon, but that alone does not satisfy his curiosity. Now that he has gone so far, he strives to reach heaven itself. Taking off from the moon and leaving the sun to his right, he flies about among the stars and arrives in heaven on the third day of that flight. But the immortals resent the intrusion. Mercury is ordered to bring him back to earth and, while Icaromenippus is not punished for his audacity, his wings are taken away from him so that he may not reach the moon and the stars again.

For more than fourteen centuries no other book like Plutarch's *De Facie in Orbe Lunae* or Lukian's *Vera Historia* was written. For much of these fourteen centuries no such book could be written, because Christian philosophy, following Aristotle, was opposed to the idea of a "plurality of worlds." In 1145 Franciscus Gratianus issued a straight *prohibitus est:* the belief in more than one world was declared to be heretical. Much hard thinking was expanded on the problem of reconciling the existence of only one world with the recognized Plenitude of God. And when the *Doctor angelicus,* Thomas Aquinas, finally evolved a convincing explanation the whole attitude was revised!

In 1277, just three years after Aquinas's death, the Bishop of Paris, Etien Tempier, with authority granted by Pope John XXI, officially condemned the proposition that there could be only one world. God's Plenitude could not be restricted; Plenitude and creative power must be unlimited. This official opinion prevailed until the end of the sixteenth century. One of the early printed books, equipped with a valid *Imprimatur,* is a kind of encyclopedia with long summaries of all the pros and cons of the much-discussed subject. Anybody who could read could make up his own mind which side he wanted to be on. The Church did not turn against the doctrine of plurality again until it was connected with the concept of a moving earth. Even then there was much hesitation on the part of the Church, and presumably much prodding by astrologers who feared for their business, and much wire-pulling by philosophers of the old school who just did not want to learn new lessons. The hesitation at least is proved by the dates. What is now often called the astronomical revolution consisted of three books and the invention of the telescope. The first of the books was that of Nicholas Copernicus of Thorn. It bore the title *De Revolutionibus Orbium Coelesticum* (*On the Revolutions of the Celestial Orbs*) and appeared in 1543. The second was Johannes Kepler's *De Motibus Stellae Martis* (*On the Motions of Mars*), which appeared in 1609. The third—the direct result of the invention of the telescope—was Galileo Galilei's *Sidereus Nuncius* (*Messenger of the Stars*) which appeared in 1610. But Copernicus's book was not placed on the *Index librorum prohibitorum* (Index of forbidden books) until 1616. And more than ten years afterward there was still hope of getting it off again; Galilei tried to do so until he himself was forced to recant in 1633. Copernicus remained forbidden until 1835.

The result was the obvious one: the rapid development of astronomy during those two hundred years took place chiefly in Protestant countries.

Each of the three books which together constituted the astronomical revolution was of almost incredible importance, even taken by itself. Together they formed, in logical progression, a new picture of the world. In the Ptolemaic, or rather Hipparchic, system the earth had been in the center. The planets moved around it on large circles, but not directly. They moved on smaller circles, so-called epicycles, and the center of the epicycle moved along the large circle. Also the center of the large circle did not coincide completely with the center of the universe, the earth. It was close, but not in the same point. Copernicus, with many references to

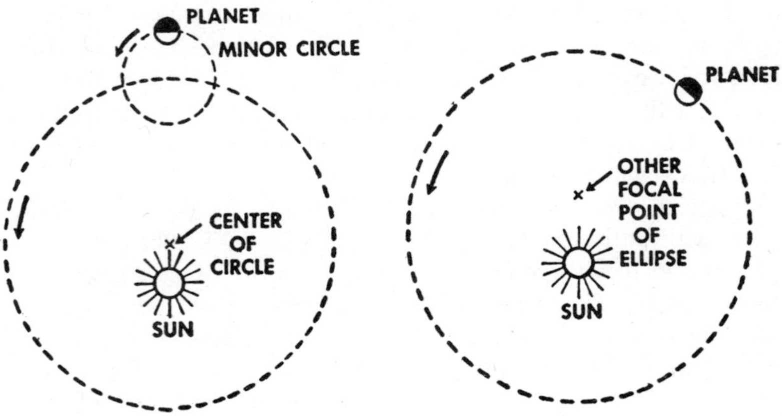

Fig. 1. Copernicus's and Kepler's conceptions of the movement of the earth around the sun.

Aristarchus of Samos, reversed the positions: he put the sun in the center and the planets, earth among them, on the epicycles (Fig. 1).

Kepler took the next step. Investigating Tycho Brahe's many observations of the positions of Mars, he ran again and again into an impossibility. No matter what one did with the major circle and the epicycle, Mars just did not conform. But if one entered its positions on a large sheet of paper another figure seemed to emerge. Kepler was doubtful at first; it should be the "ideal" figure, the circle. But it was not; it was another figure, an ellipse, which does not have one center, like the circle, but two focal points (Fig. 2). Kepler may have consoled himself about the loss of the "ideal" circle by reminding himself that the circle is only a special case of an ellipse, an ellipse with the two focal points merged. In any event he realized that

Fig. 2. An ellipse of the eccentricity 0.5 showing some of the geometrical relationships. The main characteristic of an ellipse is that the sum of the distances from the two focal points F_1 and F_2 to a point P on the periphery must be a constant value no matter which point along the periphery is chosen. The closer F_1 and F_2 approach each other, the less eccentric the ellipse will be. Least possible eccentricity occurs when F_1 and F_2 are the same point; the ellipse in that case has become a circle.

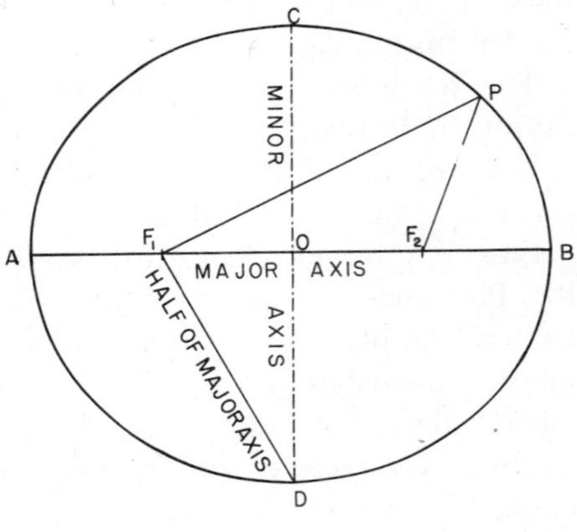

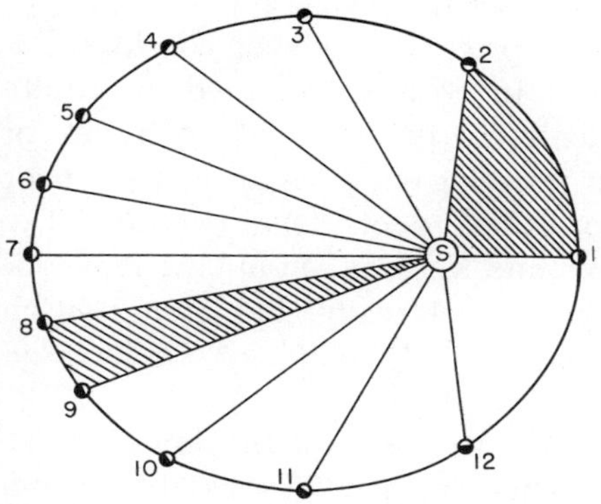

Fig. 3. Kepler's Second Law. The real orbit of the earth is so nearly circular that if it were drawn as a circle 40 inches in diameter the ellipse would deviate from the circle by less than the thickness of the line. The figure shows earth's orbit drawn as an ellipse of eccentricity 0.5; the numbers 1, 2, 3, etc., indicate when the months of the year would begin. The planet would need the same time to move from 1 to 2 or to move from 8 to 9.

the planets move in elliptical orbits, with the sun in one of the two focal points of the ellipse. That statement is now known as Kepler's First Law.

Investigating the rate of movement along the ellipse he found what we call his Second Law. The line connecting one of the focal points (the sun) with a point along the ellipse (the planet) is called the radius vector. Kepler saw that Mars moved faster when near the sun and formulated this observation by saying that "the radius vector sweeps over equal areas in equal times" (Fig. 3). His Third Law, which he discovered later, deals with the relationships between the distances of the planets from the sun and the time required for one full revolution around the sun.

Copernicus and Kepler put the mechanism of the solar system in order. Galileo Galilei, by means of a homemade telescope which he had virtually invented himself (after studying the principles on which the first "miracle tube" of the Dutchman Jan Lippershey was based), put worlds into the solar system. He saw what had been debated fruitlessly for dozens of centuries—he saw that all the planets showed disks when seen through a telescope. This proved that they were worlds comparable to earth and moon, even though they all differed in size.

All this, taken together, provided a new and solid foundation for dreams about space travel. What happened first was that Lukian was reprinted, for simplicity's sake in the original Greek, and not less than five times. For the less erudite the book was translated into Latin (by Kepler himself, among others) and for the still less erudite translations into the "vulgar languages" were made; the first English edition appeared in 1634. But meanwhile Kepler had contributed a work of his own, the one which is called *Somnium* (*Sleep*) for reasons which will become apparent soon. It also was printed in 1634, but otherwise it cannot be dated accurately because Kepler did not write his *Somnium* at any specific time. The book is the result of idle moments or, possibly, periods of sickness covering many years of his life. There is reason to believe that several hand-written copies of the manuscript existed, and one of them seems to have found its way to England. But Kepler hesitated to have it printed; he never considered it quite finished and ready for the printer. In 1630 Kepler died, not quite sixty years of age. Death came suddenly, in the form of an attack of pneumonia which he contracted while traveling long distances over poor roads and in bad weather, trying to collect debts long overdue. The *Somnium* was still not quite finished then and Kepler's son-in-law, Professor Jakob Bartsch, undertook the task of making it ready. But then Bartsch also died suddenly and the manuscript fell into the hands of Kepler's son Ludwig.

Ludwig considered it his duty to "leave to posterity the fame of my father without lacunae" and Kepler's heirs, impecunious as they were, shared the expense of the printing.

The *Somnium* is a fantasy, a fantasy with the little sad smile of a man whose good sense of humor is somewhat incapacitated by physical pain. It is full of interesting sidelights, little humorous thoughts about Tycho Brahe, the garrulous, boisterous, conceited creator of the observatory of Uraniborg, the drunkard whose nose was made of silver (the original having been lost in a youthful brawl), the ox-strong nobleman without a home, the man who made the best and most reliable observations anybody ever made up to his time—observations that enabled Kepler to throw out the artificial circles of antiquity and of Copernicus and substitute his own ellipses. It is also tinged by Kepler's own desire for sleep and rest and it shows influences of the enforced restlessness of a Protestant in those times. "If they finally chase us from the Earth," he wrote in 1629 to a friend, "my book will be useful to emigrants and pilgrims as a guidebook of the moon."

The *Somnium* begins with Kepler lying in bed, asleep, resting from incessant labors over sheets filled with computations. Soon he is dreaming. He dreams that he has bought a book in a bookstall and is now reading it. That book tells of the young Icelander Duracotus who makes the long pilgrimage to Tycho Brahe to learn what astronomers know about the moon. After several years he returns to his native island of glaciers and boiling springs. He finds his mother Fiolxhilda,[2] who spends her time in collecting strange herbs—a witch, as the son now realizes. He tells her about Tycho and his teachings and he learns, to his utmost surprise, that his mother knows much more about the moon than all astronomers in the world taken together. She has had a better teacher than her son, because her teacher is a demon—a demon from Levania.

It becomes clear immediately what "Levania" is; the demon himself provides the clue in the first sentence he speaks: *"Quinquaginta millibus miliarium Germanicorum in aetheris profundo sita est Levania insula"*

[2] The strange name Fiolxhilda was based on Kepler's memory of an old map of Europe he had seen in Prague. On that map Iceland was labeled Fiolx.

("Fifty thousand German miles away in the deep ether there is the island of Levania").

Levania is not inaccessible, but there are certain difficulties. Little demons are rampant both on Levania and on Volva (the earth). But they shun the light of the sun and they cannot cross space because the rays of the sun would catch them there. They cannot, that is, under ordinary circumstances. But when the shadow of the earth touches the moon, they are able to race across this temporary bridge. And they return when the shadow of the moon touches the earth.

It is clear what Kepler meant. The demons, the spirits of astronomy, must shun the sun's glare, for you cannot observe while the sun is blinding you. The spirits of astronomy thrive during the night when it is dark, and observations that are impossible otherwise can be made when the shadows of earth and moon reach across the void: during eclipses.

The spirits *can* take a mortal along, but it is hard work and it cannot be done at all if the mortal is too big and paunchy. (Apparently Kepler did not trust Tycho's abilities beyond making good observations.) Nor is the trip pleasant for any mortal—when the spirits pull him rapidly along the temporary bridge of the earth's shadow, he suffers from cold and lack of air. At the point where the "magnetic influence" of the moon becomes stronger than that of the earth the spirits let go and the mortal will then fall to the moon without their help. But at this point something strange takes place; the body of a mortal will curl up into itself "like a spider." Kepler had seen how certain spiders, when disturbed, pull all eight legs in and let themselves drop. He even explained in more detail what he meant: "When the magnetic attractions of the moon and of the earth equalize each other, it is as if neither of them exerted any attraction. Then the body itself, being the whole, attracts its minor parts, the limbs, because it is the whole."

Delete the word "magnetic" and you have a description of gravitational effects, long before Newton. As a matter of fact, Newton merely reported in sharp and clear language what Kepler describes in a playful mood, and I have often wondered whether Kepler may not have had such a sharp description with formulae somewhere among his notes.

As soon as the spirits arrive at the moon they hide in dark caves, knowing that the sun's rays will beat down upon the moon again at any moment. This, Kepler added, is a habit of astronomers too. "When I lived in Prague I had an apartment where there was no better place to observe the sun crossing the meridian than the beer cellar."

At this point Kepler stops playing. We now are on the moon and the astronomy lesson is about to begin. The Volva (earth) is large in the sky and the continents and seas are clearly visible. Because of the astronomical fact that the moon always presents the same face to the earth, Levania is divided into two permanent hemispheres and its inhabitants refer to themselves as Subvolvani, those who always have the Volva in the sky, and Privolvani, those who never see it unless they travel half around their world.

The mountains of Levania are generally higher and more rugged, the valleys and chasms deeper, than on earth. Since the days and nights are so long, each one lasting fourteen of our days and nights, the temperature differences are enormous. But fortunately there are many caves on the moon and they protect the inhabitants. Besides, Nature helps; the bark of the trees and the fur of the animals—or what serves in lieu of fur—make up the larger part of the mass of their bodies. Sometimes an animal is caught by the sun and seems to drop dead. In that case the outer layer of its fur or skin is singed by the sun and becomes hard. At night the animal revives and the singed portions slough off. In shape the inhabitants of the moon are mostly snakelike; a frequent variety looks like a singed pine cone during the day, while at night the pine cone opens and the animal emerges. But while the Endymionides, as Kepler terms the inhabitants of the moon, have the shape of strange animals, their mental qualities and capacities resemble those of human beings.

> Those hollows of the moon first seen by Galilei are . . . portions below the general level, like our oceans. But their appearance makes me judge that they are swampy for the greater part. It is there where the Endymionides find the sites for their fortified cities which protect them against the swampiness as well as against the heat of the sun, possibly also against enemies. They do it in the following manner: in the center of the chosen site they put a stout pole to which they attach ropes, their length depending on the size of the fortress to be built; the longest [rope] measures five German miles [about twenty miles]. Then they mark the periphery by walking around at the end of the rope. After that they amass to build the wall. . . . Whenever the inhabitants feel annoyed by the power of the sun those who live near the center move into the shadow of the outer wall . . . following the shadow for fifteen days they wander about and by this means endure the heat.
>
> (*Somnium*)

The new telescopic era which dawned during Kepler's lifetime shows in this description. Through the new "optick tube," crude and poor as it was, astronomers had now seen the round "craters" in the moon and Kepler, like everybody else, had wondered what caused them. This was his answer: the circular craters were artificial, erected by the inhabitants of the moon for the purpose of producing shaded places in which they could keep alive.

Lukian's influence went hardly beyond giving rise to the term Endymionides. Of course Lukian had not been serious when he invented his astonishing array of incredible creatures in the army of the moon. But Kepler was serious, in the sense of making a serious guess, and the biological features he invented showed greater biological knowledge than most biologists of his time possessed. As far as mathematics went Kepler was accurate, needless to say, and his conception of gravitational influences was virtually Newtonian.

Even the dream device of getting to the moon was realistically conceived. To Lukian, and also to many writers even long after Kepler, the problem of flying through the air and the problem of journeying to the moon were one and the same. The trip to the moon differed from a flight from one mountain peak to another only in being a much longer trip, requiring preliminary practice, as explained for the first time in Lukian's *Icaromenippus.* But Kepler knew that they were not the same. To make flying and space travel alike there had to be a common atmosphere to earth and moon, an atmosphere which, probably, was denser near the surfaces of the two bodies, but an atmosphere which extended all the way from one to the other just the same. And Kepler realized what everybody else overlooked, probably without thinking about it at all: that this meant friction. Friction was the one thing Kepler, who had found the true laws of planetary motion, could not use. There could be no friction. Hence there could be no common atmosphere. Consequently man could not fly to the moon, even granting that he could fly at all. The only way out was a device which did not violate the laws of Nature since it was not subject to these laws. Kepler, just because he had a better conception of the problem than anybody else for centuries to follow, had to use the dream method.

Even before the *Somnium* saw print, an English book had appeared which also dealt with a voyage to the moon. Unlike the *Somnium,* that book, the first interplanetary voyage of English literature, saw several editions and translations. The first edition was printed in 1638 under the title: *The Man in the Moone: / or / A Discovrse of a Voyage thither /* By / Domingo Gonsales / *The speedy Messenger.*[3]

The author of the book was Bishop Francis Godwin, known to students of English literature mainly as the compiler of a voluminous and boring catalogue of English bishops. Bishop Godwin's book, which still affords amusing reading, shows many influences of the *Somnium,* and he must have read a manuscript copy of Kepler's posthumous work. In fact some of the early adventures of the fictitious hero Domingo Gonzales suggest that Bishop Godwin had the German astronomer in mind as a prototype for his hero. If true, it would be a nice little literary joke that the author of one moon journey is used as hero by the author of another moon journey. Taken purely as "literature" *The Man in the Moone* might outrank the *Somnium;* taken as a work mirroring the scientific knowledge of its period it is inferior.

[3] The French translation appeared in 1648 under the title: *L'Homme dans la Lune,* ou *le Voyage chimérique fait au Monde de la Lune, par* Dominique Gonzalès, *aventurier espagnol.* The German translation followed soon after under the title *Der fliegende Wandersmann.* A reprint of the first English edition appeared in 1937, in lieu of a tercentenary celebration, as vol. xix, no. 1, of *Smith College Studies in Modern Languages.*

Bishop Godwin was careful in his work. His hero, being a Catholic Spaniard, uses the Gregorian calendar, while the bishop himself still used the older Julian calendar. But Godwin was not a follower of Copernicus. He accepted the daily rotation of the earth around its axis, an old idea which had been accepted and rejected a few score times in the interval between Pythagoras and Copernicus. But he did not accept the movement of the earth around the sun. Nor did he follow Kepler as regards the action of gravity and the presumed airlessness of the space between earth and moon. He termed the earth a big lodestone, and assumed that the "attractive force" does not reach very high, not much beyond the cloud zone. On the other hand, he stated quite explicitly that the attraction of the moon is much smaller, since the moon itself is smaller. As for the air above the zone of attraction, he described it as especially mild and pleasant, neither hot nor cold, with the miraculous property of preventing the sensations of hunger. But it is also full of "devils and wicked spirits"; Kepler's spirits of astronomy return here in a disguised form.

The trip itself is a flight. Domingo Gonzales, the hero, finds a "certain kind of wild Swans" on St. Helena. It may be coincidence, but just during the year the book appeared, a dodo from the island of Mauritius was on exhibit in London. Perhaps this large if flightless bird from a tiny island had put the thought of other large and strange birds on other islands into the bishop's mind. At any event those "wild Swans," called *gansas,* can fly and Domingo Gonzales trains them to carry weights and to obey orders. Finally he builds an "engine," which means a seat for himself, carried by a number of *gansas.* Still he only intends to fly, and the trip to the moon comes somewhat as a surprise. The birds fly in the direction of the moon in a straight line, as if it were the place to which they migrated every year. Their speed is great; Domingo estimates it as 50 leagues, about 175 miles, every hour. The flight takes twelve days—but twelve days at that rate of speed is only about 50,000 miles, about one-fourth the actual distance to the moon. Godwin probably wanted to arrive at the figure 50,000 given by Kepler, but either he overlooked the specifying *Germanicorum* in his copy of the *Somnium,* or else the copyist had omitted the word.

The moon itself is a perfect paradise, and the rigors of climate that Kepler described are not affirmed. It is a place without want, unrest, or war. The inhabitants are human, but tend to be of larger size than earth men. Larger size denotes higher rank and, presumably, greater wisdom. Their language is so musical that it cannot be written in letters; notes have to be used. The bishop did not fail to give a few samples.

The return trip to earth—Domingo happened to land in China—was made in the same manner as the ascent, after enough time had been spent on the moon to make the *gansas* eager to fly back.

If Bishop Godwin's book influenced literature, especially English literature, it shared that influence with another book by another English bishop, which appeared within a few months of the first edition of *The Man in the Moone.* This was Bishop John Wilkins' *The Discovery of a World in the Moone: or, A Discourse Tending to Prove, that 'tis probable there may be another habitable World in that Planet.*[4] This was not fiction, or a discussion disguised as fiction. It was a straightforward work on the moon, its similarity to the earth, and the possibility of its being inhabited.

> Two years later he added to the third edition of this work "The Discovery of a New World," a final chapter in which he seriously affirmed that it was possible to make a flying-chariot, wherein several men might sit, and, by means which he felt science would soon discover, might so motivate their ship that it would fly to any height and might at last reach the moon. His words were a stimulus to both science and literature and his influence in the Royal Society led that great body to turn its attention to the principles lying behind the flying-machine.[5]
> (Nicholson)

It seemed as if the good bishop's optimism was soon to be justified. Not even half a century after the first appearance of his *World in the Moone* the "flying chariot," the airship, was actually invented, though only on paper at first. The inventor was a Jesuit priest who was professor of mathematics at the University of Ferrara from 1677–79. His name was Francesco de Lana-Terzi and the description of his airship can be found in the second volume of his main work.[6]

Francesco de Lana-Terzi could not have made his invention if others had not worked out experiments which dispelled one of the most unfounded but also most persistent beliefs of older philosophers, the *horror vacui.* "Nature abhors a vacuum" had been handed down like a sacred tradition, just as had the other unfounded belief that a body falls faster the heavier it is. That latter belief had been demolished by Galilei in several ways. He first inquired of one of his teachers whether a 20-pound stone would fall twice as fast as a 10-pound stone. That was asserted. Now, Galilei continued, let's tie a 20-pound stone and a 10-pound stone together; they should fall more slowly, because the lighter stone hampers the movement of the heavier one. That was admitted too. And then Galilei sprang his trap: Tied together the stones weigh 30 pounds and therefore should fall faster than the 20-pound stone alone. After that he dropped a cannon ball and a musket ball simultaneously from the leaning tower of Pisa, demonstrating that they struck the ground at the same time.

[4] The French book *Le Monde dans la Lune, divisé en deux Livres: le premier prouvant que la Lune peut être un Monde; le second que la Terre peut être une planète,* Par le Sieur de la Montagne (Rouen, 1655), is the same, but Wilkins' name is not mentioned. Another French edition appeared in London in 1640, a German translation in 1713.

[5] Marjorie Hope Nicolson, "Introduction to *Navis Aeria* by Bernardo Zamagna."

[6] This was *Magisterium Naturae et Artis,* in three volumes which were published in 1684, 1686, and 1692, the third posthumously. (Lana-Terzi lived 1631–87.) The airship appears as *Artificum XLVI* in volume II.

This famous story has been doubted on the grounds that Galilei never wrote about it himself, we know it from the biography of Galilei by his pupil, Vincenzio Viviani, who even said that Galilei did it "repeatedly." Galilei's silence is probably due to the fact that the experiment itself was already described in books by Hieronymus Cardanus and Simon Stebin. To Galilei this was only a demonstration for which he made no special claims.

The tradition of *horror vacui* had been wounded severely by Evangelista Torricelli when he invented his "Toricellean tube," the first barometer. Blaise Pascal had developed the theory of this instrument; Monsieur Perier, Pascal's brother-in-law, had carried one to the peak of the 4790-foot-high Puy-de-Dôme in Auvergne, France, to show that there was less air pressure on mountain tops; and in 1650 the forty-eight-year-old burgomaster of Magdeburg, Otto von Guericke, had invented the air pump and demonstrated that a vacuum is possible.

Lana-Terzi presented his ideas in a very logical and systematic manner. First of all he asserted that air has weight. Then he stated that it was possible to exhaust the air from a vessel. He emphasized that the cubic content of a given vessel, say a globe, increases at a much faster rate than its surface area, and arrived at the conclusion: "It is certain that one can construct a vessel of glass or other material which would weigh less than the air contained therein; if, then, one exhausted all the air . . . this vessel would be lighter than the air itself and . . . would float on the air and ascend."

The flaw in his reasoning was of a purely constructive nature; a vacuum globe light enough to weigh less than the air it replaces would be crushed by air pressure. The real solution consisted in substituting a lighter gas for the air inside the globes, thereby eliminating the unmanageable pressure difference—a solution found a few months after the brothers Joseph Michel and Jacques Étien Montgolfier had built their first hot-air balloons in 1783. Hydrogen gas, not known during Lana-Terzi's lifetime, was used by Professor J. A. C. Charles, and Bishop Wilkins' "flying-chariot" had finally been invented.

But, contrary to all hopes, it could not carry anybody to the moon.

While Perier climbed the Puy-de-Dôme to prove that air pressure diminishes on high mountains, while Otto von Guericke heaped glory on the city of which he was burgomaster by proving that not even several teams of horses could pull two well-fitting exhausted hemispheres apart, and while Lana-Terzi deduced the principle of lighter-than-air flight, philosophers and poets were busy acquainting the world with the rapidly spreading doctrine of the plurality of inhabited worlds and entertaining it with tales that were based on this doctrine.

Cyrano de Bergerac wrote two novels of that type, *Voyage dans la Lune*

(1649) and *Histoire des États et Empires du Soleil* (1652). Some of his work drew heavily on predecessors—the inhabitants of the moon talked music, not words. As for the means of transportation, Cyrano apparently could not make up his mind which should be given preference; consequently different heroes use different means. One is semimystical—bottles filled with dew lift the traveler; since dew disappeared in the rays of the morning sun, it was evidently drawn toward the sky. Two others were "flying-chariot" devices: one an iron car lifted by means of pieces of lodestone thrown upward continuously; the other simply a box with a number of large powder rockets attached to it. Cyrano, by accident and of course without realizing it, had guessed the proper principle—the principle of reaction. But another half century was to pass before Sir Isaac Newton even stated what reaction really means.

Bernard de Fontenelle's *Entretiens sur la Pluralité des Mondes* (*Discourses on the Plurality of Worlds*) followed in 1686 and took Europe by storm. It was issued and re-issued, translated into several languages, and it even brought its author the honor of being asked to take up residence in the palace of the monarch. He accepted and spent his time writing a book on geometry of which he said that it might be fully understood only by seven or eight mathematicians in Europe, he himself not being one of them.

De Fontenelle's book, although intended as a popular astronomy, has to be mentioned here because of its strongly speculative character. Its leading idea is that every planet must be inhabited, but by beings with constitutions conforming to their surroundings—a rather modern thought.

De Fontenelle, not handicapped by such trifles as instruments which can measure the surface temperatures of other planets or tell the chemical composition of the top layers of their atmospheres, had only two things to go by: the temperature resulting from the distance from the sun, and the sizes of the planets. As for Mercury, he asserted that the heat is so enormous that there are rivers of molten gold and silver (actually a river of molten lead is the best that temperature would permit). Its inhabitants cannot imagine a world where little disks of these metals can serve as money. Still, the inhabitants of Mercury endure the heat only because the planet spins rapidly around its axis (actually it turns at the rate of once in 58.6 days, or two-thirds of its orbital period) and they enjoy themselves in anticipation of the coolness of the night which is soon to come. In general they are all hotheads and fools and enormously active because of the heat.

At this point de Fontenelle's German translator, Bode, could not help interpolating: "Very strange! for with us in Berlin we find that great heat makes people lazy and sleepy instead of lively and active."

As for Venus, it is the planet of love, even though its inhabitants are not very beautiful. They have no time for the serious sciences because their time is taken up by constant flirtations. They do practice music, poetry, and dancing since these arts have their amorous uses. But, de Fontenelle added sadly, they cannot cook, because they live almost exclusively on air.

Mars is passed over and so is Jupiter to a large extent. De Fontenelle is somewhat embarrassed by the fact that all the inhabitants of Jupiter, because of the size of that world, cannot possibly know one another, while on small Mercury everybody knows everybody else. However, the inhabitants of Jupiter have telescopes and one day somebody discovers a hitherto unknown small planet near the sun, the earth. "This is reported by their astronomical journal, but the population either does not hear about it or just laughs, the philosophers whose systems would be disturbed decide not to believe it, and all 'sensible people' harbor great doubts about this announcement."

As for Saturn, life is no pleasure there because of the extreme cold. A Saturnian, brought to earth, would die of the heat, even in the Arctic. Because of the cold, the Saturnians are slow and phlegmatic and remain all their lives in the places where they were born, just like oysters.

With Saturn, de Fontenelle had reached the limits of the solar system as it was then known, but he had to add a few remarks about the moon. Mercury and Venus, because of their proximity to the sun, do not need moons and therefore do not have any. The earth has one, Mars again none (the two small moons of Mars were not known until they were discovered by Asaph Hall in 1877), but luminous birds and luminous mountains illuminate the nights on Mars. Jupiter has, of course, its four large moons, and Saturn gets light from its rings. As for our moon, it might not be inhabited *à cause de la rareté de l'air.*

With all his nonsense, de Fontenelle did reflect some recent progress in astronomy. The moon, once loudly acclaimed as a sister planet of the earth, so similar that it might be difficult to tell the difference at first glance, had fallen into disgrace. The main reason for this remarkable and rapid cooling off was the great *Selenographia* of the astronomer Johannes Hevelius of Danzig, which had been published in 1647. It was the first systematic work about observations of the moon and it abounded with fine and detailed maps, showing a rather strange and different world. The rarity of the air was recognized and there were great doubts about the water supply. The moon no longer looked promising.

On the other hand, the close approach of Mars in 1672 had been used to determine the true size of the solar system. While this was a truly international venture, most of the mathematical work was done by Giovanni Cassini, who found, to his amazement, that the distance from the sun to the earth had to be more than 80 million miles (actually it is 93 million miles), which made the solar system at least twice as large as the most daring figures that had been advanced speculatively at earlier dates.

Things had suddenly grown too large, too impressive, for light-headed speculation on actual travel. And it also became clear to everybody what

Kepler had taken for granted more than a century earlier: that the atmosphere formed only a thin shell around the earth; that even the possession of a flying chariot did not mean the possibility of a trip to the moon. It is significant that the first great poem on aviation, Bernard Zamagna's *Navis Aeria,* published for the first time in Rome in 1768, abstains from any mention of space travel.

The poem describes the building of an airship of the type invented on paper by Lana-Terzi, but the completed ship stays close to the ground; in fact, Zamagna even warns that it should not fly too high in the atmosphere because the air may become too thin.

Consequently, the two works of this period which deal with inhabitants of other worlds are carefully "non-material." These two works, as different from each other as is humanly possible, are Voltaire's *Micromégas* (1752) and Emanuel Swedenborg's *Arcana coelestia.*

Micromégas relates the travels of a gigantic inhabitant of Sirius who is later accompanied by an inhabitant of Saturn. It is straight philosophical satire. To be a philosopher and a satirist at the same time is not at all difficult, but Emanuel Swedenborg succeeded in combining the somewhat contradictory features of research scientist and mystic. As a research scientist he contributed a great deal to various sciences, especially mineralogy and geology. As a mystic he accepted the fundamental theorem that all worlds are inhabited. It seems, of course, that his "astral men" have to be taken "spiritually," which means, in plain language, that they are not to be taken literally. Nor are they very important in themselves; their importance lies in the fact that they prompted Immanuel Kant to devote a very large section of one of his works to an examination of these beliefs.

The realization that a means of flight was not simultaneously a means of cosmic travel[7] suggested that the latter would require a new principle, possibly the application of a new force. It is only natural that the vaguely known force of electricity suggested itself.

The same Otto von Guericke who had invented the air pump had also invented an "electric machine," consisting mainly of a large ball of sulfur which could be turned by means of a handle. The experimenter himself furnished friction by placing his hand upon the rotating sulfur sphere, and it was a beautiful toy for drawing-room demonstrations and perfumed philosophical discussions with Madame la Marquise.

Guericke's sulfur ball was expanded into a spaceship by a Frenchman, Louis-Guillaume de la Follie, whose *Philosophe sans Prétention* was published in Paris in 1775. But the invention was not made on earth; it took place on Mercury where a young scientist asked for admission to the Mercurian equivalent of the Académie on the grounds that he had invented

[7] Edgar Allan Poe's "Unparalleled Adventure of One Hans Pfaall" (published in 1835 in the *Southern Literary Messenger*), which uses a balloon as a means of flight to the moon is, of course, in the spirit of banter.

a flying machine. One of the members declared that such a machine could not possibly work and vowed that he himself would use it for a flight to Hermione (earth) in case that he should be wrong and the inventor right. Of course the machine does work, and the conservative philosopher, ashamed of his rashness, does mount it for a flight to the planet earth. The trip is uneventful, but an irreplaceable part breaks during the landing, marooning the Mercurian philosopher on the third planet from the sun.

Louis-Guillaume de la Follie's short story, which was hidden in his volume of philosophic and scientific essays, is significant in several respects. It appeared thirteen years before the invention of the balloon, but de la Follie was well aware that mere flying in the air is not the same as space travel. He might have read Lana-Terzi's book. His thought about the workings of such a machine was probably inspired by observation of magnetic attraction and repulsion. In any event he was the first in more than a hundred years to "invent" a material spaceship for story purposes, foreshadowing the stories to be written in the nineteenth century.

2

The Decades of the Great Dreams

In fact the two planets earth and Mars are the middle links of the planetary system, and it may be suspected with fair probability of their inhabitants that they stand in the center between the extremes as regards physiology as well as it regards morals.

—Immanuel Kant, *General Natural History and Theory of the Heavens* (1755)

On November 13, 1833, Sir John Herschel, the son of Sir William Herschel, embarked for Table Bay. He arrived at Capetown on January 15, 1834, and on March 4 he began the astronomical observations for which the long trip had been made. It was the first systematic attempt at exploration of the southern sky, and not only professional astronomers but also the public knew well that Sir John had left for South Africa and why.

On August 25, 1835, the New York *Sun* carried a front-page story about the discoveries made at the Cape by Sir John. It was, as may well be stated at the outset, the biggest and most successful scientific hoax ever perpetrated. While purporting to be a reprint from a special supplement of the *Edinburgh Journal of Science,* it had been written in New York by a staff member of *The Sun,* Richard Adams Locke, who by the very invention of this story was proved to be a genius in his own right.

The hoax began with an editorial which called the attention of the readers to the first installment of the serialized reprint from the Scotch journal and promised the announcement of astounding discoveries. But the first installment, contained in the same issue of *The Sun,* did not relate any of these discoveries. It merely spoke of the instrument which had made them possible. Even that, if true, would have been enormously important. In a dignified scientific style it was explained that opticians could easily increase the magnification achieved by a powerful telescope.

As far as optics and mechanics go, there is almost no limit, but there is one annoying fact which renders such an undertaking useless. That fact is that the image grows weaker the larger it is so that in the end it becomes so faint that nothing can be distinguished. Actually this is only part of the truth. If it were the whole truth concerning the reasons why magnification beyond a certain level cannot be used, the problem would have been solved in the meantime by the invention of the photographic plate which accumulates impressions. The most important reason is that an astronomer on earth has to look through the atmosphere and that this atmosphere is never quiet all the way through. The sickle of Venus is apt to look like the flame of a candle in a draft because of that, and the faint image of a distant star may jump about in the field of vision. The trouble can be remedied somewhat by placing the telescope on top of a high mountain so that the densest and most disturbing section of the atmosphere is beneath, but there is always enough atmosphere left to annoy astronomers and to ruin many nights.

But *The Sun,* or Mr. Locke, went on to say that Sir John Herschel, while discussing this problem with Sir David Brewster, had found a way out. They discovered a method of "transfusing artificial light" into the image and thus brightened it sufficiently to allow for larger magnifications. Just in case some reader might happen to see through this fallacy, several paragraphs were devoted to loose talk about microscopic reflectors, angles of incidence, properties of rays, etc., until the reader felt that he could not follow anyway and was, therefore, ready to accept anything he was told. Of course, the story continued, a powerful telescope was needed to begin with, and Sir John Herschel had a 24-foot mirror cast for this purpose. (The 200-inch mirror on Mount Palomar measures 17 feet.) Theoretically this supertelescope, in combination with the new instrument for the "influx" of artificial light, should be powerful enough to see a butterfly on the moon, provided there were butterflies present.

The second installment described at length the journey to the Cape of Good Hope, the setting up of the telescope, and the observations of some distant stars and star clusters. Then, on January 10 (1835) the huge instrument was directed at the moon. The first thing that appeared on the projection screen was a huge basaltic formation. Everything was as clear as if the observers were floating over the moonscape in a balloon. Then they saw trees resembling yew trees, forests of firs, level grassy plains, and herds of animals. "Brown quadrupeds having all the external characteristics of the bison, but more diminutive." Soon after, another animal appeared on the screen, of bluish lead color, about the size of a goat, but with a single horn.

The unicorn climaxed the second installment.

As far as business was concerned, *The Sun* had every reason to be exhilarated. It had existed for only two years then, and it was the newspaper with the largest circulation in the world. It sold 19,360 copies, while the circulation of the London *Times* was only 17,000 copies!

The following installments described additional animals and winged batlike creatures. The latter, when observed more closely, turned out to be the inhabitants of the moon, bat-men and bat-women, flying, sitting around, talking with many gesticulations, and in general showing rather human behavior.

Later installments told of temples with roofs "consisting of a yellow metal" and the last one wound up with a quick and not particularly inspired survey of the other planets of the solar system.[1]

Business remained splendid until the bitter end. A sheet of drawings of the things observed on the moon sold well. A reprint of the whole story in pamphlet form found very willing buyers. But the hoax could not be maintained indefinitely. The first editorial had stated, and so did a postscript to the last installment, that the original report continued for another forty printed pages consisting of mathematical data and technical discussions of little or no interest to newspaper readers.

This statement brought Professors Olmstead and Loomis from Yale to New York. They did not blame the editor of *The Sun* for not printing these forty pages, but they were obviously in his possession and they wanted to see them. They were held at bay by Locke who told them to go to the printer. Locke himself raced ahead and instructed the printer where to send them next, racing ahead again, and again, and again, until the professors gave up. Then the *Journal of Commerce* wanted to reprint the whole story and sent a reporter to Locke to get the *Edinburgh Journal.* Locke at first tried to talk him out of reprinting the old stuff, then admitted that he had written it. The next day the *Journal of Commerce* gleefully informed its readers about the truth of *The Sun's* marvelous story.

For a while New Yorkers were indignant, then they began to laugh—as did Sir John Herschel who learned about it some time later—and the "Panorama" exhibits and the stage began capitalizing on Locke's great moon hoax.

It has been claimed that Locke had a collaborator when he wrote the moon hoax, a French astronomer named Jean Nicolas Nicollet who had left his native country because of financial difficulties and who might have been susceptible to a profitable suggestion of that kind. Only Locke himself could have told whether Nicollet helped him or not, but Locke kept quiet about that point.

But whether Locke had been helped along by a professional astronomer or not, the moon hoax portrayed better than any learned treatise the general attitude of the public—and of a large number of the scientists—toward the problems involved in that theme. The nineteenth century had begun with a number of important scientific discoveries and more were confidently expected, an expectation which, as we know now, was realized to an almost unbelievable extent. New astronomical discoveries were expected and there was a feeling (well exploited by Locke) that they might follow the invention of new instruments. This also came true in the latter part of the nine-

[1] The complete text of the moon hoax was reprinted in *The Sky* in several installments, beginning in February 1937, with an introduction by William H. Barton, Jr.

teenth century with the introduction of the camera and the spectroscope. Herschel's alleged "light infiltration" was a guess at what might be invented. Animals and inhabitants and dwellings on the moon were a guess at what might be discovered, and here again Locke merely expressed a prevalent belief, although the moon was not too well chosen as locale. One of the planets, except for the greater distance, might have been more convincing.

The "plurality of worlds" was no longer a matter of dispute. Venus and Mars and the other planets were acclaimed as "other earths," while the stars were now definitely taken as "other suns," each one presumably with a system of planets around it. And all the suns together formed the Milky Way, the galaxy. In addition, there was reason to suspect that the so-called nebulae, especially the big one in Andromeda, were other galaxies like ours. Sir John's trip to the Cape had as one of its main purposes the collection of more data on the existence of these additional galaxies.

But this growth of astronomical knowledge had also resulted, as we have seen in the preceding chapter, in a literal "growth" of the universe. The dimensions were now such that even a million miles were no more than in inch as applied to the dimensions of a city. Astronomers began to use bigger yardsticks. The distance between earth and sun, 93,000,000 miles, was chosen as an astronomical unit (A.U.), and the velocity of light, some 186,000 miles per second, furnished another yardstick. Astronomers began to say that the moon is a little over one light-second from the earth, and the sun some eight light-minutes, meaning, of course, that the light from the sun needed about eight minutes to reach the earth. But even on that scale the *nearest* star was several light-*years* away.[2]

The feeling was that the universe was full of suns, each accompanied by planets most of which were probably inhabited. But all the various inhabitants could do, even those closest to one another in the same solar system, was to look at one another. A vacuum separated them and nothing could bridge it except gravitation—which was unchangingly ever-present and therefore quite useless—and light.

While the nineteenth century became the era of the greatest astronomical dreams, it was some time before the dreams included direct communication. There were many books and articles on the inhabitants of other worlds, but at first nobody suggested that they might visit us or that we might visit them. The early enthusiasm about the invention of the balloon had quickly changed into disappointment when balloons proved unable to carry their pilots more than a very few miles up, and even that distance with great danger to life and limb because of the cold and lack of oxygen experienced at higher altitudes. The highest balloon ascent was no more of a departure

[2] In modern astronomy the parsec is used along with the light-year. The term means a distance which produces a *par*allax of one *sec*ond of arc, an apparent yearly shift of 1/3600 of one degree. A parsec works out to 3.259 light-years; it is not much bigger as a unit, but it is more convenient to handle. A light-year is 5880 billion miles.

from the native planet than the distance a fish may jump out of its tank. And it was fraught with the same dangers—suffocation and inclemency. As a matter of fact, the planets began to look like cosmic fish tanks and the various presumed humanities on them like self-contained little worlds in the well-kept tanks. They had been put there by the act of creation and sealed into their particular surroundings. Subject to catastrophes even in their own worlds—Monsieur Cuvier told terrible things about such catastrophes—they could do nothing but look out. Dependence on their surroundings forced them to remain at home, and the most they could hope for was to see other fish tanks and guess about those other surroundings.

That, in approximation, was the answer of the year 1830 to the problem of space travel.

But even if those fish could never possibly reach one another, they might—assuming that they had the necessary intelligence—signal one another. In fact this was the way to find out whether the fish in another tank were intelligent or not. If they understood a signal and replied to it, they were. If they did not reply, either they did not exist or, if they did exist, they were not quite bright enough. Of course one had to use a language for the signal which could not possibly be misunderstood.

This was, of necessity, the language of mathematical symbols. For example, if the figure known as the "Pythagoras"—the figure which expresses the fact that the square of the longest side of a right-angled triangle is equal in area to the sum of the squares of the other two sides—were produced on earth on a scale large enough to be seen from another planet, intelligent beings of that other planet—provided they existed and provided they were intelligent enough to know elementary geometry—would know about us, and they would reply in a similar manner.

The Siberian tundra offered a "blackboard" of sufficient size. To "draw" the symbols was a job for farmers and foresters. The outlines would have to be planted in dark pine forest, each line 10 miles wide. And for contrast the interior of the triangle and of the three squares could be seeded with rye or wheat.

This was not a crackpot's idea; it was advanced by Karl Friedrich Gauss, the "Prince of Mathematicians."

And Littrow of Vienna, a name well known to any astronomer, professional or amateur, followed up with another similar idea. Use the Sahara as a blackboard and use large fires for signs. Arrange them in a circle, for example. Dig a circular trench several hundred yards in width and 20 miles or so in diameter, fill it with water, and pour kerosene on the water, enough kerosene to burn for six hours. Then ignite it. Use a quadratic trench the following night and a triangular trench the night after. This regular succession of regular mathematical symbols ought to convince the inhabitants of the nearest planets not only of our intelligence, which grasps the mathematics of simple figures, but also of our intention to give signs to them, since we would never need flaming signs of that size for teaching our own kind.

While Gauss and Littrow evolved these schemes for signaling the other planets, other astronomers frantically searched for signs their inhabitants might send to us.

In July 1822 a Munich astronomer with the strange name of Franz von Paula Gruithuisen (*Khroyt-hoyzen*), who was also a somewhat strange man all around, announced that he had found a "walled city" on the moon. It is located near Schröter, a crater almost in the center of the half of the moon we can see. Gruithuisen described it as a regular grid of high walls and drew a picture which looks like a bird's-eye view of a multiple-walled fortress, partly submerged in a flood of mud which later hardened.

The announcement created a stir which can be traced through the whole literature of that time—it is even mentioned in some love stories written about 1830. Any definite building on the moon or elsewhere would be a sensation of the first class. And a kind of fortress would be a solid basis for quite a number of justified conclusions. It would be unmistakable evidence for the existence of intelligent beings—at least the former existence of intelligent beings, if you want to be very careful. It would also prove that there is or was another and hostile life form about, presumably intelligent too. It would prove a lot of things—provided only that its nature were established.

Gruithuisen's walled city is *not* established. The German astronomer Mädler who, some time after Gruithuisen, produced maps of the moon which remained the best for many decades, drew the famous spot as an irregular pattern of small mountain chains. Modern observations seem to indicate that both exaggerated. The mysterious spot is not as regular as Gruithuisen drew it. Nor is it as irregular as Mädler made it out to be. At any event, it is something for which we don't know a counterpart on earth. If it is not a work of intelligent beings, there is no natural form on earth which is strictly comparable.

Nor, it may be added, do we have anything on earth which is comparable to the Great Valley of the Lunar Alps, or to the 70-mile-long Straight Wall near the crater Thebit in Mare Nubium. As for the Great Valley, the only possible explanation is that it was caused by a very large meteorite, grazing the surface of the moon, while the Straight Wall, also called the Railway, is presumably a rocky fault, with one ledge higher than the other. The fact is that we don't have such things on earth. We don't have such a valley at all and our rocky faults are not only much smaller but also not as straight. Which, when applied to Gruithuisen's "walled city," means that the mere fact that no comparable natural form exists on earth does not imply that the lunar form has of necessity to be man-made.

The same Gruithuisen also found "life on Venus" in rather extreme manifestations. But before we go on to this, we must mention what is probably the wildest astronomical hypothesis ever advanced.

After Hevelius, as we have seen, the moon gradually fell into disrepute as an inhabited planet. The steadily improving telescopes left no doubt that there were no clouds on the moon. The darker areas which had been called *maria* (seas) with early enthusiasm turned out to be just darker and generally smoother areas. They might still be the bottoms of old seas—although even that is by no means sure—but they certainly were not seas any more. There was, in short, no water on the moon. And there was no air to speak of; when the moon's disk covered a star, that "occultation" occurred precisely at the proper second. If the moon had an appreciable atmosphere there would be a delay. Without air and water there could be no life as we know it. The lunarians, or Selenites, or Endymionides, if they ever existed, belonged to the past.

The nineteenth century made an attempt to "save" them.

The moon, by turning once around its axis while moving once around the earth, succeeds in keeping its opposite hemisphere permanently hidden from the interested eyes of our astronomers. That is, the moon almost succeeds, because there is a slight wobble technically known as libration which permits small portions near the rim to be seen on occasion. By accumulating such observations astronomers have succeeded in seeing about four-sevenths of the lunar surface. But the remaining three-sevenths simply have never been seen, nor can they ever be seen by observers stationed on earth.

Now the motion of the moon is peculiar; in fact it is one of the really difficult problems of mathematical astronomy. For a while it seemed as if the evidence pointed at a very peculiar solution. It seemed that the moon was not spherical as all other planets are—spherical, that is, within the customary deviations from the true mathematical sphere—but egg-shaped. The sharper end of the egg was the one pointing toward the earth, and it seemed as if the visible hemisphere was the enormous bulge of an irregularly shaped body—as if the whole hemisphere we can see, the whole hemisphere with all its craters and maria and mountains, was in itself an enormous mountain.

Then, if that were true, the other side would certainly be very different. The waterless and airless moon would be waterless and airless only where we could see it. It would be waterless and airless because it is an enormous mountain. The "back" might resemble earth much more; it might be an enormous tropical jungle with dense and steaming air. It would have to be a watery jungle, it was reasoned, because naturally all the water of the moon would be concentrated on that one hemisphere.

Suddenly it seemed as if the old Pythagorean idea of the "uninhabitable hemisphere" of earth, the hemisphere which cannot be inhabited because it has the "central fire" overhead, were illustrated in the sky. (The uninhabitable hemisphere was, of course, the Western Hemisphere.) The moon seemed to conform to this old fantasy with an enormous bulge sticking out into cold and airless space. From declaring that the other hemisphere was likely to be much more earthlike and inhabitable, it was only one step to saying that it was inhabited.

That hypothesis was never generally accepted by astronomers, but for some time it was an acceptable hypothesis; things *might* be that way. The moon "bubble" was pricked by Simon Newcomb who spent a great deal of his long and busy life investigating the orbital motion of the moon. It turned out that the moon is not precisely spherical, but the difference is too small to matter much. The gigantic earth-pointing "mountain" collapsed and with it collapsed the interesting idea of the watery and inhabitable "other hemisphere."

Decades later, when the dream was almost forgotten in astronomical circles, it was resurrected for literary purposes by the Polish poet, Jerczy von Zulawsky, in a novel *On Silvery Plains,*[3] in which the crew of a spaceship from earth sets out on a long trek across the visible hemisphere of the moon to reach the other hemisphere. Some of them succeed and a strange race grows up there, the children of the arrivals from earth.

As photographs with moon-orbiting devices (the Lunar Orbiters, see Chapter 12) have proved that the "other hemisphere," with one striking exception, looks like the hemisphere we can see. The exception is that the far side of the moon is heavily cratered all over, without the large dark *maria* that are naked-eye objects on the visible hemisphere.

If Gruithuisen felt certain of life on the moon—which, of course, meant intelligent life—he was more than certain of life on Venus.

It was well known in Gruithuisen's time that Venus is almost precisely a second earth as far as size and mass go. But Venus, circling the sun inside the orbit of earth, would be generally warmer. It would be a tropical planet with a vegetation, to use Gruithuisen's own words, "incomparably more luxuriant than even the virgin forests of Brazil."

Now Venus exhibits a strange phenomenon from time to time. The dark parts, the parts where it is night on Venus, are not really dark, but glow with a faint luminosity, reminiscent of the so-called earth-light of the moon. The earth-light of the dark parts of the moon is, as the name implies, caused by the earth; these parts have night as far as their position in relation to the sun is concerned but they are illuminated by the Volva, to fall back on Kepler's name for the earth in the lunar sky. It is possible that the night side of the earth shows a similar faint light to an outside observer when there is a full moon illuminating it.

The very first observers of the Venus-light explained it just that way—

[3] This is a translation of the title of the German edition, *Auf silbernen Gefilden.* As far as I know there are no other translations of Zulawsky's works.

there was night on Venus with a full moon in the sky. But these first observers had a moon of Venus at their disposal. We haven't. The "satellite of Venus," reported by several early astronomers, does not exist. We are perfectly certain now that Venus, like Mercury, does not have a moon; those early observers mistook a fixed star for a moon of Venus. There was, consequently, no explanation for the faint luminousness of Venus's night side.

Gruithuisen found one.

Going over astronomical records he noticed that "the principal observations of the ashen light on Venus are those of Mayer in 1759 and of Harding in 1806." There was an interval of 47 years or of 76 Venus years, an interval which makes no sense whatever from an astronomical point of view.

Gruithuisen agreed with this: "If this period has a religious character we cannot see any justification for that number of years. But," he continued, "it becomes more comprehensible if we assume that some Alexander or Napoleon then attained universal power. If we assume that the ordinary life of an inhabitant of Venus lasts 130 Venus-years, which amount to 80 earth-years, the reign of an Emperor of Venus might well last 76 Venus-years."

The ashen light of Venus, to Professor Franz von Paula Gruithuisen, was the astronomical result of a general festival illumination in honor of the ascension of a new emperor to the throne of the planet. Later Gruithuisen was apparently frightened by his own hypothesis and amended it by saying that it might simply be the burning of large stretches of jungle to produce new farm land—he just missed saying "new *Lebensraum*"—for the growing population of the evening star. "Large migrations of people would be prevented and the resulting wars would be avoided by abolishing the reason for them. Thus the race would be kept united."

To think of a phenomenon like our aurora borealis and aurora australis in order to explain the ashen light would have been *much* too simple.

It was, of course, difficult to surpass Gruithuisen, but the French inventor Charles Cros did his best. His *Moyens de communications avec les planètes* was published in Paris in 1869. The fact that luminous pin points had been occasionally observed on Venus and Mars—in all probability high clouds which were still struck by the light of the sun while the surface was already in darkness—was sufficient for him to state that the inhabitants of these two planets were trying to communicate with the earth. Cros proposed to answer by means of an enormous mirror, similar to the mirrors used in reflecting telescopes, but many many times as large and of a very shallow curvature.

It was to be so shallow that the focal point was no longer on earth at all but on the surface of the planet with which communication was to be established. If the mirror were large enough to collect a sufficient amount of sunlight, Cros reasoned, and if the focal point of it were near the surface of the Martian desert, the power of the sun's rays would fuse the sand there. Then it would be possible to "write" gigantic figures in the sands of the deserts of Mars: first simple symbols, like triangles and circles, then more complicated mathematical figures, and finally even simple pictures: the head of a man, the outline of a house, and so on.

Cros wanted the French government to pay for such a mirror and he spent many years of his life making petitions, writing memoranda, and pulling what strings he could. Nobody ever succeeded in convincing him that the existence of the Martians was, after all, not assured. Nobody ever succeeded in convincing him that no optician could build a mirror of sufficient size and precision. Cros stubbornly clung to his idea. He died disappointed and in extreme poverty.

The theme of devising means of communications with other planets has been a recurrent one ever since, although more in the spirit of intellectual curiosity. The German astronomer Plassmann once devoted a paper to the question of whether the Martians could see the light of our large cities. It was written in the early nineteen-twenties, when cities were much larger and much brighter than in Gruithuisen's and Cros' time. He came to the conclusion that the Martians, provided their eyes were as good as ours and provided that their instruments were as good as ours (and no better), would never be quite certain whether they did or did not see pin points of light in those spots where terrestrial maps show the names of Moscow, Berlin, Paris, London, New York, Chicago, and Los Angeles. With somewhat better instruments they could be certain that they did.

Plassman disregarded the Venusians completely in his article, and with good reason. Blanketed as they are by their eternal cloud layer, they certainly would not be astronomers and probably would never have invented the telescope. As a matter of fact, it would be easy to "prove" that the presumed Venusians in their presumed environment would not have been able to attain any intellectual level worth mentioning because their environment was not suitable for intellectual development. Intellectual or at any event scientific activity begins with the realization that effect follows cause, that the same cause always produces the same effect, and that there are natural laws. On earth, as far as we can find out, such activity was stirred or at least encouraged by the regularity of certain astronomical features. The moon went through its phases regularly, the seasons repeated regularly, the stars seemed to wander across the sky in a measured tread. The orderliness of the sky suggested orderliness in nature in general and investigation of nature was begun and finally progressed, in spite of many handicaps in the form of faulty analogies, mistakes, and misunderstandings. Without the starred sky as a starting point nothing like that might ever have taken place.

As for the Selenites, if they exist, it may be said that their knowledge of the earth would be large in extent. Again assuming that their eyes and instruments were as good as but not better than ours, their maps of the earth should be excellent. In fact, because of their favored position for the observation of the surface of the earth, their maps would surpass our own for a number of areas, including some parts of Siberia, of Inner Australia, of Antarctica, and even sections of Alaska and South America. They would know the location of any town of more than a few thousand inhabitants and they could follow the maneuvers of a battle fleet and even of single large liners like the *Queen Mary.*

Some other aspects of interplanetary communication have also been re-examined more recently, of course again in the spirit of intellectual curiosity, not in the form of "projects," even if the newspapers occasionally called them that. An English engineer, for example, calculated the amount of flashlight powder needed to produce a flash that could be seen from Mars, provided the Martians knew where and when to look for it. I don't recall the figure, but it was enormous. Things would be somewhat better if the flashlight powder were not exploded on the ground but raised above the densest layers of the atmosphere by means of captive balloons which rise as high as they can possibly go and are exploded by electric ignition along with the flashlight powder. Under that assumption the figure for the amount needed looked somewhat better but it was still on the order of a trainload.

And finally some people have amused themselves in working out a "letter to the Martians." The purpose of that mental exercise was to find out whether it would be possible at all to establish a mutual understanding with beings of intelligence but of a completely alien race and environment. The "language" would have to be partly pictures and partly mathematics. The letter would begin with a single dot, accompanied by the figure 1. Then two dots accompanied by the figure 2. Then three dots with the figure 3, etc., until the meaning of the figures would be explained. Then the "letter" would progress to simple arithmetic and algebraic operations and to elementary geometry. The meaning of the symbol of an arrow (direction) would have to be explained and then a picture of the earth could be shown, with an arrow pointing to the picture of Mars. Of course there would be a diagram of the solar system with more arrows drawn in. It seems likely that a message of this type would show to anybody who can think and see that it came from earth.

All these recent amusements, you may have noticed, spoke about the Martians as an example. This in itself is a modern thought. Around the middle of the nineteenth century Mars was still generally disregarded. It had not yet become important as a planet about which to speculate. If interest centered upon any specific planet, it was, thanks to Gruithuisen's efforts, Venus. Naturally such an exciting thought could not fail to leave its mark upon literature and the first novel about a trip to Venus was not long in coming.

Its author was a Frenchman, Achille Eyraud. It was printed in Paris in 1865 and its title was simply *Voyage à Vénus.* It was a small book without much literary value, which completely failed to create a stir and never got beyond the first printing. In retrospect, however, it is important as signifying a change in attitude, for Eyraud did something from which writers had shied away ever since Cassini finished his calculations about the size of the solar system.

Eyraud "invented" a spaceship for his novel and described its working principle. And he must have had a good working knowledge of theoretical physics or else the help of a professional physicist because he described the right principle: his ship was propelled by a *moteur à réaction.* He explained to his readers that a fireworks rocket rises by means of reaction, that the recoil of a gun was the same thing, and that his *moteur à réaction* was a refinement of a gun's recoil.

For many years I knew about Eyraud's novel only by way of a summary in the appendix of a late edition of Camille Flammarion's *Les Mondes imaginaires et les mondes réels.* When I finally found a copy in the Library of Congress, I saw to my great surprise that Eyraud had not only proposed a rather modern thought, but that he had also made a "modern" mistake. His *moteur à réaction* was to operate with water as "reaction mass." But he wanted to avoid the loss of the water. So he reasoned that the water should not be thrown out into space but into a big container from which it could be collected to be used over and over again. That, as we shall see later, is nonsense. But a reaction motor of just this type was proposed in 1927 by the Austrian engineer Franz Abdon Ulinski. If an engineer could make such a mistake in 1927 one may forgive a novelist of 1865 for catching himself in a self-constructed trap. Actually, it is not important that Eyraud made this mistake. It is slightly more important that he at least used the theoretically correct principle. What is really important is that, by assuming such an invention, Eyraud expressed the belief that science, after having made the whole problem impossible, would produce a solution too. If one branch of science had separated the worlds hopelessly by finding enormous distances and airless space, another branch of science might be able to discover means of bridging those distances and of braving the nonexisting environment of a vacuum.

Seen from this point of view, Eyraud's novel simply expressed confidence in the powers of science and discovery, a confidence well fortified by the fact that during the preceding half century a number of important discoveries had been made and a number of important inventions had been put to work.

It was the same feeling which produced the work of Jules Verne.

Kenneth Allott, in his excellent biography, has taken great pains to point out that Jules Verne's writings are romantic in sentiment. Of course they are, since explorations of the unknown and possibly also of distant things are by their very nature "romantic." But the more important thing is that Jules Verne represents a new attitude. Consistently his heroes (who are, of course, merely personifications of the scientists, engineers, and explorers of the nineteenth century) do things themselves. They do them in a novel way. They don't do things in a traditional and poor and inefficient manner for the sake of tradition. Nor do they look for "lost arts."

They are not even very romantic in behavior; they are mainly truthful. Instead of yielding to the traditional modesty of being "insignificant sons of great ancestors," they act with the full knowledge that their time has surpassed any preceding time. They know that they know more than their ancestors did.

They expect their sons to be better than they are and they expect the future to be greater than the present. They don't hesitate to cruise under the seas or to fly through the air. And to them the problem of reaching the moon is what it really is: a question of attaining a sufficiently large velocity in the right direction at the proper time.

Jules Verne's *De la Terre à la Lune* appeared in 1865, the same year that Achille Eyraud's novel saw print. It was a prolific year for that sort of literature. Alexandre Dumas published his *Voyage à la Lune,* Henri de Parville published his *Un Habitant de la planète Mars,* an anonymous French author published another *Voyage à la Lune,* an English clergyman *The History of a Voyage to the Moon,* and, as if to round out the picture, Camille Flammarion published his *Mondes imaginaires et mondes réels,* which is essentially a description of all preceding works of literature that deal with astronomy, astronomical philosophy, plurality of the worlds, habitability of the planets, and attempts at interplanetary communication.

The book by Dumas is unimportant though entertaining, the Mars story by Henri de Parville is essentially a treatise on the plurality of the worlds. This story shares with the anonymous *Voyage à la Lune* a literary trick: both begin with the finding of strange meteorites that are actually messages. The story by the English clergyman (he wrote under the name of Chrysostom Trueman, which seems to be a *nom de plume*) brings two new ideas. The ship is propelled to the moon by a substance that is not attracted by the earth but repelled by it, it was found in Arizona or Colorado weighed down by a layer of rocks. The ship lifted by this substance is a large oblong box, like a railroad freight car, but inside there is a gallery running all around the interior, giving access to quarters. The rectangular space in the middle below the gallery is filled with top soil and growing plants which produce the oxygen for the travelers—this is the first appearance of the idea of a balanced ecology inside a space vehicle.

Jules Verne gave the greatest effort to trying to make his method credible. He had decided on a cannon as the proper means of bridging the gulf between earth and moon. He had calculated—and had his calculations checked by astronomical experts—just what muzzle velocity a cannon would have to have to throw a projectile to our satellite. He threw in a number of things then new: the propelling charge was to be guncotton, invented only some fifteen years before by Schönbein; the ball (which was to be large enough to be seen in flight by a new giant telescope) was to be of aluminum, which at that time was still so rare that chemists made one another presents of little bars weighing a few ounces. In fact the whole book is the story of the preparations for the great event.

The locale of the story is America and the main characters are all members of a fictitious organization, the Gun Club, consisting of gunnery experts who served in and survived the Civil War, although not always complete in their anatomical make-up. The leaders of this club at first only want to send a projectile to the moon in order to show that it can be done. But then a young man by the name of Ardan appears and proposes to be a passenger of the projectile.

Jules Verne had his little joke with that name. Everybody in Paris at once recognized the anagram; Ardan was Nadar and Nadar was Felix Tournachon, a Paris journalist. "Nadar" had studied medicine, but then turned newspaperman and photographer. In 1852 he had opened a studio in Paris, but his puttering in the darkroom did not prevent him from being active in aviation. He built a balloon, the *Géant*, which ascended for the first time in 1863, with Jules Verne in the gondola, and later he built the first of the balloons which left besieged Paris during the Franco-Prussian war of 1870–71. "Nadar" was also Lesseps' secretary for some time.

Well, Ardan-Nadar compels the Gun Club to revise the plans. The round ball has to be redesigned as an elongated projectile, padded and furnished inside, and equipped with "water buffers" to enable the passengers (they are three, in the end) to survive the shock of firing.

Two years after *De la Terre à la Lune* came *Autour de la Lune*, the story of the three heroes imprisoned in the aluminum shell. As soon as they recover from the shock—actually it would have done more to them than just kill them—they begin to spout astronomical and mathematical wisdom in large doses. They get an additional reason for doing so because their projectile passes earth's "second moon," which slightly deflects its course. This proves to be their salvation for, instead of striking the moon, the shell misses and, caught by the force of lunar gravitation, circles around it. Unfortunately for Jules Verne's readers, the side of the moon which cannot be seen from earth is in darkness, and all that the three heroes can see is what appears to be a volcanic eruption. Finally they cross into the field of terrestrial attraction and fall back to earth, landing safely although tumultuously in an ocean from which they are rescued.

Although these two books are not easy to read, Jules Verne's two-volume story of a shot to the moon has become the classic moon-trip story and its fame is universal. This has had the rather unfortunate result of associating space travel with shooting in the minds of most people. But what Jules Verne really accomplished was to acquaint the public with the fact that a trip to the moon is a question of velocity. That he used a gun to produce that velocity is merely a literary device which is not to be taken seriously.

Just the same, this method has been examined critically more than once, and such an examination of Jules Verne's cannon shot is very interesting and teaches a lot about the actual problem. But because it involves too long a scientific discussion I am going to postpone it until it can be compared with modern conceptions. (See Chapter 10.)[4]

Possibly because in 1865 there were so many novels about visits to other planets, it was some time until the next worthwhile piece of fiction of this type was published. Its author was an Englishman, Percy Greg, and his long novel (two volumes) bore the title *Across the Zodiac.* It was published in 1880, and a German translation followed in 1882. It seems that the English edition is now much rarer than the German translation and for this reason I was misled for quite some time into believing (as I said in the first edition of *Rockets*) that the German version was the original and only *claimed* to be a translation from the English.

In any event it is a fairly interesting book. It too begins with the fall of a meteorite which buries itself in one of the South Sea islands. When the wreckage is examined, a strong box of an unknown metal is found in the meteor crater, containing a fat manuscript written in unknown characters. At first nobody believes that these characters can be read. But the finder of the box reasons that the writer may have been an earth man who did not know, of course, in what country his vessel might land or crash. In that case he would not write in his native tongue, whichever that was, but, if he was able to do so, in the one language which can be read by a number of people in every nation: Latin. Decoding the manuscript on this assumption proves easy. It is Latin, written in a different alphabet, no more difficult than English would be in the Greek or the Russian alphabet.

It turns out that an engineer who pondered about gravitation had the thought that there should be something like "negative gravity" which he calls "apergy." After some experimentation he succeeds in discovering this apergy and sets out for a trip to Mars, where he finds people greatly resembling those of earth, only somewhat smaller in stature. The remainder of the story is devoted to a description of Martian society, but the novel

[4] Jules Verne's cannon shot is discussed in detail by Max Valier in *Der Vorstoss in den Weltenraum* and in *Raketenfahrt*. Incidentally, the calculations, although not credited by Valier, were made by Hermann Oberth. Another discussion is that by Count Guido von Pirquet in my book *Die Möglichkeit der Weltraumfahrt* (1928), and a list of the mistakes may be found in "Marvelous Voyages—III" by Dr. Laurence J. Lafleur, *Popular Astronomy* (April 1942), vol. L, no. 4.

is not a Utopia. That society is by no means an "ideal state"; there are a lot of things wrong with it. It is a different society, logically constructed.

The date of that novel is, as has been said, 1880. And we are now entering the period of speculations about Mars. After the "moon mountain" of Hansen's hypothesis had collapsed, the moon had finally been given up as dead. Venus was neglected and Mars was to rule supreme.

This shift in interest was not accidental; it was the logical result of scientific developments. The progression of these developments was about as follows: In 1828 Wöhler had shown that living substances consist of the same elements as nonliving matter. In 1859 Kirchhoff and Bunsen had developed the spectroscope which, within its natural limitations, proved that the other stars and planets consisted of the same elements as earth, although the distribution of elements varied.

As if to make the case airtight as soon as possible, Dimitri Mendeleyeff and Lothar Meyer—working independently—produced the periodic system of the elements, showing that only a limited number of elements was possible. Mendeleyeff's work was published in 1869. Charles Darwin's *Origin of Species*, published for the first time in 1859, had to be added to these developments in the fields of physics and chemistry. The controversy about Darwin did not begin immediately, but by 1880 it was already clear who would win. Now Darwin's theory meant that all life was interrelated, not only in a symbolic manner of speaking, but actually. Wherever life started at all it was likely that it, given enough time, would produce intelligent beings in the end. And it seemed probable that life would start wherever conditions were suitable and that it would start soon after conditions became suitable.

On top of all this, astronomers believed they had a correct explanation of the formation of a planetary system of a sun. It was the so-called Kant-Laplace hypothesis—in spite of its name it was mainly Laplace's conception—which stated that the planets had condensed out of matter that had been thrown off their suns. Obviously, then, the relative ages of the various planets could be read from their distances—those farthest away were the oldest.

Mars, therefore, was older than earth by an unknown number of years, probably a large number. Consequently life had started that much earlier on Mars; consequently Martian humanity was older and wiser than we are. By the same token Venus was younger and had probably not developed intelligent beings yet; Venus was a replica of the earth at the time of the carboniferous forests or at the time of the great dinosaurs. But Mars was a replica of the earth as it was to be later; looking at Mars one looked, actually and almost literally, into the future.

The year 1877 approached.

It was the year in which Asaph Hall discovered the two small moons of Mars. But it was also the year in which still more exciting news about

Mars came from Italy, originating from an astronomer by the name of Giovanni Schiaparelli. The news said that Schiaparelli had seen *canali.*

Now the Italian word *canali* simply means "grooves" or "channels." The meaning of "canals" is secondary, although *canali* is also used to mean artificial waterways. But in all other languages the word "canal" means only that, and Schiaparelli's *canali* was quickly, wrongly, and emphatically translated into "canals." The jubilation was great; everything seemed to fit perfectly. The planet which had logically been pronounced the most likely abode of intelligent life in our solar system now showed by gigantic works of engineering that it really was.

Mars, as had been suspected from the combination of Kant-Laplace and Darwin, was older than the earth. Because of its age it had lost most of its water. But its inhabitants, also because of their greater age, had progressed in technological knowledge and the spirit of mutual cooperation to the point where they could overcome that handicap. They had covered their planet with a network of canals which led the water from the melting polar icecaps over all the continents. Those Martian canals were waterways, but they served as means of transportation only secondarily; primarily they served as gigantic irrigation ditches, as center lines of irrigated farm land.

This conception also disposed of the first critical argument—the fact that a Martian canal, to be visible from earth, would have to be at least 30 miles wide and that many of the *canali* were three, five, and even ten times that wide. Of course the lines we saw were that width or else we could not see them. But what we saw was not the canal itself; it was the wide ribbon of vegetation that stretched through the deserts alongside the canals. There was one river on earth which probably looked to the Martians like one of their canals, for the same reason and especially because it happened to be almost straight: the Nile!

Everybody knows that Schiaparelli's first announcement was followed by three decades of Mars enthusiasm, three decades during which reports from astronomical observatories were awaited and read as avidly as reports from the front in the middle of a war. Several amateur observers reported light signals from the Martians, most of which turned out to be high-flying whitish clouds. Astronomers made careful maps of the surface of Mars, and many who studied them looked for something on the order of the mathematical forest in Siberia proposed by Gauss. Such convictions were so strong that they led to a qualifying paragraph in the famous *Prix Guzman* which was deposited in Paris some time around 1900. The prize amounted to 100,000 francs, then $20,000, and the award was to go to the person or persons who established interplanetary communication. But Madame Guzman, when founding the award, stipulated that Mars be excluded because communication with Mars would be too easy to deserve a prize.

It goes without saying that literature did not fail to contribute to these decades of the great dreams. Much later, in 1927, the popular astronomical museum at Treptow near Berlin staged a "Mars Exhibit," consisting of maps of Mars, greatly enlarged photographs, and a large collection of books. And at that exhibit I met an elderly gentleman who swore that he had originally learned German for one reason only: to be able to read the novel *Auf zwei Planeten* by Kurd Lasswitz.

Kurd Lasswitz was born in Breslau in Silesia on April 20, 1848, the son of a well-to-do merchant with some, for his time, highly unmerchant-like characteristics. He had a small observatory on the roof of his suburban home and later had himself elected to the Prussian parliament (*Landtag*) as representative of the Democratic Party of his district. The son studied philosophy, natural history, mathematics, and—naturally—astronomy. In 1876 he became professor of mathematics in Gotha, where he spent most of his life. The novel *On Two Planets* was neither his only book nor his only work of fiction, but it was the one that made him famous. It has been a permanent part of German literature since its first appearance in 1897. The Nazis banned it as "democratic," but it was re-issued in 1948 in celebration of the hundredth anniversary of Lasswitz' birth.

Lasswitz was almost fifty years old when he wrote it and by that time he was also a privy councilor in addition to being professor of mathematics. Presumably he kept an appointment book in which he also jotted down other things, for it is on record that he began writing *Auf zwei Planeten* in November 1895 and finished it on April 11, 1897. It appeared during the last week of October, 1897, and within ten years had been translated into Swedish, Norwegian, Danish, Dutch, Spanish, Italian, Czech, Polish, and Hungarian.

Kurd Lasswitz had carried that logic about the older Martian race a few steps farther. If the Martians were so much older it would be a succession of Martian inventors and scientific institutions that had gradually solved the difficult problem of space travel. Consequently it would be the Martians who would be the first to visit earth, not the other way round. Also, these *canali* were really strips of vegetation through the deserts, wide bands of tall forests. (Under the lighter gravity of Mars trees would be higher.) Those forests shadowed the buildings of the loosely built cities and the moving roadways. (When traffic is too heavy in volume a moving roadway is superior to independent vehicles.) The precious water was carried in pipes to prevent evaporation, and none of the carefully preserved water was wasted on edible plants; food, under the Martians' advanced chemistry, would be synthetic and so inexpensive as to be virtually free.

It was partly a Utopia portraying a higher society; partly it was simple prediction: engineers have often considered rolling roads for specific traffic problems. And there cannot be a better solution for the basic problem of sociology, eliminating hunger, than synthetic food which can be produced in any required volume at very short notice.[5] It was also basic psychology to show that the highly ethical Martians, when confronted with terrestrial stubbornness, quickly revert to war, fought in a highly efficient manner with a minimum of actual killing.

But the most interesting thing in Lasswitz' novel was his solution of the problem of space travel. Lasswitz saw to it that his readers got acquainted with that problem early in the book.

The novel begins with a flight to the North Pole in a free balloon; the tragic attempt by Andrée to reach the Pole in this manner was still fresh in everybody's memory. When the Pole comes into sight, the men in the gondola see to their great surprise that there is an unmistakable building on the Pole, strange in shape, but not an Eskimo hut. At the same time the balloon begins to circle as if caught in a whirlpool, but it also begins to rise. Minutes later the balloon rises vertically, and the gondola rises faster than the gasbag. They are caught in a field of reversed gravity, induced by the connection between the building on the Pole and a structure high above the Pole, one earth-diameter from the center of the earth.

These are structures of the Martians, their first foothold on earth. The Martians get the travelers out of the field by turning it off as soon as the car which happened to be in transit between the Pole and the "outside station" has arrived on the latter. Revived, the three earth men are informed about the accident that has happened to them. Later on they are taken to the outside station and asked to point out their homes; the Martians do have that kind of telescope that Locke ascribed to Sir John Herschel. But Lasswitz explains it better. The original image falls on a small screen consisting of millions of tiny photoelectric cells. Each cell operates an "optical relay," releasing visible radiation of precisely the same color but of much higher intensity, and these rays form the image on the large screen. (Lasswitz, incidentally, had astronomical experts check the precise range of such a presumed instrument in that presumed position.)

And then the earth men learn how the Martians crossed space. They succeeded in developing a material which, although not weightless in itself, had the quality of becoming so as soon as it formed an unbroken shell. Lasswitz vaguely indicates that gravity will "flow around it" in this case. And all the things inside become "inaccessible" to gravity. Thus a Martian spaceship—they are spherical—becomes weightless as soon as the last door is closed. If that were done near earth the weightless ship would at once move away from earth's orbit along a tangent and in time reach the orbit of Mars, to land on Mars if the planet happens to be near that

[5] Later, when synthetic food is introduced on earth, it becomes more than just a chemical problem. Lasswitz compresses the whole struggle in one scene, the scene where a man who has been out of touch with events, orders supper in a small restaurant, deciding on a liverwurst sandwich. The waiter asks whether he wants natural or synthetic.—" 'What, synthetic?'—'Yes, sir, Martian style, sir. You see, sir, "grown" is considered patriotic, but synthetic is much less expensive and, really, it's much better.' "

point of its orbit at that time. Or else the Martians can take off from the home planet in the same manner, wait until their planet is sufficiently far away, fall toward the sun until they have reached the orbit of earth, and then land on earth.

The first attempts, the earth men are told, were unsuccessful and ended in disaster, partly because of the clumsiness of the method, partly because that marvelous substance, developed on Mars, disintegrated in the moist atmosphere of earth. They had to have two improvements. One was the station over the Pole—they have similar stations over their own poles because "you do not jump onto a train when it is moving but when it is standing still"—and secondly a method of steering, accelerating, and stopping their ships in space. The solution to the latter problem was "repulsit," a substance somewhat beyond the grasp of terrestrial physicists, but essentially a highly explosive substance. The reaction of such an explosion would steer or stop or accelerate their ships, but they always needed quite a number of such "shots," each releasing a small dose of repulsit, since the jolt resulting from a dose of the proper magnitude would have killed every man aboard. Still, they could and did travel that way, finding only the earth inhabited by a race similar to themselves. But just recently a Martian engineer named Fru has made the final invention they need—Fru's "repulsor" which releases the repulsit charge in a prolonged push instead of a sudden explosion.

If we disregard the marvelous but impossible—and actually useless, as we see later—substance which makes things inaccessible to gravity, Lasswitz presented a solution of the problem of space travel as mathematicians see it. Once free of the immediate gravitational influence of a planet, a spaceship would drift to the orbit of another planet in a time interval which can be calculated, and it depends on the timing of the departure whether you meet the other planet at that point of its orbit or not. The reaction of a shot—preferably a pushing instead of jolting reaction—can be used as means to decelerate or accelerate in airless space. All this was worked out so well that most readers, as one critic put it, did not even notice that these pages of the book (only a score or so) presented a number of new thoughts.

One of the results of Lasswitz' novel was the re-issue of Kepler's *Somnium,* and the *Somnium* greatly influenced H. G. Wells when he wrote *The First Men in the Moon* a year or two later. In this novel, as most readers will remember, Wells uses a substance very similar to Lasswitz' Martian invention, but ascribes it to an earth man, the engineer Cavor, and consequently refers to it as Cavorite. Cavorite gets the two space-travelers to the moon and they are finally captured by the Selenites who live in the numerous hollows and caves under the volcanoes.[6] The Selenites have evolved from insects, and Wells infuses these beings with the spe-

[6] Wells probably misread "subvolvani" as "subvolcani," converting the "dwellers under the Volva" into "dwellers under the volcanoes."

cialized nature of members of a termite colony. Finally one of the travelers escapes and returns to earth.

Wells' other early interplanetary novel, *The War of the Worlds,* which appeared almost simultaneously, describes an invasion of the earth by the Martians. In this novel Wells used Jules Verne's method of shooting, but the book itself is devoted to the description of the attempted Martian conquest. The Martians are described as octopus-like creatures, rather frail, at least under the pull of terrestrial gravity. Consequently they move about mechanically, sitting in towering fighting machines equipped with heat rays and poison-gas nozzles. Finally the Martians succumb, not to human opponents but to the "invisible devils" of which the terrestrial atmosphere is full. They die of bacterial infection.

After these early works of H. G. Wells there followed a comparatively barren period. The main and most obvious situations had been used; technological imagination was restricted to side issues. Consequently the imaginative novel quietly dropped out of the history of the space-travel idea, of which it had virtually formed the mainstream for a considerable length of time. Its development slowed down and for many years it was almost an oddity, until it emerged again as the modern science-fiction story, equipped with fresh concepts, greatly expanded technological scope, and a new kind of plot structure. But even with all these improvements it could not regain the position it had held during the second half of the nineteenth century.

For in the meantime science itself had taken over.

3

"The Rockets' Red Glare"

Actioni contrariam semper et aequalem esse reactionem; sive corporum duorum actiones in se mutuo semper esse aequales et in partes contrarias dirigi.

Reaction is equal (in power) but opposite (in direction) to action; or the actions of two bodies are equal (in power) but point in opposite directions.

—Sir Isaac Newton's Third Law of Motion

The story of the rocket is part and parcel of the story of flight as well as of that of space travel. While this connection may have been obscured at times in the course of history it has always existed. And modern rocket theory has its ultimate foundation in the simple powder rocket which originated as a weapon of war, then became a device for amusement, turned weapon again, changed back to amusement, and has now turned weapon once more.

If it were only a question of theory, most of the history of the old blackpowder rocket could safely be neglected in a modern discussion. But because the application of rocket power is a problem of engineering design, the history of the rocket is important. Through the centuries rockets slowly grew and acquired refinements. Their history, therefore, mirrors a development consisting of many successive small achievements by the trial-and-error method and showing the limitations of the devices developed by the traditions of families of artisans. I refer, of course, to the later European tradition of which we have knowledge. But the rocket is not a European invention; the Europeans got it from the Arabs and the Arabs, in turn, had it from the Chinese.

There has never been any doubt that the rocket, the ordinary skyrocket, was originally a Chinese invention. But for want of definite knowledge, its origin has often been shrouded in legend and its age has been vastly exaggerated. Older books often state in an offhand manner that rockets and similar fireworks have been known to the Chinese since very early times, probably since 3000 B.C. Where the authors of these books got their information is something a historian with a talent for languages should investigate some day. There is no reason, and there never was a reason, to believe in such a great age. The oldest known

Chinese source in which rockets are mentioned is a chronicle known to sinologists as the *T-hung-lian-kang-mu,* and that chronicle dates the first use of the invention—though not the invention itself—as having been made in A.D. 1232. The occasion was a siege laid to the city of Kai-fung-fu (Pien-king) by the Mongols. But the Chinese had two new weapons which the Mongols disliked intensely. One of those weapons was a kind of bomb dropped from the walls of the city on the heads of the besiegers. The Chinese name for that weapon is *tchin-tien-lui* or heaven-shaking thunder. The other new weapon goes under the name of *fe-ee-ho-tsiang* or arrow of flying fire. These "arrows" are taken to have been rockets and the description supports that supposition.

That description, according to the French sinologist Stanislas Julien, reads as follows:

> The defenders also had "arrows of flying fire." They attached an inflammable substance to the arrow. The arrow suddenly flew away in a straight line and spread its fire over an area measuring ten paces. The Mongols dreaded these arrows of fire very much.[1] (Julien)

Those "arrows" are taken to be rockets because no mention is made of a bow or any other instrument for shooting them. It is merely stated that they flew away suddenly and in a straight line. Recently this explanation has been criticized on the grounds that a rocket with a range of only ten paces would be ridiculous. I rather think that the objection is ridiculous, for the book *T-hung-lian-kang-mu* doesn't say a range of ten paces, unless one assumes that Julien could not read Chinese. The ten paces refers to the area over which the weapon spilled its fire; whether that was 50 or 100 or 200 paces from the point of origin is simply not mentioned.

It is unfortunate that this book does not show a picture of one of the weapons. In all probability they were real arrows with rockets tied to them. In later Chinese manuscripts pictures of such arrows with rockets tied to them are quite common, as was pointed out for the first time by a French Jesuit missionary, Father Amiot. And travelers who have paid attention to such minor matters have asserted that Chinese fireworks rockets, at least as late as the year 1900, were feathered at the lower end of their guiding sticks, just as if they were arrows and notwithstanding the fact that the feathers burned away very quickly when the rocket was ignited. The arrows of Pien-king were literally "arrows of flying fire," a term which, to a European, would mean something entirely different: a fire arrow.

A fire arrow or incendiary arrow consisted of a small package of incendiary materials tied to an arrow. Such arrows were dangerous weapons in times when most fortifications consisted of wood, and they were even

[1] In *Journal Asiatique* (1849). The original reads: *"De plus, les assiégés avaient à leur disposition des 'flèches à feu volant.' On attachait à la flèche une matière susceptible de prendre feu; la flèche partit subitement en ligne droite, et répandait l'incendie sur une largeur de dix pas. Les flèches à feu volant étaient très redoutés des Mongols."*

more dangerous in later naval battles when sailing vessels had extensive tarred riggings. But rockets often substituted for fire arrows in these later naval battles and it seems that the two, the fire arrow and the fireworks rocket, are actually related to each other—one is the ancestor of the other.

The story of the fireworks rocket prior to the siege of Kai-fung-fu is not known, but more than half a century ago S. J. von Romocki evolved a brilliant conjecture about this unknown part of the rocket's history, a conjecture which has never been contradicted and which seems to be as close to actual truth as may be hoped for under the circumstances.

The fire arrow, the real fire arrow that was shot from a bow, was always an effective weapon in the wars of those days and it is only logical that many men sought to improve upon it. The main drawback of the fire arrow was its short range; it could not be shot from a fully drawn bow because that would extinguish the flame of the small package of incendiary materials tied to the shaft. Obviously it was necessary to find a material that would not be extinguished that easily. Now Greek philosophers held that salt, when added to an incendiary mixture, would make the flame hotter. The simple truth of the matter is that it merely makes the flame brighter, but since there was no way of measuring flame temperatures—or temperatures of any kind—in those days, the idea that the brighter flame was also the hotter flame was as obvious as it was beyond proof or disproof.

This "secret," the addition of salt to incendiary mixtures, seems to have traveled to China. And the Chinese seem to have substituted saltpeter for salt, either because they were not satisfied with the results of the "secret" or else because they accidentally discovered the properties of saltpeter. But when they added saltpeter they had a real improvement, because then their incendiary mixture was well on the way to becoming gunpowder. Gunpowder, as every schoolboy knows nowadays, consists of a mixture of saltpeter (about three parts) and pulverized charcoal and sulfur (constituting the fourth part in about even proportions). By about A.D. 1200 that process seems to have been completed. Hence the defenders of Kai-fung-fu had bombs which exploded with noise and violence and they also had fire arrows which had grown powerful enough to depart on ignition, without requiring a shooting instrument.

The real secret—namely what saltpeter would do to or for an incendiary mixture—was not confined to China for very long. Just eight years after the battle of Kai-fung-fu a learned Arab compiled a book in which saltpeter is mentioned. The name of that Arab was Abú Mohámmad Abdallah ben Ahmad Almaliqī, the last word meaning "from Malaga." Since he was also called Ibn Albaithar, meaning "son of the horse doctor," it is not too surprising that he was later quoted mainly by that designation and not by name. He called saltpeter "the flower of the stone of Assos" but added that this substance "is now called 'Snow from China' by the Egyptians." The Chinese influence is obvious; one may infer also from his statement that saltpeter may have been known before travelers returning from China told what could be done with it. As for the "people of the West," meaning the Arabs, Ibn Albaithar continued, they call the same substance *barood.* But at that point he stopped, without saying what they, or the Egyptians for that matter, did with *barood.*

We get more complete information about the things Ibn Albaithar may have been hiding from another Arab manuscript, *The Book of Fighting on Horseback and with War Engines.* It was written by one Hassan er-Rammah, described as a brilliant hunchback, whom his contemporaries adoringly called Nedshm Eddin (Star of Faith), and who seems to have completed his manuscript in A.D. 1280. The book contains recipes for making gunpowder and directions for making rockets. And the rockets are still called *alsichem alkhatai,* Chinese arrows.

But Hassan adds one unsuspected novelty: a rocket-propelled torpedo consisting of two flat pans, fastened together and filled with powder or an incendiary mixture, equipped with a kind of tail to insure movement in a straight line, and propelled by two large rockets. The whole was called the "self-moving and combusting egg," but no instances of its use are related (Fig. 4).

Some thirty years before Hassan finished his book, rockets and powder (not yet gunpowder, because the guns were still to be invented) had been introduced into Europe. The rockets bore the Latin name of *ignis volans* (flying fire) or the Germanic name of "wildfire," the latter being used later for "Greek fire," which was a pure incendiary without explosive qualities.

Thanks to the patient work of several historians the story of the introduction of rockets into Europe is now reasonably well known. One of the most important sources is the *Epistola* of the English monk Roger Bacon, whom his contemporaries, with much awe and a little fear, called the

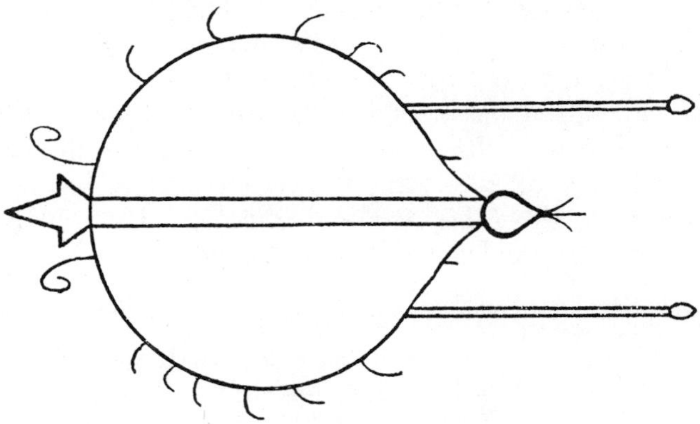

Fig. 4. The "self-moving and combusting egg" of Hassan er-Rammah.

Doctor mirabilis. This *Epistola* is also the earliest source that can be dated with any accuracy. It must have been written a year or two prior to 1249 and contains information about the making of powder and of rockets. Because Bacon did not want everybody to understand what he was saying he wrote three chapters in a cipher consisting of misleading terms and much senseless "padding." The most important thing then was to get refined saltpeter. Hassan er-Rammah wrote about that openly, though in Arabic; Roger Bacon wrote in easy Latin, but behaved as if he were describing the making of gold or of the Philosopher's Stone, confusing not only his contemporaries but also countless later historians. The most mysterious part was an anagram in the eleventh chapter, reading: *sed tamen salis petre* LURU VOPO VIR CAN UTRIET *sulphuris.*

Since the first legible words refer to saltpeter and the last word is "sulfur" it looked as if the jumble of letters in between had to refer to the third ingredient for powder, namely powdered charcoal. Hence some historians rearranged the letters to CARBONUM PULVERE. That this requires at least one letter (M) which is not in the anagram and also leaves a few letters over was generously disregarded. After all, everyone knows how careless copyists often were. The real solution of the anagram was not forthcoming until the British military historian Lieutenant Colonel Henry W. L. Hime found another arrangement of these letters which made more sense and did not involve changes. His rearrangement was R VII PART V NOV CORUL V ET, so that the whole sentence, with these abbreviations expanded, reads *sed tamen salis petre recipe vii partes, v novelle coruli, v et sulphuris.* This is not the very best classical Latin, but it has a meaning: "Take 7 parts of saltpeter, 5 of young hazelwood and 5 of sulfur"—this produces a weak variety of blackpowder, well suited for simple rockets.

Roger Bacon's German counterpart, Albertus Magnus, informed his readers about propelling charges for rockets without any cipher or secrecy. In his book *De Mirabilibus Mundi* (*On the Wonders of the World*), written sometime between 1250 and 1280, he said, under the heading *Ignis volans,* without any circumlocutions: *recipe libram unam sulphuris, libras ij. carbonum salicis, libras sex salis petrosi,* "take 1 pound of sulfur, 2 pounds of charcoal, 6 pounds of saltpeter." And this recipe, in turn, was simply copied from another book, the *Liber Ignium,* or *Fire-Book,*[2] written somewhat earlier by an otherwise unknown Marchus Graecus. The discussions about

[2] The *Fire-Book* says, under *Ignis volans: Accipias libram I sulphuris vivi, libras duas carbonum vitis vel salicis, VI libras salis petrosi;* one does not have to know Latin to see that it is the same recipe. But even at the time of Albertus Magnus one could find a powder recipe without knowing Latin. A German writer named Hanns Hartlieb had quickly utilized this new information for a book of his own. "*Nym lebendigen swebel .j. tail / velber koln .ij. tail / Saliter .vj. tail,*" he copied, but added, "*anderschwo stet .ij. tail*" ("elsewhere it says 2 parts"), proving two things: first, that he merely compiled without having any practical experience with the things he was writing about, and, second, that powder recipes must have been available for copying around the middle of the thirteenth century.

Marchus Graecus and the *Fire-Book* which can be found in print are tedious and inconclusive, but there is fair agreement on the assumption that this undatable book is essentially a translation from the Arabic.

That the introduction of rockets was not just a literary phenomenon is proved by occasional references to the thing itself. The chronicle of Cologne speaks about them in 1258 and the Italian historian Muratori credits a rocket with an important victory in the battle for the Isle of Chiozza in 1379. One defending tower resisted all assaults until it was set afire by a lucky hit.[3]

At that time firearms were already in existence, having been invented—according to the only source which gives a date—in 1313, but they were still crude and of poor performance, so that rockets were a serious competitor. Naturally the military engineers of that period experimented with both, guns as well as rockets, trying to improve their power and range. Consequently more than one type of rocket existed in 1400. The German military engineer, Konrad Kyeser von Eichstädt, in his book *Bellifortis* (finished in 1405), mentions three types: vertically rising skyrockets, floating rockets, and rockets running along taut strings. The sketchbook of an Italian military engineer, Joanes de Fontana, which can be dated as having been finished in 1420, is full of much more ambitious projects.

This sketchbook, later called the *Bellicorum Instrumentorum Liber,* contains drawings of rockets disguised as flying pigeons, as floating fish, and as running hares, all of them supposed to set fire to the enemy's fortifications. The running hare, mounted on a wooden board, does not run on wheels but on wooden rollers—de Fontana was looking for a device that would enable his disguised rocket to cross uneven ground. The same idea, magnified in size, resulted in the sketch of the first rocket car, apparently expected to breach walls or at least gates. And finally de Fontana also sketched a wooden rocket torpedo, painted to resemble the head of a large sea monster (Figs. 5 and 6).

Most of these sketches were probably nothing but ideas which were never tried. However, in the famous *Chronicles* of "Messire Iehan Froissart" there is a suggestion which would have been worth trying; we now know that it works. Froissart (who died about 1410) said that one should fire rockets from tubes to give them direction. And some innovations resulting from practical experimentation were added soon. Rockets with crude parachutes are described in a heavy manuscript—it never reached print—compiled by Count Reinhart von Solms during the first part of the sixteenth century. A little later the Count of Nassau described a rocket which would dive and finally explode under water (Fig. 7). Like that of the Count von Solms, the manuscript of the Count of Nassau was never printed. It is an enormous folio tome of 725 pages, completed in 1610.

[3] The original reads: *"Pure una rocchetta fu tirata nel tetto della torre de si fatto modo, que il tetto s'accese."*

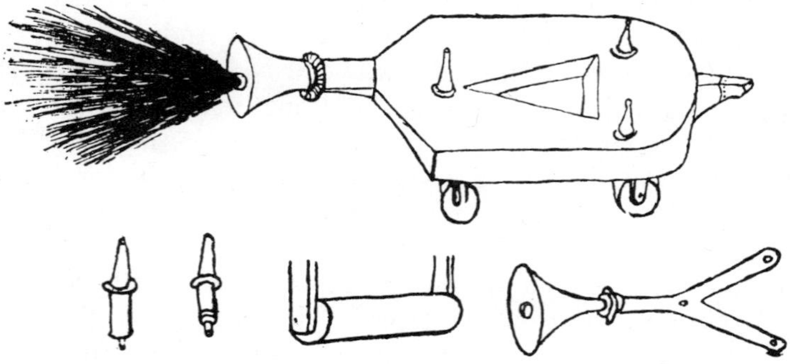

Fig. 5. Joanes de Fontana's rocket car.

Some two decades later the architect Joseph Furttenbach of Ulm wrote two books, entitled *Architectura navalis* and *Halinitro-Pyrobolia* (they are in German, in spite of their Latin titles) which contain some interesting information. At sea, Furttenbach reported, rockets were not only used for signaling, they were also employed as incendiary weapons, aimed at the tarred rigging of other craft. Pirates, especially, used them that way and Furttenbach suggested that other vessels use the same means for fighting back. By that time military use of rockets in land warfare seems to have ceased. Main witness for this assertion is the chief armorer of the city of Frankfort-on-the-Main, Leonhart Fronsperger. He wrote a book, published in 1557, which is of great importance to historians of artillery and most of which is devoted to firearms. But there is also a section on fireworks for amusement and it is there that the "roget," as Fronsperger calls it, appears. The description is short enough to be quoted in full:

> ROGET is the simplest of the fireworks, made of powder, saltpeter, sulfur and charcoal, pounded tightly into paper, is to go high into the air, to give off beautiful fire, to use itself up in air and to disappear without any harm. And while the roget is of small power and does not last long, many beautiful fireworks can be made from it, when several are combined in balls or wheels or are shot from mortars; they are an ornament and propulsion to other fireworks and rise into the air from their own fire, without shooting. (Fronsperger)

These two admittedly somewhat long sentences are all Fronsperger has to say about rockets.

Fronsperger's failure to go into more detail may also have been due to consideration for the fireworks makers; possibly he did not want to give away their trade secrets. But 34 years later one Johann Schmidlap published the first book dealing exclusively with nonmilitary fireworks. In the introduction to the thin folio volume, printed in Nuremberg in 1591, the author states that other fireworks makers will probably "speak ill" of him.

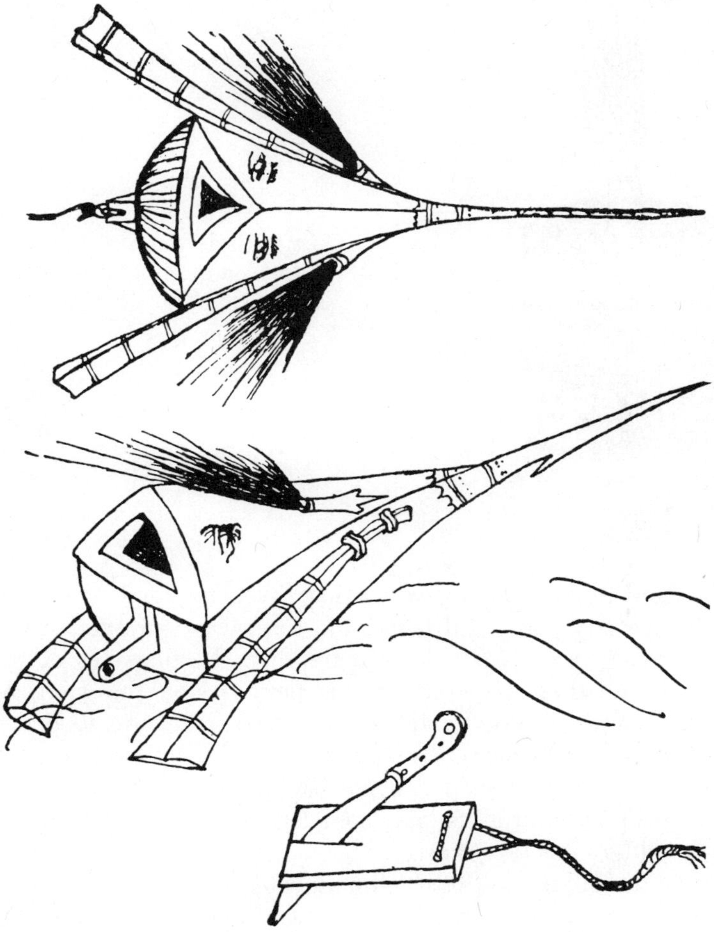

Fig. 6. Fontana's rocket torpedo. The dagger thrust through the tail piece is to indicate that the whole is made of wood.

Apparently this did not worry him too much, because he proceeded to tell everything in great detail. The making of fireworks rockets had become a standard practice by then and has remained very much the same since.

The raw material for the making of a rocket was "lazy" gunpowder; that is, gunpowder which had been made relatively slow-burning by the addition of extra charcoal. This was a comparatively recent innovation, because prior to, say, 1500 all gunpowder had been weak enough to be used in rockets without modification. The rocket case consisted of pasteboard; the fireworks maker's chief tools were cylinders made of hardwood and a mallet. The first step was the pasting of the tube which was shaped

Fig. 7. Underwater rocket, as pictured in the manuscript of the Count of Nassau (1610).

around a cylinder of hardwood which had been rounded off at one end. Then, while the paste was still moist, the tube was "throttled" by pulling the wooden cylinder out for a little more than the diameter of the tube so that the rounded-off end was at this point inside the tube. Another rounded-off plug was inserted and then a soaped string was looped around the moist tube at the point where the two rounded-off cylinders of wood met inside. By pulling the soaped string back and forth the tube was constricted to about two-thirds its full diameter. After that it was permitted to dry thoroughly.

When dry, the tube was filled with powder which was hammered in solidly, layer after layer, until the tube was filled to the top. The end which had been constricted formed the lower end of the rocket and the fuze was put in through the constriction or nozzle. The important point was that the powder was hammered around a "thorn" (Fig. 8), so that the powder charge had a center hole, usually conical in shape, which reached almost all the way through. The purpose of this center hole—which Konrad Kyeser was the first to mention—was to provide a large burning surface for the powder so that a large quantity of gases would be generated. For some reason this center hole was called the "soul" of the rocket, which set mystically inclined characters to speculating about the "power of the rocket soul."

This standard practice, which may even be an Arab invention, has survived until now—most fireworks rockets are still made in this manner. Only the larger factories have adopted some modifications, for example, hydraulic pressure in lieu of hammering. Such a "pressed rocket" is usually

Fig. 8. Mechanism for making powder rockets. The cardboard tube was pushed over the "thorn" and the powder charge was then hammered into the tube around the thorn. For large rockets a metal tube was slipped over the cardboard tube during the hammering to prevent the cardboard tube from splitting open.

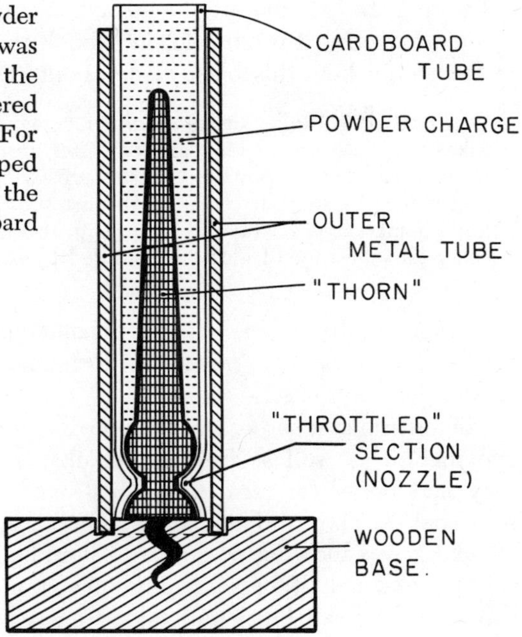

made solid, with a layer of clay at one end. The "soul" is then bored out through the layer of clay, the remains of which serve as a primitive nozzle, obviating the need for the tedious "throttling" of the tube.

The finished rocket, according to Schmidlap's description, is then tied to a straight stick "which must be about seven times as long as the rocket itself. You then balance your rocket on your finger or on the back of a knife blade, placed just below the firing hole. If it balances well the stick is just right; if the stick be heavier, cut it until the rocket balances, but cut from its thickness only so that it have its full length." This test, also, is still in use for handmade rockets.

Among Schmidlap's elaborations one can find the first multiple rockets, or, as they are now called, step rockets or multistage rockets. One of his pictures shows a large rocket carrying a second smaller one, and the smaller one carries a still smaller third rocket in its head. Such step rockets are the means for extreme altitude or range, but in Schmidlap's case they were used only for the purpose of producing an interesting display.

At this point, I think, an explanation of how a rocket works is in order. All the time, from the Chinese and Arabs on, none of the countless artisans who "throttled" and "hammered" rockets had the faintest idea of just what they were doing. They only knew that if they did thus-and-so it would work.

Burning gunpowder produced a blast, that much was certain. And an Italian, Vanoccio Biringuccio, in a book published in 1540,[4] had explained in fair detail how this blast came about. This is what he said:

> One part of fire takes up as much space as ten parts of air, and one part of air takes up the space of ten parts of water and one part of water as much as ten parts of earth. Now powder is earth, consisting of the four elementary principles, and when the sulphur conducts the fire into the dryest part of the powder, fire and air increase . . . the other elements also gird themselves for battle with each other and the rage of battle is changed by their heat and moisture into a strong wind. (Biringuccio)

Biringuccio had observed the phenomenon correctly enough; it was not his fault that he had no better technical language at his disposal than Aristotelian nonsense.

The artisans who fashioned powder rockets knew, from Biringuccio's explanation as well as from direct observation, that something for which they had no better term than "a strong wind" emerged from the nozzle of a rocket. They did not know the reason why this strong wind, which evidently was blowing downward, should carry the rocket upward. Another century and a half passed before Newton found out why this happened. But it could have been understood even without Newton's Third Law of Motion by thinking of a rocket as a vessel filled with gas under pressure; gas meaning anything that is gaseous—compressed air, if you will, or steam under pressure, or carbon dioxide—any gas will do. As long as the vessel containing the compressed gas is without leaks or holes, nothing will happen; the gas will press against the walls of the container with equal force in any direction, up and down and left and right. But the picture changes when the container gets a hole through which the gas can escape. We'll assume that the hole appears suddenly in the bottom of the container. Then the gas still presses against the walls and the top, but it cannot press against the bottom because there is nothing to press against. Consequently the pressure pointing upward is not counteracted and the container rises (Fig. 9).

Abandoning historical sequence, we find a perfect illustration of this principle in the steam rockets for which James Perkins of London was

[4] *De la Pirotechnia.* An English translation was published in 1943, a German edition some twenty years earlier.

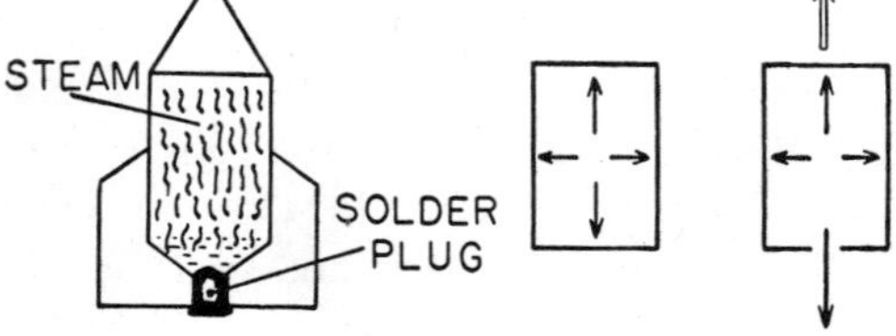

Fig. 9. Probable appearance of Mr. Perkins' steam rocket, with diagrams of the elementary principle of rocket motion.

granted a British patent on May 15, 1824. Mr. Perkins' rocket consisted of a metal container which was partly filled with water. A round hole in the bottom was closed with a plug of metal alloy which melted easily, probably a lead-tin alloy. Then the whole thing was put into a blazing bonfire. The heat soon changed the water into steam which tried to get out of the confined space. But the container held together until the whole rocket had become hot enough for the plug to melt. Then the steam could push it out, and suddenly, with a hole through which the steam could escape, the rocket shot up into the sky. I don't know for what purpose Mr. Perkins constructed these steam rockets and I know even less why he went to the trouble and expense of acquiring a patent, but I feel grateful for his efforts because he provided such a simple demonstration of the basic principle of rocket motion.

The chief difference between his steam rocket and a powder rocket (or even a modern liquid-fuel rocket) is that his actually carried a compressed gas, while the other rockets carry the gas in a different form, by carrying a fuel which as a result of combustion changes into a gas.

And now we are ready for an explanation of the Third Law of Motion itself. It states that "every action is accompanied by an equal but opposite reaction." The simplest example, I believe, is provided by imagining a frog sitting on a piece of wood, floating on a quiet lake. The frog weighs 1 ounce and so does the piece of wood and nothing happens for some time. Then the frog sees a fly a short distance away and jumps violently off the piece of wood. The frog, of course, gets to the fly and the piece of wood gets elsewhere—in the opposite direction. Assuming that the resistance of the water did not exist in this case, the wood gets as far as the frog does. If the frog, in pushing himself off the piece of wood, acquired enough speed to sail through the air a distance of 4 feet, the wood also got a sufficiently strong push to move 4 feet in the opposite direction, both (because of their equal weight) moving with equal speed (Fig. 10).

When everything is over we will have the picture of a 1-ounce frog 4 feet from the original resting place due north, and a 1-ounce piece of wood 4 feet from the original position due south. If both had moved along the balance of a scale, the scale would not even have quivered because of the perfect distribution of the two weights to the right and to the left of the original center. This is the important point—the two components balance each other perfectly at any given moment.

In the case of a rocket, the rocket itself is the piece of wood of our example and the powder gases are the frog. The gases, in "jumping" away from the rocket, "kick" it back—something that would happen in airless space as well as in air and that has nothing at all to do with "pushing" against the air.

It is precisely the same force which kicks a gun barrel backward when the projectile and a stream of violently expanding explosive gases emerge

Fig. 10. The Third Law of Motion.

from the muzzle. It is the same force which makes a chair topple over when a cat jumps from its back onto the bookshelf, the same force which kicks your canoe back into the river when you jump ashore.

The reason this sounds surprising to many people is that it is not noticed much in everyday life. Not because the force is not there, but because it is not wanted. Ordinarily the action is the desired part of the whole procedure; the reaction is a nuisance and is therefore overlooked. But it can be utilized. You can, for example, propel a light canoe by throwing heavy rocks with all your might over the stern. It isn't very efficient and it is exhausting too, but it can be done, and it shows, incidentally, that propulsion by reaction has nothing at all to do with "pushing against the air." You push against whatever it is you throw out, against the "reaction-mass" which is expended for this purpose.

Now it can be shown that your reaction-mass should consist of very small particles. In the example of the frog, each part of the whole system that separated was a half. The result was that one-half of the total mass attained one-half of the jumping speed. If there had been several smaller frogs on that piece of wood, the final result would have been better; the same speed could have been attained with the expenditure of less reaction-mass. If you want a system to attain the full jumping speed—let's call it "exhaust velocity" from now on—you can always throw away one-half of what the mass happens to be at a given moment, or you can throw away one-quarter of the mass each time. Or a smaller fraction. In each case the mass that is left shall be called "1." The question now is how much mass you need to begin with. The answer to that question is provided by the following table:

If	are expelled, the original mass must be
halves	4.000
quarters	3.165
tenths	2.868
hundredths	2.729
thousandths	2.723

Of course, each of the gas molecules that is expelled by a real rocket is a great deal less than one-thousandth of the rocket mass. Hence one has to add one more line to that table, for "infinitely small" particles, and the answer then is that the original mass must be 2.7182. This figure is well known to mathematicians, who have designated a letter to represent it: the letter *e*, which crops up often in the discussion of modern rockets.

One thing which is in itself obvious should be mentioned at this point, however. There is, of course, no reason why the ejection of particles should not continue even after the rocket has reached full exhaust velocity. Naturally it is possible to go on, provided only that there is still some "ammunition" left. We can now summarize what has been said so far in four statements:

1. The motion of a rocket is independent of the surrounding medium; in fact the surrounding medium is effective only in offering resistance to the movement of both the rocket itself and its exhaust. Hence the thinner the surrounding medium, the more advantageous. Most advantageous of all is, of course, the complete absence of a surrounding medium, in other words, a vacuum.
2. The exhaust of a rocket should consist of very small particles; this is normally an accomplished fact because the exhaust is normally gaseous.
3. The velocity of a rocket can be increased either by increasing the mass of the exhaust or by increasing the velocity of the exhaust. (If a choice is possible the latter is superior.)
4. The velocity of a rocket can exceed its own exhaust velocity. The velocity of the rocket is limited (aside from external resistance) only by the amount of reaction-mass (fuel) available.

All this is actually contained in the simple statement called the Third Law of Motion. Other physicists began to understand the problem after Newton had pointed the way. What seemed of possible practical importance was that rocket motion did not depend on some mysterious quality of gunpowder, but was solely a question of producing an exhaust of some sort in some convenient manner. And this implied that one might propel a vehicle that way—a thought which was highly intriguing when no engine of any kind existed. But actually it only came to a classroom demonstration.

A Dutch professor, Jacob Willem s'Gravesande, actually produced a small steam-reaction car for classroom use. A picture of it in the second volume (printed in 1721 at Leyden) of his book *Physica elementa mathematica* shows a small spherical metal container mounted on a small flat carriage. There was a small fire under the container and the steam produced escaped through a long tube pointing rearward. The model moved.

There are rumors that Professor s'Gravesande later tried to build such a steam-reaction car on a large scale and there even exists a drawing of such a vehicle (Fig. 11). But since this drawing looks as if it had been made later and since no direct report about the attempted construction is known, it is doubtful whether s'Gravesande ever really tried, although he may have toyed with the idea.

Fig. 11. A late drawing of s'Gravesande's steam-reaction car.

Meanwhile people tried to build large rockets.

In 1718 there appeared a book by a Colonel Christoph Friedrich von Geissler, commander of the field artillery of the Elector of Saxony and King of Poland, who revealed that he had made some interesting experiments near Berlin many years earlier, in 1668. Colonel Geissler's book is an outstanding example of the writing of an inarticulate person with a complete absence of style, but the facts are clear enough. The rockets were of two sizes, "50 pounds, also 120 lbs" (for U.S. weight add 10 per cent). Their cases were wood, covered with sailcloth soaked in hot glue. The propellant was a mixture of 36 pounds of saltpeter, 16 pounds of sulfur, and 12 pounds of charcoal of linden wood. The charge was pressed, not hammered, into the case. Payload was a 16-pound bomb. I'm now quoting, as well as that can be done: "When said rocket burned, the flame was so forceful that it made a deep hole in the sand, before and until said rocket rose and went on its way; smoked fearfully and the press on which it was pressed, cost much money, wherefore in a case of distress a drop weight might be used."

That Colonel von Geissler spoke from experience is also shown by the advice he gave: "When firing . . . take care not to have a hampering wind, for wind will always turn the guide-stick of the rocket around so that generally and especially at somewhat stronger wind the rocket will straightways turn into it."

The appearance of this book apparently served as a reminder and a challenge to the younger generation of ordnance officers in Berlin, because in 1730–31 more experiments with 100-pound rockets were made. According to the book by von Malinowsky and von Bonin (see Bibliography), the fireworks men compounded four different propellant mixtures for such heavy rockets. They were:

	PARTS BY WEIGHT			
	NO. 1.	NO. 2.	NO. 3.	NO. 4.
Saltpeter	36	48	72	28
Sulfur	10	15	22.5	7
Charcoal	10	15	22.5	7

"Mixture No. 1. was tested in June 1730 by Capt. Holtzmann and found good. Mixture No. 2. was tested on July 17 of the same year. The rocket case weighed 33 lbs., the charge 23 lbs., the guiding stick 33 lbs. and the cap and payload 4 lbs., the whole rocket, therefore, 93 lbs. It rose to an extremely high altitude. Mixture No. 3. was also tested in 1730 and mixture No. 4. in 1731 near Sheep's Bridge, Berlin."

If history progressed as logically as one has a tendency to assume, the great wave of military rocket enthusiasm which was soon to come should have been the outcome of these experiments and other experiments succeeding them. However, in the history of technology, the chronological sequence of individual work and the true "line of evolution" of an invention do not always coincide. The period of military rocket enthusiasm which began shortly after the year 1800 and which is now generally referred to as the "Congreve Period" after Sir William Congreve, the creator of the British war rockets, was neither caused nor even influenced by this earlier work.

The direct cause for that revival of war rockets did not even lie in Europe. It was an unsuccessful military campaign in faraway India, related after its conclusion in a book with the title, *A Narrative of the Military Operations on the Coromandel Coast,* etc. etc., by one Innes Munroe, Esq., published in London in 1789. This book confirmed stories which had reached England earlier and which said that the Indians used rockets against the British troops. They were said to be similar to the rockets used for amusement in England, but much larger. Instead of having a pasteboard tube they had an iron tube for the charge, their weight ranged from

6 to 12 pounds, and they were guided by a 10-foot bamboo pole. Their range was estimated from a mile to a mile and a half and while they were not very accurate, they caused much damage by mass application, especially when used against cavalry.

The man responsible for all this was a native ruler, Hydar Ali, Prince of Mysore, who had attached a corps of rocket gunners to his regular army. Originally that corps numbered some 1200 men, but when the "new" weapon proved successful Hydar Ali's son, Tippoo Sahib, increased the strength of the rocket corps to 5000 men. The British suffered badly from these rockets, especially at Seringapatam in 1792 and 1799.

When this news was digested, military men remembered that during those same years a Frenchman, Citizen of the Republic Chevalier, had made similar experiments. He had fired incendiary rockets of a large size against a large target fashioned from an old sail which he intended to set afire. It had not worked simply because the rockets had torn through the old sail so fast that the target had no time to catch fire.

But perhaps there was something to Chevalier's idea after all.

Several military men and scientists began to experiment in a modest way, but only one of them was really successful, the British Colonel William Congreve, M.A. (the "Sir" was not added until several years later). Contrary to a statement that can be found in many books, Congreve had never seen the Indian war rockets in action, for the simple reason that he never went to India. His only information came from books and journals, available to anybody interested in the matter.

In 1801 or 1802 Congreve bought the biggest skyrockets available in London, paying for them out of his own pocket, and began to indulge in some private long-range practice, or rather in firing practice which had the purpose of establishing the possible range. He found it to be between 500 and 600 yards, less than half the range of the Indian war rockets. Then he took the idea to the authorities, apparently aided by his father, Lieutenant General Sir William Congreve, comptroller of the Royal Laboratory at Woolwich. Lord Chatham granted permission to use the laboratories and the firing ranges, and Congreve soon obtained a range of 2000 yards. In 1805 the new weapon was demonstrated to the Prince Regent, and later in the same year Congreve accompanied Sir Sidney Smith in a naval attack against Boulogne.

What happened during that first expedition is not quite clear; some military historians say that the weather prevented the use of the rockets, others state that about two hundred rockets were fired which damaged only three houses and that the French soldiers marched around in the city after the attack, carrying the empty shells of the rockets and making unprintable jokes about them. It is unimportant whether the one or the other claim is correct, because the rockets proved their effectiveness in the years to come. Boulogne underwent a terribly devastating fire raid in 1806, and in 1807 the greater part of the city of Copenhagen burned to the ground, set aflame by a mass expenditure of about twenty-five thousand rockets.

Danzig, in 1813, saw very much a repetition of the Boulogne bombardment. The first barrage, fired on August 26, failed to do any harm and the empty burned-out rockets were carried around in the streets and bought by people who exhibited them in stores for a small fee. The second bombardment in September set twenty-three buildings afire, among them the hospital of the Dominican monks. And the third bombardment on October 20 fired the food stores so that the city had to surrender on November 27.

During the days immediately preceding the effective October bombardment of Danzig, the British "Rocket Corps" distinguished itself during the Battle of Leipzig (October 16–19, 1813) which broke Napoleon's power. The Corps had no part in the actual taking of the city, but it had served effectively during the preliminary actions and was granted permission to display the word Leipzig on its standards.

Congreve had started out with incendiary rockets of a caliber of 3.5 inches. The case was heavy sheet iron, 40.5 inches long; the 16-foot guiding stick was attached to the case by means of a copper hoop and held in place by two smaller iron hoops, soldered to the case. Although these rockets had decided several sieges, Congreve seems to have been attacked by officers of other branches of the service because he hit back angrily in one of his books. "One of my 32-lb. incendiary rockets contains about 7 lbs. of carcass composition which is the same as that contained in the bombs of the 10-inch mortar." But, he added, his rockets had a maximum range of 3000 yards, the 10-inch mortar a maximum range of 2000 yards. And his rocket bombardment groups were highly mobile units, while the 10-inch mortar was the "most ponderous piece of ordnance in the services."

Others must have said something about the high cost of rocket weapons, while still others must have suggested that the incendiary rockets lost most of their terror as soon as the enemy got acquainted with them, and that soldiers, after having learned that there was no immediate danger to them, could easily extinguish incendiary rockets that fell close to them. Congreve's reply was very definite:

> A certain portion of the rockets have a 6-pounder shell or a cylinder, holding nearly one pound of powder, contained in the carcass which bursts during the combustion; nor can the enemy distinguish the one sort from the other, and will, therefore, avoid all. But in addition to the above precaution, every rocket contains a certain quantity of smoke-ball composition; and the suffocating quality of this matter is well known to be such, that no person can exist in the room where it has burnt but a few seconds. (Congreve)

The question of which type of weapons, rockets or artillery, was the more expensive, was answered by a little expense acount for each:

ROCKETS

	£	s.	d.
Case complete	0	5	0
Cone	0	2	11
Stick	0	2	6
Rocket charge	0	3	9
Carcass charge	0	2	3
Labour, paint, etc.	0	5	6
	£	1/1/11	

TEN-INCH MORTARS

	£	s.	d.
Carcass charge	0	15	7
Powder charge	0	6	0
Cartridge, etc.	0	1	0
	£	1/2/7	

(Plus the cost of the mortar)

Beginning in 1813, Congreve changed the design of his rockets somewhat. The case holding the propulsive charge was no longer cylindrical with a conical cap; it was conical too so that the rocket's largest diameter was just behind the cap. At that point it measured 6.4 inches; at the rear end the diameter was 4.5 inches. With such a slanting body the guiding stick could no longer be attached in the traditional manner and Congreve introduced a new method: the guiding stick continued in the center line of the rocket body, being held in place by a three- or four-pronged fork which gripped the rear end of the rocket. The introduction of the "center-line stick" may have been due only to the changed body shape, but it also improved the accuracy. Congreve also began introducing many different types of bombs, etc., to be carried by the rockets. In 1817, according to a list published by Congreve himself, the following types existed:

DESIGNATION	TYPE OF PROJECTILE CARRIED	EXTREME RANGE IN YARDS	MINIMUM ELEVATION FOR EXTREME RANGE
42-lb carcass rocket	Large: 18-lb carcass Small: 12-lb carcass	3000	60°
42-lb shell rocket	Large: 12-lb spherical bomb Small: 5.5-lb spherical bomb	3000	60°
32-lb carcass rocket	Large: 18-lb carcass	2000	60°
	Medium: 12-lb carcass	2500	55–60°
	Small: 8-lb carcass	3000	55°
32-lb shell rocket	9-lb spherical bomb	3000	50°
32-lb case shot rocket	Large: 200 carbine balls	2500	55°
	Small: 100 carbine balls	3000	50°
32-lb explosive rocket	5–12 lb powder	2500–3000	55°
12-lb case shot rocket	Large: 72 carbine balls	2000	45°
	Small: 48 carbine balls	2500	45°
Flare rockets equipped with parachutes			

These rockets represented all types of artillery ammunition then in use, save for the solid round shot. It was Congreve's firm belief that his rockets would replace artillery, except for naval warfare, within a few decades. His favorite term in speaking about his rockets was "the soul of artillery without the body" and in a sense he was right. His rockets surpassed all the easily movable artillery pieces of his time in range. As far as accuracy was concerned, which, from our point of view, was nothing to boast about, they were about even.

But the main advantages of rockets were the absence of recoil, which permitted their use from small boats (Congreve himself always disliked that particular application) and the associated fact that they did not need strong and heavy barrels to give them direction. He used thin-walled launching tubes made of copper. And for mass bombardments they needed only collapsible wooden frames which looked like very wide stepladders.

What Congreve did not know—and what he could hardly be expected to know—was that the artillery pieces of his time were crude and clumsy devices which, in spite of their age of some five hundred years, were still in the earliest stages of their development. His rockets, on the other hand, were just about as large and as powerful as a blackpowder rocket can be. Since he could not know this, it was only natural that his conclusions were wrong. Artillery soon began to outrange, outshoot, and out-perform rockets in every respect, and the later invention of the internal-combustion engine made even the weight of the guns unimportant.

Congreve died May 16, 1826. Among his papers were found plans for a rocket with a diameter of 8 inches, as well as some notes relating to rockets weighing 500 and even 1000 pounds.

Congreve's influence was enormous. Denmark, Egypt, France, Italy, The Netherlands, Poland, Prussia, Sardinia, Spain, and Sweden attached rocket batteries to their artillery. Austria, England, Greece, and Russia had rocket corps which were independent units. So had Switzerland, but only on paper. The United States formed rocket units—there is some doubt whether they were independent or artillery units—some time later when William Hale had made a new invention which will be described shortly. At least twenty books on war rockets appeared during as many years, written in English, French, and German (the Russians wrote French at that time).

Congreve's war rockets gave rise to two or three useful, and ten times as many useless, inventions. And they are responsible for allusions which are still known to practically everybody, even though people do not know the reference any more. Everybody knows, for example, that one of the famous early locomotives was named the *Rocket*. No direct allusion to the speed was intended; the name grew out of an article in a British engineering journal. George Stephenson had promised a locomotive of twice the speed of a mail coach. And a critic exploded into words like "silly and ridiculous . . . twice the speed of a horse-drawn coach. It may just as well be expected that the inhabitants of Woolwich will consent to a ride on a Congreve war rocket than trust their lives to such a machine." Whereupon Stephenson called his new locomotive *Rocket*.

And every American sings about "the rockets' red glare, the bombs bursting in air . . ." which refers to the bombardment rockets launched in quantities against the bastions of Fort McHenry.

> On that memorable night in September 1814, Congreve's brainchild was given a real tryout. . . . Indeed, so intensively was this new agency employed on this occasion that several vessels especially fitted out as rocket ships, accompanied by a number of naval barges, were delegated to ascend the Middle Branch of the Patapsco and take the fort in the rear. Fired upon, however, by three batteries of whose existence they were apparently ignorant, they were forced to retreat after the loss of one barge with its full complement of men. But though the rocket failed in this instance to accomplish its objective, it already had left its mark, and that an indelible one, on American history. For, only some three weeks previously at the battle of Bladensburg (August 24, 1814) a flight of these ungainly projectiles directed against Stansbury's brigade had caused the regiments of Schultz and Ragan to break and flee in wild disorder. As a result, the American flank was turned, and despite stout resistance on the part of the Fifth Maryland Infantry under Sterrett, the day was lost. Thus we may indirectly (or perhaps directly) thank Congreve and his invention for the capture and burning of Washington which followed. . . .[5] (C. Goddard)

The activities of the other European armies followed an easily discernible pattern. The first job was to learn everything that could be learned about Congreve's rockets and to obtain specimens, if possible. The second was to duplicate the British achievement. And the third was to improve upon it, if possible. There seemed to be one point especially where improvement was possible: the 16-foot pole the rocket had to carry to keep it from moving erratically. The pole was necessary, no doubt, but it also looked like a lot of dead weight.

Fig. 12. Demolition of a target by a Congreve rocket. The target consisted of a double wall of oak timber, each piece of timber being 6 by 6 inches. This experiment was made by the École de L'Artillerie at Metz in 1817. The picture was published in the first French translation of Congreve's book which was for distribution among officers of the French Army only.

[5] Lieutenant Colonel Calvin Goddard, in *Army Ordnance,* May–June 1939.

The Dutch Army, for example, began by buying a number of Congreve rockets. But when things had progressed to actual experimentation the rockets misperformed, in all probability because they had been stored for over a year. Experimentation continued with Dutch-made rockets which did not carry a guiding stick. A Captain de Boer had proposed to eliminate it and to stabilize the rockets by means of three metal vanes of only a fraction of the weight of a long pole with metal attachments. This should have worked, but apparently the Dutch were not satisfied, for two years later they again ordered Congreve rockets from Great Britain. After some more experimentation they decided not to introduce rockets at all, except in the colonial services. And then they won an apparently hopeless battle against 6000 native warriors on Celebes with the aid of war rockets (1825).

In France, the ordnance experts were at first doubtful; the *Aide-mémoire à l'usage des officiers d'artillerie de France* for the year 1819 almost stated that war rockets were imaginary. It did not quite put it that way, but it came close. However, an enthusiastic artillery officer translated Congreve's book and a special commission for rocket research did some experimental work, mostly near Metz (Fig. 12). They developed types of their own of a weight of some 40 pounds, still following the Congreve pattern. And then they tried to eliminate the guiding stick. Since the Dutch had not succeeded in doing without it, the French reasoned that it could not be avoided completely. But it might be possible to utilize it in another way, by making it do double duty as guiding stick *and* as projectile (Fig. 13).

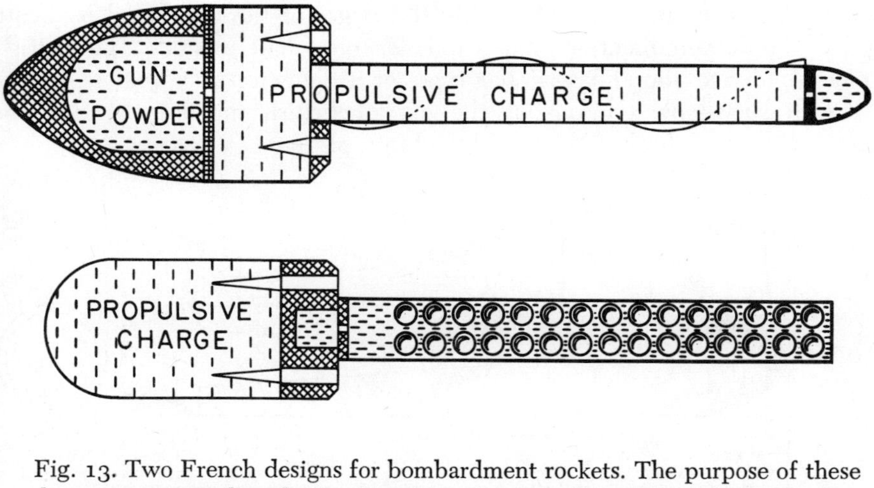

Fig. 13. Two French designs for bombardment rockets. The purpose of these designs was to utilize the dead weight of the guiding stick in some manner, either for the propulsive charge (top) or as a projectile (bottom). Picture from Montgéry's *Traité des fusées de guerre.* It is doubtful whether such rockets were actually built.

When the stickless war rocket actually appeared, around the middle of the nineteenth century, it turned out to be a British invention. The inventor, William Hale, had solved the problem of stabilizing stickless rockets by rotating them. He did this by inserting three metal vanes in the exhaust nozzle, slightly inclined so that the exhaust itself caused the rotation (Fig. 14). But by the time the invention was ready, most of the rocket corps of the various armies had been dissolved. The one that lasted longest was the Austrian rocket corps, which used Hale's rockets exclusively from the moment they became available. The Austrians recorded a long list of engagements in Hungary and in Italy which they claimed had been decided in their favor by the war rockets. Most of these engagements were fought in mountainous country where the enemy probably did not have much artillery. For when the Austrian rocket corps faced the Prussians in reasonably flat country in 1866 it was so unsuccessful that it was dissolved a year later.

But Hale's rockets were introduced in the United States Army. Scott's *Military Dictionary* stated in 1861:

> The rockets used in the United States service are Hale's . . . The two sizes in use are the 2¼ inch [diameter of case], weighing six pounds, and the 3¼ inch [diameter of case], weighing 16 pounds. Under an angle of from 4 to 5 degrees, the range of these rockets is from 500–600 yards, and under an angle of 47 degrees the range of the former is 1760 yards and of the latter 2200 yards. War rockets are usually fired from tubes or troughs, mounted on portable stands, or on light carriages. (Scott)

Apparently one size was dropped quickly, for the *United States Ordnance Manual* of 1862 described only the larger of the two listed by Scott and did not even mention that a smaller type had existed. The British Colonial Service adopted a similar type, of a diameter of 3¾ inches. It is listed in successive editions of the official British manual, *Treatise on Ammunition*, for the last time in the edition of 1905.

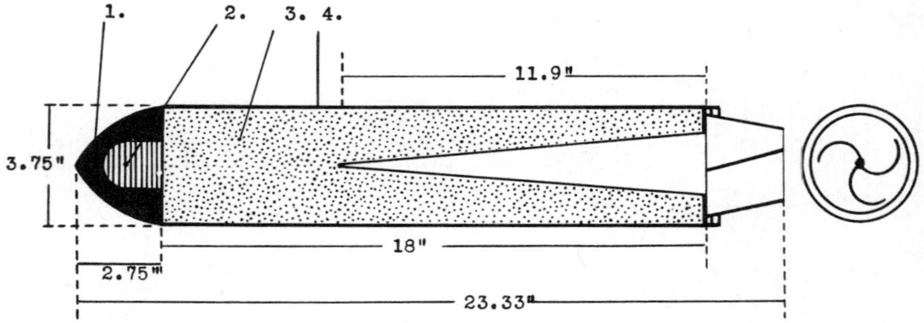

Fig. 14. Late British war rocket of the Hale type (1905). 1. Cast-iron head; 2. hardwood core; 3. propulsive charge; 4. steel tube.

The last actual use of war rockets during the nineteenth century is reported from Russia. It took place during the long-drawn-out Turkestan War which was directed against a large number of more or less independent tribal chiefs, each with his own army, region, and mountain fortress. The published reports of a Colonel A. Serebrennikov contain many references to "rocket stands" but with little information about them. From a Russian "technical encyclopedia" published in 1897 one can gather that the rockets were of the Congreve type, with a diameter close to 2 inches and a weight of about 10 pounds. The "rocket stands" look like surveyor's tripods, except that they carry a launching tube instead of a small telescope. First mention of the use of rockets in that series of campaigns is for the year 1864; last mention is for January 12, 1881, during a battle near a locality called Glok-tepé.

As a weapon of war the Congreve rocket lasted less than a century, but in the meantime the rocket had gained a place in naval life. The life of a seaman often required something resembling shooting but less violent than real shooting. The Congreve rocket fitted perfectly into this picture. And in 1821 Captain Scoresby, on the good ship *The Fane,* began to hunt whales with rocket harpoons. It was a good idea and it is likely that rocket harpoons would have endured to this day if it had not been for the fact that a rocket is not quite accurate enough. Because of this the first real harpoon gun replaced the rocket harpoon.

But rockets could throw lines, and that feature was utilized. Small vessels the size of lifeboats that had to negotiate a heavy surf did so by using a rocket to throw a small anchor across the surf. Then they pulled themselves to safety along the anchor rope.

This method, of course, worked both ways and resulted in the most useful rocket invention that was to follow the Congreve war rocket. This was the so-called lifesaving rocket, used to rescue the crews of stranded vessels. It works in this fashion: a rocket carries a thin line from the shore to the stranded vessel. By means of this thin line the crew of the vessel hauls a heavier rope across and the men leave the vessel one by one in a so-called breeches-buoy, a saddlelike affair suspended from the rope and pulled back and forth by means of thinner lines.

The first plan of this kind originated in the mind of a tradesman, the Prussian Master Weaver Ehrgott Friedrich Schaefer. This was before Congreve, and Schaefer did not think of rockets; he wanted to shoot the first line from a small siege mortar. He drew up his plans and submitted them in 1784 to the commander of the artillery of Frederick the Great. A committee of ordnance officers was entrusted with the evaluation of the idea, but most of them seem to have known the sea only from books; they decided that the invention was impractical.

Thirteen years after Schaefer a British Lieutenant of the Royal Artillery by the name of Cell advanced the same suggestion. The performance was the same, with the minor difference that the idea was not actually rejected. Instead no action was taken. A few years later, however, George William Manby, Royal Military Inspector, witnessed the stranding of a ship on which 67 people perished in plain sight of the horrified but helpless spectators ashore. After that Lieutenant Cell's suggestion was revived, Manby himself constructed the mortar, and the invention saved 332 lives on the coast of Norfolk during the years 1807 to 1823.

When Manby's mortar proved successful the Prussians remembered Schaefer's suggestion and went to work to build a line-throwing mortar. It was tried for the first time in the harbor of Pillau on the shores of the Baltic Sea and in July 1819 the provincial government officially permitted "the use of this new means of saving lives, although the Government is aware of the fact that this means might be dangerous in itself."

Meanwhile a captain named Treugrouse had substituted Congreve war rockets for the line-throwing mortar in private experiments made in 1807. The experiments were successful, but remained almost unknown until John Dennet of Newport on the Isle of Wight repeated them in 1824. It was Dennet who applied for and was granted the first British patent for such an apparatus. It is dated August 2, 1838. Four years after Dennet's first experiments the Prussian Major Stiehler demonstrated line-throwing rockets for the same purpose near Memel, using a continental (possibly Polish) imitation of the Congreve rocket as a missile. Major Stiehler's experiments attracted much attention and led to the introduction of that method on the European continent. The rockets used were mostly straight Congreve rockets; a special design for this purpose by one Colonel Boxer is the first adaptation of the step principle for serious use (Fig. 15).

But the nineteenth century, which saw the rise and fall of the Congreve war rockets—both literally and figuratively speaking—did not close without witnessing attempts to revive the old ideas of Hassan er-Rammah and Joanes de Fontana about rocket-propelled torpedoes. Early in the nineteenth century Robert Fulton had built a small submarine and had demon-

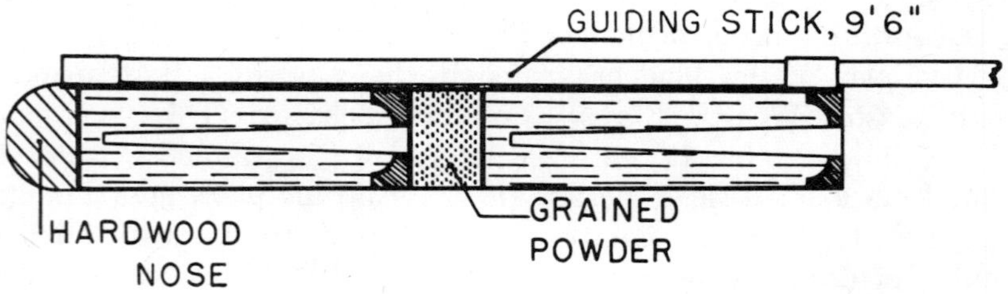

Fig. 15. Colonel Boxer's line-carrying rocket (1855). The first example of utilization of the step principle for purposes other than display.

strated that an underwater explosion, because of the incompressibility of water, had far greater power than the same explosion in air.

Fulton, in turn, had derived the germ of this idea from a book on physics, written in French but published in London in 1734. That book contained the description of an accidental discovery by its author, the French refugee physicist Doctor Desaguliers, during a pleasure trip on the river Thames. The party of which Desaguliers was a member amused itself with fireworks of various kinds, among them "water rockets," which are still manufactured by pyrotechnicians on occasion. Such a water rocket consists of a waterproofed cardboard tube weighed down at one end to make it float in a vertical position. It is charged with alternating layers of strong and weak powder mixtures. The weak layers do no more than produce a shower of sparks when the rocket floats at the surface. The powerful layers push it down so that it reappears at another point when the next weak layer catches fire. The spectacle ends with the explosion of a small powder charge which goes off at the surface when the last layer is weak, or under water when the last layer is powerful.

One of these water rockets happened to rise under the bottom of one of the pleasure barges when the last powder charge went off. Desaguliers emphasized how small it was, "much less than an ounce" (*"beaucoup moins d'une once de poudre,"*) but that small charge pushed the bottom of the barge in, so that the people aboard did not even have time to change over to another boat which was very close, but fell into the river. Doctor Desaguliers thought of the right explanation at once, that because of the rapid expansion of the powder gases the water acts like a solid body so that the bottom of a ship offers less resistance than the water does.[6]

The submarine mine, based on this principle, was evolved independently by the Russians during the Crimean War and by Confederate engineer officers during the Civil War. The problem was to give mobility to these mines which, incidentally, were then called "torpedoes," a name which was later transferred to the mobile torpedo.

Several dozen torpedo types were invented and mostly tested during the forty years between 1860 and 1900, the only surviving type being that of the Scotsman Whitehead, which is really a model submarine, complete with engine and steering devices. Some of the early rivals of the Whitehead torpedo, and the only ones which interest us here, were rocket-propelled.

The first experiments of this kind of which we have definite knowledge were made in 1862 from a submerged caisson near Red Hook, in New York harbor. The experimenter and inventor of the device was one Major Hunt

[6] *"Dans l'explosion de la poudre, dont la vitesse, dans son expansion, est excessivement grande, l'eau resista comme un solide, et par conséquence la poudre se fit un chemin par le fond de la barque dont let planches étoient peu épaisses et moins denses que l'eau."*

of the Engineer Corps, U.S. Army. His torpedoes had the shape of a projectile and consisted of three parts: a warhead filled with cannon powder or possibly guncotton, a powerful powder rocket for propulsion, and a wooden "shoe" which surrounded the rocket and carried inclined ridges supposed to produce rotation of the torpedo.

An 11-inch Dahlgren gun was used as a launching tube at first, but it was soon replaced by a specially constructed 12-inch tube with two sliding valves.

It was found that the direction was not very reliable but that the rocket torpedoes could be propelled in this manner for several hundred feet, which made it appear that such submerged caissons would be a valuable adjunct to shore batteries in the defense of harbors and inlets. Unfortunately Major Hunt's experiments found a tragic end. Once, after the rocket had been ignited, it was found that the outer valve had not been raised. Major Hunt feared that the gun might burst and descended into the caisson. The gun did not burst, but the Major, as well as an assistant who tried to come to his rescue, was smothered to death in the caisson by the gases escaping around the rear valve.

But while this accident ended American experimentation with such a device, it did not eliminate the idea altogether.

> Mr. Quick, an English engineer in the Royal Navy, somewhat later proposed an invention on the same principle as Major Hunt's, and an experiment was recommended by the Royal Engineer Committee, which took place at Shoeburyness in 1872.
>
> The torpedo was a cylinder, something over 5 feet long, with a sharp-pointed head, and immediately behind it was a space intended to be filled with guncotton. The after part of the body consisted of four rockets, which were in communication with the igniting charge, and whose gas escaped on ignition through spiral vents designed to give rotation to the rocket and keep its axis steady while projecting it through the water.[7] (Barber)

The invention was carefully thought out, as was the experiment devised to test it, but the latter turned out unlucky. As Lieutenant F. M. Barber said in his lecture:

> The torpedo was fired from a 10-inch gun, which was laid on the beach at about 5 degrees elevation, at a spot where the incoming tide would cover it to a depth of about 3 feet. The muzzle was closed by a disc of glass fixed in a wood washer tightly sealed round the edge, while an electric wire led through the vent of the gun to a small igniting charge in the centre of the base of the torpedo. When the tide rose and covered the gun the experiment took place. On firing, the torpedo burst open close to the muzzle of the gun, two rockets rising high into the air, one of which descended again almost immediately. . . . No further experiments have been made that I am aware of on Mr. Quick's plan. . . . (Barber)

William Hale, the inventor of the stickless rocket for land warfare, is

[7] Lieutenant F. M. Barber, *Lecture on . . . Torpedoes*, Newport, R.I., 1874.

known to have been opposed to the idea of a rocket-propelled torpedo. Hale pointed out that the main value of a torpedo consists in striking the enemy vessel below the waterline and exploding there. But this condition, he said, cannot be fulfilled in the case of a rocket-driven torpedo, since the rocket, due to the steady combustion of its propelling charge, is steadily losing weight.

At about the same time that the British Navy made the single ill-fated experiment with Mr. Quick's rocket torpedo an American engineer who is identified only as "Mr. Phillips of Indiana" submitted a novel idea for a rocket torpedo which was designed to circumvent the difficulty pointed out by Hale. Phillips' rocket torpedo was really a torpedo-towing rocket. The exceptionally large rocket, 6 or 7 feet long and 1 foot in diameter, was to float at the surface, towing under water the comparatively small and generally cigar-shaped torpedo which was to be nonbuoyant. "When the rocket strikes the enemy it explodes at the water's edge (carrying a separate bursting charge of its own) while the torpedo dives and is carried by its momentum under the ship where it explodes on striking bottom." Apparently Mr. Phillips' rocket torpedo was never built and tested. It would have been interesting to see whether such a double explosion would have taken place in the manner predicted by the inventor. It would also have been interesting to see whether a rocket of such dimensions could have skimmed across choppy waves without being deflected from its course.

In 1873 the British Navy also experimented with another rocket torpedo, invented by Charles Meade Ramus. The first experiments were not very successful, but the inventor did not give up hope and improved his invention greatly so that he could demonstrate it with much better success again in 1879. According to a now defunct German military magazine,[8] the torpedo traveled 15 feet below the surface with a speed of about 50 feet per second. Its range was about 1000 yards. The bursting charge consisted of guncotton. The torpedo also carried a device for cutting antitorpedo nets. Ramus' offer was rejected again, however, because a range of 1000 yards was deemed insufficient.

To return to America, a rocket torpedo (Fig. 16) was submitted in 1874 by Lieutenant Barber for a contest. Another entry received only weeks earlier was the rocket torpedo of "Mr. Weir, a civil and mechanical engineer of New York." Because Mr. Weir had a slight priority and also because his design was supported by experimental evidence it should be mentioned first.

The inventor stated that the over-all length of his torpedo would be 7 feet 7 inches, the diameter 12 inches. The rocket composition was to weigh 78 pounds, the explosive charge in the head 74 pounds, the over-all weight was to be about 245 pounds. The inventor also reported on tests

[8] *Archiv für die Offiziere der Ingenieur und Genietruppe,* vol. 85.

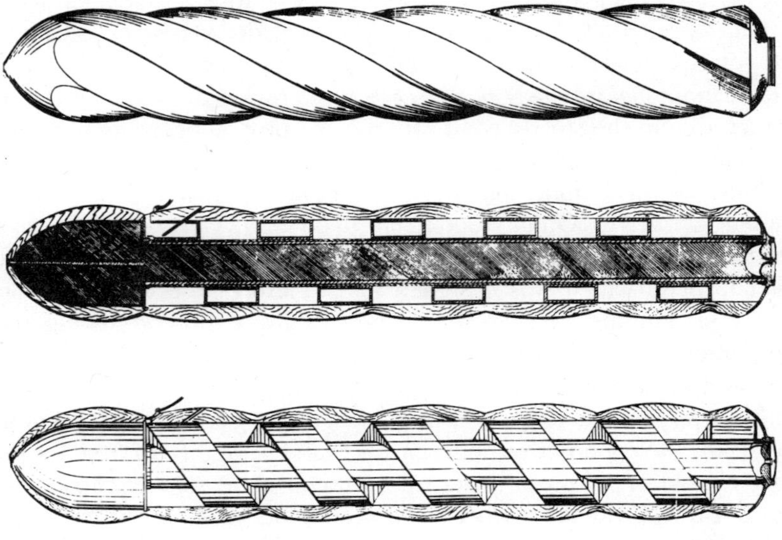

Fig. 16. Lieutenant Barber's rocket torpedo. Among the many efforts of the late nineteenth century to produce self-propelled torpedoes were a number of rocket torpedoes. Lieutenant Barber's type represents one of the more complicated designs.

which he had made privately. The rocket torpedoes were fired under water, 4 feet below the surface, and aimed at a line 4 feet below the surface on a target 63 feet distant. They struck the target after 4.5 seconds, 5 to 7 inches below the 4-foot line at which they were aimed.

At the same time as Mr. Weir, Lieutenant Barber, from whose published lecture many of the preceding data and Fig. 16 have been taken, submitted to the Bureau of Ordnance plans for a rocket torpedo of his own. The length of the device, which apparently was never built, was to be 7 feet, the diameter 12 inches, with a displacement weight of 287 pounds. The rocket charge was to weigh 51 pounds, the bursting charge 48 pounds. A tube containing rocket composition formed the center of the torpedo. Around it was wound a light metal spiral tube which could hold 52 pounds of water. The water, which, of course, had the purpose of replacing the weight of consumed rocket composition, was to enter in front. The rotation of the rocket, brought about by the shape of the outer wooden cover, would keep the water at the outer walls of the spiral tube with a tendency to flow toward the rear end of the rocket. That way, it was hoped, the weight of the propulsive charge would not only be replaced by water at approximately the proper rate, but the water would also distribute itself to the right place. The explosive charge was to be ignited by a percussion fuze on impact; as a precaution a powder train ran from the propulsive charge to the explosive charge so that the torpedo would be exploded in any case soon after the propulsive charge was exhausted.

No rocket torpedo was ever introduced as a naval or coast defense weapon in any country. None of the many designs proposed ever looked as if it would be able to overcome the fundamental drawback pointed out by Hale. And so, with the end of the nineteenth century, the war rocket was once more a thing of the past.

But just at that time some people began to expect different and much bigger things of rockets than that they might be battle-winning weapons.

4 Prophets with Some Honor

> It is not the gray, cold, naked objective truth that counts in the history of mankind and will advance the cause of civilization, but it is the flight of human imagination, the impulses and visions of a genius, very often his errors and miscalculations, which have stimulated inventions and progress.
>
> —Berthold Laufer, in *The Prehistory of Aviation*

> The history of science is science itself, the history of the individual, the individual.
>
> —Johann Wolfgang von Goethe, in *Mineralogie*

Parallel to the story of the use of rockets runs another story: that of the application of rocket power to machinery in general and to vehicles in particular. No systematic attempts were made prior to the days when Professor s'Gravesande astonished his pupils by demonstrating a steam-spitting small vehicle that scampered across the floor. Consequently some of the early instances in that story stand alone in splendid isolation in space and time.

The earliest of all seems to have been the apparatus which later became known as the Flying Pigeon of Archytas, dating back to about 360 B.C. For many centuries the wooden pigeon has been praised as one of the most ingenious inventions ever made by man and practically every writer who praised it added that the secret had been lost. While the praise was justified, in view of the early date, the lament was not. A Roman author, Aulus Gellius, had described it quite clearly, clearly enough, at least, so that we know the general principle. We could build a duplicate that would work, but we could not be sure that it would also be historically correct.

According to Aulus Gellius the pigeon of Archytas was suspended from strings with counterweights "which preserved the equilibrium and the pigeon was propelled by the blowing of the air mysteriously encased therein." This means that the wooden bird model did not fly at all, as is usually asserted, but merely moved in circles around the point of suspension, motivated by air blowing from its body. If such a model were to be constructed today, the simplest way of accomplishing such a performance would consist in placing a small container of compressed air inside the body. Archytas must have used another method, easier to handle with the means at his disposal. In all probability he used steam. Unless all the

Fig. 17. Heron's aeolipile. The oldest reaction machine and the direct ancestor of the modern lawn sprinkler.

descriptions we have are misleading it may be assumed that the bird model was moved around by the reaction of a steam jet.

This would make it an earlier edition of another ancient invention, models of which are still shown and operated in elementary physics classes: the aeolipile, invented, according to tradition, by Heron of Alexandria.

Unfortunately we don't know just how the original aeolipile looked. Nor do we know just when it was invented. The one definite clue is that the inventor, Heron, is stated to have been a pupil of Ktesibios. Just to make things easier, there was more than one Ktesibios, but Heron's teacher most likely was Ktesibios Mechanicus who lived in the third century B.C., fairly soon after Archytas.

The aeolipile probably consisted of a bowl-shaped container which was filled with boiling water and then placed over the fire on an altar. The steam rose through a tube into a spherical chamber and escaped through two narrow tubes bent backward near their tips. The reaction whirled the tubes and the spherical chamber around as long as fresh steam came from the boiler (Fig. 17).

About two thousand years after its original invention, a physicist by the name of Segener had the idea of turning such an aeolipile upside down and forcing water into it under pressure. This invention is now in general use, turned right side up again and known to everybody as a rotating lawn sprinkler.

Another such isolated instance of the application of rocket power is a story which may be legendary or may be true—there is no way of telling.

It centers around the otherwise completely unknown person of a Chinese official whose name is given as Wan-Hoo.

This Wan-Hoo, the story goes, committed a rather spectacular suicide in or around A.D. 1500 by inventing and testing a rocket airplane. He took two large kites and connected them with a framework in the center of which a saddle was fastened. Forty-seven large powder rockets had been attached beneath the kites in strategic places and forty-seven coolies stood ready with flaming torches to ignite these rockets at a prearranged signal. When everything seemed ready, the learned and daring Wan-Hoo seated himself in the saddle and finally signaled to the waiting coolies. They rushed at the machine, each one applying his torch to the rocket he was to ignite, and Wan-Hoo and his machine disappeared in a noisy cloud of black smoke.

Precisely the same can be said about the second known attempt to combine the principles of flight and of propulsion by reaction.

In 1783 the Montgolfier brothers succeeded in making their first hot-air balloons rise. Just one year later two Frenchmen, the Abbé Miollan and one Monsieur Janninet, announced that they had solved the problem of making such balloons fly in the desired direction. Their idea was simple. The bag of the balloon was filled with hot air or smoke, which had the tendency to expand if permitted to do so. If there were a hole in the side of the bag the hot air would escape through that hole, naturally causing a reaction in the opposite direction which would propel the balloon sideways. All that was required was to have several holes around the circumference of the bag, kept shut by valves which could be operated from the gondola.

It was decided to make the first attempt with a balloon which had only one exhaust hole. Subscriptions for this balloon were offered and were sold to an eager populace. The ascent, of course, was public and, equally of course, an admission was charged for the privilege of witnessing the epochal event.

The balloon was ready in midsummer 1784, the attendance was good, but Miollan and Janninet had the particularly tough luck to pick a day which turned out to be unusually hot. Because of that the most frantic stoking resulted in only a partial inflation of the bag. It looked weak and wrinkled and did not even have enough lifting power to lift itself and the gondola, to say nothing of two passengers. Finally two things happened almost simultaneously: the balloon burst into flames and the spectators lost their patience which was strained by the hot weather anyway. They rushed to the balloon and tore off smoldering pieces to be kept as souvenirs, shouting, meanwhile, for the two unlucky inventors.

Janninet merely disappeared, it seems, while Abbé Miollan proved a little more resourceful: he grabbed the box containing the admission money and left Paris with a speed that far surpassed anything he had promised as the probable speed of his balloon.

The third attempt at rocket flight ended even more ingloriously: it was forbidden by the police.

In the early decades of the nineteenth century there lived in Paris a rocket maker by the name of Claude Ruggieri, obviously an Italian. It was a period which was full of stories of balloon ascents on the one hand and of Congreve war rockets and their exploits on the other hand. Ruggieri profited from the news by staging demonstrations in which small animals like mice and rats were carried aloft by large rockets and brought back to earth safe and sound by small parachutes.

Claude Ruggieri's rockets grew in size and power and about the year 1830 the resourceful artisan announced that a large combination-rocket, as it was called, would carry a ram aloft. On the day after the announcement Ruggieri received an offer from a young man who wanted to take the place of the ram. Ruggieri accepted the offer and the rocket ascent was announced to take place on a certain date from the Champs de Mars. But the police intervened, probably because the "young man" turned out to be a small boy. I don't know whether the whole project collapsed because of this or whether Ruggieri used his ram after all. Dr. Robert Paganini of Zweisimmen, Switzerland, informed me several years ago that the boy's name was Wilfried de Fonvielle. He must have been at the most eleven years of age when he offered his person for the venture. Later he became a balloonist and was one of those who left Paris by air when the city was besieged by the Prussians in 1871. Wilfried de Fonvielle died in 1914, just a little over ninety years old.

In 1843 the daily press of Russia devoted a great deal of space to the invention of one Emil Jir who claimed to have solved the problem of the dirigible balloon by the invention of a mechanism which would permit balloons to find favorable winds by rising or falling, without the necessity of sacrificing either ballast or lifting gas. The mere mention of this project brings to mind *Five Weeks in a Balloon* by Jules Verne, but Verne's and Emil Jir's methods were not alike. Jir planned to accomplish the rising and falling by means of reaction; he wanted to carry compressed air along for this purpose, probably also a hand-operated compressor. The modern verdict on this idea would be "theoretically sound but quantitatively insufficient."

This verdict applies to most of the "daring inventions" of this type.

Six years after Emil Jir, a member of the engineer corps of the Russian Army submitted a manuscript of 208 pages to the governor of the Caucasian provinces, Prince Vorontzov in Tiflis. The manuscript was entitled "On the methods of guiding an airship," and was signed by "Engineer Tretesskij."

Tretesskij planned to equip the airship with exhaust nozzles pointing in all directions; if movement in a certain direction was desired, the proper nozzle was to be connected with the "blast generator," to use a modern term. The blast was to consist of compressed air or steam or, possibly, of air heated by an alcohol blowtorch.

Both these projects may have been influenced by a British patent which had been granted on January 4, 1841, to one Charles Golightly. Unfortunately the patent itself is lost and no description or picture of Golightly's project is known. All we have is a cartoon on Golightly's project which bears the jeering title: "Steam-Horse, on which one may ride in one hour from Paris to St. Petersburg." If the caricaturist expressed the idea correctly the Golightly project was a steam-reaction airplane (Fig. 18).

Strangely enough, this cartoon had more influence than the original invention. It set quite a number of people to thinking about reaction planes. The following quotation from a German book seems to express the state of mind of the leading mentalities of the middle nineteenth century quite significantly.

> In the diaries of my mother, Elisabeth Lepsius, née Klein, I find under the date of November 12, 1846, a description of a house party where a friend of the family, the naturalist Ehrenberg, was one of the guests. He was the famous professor Christian Gottfried Ehrenberg (1794–1876) whose discovery of the *infusoria* had populated all the waters and even the air with the smallest living things then known. . . . A great deal of the conversation was devoted to guncotton. Ehrenberg was highly excited about the discovery and declared that it would exert an enormous influence on the development of aviation and that we would soon come to the point where we are able to propel airships vertically as well as horizontally by means of rockets.[1] (Lepsius)

Guncotton, it may be explained here, had been discovered just one year earlier (1845) by the German chemist Christian Friedrich Schönbein, who had tried to dissolve cotton in a mixture of nitric and sulfuric acids. Of course the cotton had refused to dissolve and Schönbein, writing off the experiment as a failure, had gone home for supper, after putting the still

[1] B. Lepsius, in *Lili Parthey, Tagebücher aus der Berliner Biedermeierzeit* (Berlin, 1926).

Fig. 18. Caricature of Golightly's projected "aerial steam-horse."

wet strands on top of the hot stove to dry. Schönbein lost his laboratory, but discovered guncotton in the process.

Another well-known man who became interested in the idea of guncotton-propelled airplanes was Werner von Siemens, the founder of the enormous Siemens Works near Berlin. Siemens drew a sketch of such a plane and had it published. Because he was then holding an army commission he did not use his name but labeled the sketch "a proposal coming from an officer." (In Russian books on rockets this project is erroneously ascribed to a Nuremberg mechanic named Rebenstein.) (Fig. 19.)

And finally there was the Phillips helicopter. Phillips built a functioning model, running on steam. It was, in fact, an aeolipile with helicopter blades. The model was demonstrated to learned societies in Paris in 1842 and it worked. According to contemporary descriptions, it was made of metal and weighed about two pounds, the center of the apparatus being, of course, the boiler. The rotor had four blades, tilted at an angle of 20 degrees and supported by four struts. The four exhausts of the aeolipile part were connected with these four struts. The reports do not give figures about the performance, but state that the small helicopter "rose to considerable height and traveled a long distance horizontally until it touched the ground again." As did the Golightly patent, the demonstration of the Phillips helicopter gave rise to a caricature (Fig. 20).

Three other names that must be added to the list of the early inventors of rocket-propelled aircraft are those of General Russell Thayer of Philadelphia, engineer Nicholas Petersen of Mexico City, and the inventor Sumter B. Battey. The dates are, in the same order, 1884, 1892, and 1893. All three inventions relate to the propulsion of elongated airships, not round balloons.

General Thayer had, of course, a military application in mind. His air-

Fig. 19. German drawing of a rocket airplane. Probably by Werner von Siemens, drawn some time between 1845 and 1855.

Fig. 20. Cartoon of a steam-reaction helicopter (about 1860).

ship was a war vessel, equipped with cannon. Propulsion was to be accomplished by a jet of compressed air, and the air was carried along in steel cylinders, but the compressors were stationary on the ground.

The other two suggestions were similar to his in several respects. Nicholas Petersen equipped his vessel with a mechanism closely resembling the old but then very common Colt revolver except that it was much larger and large rockets were to be used instead of cartridges. Sumter B. Battey, who was granted a U.S. patent on his invention, had in mind a mechanism more like that of a machine gun. His reaction motor worked with a large number of small explosive fills which were fed into the chamber automatically (Fig. 22).

In judging all these ideas and projects, one thought has to be kept paramount. The choice of rocket or steam reaction as a means of propulsion was not a free choice, it was a matter of necessity. At that time no other prime movers except the steam engine and the clockwork existed. The one was obviously too heavy and the other was, equally obviously, too weak. Rockets, on the other hand, had proven successful in certain applications

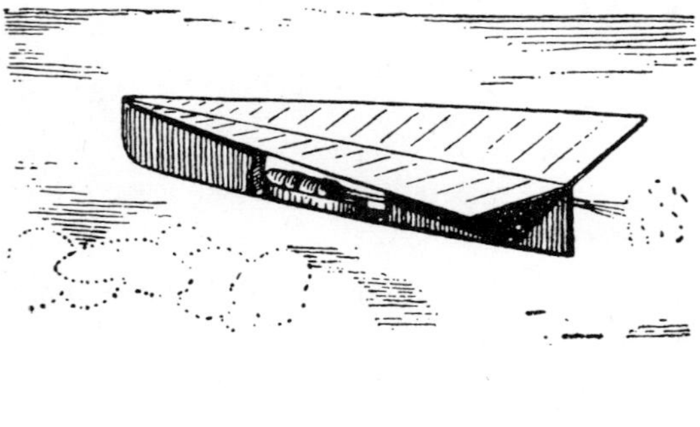

Fig. 21. Steam-reaction airplane of 1867. This one may prove to have been prophetic, with the addition of a vertical swept-back fin.

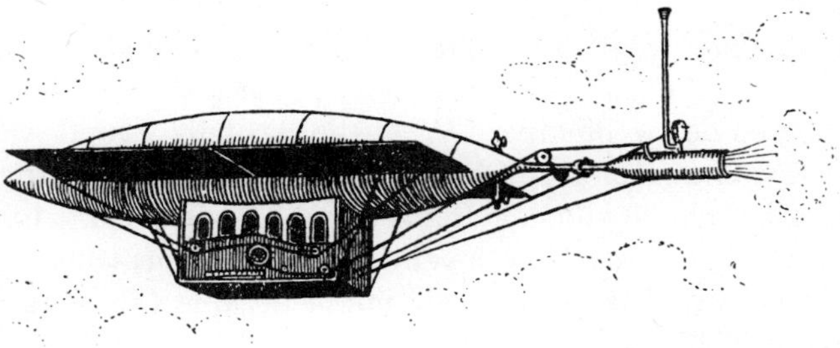

Fig. 22. Battey's rocket airship of 1893.

and, as long as there was no proof that they could *not* perform as a prime mover for aircraft, they were worth considering.

It was precisely this kind of reasoning which caused the conception of a project that, because of the circumstances under which it was born, remained unknown for more than thirty years. It was safely locked away in the archives of the Imperial Russian Secret Police. The inventor was one Nikolai Ivanovitch Kibaltchitch and the invention had been made, or at least put on paper, in a cell of the Peter-Paul Citadel while the inventor was awaiting the day of his execution.

Nikolai Ivanovitch Kibaltchitch was one among six members of the Narodnaya volya, the revolutionary party called "Will of the People," who were charged with the "major crime," the assassination of Czar Alexander II, on March 13, 1881 (March 1, Russian style). The trial[2] took place at St. Petersburg, then capital of the Russian Empire, April 7 to 9, 1881, ending, as was to be expected, with death sentences for all six defendants, two of whom were women. (The death sentence for one of the women was commuted to lifelong imprisonment because she was pregnant then, but she died only one year later.)

The leader of the group was Alexander Shyelyabov who, when in court, missed no opportunity of inserting a political speech into the record. The actual assassin, the man who had thrown the bomb that killed the Czar, was one Nikolai Ryssakov, a young man who was extremely nervous all through the proceedings and hysterically voluble when asked to answer a question. Kibaltchitch had taken part in only one respect: he had made the bombs and had instructed Ryssakov and others how to use them.

Kibaltchitch had been arrested on March 29, 1881. Two days later he had been confronted with Ryssakov who recognized him at once as the bomb

[2] For a complete account of the trial, see David Footman, *Red Prelude* (New Haven, Conn.: Yale University Press, 1945).

instructor. According to court procedure an attorney was then assigned to defend him.

When the attorney went to Kibaltchitch's cell on one of the first days in April, he probably expected to find either a fanatical revolutionary or the proverbial broken-down offender. What he actually found was a well-dressed, quiet young man who was deep in thought when the cell door opened. But Kibaltchitch did not think about his personal fate, which he knew was decided; he was busy inventing a rocket airplane. Literally his first question was whether his attorney might be able to obtain permission for him to write in his cell.

The attorney realized soon that the case was hopeless from the legal point of view. Kibaltchitch not only refused to go through the legal routine, he also did not comply with any of the established rules. When in court he sat as if he were paid for attending, but were not obliged to listen to the proceedings. He came to life only when the explosives experts were called to testify. Then he would enter into a discussion, argue about fine points of the manufacture of a detonator, ask questions about facts which he had failed to find in literature, and be deeply gratified when the experts asserted that the bombs he had made had been good, professional, reliable bombs.

He also called public attention to the fact that he had just finished a manuscript which bore the title "Preliminary Design of a Rocket Airplane" and that he had asked his attorney to submit it through the prison authorities to a committee of technical experts.

That had actually been done, but the manuscript had not gone beyond the prison administration. The officials in charge decided that a public discussion of Kibaltchitch's manuscript would stir up too much interest in the person of the defendant—the newspaper accounts of Kibaltchitch's short remarks in court had done too much harm already in their opinion. The manuscript was simply attached to the documents of the trial and locked away in the archives, until the latter were systematically searched by the Tcheka in 1918. It was then published, with comments by historians and engineers, in the Russian journal *Byloje* (*The Past*), October–November 1918.

Kibaltchitch had visualized his rocket airplane as a platform with a hole in the center. Mounted above this hole was a cylindrical explosion chamber into which "candles" of compressed powder were to be fed. The apparatus was to rise vertically at first, then the chamber was to be tilted for horizontal motion; the speed was to be regulated either by the dimensions of the powder "candles" used, or by their number.

Kibaltchitch died in the belief that a committee was still debating his plans.

None of all these inventors of rocket aircraft thought of his projected vehicle as a possible means of leaving the earth. Although this idea had been expressed in novels by that time, as we have seen, it was still not a subject of serious consideration. But the first project for a rocket-propelled vessel for ascents beyond the outer layers of the earth's atmosphere was not long in coming.

As such things often go, the idea was conceived by two men almost simultaneously, two men of different backgrounds and different characters, of different habits and different nationalities.

Both were strange and each in his own way was a colorful character. Both were independent in their thoughts and stood alone with them. Both were, which is also interesting, self-taught and both failed to be very successful. But their similarities end with these statements; everything else about them was different. The Russian school teacher Konstantin Eduardovitch Ziolkovsky (pronounced *Tsee-oll-kóvs-key*), handicapped by the same physical ailment that did not handicap Thomas Alva Edison at all, lived a long quiet life in a rather unimportant and, to the outside world, practically unknown corner of the Russian Empire, near Kaluga. The German law student Hermann Ganswindt, who never got to be a lawyer or a judge as his parents had dreamed he would be, also lived a long life, but a stormy one, battering his way through almost four-score years in a manner which could not possibly be used as a plot for a story. Any self-respecting editor would reject it as "unnatural and improbable."

Of the two, Ganswindt was more impetuous and less scientific. And, since he was Ziolkovsky's senior by the narrow margin of one year, I'll begin with him.

He was born in a small town in East Prussia on June 12, 1856. His parents, well-to-do people, decided that little Hermann was to study law, acquire a doctor's degree, and then settle down to a successful and "reputable" life in the legal profession. Filial rebellion was not a part of the program, but young Hermann rebelled. He decided in an upsurge of revolutionary spirit that "justice" and "right" were not always identical, and this decision quickly condensed into the belief that the courts by administering the law could not do other than dispense "injustice and law" instead of "right." Hermann Ganswindt wanted no part of that, and he wrote a book about his worries, or rather a book in which these worries were expressed vehemently along with other things. It was thoroughly misunderstood, of course, but Ganswindt would never admit that he had written a book which could be misunderstood; he only blamed the others for not understanding it.

Having turned his back on the legal profession, Ganswindt turned toward his real love: mechanical gadgets. All his inventions had to do with transportation in one form or another. He invented (not in that order): bicycles, horseless carriages, "motor" boats, fire engines, airships, and spaceships. Some of his inventions remained on paper, but most of them were actually built; in many cases the inventor was his own craftsman.

His first attempt was the airship. There were some airships on record at this time: Giffard in Paris had built one with a small steam engine in the gondola; it was too weak. Dupuy de Lôme, also in Paris, had made one for propulsion by muscle power. Hänlein had completed an airship in December 1872, and it had been the largest of the three—all of 165 feet long. Ganswindt wondered why they had all been unsuccessful and found an answer which he incorporated in a patent that was granted to him in 1883 by the Patent Office in Berlin. (It was D.R.P. No. 29014.) The main mistake, according to Ganswindt, had been the semi-logical idea that you had to build a small airship before you could build a large one.

In the case of the airship that was wrong, said Ganswindt. In order to be really a "dirigible," a ship must be able to maintain some 30 miles per hour. Nothing less would do, or any breeze could change the dirigible back into a free balloon. Adequate speed would require an engine of at least 100 horsepower, but a steam engine—there was no other kind—of such power would be too heavy for an airship. Too heavy, that is, unless you built an airship of sufficient size; natural laws see to it that things improve with size in that case. If you enlarge a given airship ten times in every dimension, its volume and hence its carrying capacity are a thousand times as large as before. But its cross section and its air resistance are only ten times as large. *Ergo* an airship will become a true dirigible if only you make it large enough. The proper size for a 100-horsepower ship would be a length of 500 feet with a diameter of 50 feet.

His reasoning in relation to air resistance was somewhat crude but essentially he was right—such a ship could have been built then. Everybody knew, of course, that airships would have a great military value if they could only be built, hence Ganswindt submitted his idea, copy of patent attached, to Field Marshal Count von Moltke. The answer, signed by Major General von Regely, came quickly. It said that the "General Staff does not have any funds . . . numerous projects of aerial travel received. . . . Etc. Sorry. Period."

Ganswindt was not discouraged; in fact he had expected something like this. He sat down and wrote a book devoted exclusively to the problem of the dirigible airship—*Die Lenkbarkeit des aërostatischen Luftschiffes*—and had it printed at his own expense. It came off the presses in July 1884. Now, he felt, now he had ammunition for another attack on conservative and reluctant officialdom.

It was rather good ammunition; in retrospect one can only say that his grasp of the problem was sound. Unfortunately that was not apparent to his contemporaries. The shape of his airship was a long cylinder, both ends with curved points, like the most up-to-date artillery ammunition of the time. Ganswindt apparently did not know the terminology; he could have described the shape of the gas-envelope as cylindrical with ogival ends of two calibers radius. This envelope was to consist of sheet rubber, almost

⅛ inch thick. All the ropes from the envelope were to attach to a supporting structure some distance below the envelope, running the whole length of the ship. It was an interesting wrinkle that Ganswindt did not want to use a girder construction for this supporting structure, as others had suggested (and as was later done) but a strong metal pipe of considerable diameter. This pipe was to be filled with hydrogen under pressure, as a reservoir against seepage losses from the envelope. The gondolas, each with its own steam engine, were to be suspended one from the other at the center of the ship. Ganswindt said that in future air travel from city to city all passengers, mail, and freight for the nearest city could be put into the lowest of this vertical stack of gondolas. This would then be disconnected as a unit and either ballast or another gondola taken on. Ganswindt also visualized that a gondola might be dropped by means of parachutes without landing the ship, but admitted that it would take time and demonstrations to get passengers used to that idea. For ground use he conceived the anchoring mast, with a circular rail around it for anchoring the rear end of the ship.

Although some details may sound silly to us, Ganswindt's groping was generally in the right direction, marred somewhat by loud insistence on specific ways of accomplishing something. Still, a line of airship development could have been based upon his suggestions.

Armed with his own book, Ganswindt started his second campaign. He wrote (copy of book attached) to the Ministry of War. This happened to be a place which was very weary of airships. Everybody seemed to have ideas about the "conquest of the air" and the only ones who did not submit them to the War Ministry were those who were too busy to write them down. Not only was the War Ministry subjected to daily annoyance; it did not mind disbelieving entirely in the conquest of the air—if it were possible, which, of course, it wasn't, it would confuse tactics and ruin established strategy. Furthermore, all useful military inventions had been made in the military laboratories, and, well, maybe by Krupp engineers. And one mustn't give in to the idea of scientific research. If one did, one might get a request for a cruiser with full complement to chase the Great Sea Serpent.

As for that man Ganswindt—well, he is 28 years old and has never finished his studies. Now he wants to be an inventor and travel through the air, the crank, and not only that, he wants to travel through the air with structures twice the height of a church steeple. Nonsense! If the small airships which some Frenchmen had built did not work, how could a big one be better? The answer to Ganswindt was: No.

But meanwhile Ganswindt had mailed copies of his book to all kinds of important people. Among the recipients was, of course, the Crown Prince, the same who was later to rule for 99 days as Kaiser Friedrich III. The Crown Prince was interested in intellectual matters; he also was dutifully interested in military matters. He read the book, he agreed with Ganswindt, and he ordered an investigation. The War Ministry was displeased, not openly, of course. Ganswindt once more. He had just been rejected through channels; now he was back in the form of a "most-high personal" request of His Royal Highness the Crown Prince. One had to ask technical experts. And the technical experts, the churls, wrote guarded reports, saying that there were no obvious flaws in Ganswindt's reasoning, although it was always possible that such flaws might be revealed by actual tests. Actual tests would cost money. War Ministry money. And if such hidden flaws then showed up, the newspapers would remind the War Ministry that its money was taxpayers' money. But the same newspapers, or at least some of them, might let Ganswindt scream in print in case he was rejected. Somebody then recalled that diplomacy was the domain, but not the monopoly, of the Foreign Office. The War Ministry could be diplomatic too, if nothing else helped. And Ganswindt was informed with much politeness that "airships of a length of 500 feet far surpass military requirements."

The Crown Prince, poor of health, was satisfied. Too bad; it had sounded interesting. As for the War Ministry, it stuck to its figurative guns and rejected *all* airship proposals thereafter, including one originating from a Count Zeppelin. Only after Zeppelin had spent his own money and that of a few wealthy friends proving the possibility of large airships by building them, did the attitude change.

Ganswindt also felt that he had to show more than a copy of a patent and a small book. He tried to raise money by founding a society and soliciting donations. That was not too successful and he decided to set himself up in business as a full-time inventor. He quickly became a well-known character in Berlin and especially in Schöneberg, the suburb where he lived. It made him very happy when people began to refer to him (even if jokingly) as the "Edison of Schöneberg." There he founded a small factory for his inventions, combined with a permanent Ganswindt Exhibit with all the characteristics of a fair, including large and brightly colored posters all over Berlin. The factory built a free-wheeling mechanism for bicycles and a new type of wheel hubs which were advertised as "practically frictionless." The factory built what Ganswindt called the *Tretmotor.* It was an ersatz engine, consisting of an ingenious arrangement of levers and pieces of rope. Two small platforms, slightly larger than a man's foot, were on top of or behind the mechanism. The driver stood on these platforms and by alternately shifting his weight from one leg to the other made the mechanism whir. Ganswindt built a well-functioning "motor" boat that was propelled in that manner, adding a small artificial lake to his fair for demonstrating it. He also built a carriage for two people, with the driver standing behind them. It worked well and Ganswindt rode around the city in it.

The sudden appearance of a horseless carriage in 1894 caused so much traffic confusion that Ganswindt received a letter from the Police Commissioner: would he please be cooperative and tell the police about each trip in advance so that they could take the proper measures? Simultaneously he got a letter from the fire department. Would he lend them a carriage of this type for trials? Ganswindt thought that this would provide the needed airship money. He did not lend the existing carriage to the fire department but built one especially for them. They liked it fine; in fact, they wrote rather enthusiastic reports which Ganswindt showed me later. But no business developed.

Not that the fire department did not wish to buy the carriage. It did, but Ganswindt asked 4000 marks. The fire department could not pay that much and offered 2000. Ganswindt replied that his own expenses had been more than 2000 marks and took his carriage back. It did not occur to him that the cost price of later cars would be much lower; he had no business sense. He only demanded his "right," and his "right," in this case, was to make a profit from the first carriage on. When I asked him later why he had not accepted the price offered to him, since that would have resulted in more orders and later profits, he failed to understand my reasoning. "But I would have lost money on that car," he insisted. "If, as you say, later cars would have cost less money to build, and I agree with you on that, I would of course have reduced the sales price."

In between he built a large helicopter with seats for two people. The machine lacked only the motor, nothing else. Newspaper reports said that there was an internal-combustion engine of sufficient horsepower on display at the Paris Fair. Ganswindt went to Paris and offered to buy the motor, provided that it developed the necessary power. That was promised and weeks later a crate arrived at Schöneberg. Ganswindt tested the motor and it lagged by a full third behind the promise. He returned it and thought about a way out. He found one. The helicopter had been assembled around a central axis; Ganswindt replaced the central axis by a length of steel tubing. Through this tube he threaded a steel rope which was attached to the ground as well as to the projecting roof of the factory building. Then he wound a second rope around the central tube and attached it to a heavy weight which, when released, would drop into a well dug for this purpose. It was an adaptation of the well-known trick of winding up a top.

The helicopter actually flew with two people in it. Not high, because of the short duration of the impulse received, but it flew. It is obvious now that Ganswindt was just lucky. A modern airplane engine could take his helicopter up without any difficulty but could not prevent it from sideslipping and crashing, which it would most assuredly have done after 15 or 20 seconds of flight. Not only had Ganswindt forgotten that there is such a thing as torque, he had made a design of zero stability. On the occasion of the helicopter's short "flight," one of the early inventors of the motion picture, Max Skladanowsky, was present and took a film of the performance. Together with other similar "shots" it made one of the first newsreels. The film was taken in June 1901 and the newsreel was shown in the Wintergarten in November of the same year. I have seen a copy of the program for the night of November 5, 1901. The last feature of the program was "Max Skladanowsky's Biograph" and one of the features of the Biograph newsreel was Hermann Ganswindt's Flying Machine.

By then Ganswindt had quite a number of personal enemies, many of them, it seems to me, disappointed investors. They claimed that the helicopter was pulled up by the wire rope. They said that "any blacksmith can duplicate the contraption at an expense of 500 marks." Ganswindt's friends countered with a bond of 35,000 marks, to be paid to the blacksmith who did. Ganswindt's enemies dug up yellowed copies of his first book with its indiscreet and exaggerated remarks about justice and right and "proved" that he was a Social Democrat in disguise. And they filed a complaint with the local District Attorney.

What happened then Ganswindt later described as follows: He received a request from the police commissioner, stating that either he or his representative wanted to see the helicopter in action. Ganswindt was up early on that morning of April 17, 1902, in cutaway and top hat, ready for his visitor. But instead a plainclothesman appeared with a warrant for his arrest. The charge was that he had accepted money from trusting investors, promising to build things which could not be built. And since he had often displayed a sufficient knowledge of physics and chemistry to be aware of that, he was guilty of fraudulent claims. The trial, which included a court visit to Ganswindt's exhibits, ended with acquittal. But as far as public repercussions went, he might as well have been convicted. The company that rented poster space refused his posters; the newspapers refused his advertisements.

Ganswindt, embittered and hurt—all the more hurt, probably, because he felt that he was not entirely blameless—went berserk and fought back. He did this, ironically, by dragging his enemies into court, with lawsuits that were often based on minor technicalities. The courts were embarrassed and delayed proceedings; Ganswindt complained to higher courts about the lower ones. The performance was quieted only by the outbreak of the First World War. Then Ganswindt turned to the War Ministry again with bulky "recommendations," offering his airship once more and stating loudly that the Zeppelin ship was an inferior imitation. He asked the Ministry for a sum of several millions as damages for the suppression of his invention, but promised that he would use the money to build his airship and put it at the disposal of Germany.

The German Minister of War—von Stein was his name—voiced the general sentiment in one sentence. When, in 1917, another one of Ganswindt's innumerable petitions arrived, he wrote in red pencil on the cover page: "*Lebt denn dieser Unglücksrabe immer noch?*" ("Is that bird-of-woe still among the living?")

The question was justified in that the Ganswindt of 1917 was no longer the Ganswindt of the period from 1883 to 1901. He was no longer an inventor who wanted to create something; he was a quarrelsome person who fought only to redeem an injustice done to him. He wanted it redeemed with money and public recognition. He wanted to hear "officially" what he put on his stationery and envelopes: "Hermann Ganswindt, original inventor of the free-wheeling mechanism, the automobile, the airplane, the airship, and the spaceship."

He did get some kind of recognition in the end. Articles about him began to appear in newspapers and magazines and the general feeling was that he had been condemned too hastily. Some people even sent him anonymous gifts of money when they learned that the monetary inflation had left him impecunious and that he was supported by welfare agencies and by the elder of his twenty-three children. And in 1934 a semiofficial government agency sent Dr. Stark with a gift of 1000 marks to Ganswindt.

Ganswindt was sick with a cold on that day. He received the money and the official recognition which he thought it implied—actually it was one of the Nazis' propaganda gestures intended for domestic consumption—with deep gratitude. A week later he died (October 25, 1934). He had lived to fight for recognition; recognition achieved, he had nothing left to live for. He had received his "right."

What is significant is this: The recognition which Ganswindt finally received had nothing to do with the fact that he had once been close to inventing the helicopter. Nor did his old airship patent have anything to do with it. What renewed his fame in his old age was that he had prophesied the spaceship. He himself did not pay too much attention to this later because at one time the War Ministry had seized upon it to make him appear ridiculous. But Ganswindt's conception of a reaction-propelled spaceship is actually the first I have been able to find, except for Eyraud's fictional one. Unfortunately I am not able to date Ganswindt's idea with complete precision.

I have seen, in Ganswindt's own collection, a printed program of one of his lectures, delivered on May 27, 1891, at the Philharmonie—as the name indicates, a concert hall—in Berlin. The lecture was mainly devoted to aviation, but newspaper reviews about it proved that Ganswindt had expounded his theory of space travel, too. He had come across the principle of reaction fairly early in his career and had realized at once that reaction must take place in empty space too. Hence, if this principle could be utilized for propulsion, one could build a spaceship. I tried to make Ganswindt remember just when he had thought of that first. But while he recalled some detail, he could not mention a year. He did know that he had been "a very young man." And he added that he had thought of the spaceship "before Jules Verne wrote his novels," meaning, of course, *De la Terre à la Lune*. However, since that novel appeared when Ganswindt was just nine years old, the statement obviously could not be taken at face value. I assume that Ganswindt took the year during which he read the book as the year of its publication, and in all probability he referred to the German translation. Thus, the date of May 27, 1891, is the earliest that can be *proved.* A painting of his spaceship (reproduced in Plate I), which was hung over the piano in his home, was undated, but looked as if it had been painted in about 1890.

Although Ganswindt had intuitively grasped the principle of reaction, he was never able to understand the mathematics of the problem. He insisted that a fireworks rocket mostly worked by "pushing against the air" because a "mere gas" could not produce enough reaction. In order to get useful reaction, he said, it was necessary to throw off solid bodies, weighing at least two or three pounds apiece. Because of this preconceived notion, his "fuel" consisted of heavy steel cartridges containing a charge of dynamite. Those cartridges were to be fed into a bell-shaped steel explosion chamber. One half of the cartridge would be thrown out by the explosion of the charge inside; the other half would hit the top of the explosion chamber and, after transmitting its kinetic energy to the chamber, drop out. The chamber was rigidly connected with two cylindrical fuel drums on either side.

This assembly constituted the "motor." The passenger cabin was suspended from the motor on springs to ease the shocks. Naturally it was provided with a center well through which the steel pieces could pass. Once a sufficiently high velocity was reached, Ganswindt would stop feeding cartridges into the explosion chamber. Then, he knew, the passengers would undergo the experience of apparent weightlessness which he intended to combat by rotating the long cylindrical cabin around its center well, thus substituting centrifugal force for gravity, making the two ends of the cabin the floors. That idea was correct too, but Ganswindt lacked either the ability or the patience—probably both—to work the plan out in detail.[3]

[3] The Viennese periodical *Die Zeit* (*Time*) No. 304, July 28, 1900, carried a long and very interesting article about Ganswindt's space-travel plans, written by Professor Doctor Román Baron von Gostkowski, dean of the College of Engineering, Lemberg. Professor von Gostkowski proved that Ganswindt's ship could reach neither Mars nor Venus because existing explosives do not contain enough energy. He admitted that explosives could enable man to reach the limits of the atmosphere but pointed out that the ship would then fall back because it could not carry additional explosives for braking its fall. Apparently it did not occur to him that one might carry instruments only, without a pilot. This article is one of the very earliest discussions of space travel by a professional physical scientist.

Now for Ziolkovsky, the other early spaceship inventor.

Unlike Ganswindt, he was a patient and quiet man. Unlike Ganswindt, he never overlooked a single detail. Unlike Ganswindt, he never had any enemies. And most unlike Ganswindt, he had considerable trouble speaking about himself and his work. When, about 1930, Professor Nikolai A. Rynin of Leningrad wrote to Ziolkovsky for a complete autobiography for his forthcoming seventy-fifth birthday, he received only a thin manuscript of a few pages, accompanied by a letter stating: "This, Nikolai Alexeyevitch, is the best I have to offer." Ziolkovsky also enclosed a collection of photographs, and it is typical that he did not even mention in his autobiography the fact that he got married, although some of the later photographs show him surrounded by grandchildren.

Ziolkovsky was born September 5, 1857,[4] at Ijevsk in the District of Ryasan. His father, a Pole, was a forester by profession and an unsuccessful inventor and philosopher by avocation, while his mother came from a family of Russian artisans. Ziolkovsky remembered these facts and he also remembered that his first toy, which he got at the age of eight or nine, was a small balloon, a rare toy in those days. Soon after, at the age of ten, Ziolkovsky was dangerously ill with scarlet fever which caused deafness; his hearing never returned. "At eleven I wrote the foulest verses."

His deafness made it difficult for him to attend school—even for what little schooling there was in Ryasan in 1870—but he studied zealously, especially elementary mathematics and physics. Even that was difficult because there were only "a few doubtful books." But he managed to absorb the basic principles of science and those few fundamental facts he could find in the books he had. He not only satisfied the small requirements of a pupil; soon afterward he even satisfied those of a teacher and began to teach in 1876. In 1882 he was offered a position at the school in Borovsk in the District of Kaluga, and ten years later he became a teacher at the school of Kaluga, a position he held until he resigned because of ill health in 1920.

Most of the time he got up at dawn, went to school, and hurried home as quickly as possible to do his private work. The latter mainly concerned problems in physics, but he also worked on some inventions. At the age of twenty-three he submitted a few scientific papers to the St. Petersburg Society for Physics and Chemistry.

When his papers were received there they caused not a little consternation; the reaction was partly incredulous and partly indignant. The papers dealt with some aspects of the theory of gases. Those formulae had been worked out by others twenty-four years earlier. They dealt with methods of measuring the velocity of light, methods that had been in use decades before that. At first glance it seemed as if somebody with a restricted

[4] Old style; according to the Gregorian calendar, September 17.

mentality had tried to gain some doubtful fame by doubtful methods. But a careful reading of the manuscripts made it clear that their author, the teacher Konstantin Eduardovitch Ziolkovsky, simply had not known that these formulae and methods existed.

One of the leading men of the Society took it upon himself to write to Ziolkovsky, to inform him about his strange misfortune, and to express the interest of the Society in his future work. The man who wrote that letter was Dimitri Mendeleyeff, the famous discoverer of the periodic system of the elements.

Ziolkovsky passed over his disappointment later with the remark that it had been "good practice," but he concentrated on inventions from then on. He also became interested in airships and he came to the same conclusion as Ganswindt did—Ganswindt would have sued him if he had known—that airships were too small to work well. He began to work on a large all-metal ship.

But there was something that was not known: how much skin friction would a metallic ship create at a certain speed. Ziolkovsky thought about the problem, but it could be decided only by experiments with models. For these experiments he built a small wind tunnel in 1891, possibly not the first in the world but certainly the first in Russia. Mendeleyeff was interested in Ziolkovsky's airship plans and helped as much as he could. It was probably mainly his influence that made the Academy of Science in St. Petersburg contribute 470 rubles to Ziolkovsky's work. Ziolkovsky also had a friendly press, and newspaper readers from various parts of the country contributed small amounts, all in all 55 rubles. Ziolkovsky was perfectly aware of the fact that they were meant as alms. But he did not return them in indignation as most people of his period would have done; he used the money for his research work.

One night in Moscow, when he was still a boy, he had looked up to the sky and tried to visualize what he had learned: the spinning earth in space, racing around the sun along its orbit. Spinning suggested centrifugal force with the familiar picture of a weight whirled around on a piece of string. Such strings occasionally snap—and young Ziolkovsky found no more sleep during that night. He was, as Columbus' son said about his father in another sense, "drunk with the stars." The idea of space travel had caught him and never let go again.

Many years later, in 1895, Ziolkovsky cautiously mentioned space travel in an article for the first time. To his surprise the article was printed (in *Nature and Men*) and from then on the idea of space travel crowded everything else from his mind. After that first publication, he really began to think about the problem, or rather the complex of problems, taking them up one by one.

Between the planets there was mostly empty space, hence the space vessel would have to have a completely sealed cabin with oxygen reserves and air purification. Because of that empty space only a mode of propulsion which would work in empty space could be considered: that was the principle of recoil, the rocket. But rockets were not efficient enough; they needed higher efficiency. Higher efficiency could be obtained best by using fuels with a higher exhaust speed. Hence Ziolkovsky decided on liquid fuels of the kerosene type for the spaceship. He went over the whole problem step by step and in 1898 he had something that could be called a preliminary study. He submitted it to the editors of a journal called *Na-ootchnoye Obozreniye* (*Science Survey*) and the editors, apparently after hesitating for quite a while, finally published it in 1903.

Ziolkovsky's article was the principal section in the issue in question. But unfortunately the magazine contained among other articles one that was offensive to the Czar's secret police and the whole issue was confiscated, though a few copies seem to have reached regular subscribers. Ziolkovsky was officially presented with one copy, which he had bound in leather. In the years to come this article was somewhat expanded and partly rewritten, and it formed the first of a series of articles by Ziolkovsky which appeared from 1911 to 1913 in a technical magazine with the name *Aviation Reporter.*

This was a magazine that was widely read in Imperial Russia (it folded with the revolution) and it brought Ziolkovsky a disciple who proved very valuable, a writer on scientific subjects, mainly physics, named Dr. Jakov I. Perelman. Perelman, who was later to become the science editor of the *Krassnaya Gazyeta* in Leningrad, devoted several chapters in his *Entertaining Physics* books to Ziolkovsky and to rockets in general. But he also wrote a straight popularization of Ziolkovsky's more severe works, and it was Perelman's eloquence which made Ziolkovsky's name known to the Russians.

There are many legends connected with that early period which unfortunately acquired some permanence because a German-writing Russian by the name of Aleksander Borissovitch Shershevsky uncritically put hearsay into some of his articles and his only book. One of the legends is that Ziolkovsky and the Frenchman Robert Esnault-Pelterie (one of the French pioneer aviators who later became an ardent devotee of space-travel problems) held a debate on space travel in the presence of the Czar. Determined to set the record straight, I asked both of them by mail about this assertion. Robert Esnault-Pelterie wrote at once saying that the legend may have originated with the fact that a year or two before the First World War, he had delivered a lecture at St. Petersburg. But the Czar had not been there. Nor had he ever seen him. Nor had he ever met *le Monsieur Ziolkovsky.* And his lecture had been *pas sur l'astronautique mais sur l'aviation.* Ziolkovsky's letter, written in longhand and in German, arrived a week or two later, and also contradicted the story.

Another legend, originating from the same source, stated that the well-known Russian astronomer Tikhov had devoted much of his time to problems of space travel. Perelman assured me that all that Tikhov had ever done along that line had been to be chairman during one of Perelman's lectures.

The First World War, naturally, put a stop to Ziolkovsky's and Perelman's endeavors. But the Russian revolution did not touch them; in fact the Soviet government, before it was completely stabilized, put a protecting hand over Ziolkovsky's head. He mentioned in a letter to me that he received sums of money "quite frequently," probably by way of honorariums for reprints of his former articles which were published in the form of pamphlets by the Kaluga Branch of the Government Publishing Office. The first publication by Ziolkovsky under the new regime was not a straight scientific work, however. It was a novel with the title *Outside of the Earth,* a fictionalized account of a journey away from the earth. Such books always sold well in Russia and the commissar in charge may well have thought that Ziolkovsky's novel was just one more of that type, even though he was probably informed that every sentence was the result of painstaking research.

My reason for assuming this is the fact that none of Ziolkovsky's "straight" works was reprinted until after 1923, the year Professor Hermann Oberth published in Munich a German work on the theoretical possibility of space travel. One year later the Government Publishing Office in Kaluga reprinted the long article that had originally appeared in *Na-ootchnoye Obozreniye* in 1903. The title was changed for republication and now read: *Rakyeta v kosmeetcheskoye prostranstvo* (*The Rocket into Cosmic Space*). The octavo pamphlet of 32 pages had a bilingual title page, with a German translation of the title (*Eine Rakete in den kosmischen Raum*) at the top. It also contained a preface in German by Aleksander L. Tchijevsky, stating that the Russians, after a short review of Oberth's book had been printed in the official Russian dailies, remembered that their compatriot Ziolkovsky had expounded the same theory thirty years earlier. The whole preface was an early example of a Russian "song" which we know well now: the loud assertion "we invented it first." Aside from that, it was an emphatic appeal for Russian patriotism (writen in German) closing with the words: "Do we always have to get from foreigners what originated in our boundless homeland and died in loneliness from neglect?"

The other articles appeared in quick succession after that; Ziolkovsky found himself honored beyond his most ambitious expectations. When he rounded his seventy-fifth year in 1932, official agencies prepared articles about him which were published in all the official newspapers. The Osoaviakhim, the Russian aviation organization, arranged a large-scale celebration of his birthday. The Academy of Sciences held a special meeting in his honor, and the Soviet government awarded him the "Red Banner of Labor."

He died three years later, on September 19, 1935.

5

The Battle of the Formulae

Those not imbued with enthusiasm will, as a rule, avoid major mistakes but will also miss major accomplishments.

—Professor Dr. Wilhelm Wundt

Knowing what went on in the way of rocket development during the nineteenth century and, simultaneously, what happened later, the first two decades of the twentieth century look like a rather barren stretch in the story of the rocket. Even to a contemporary onlooker it must have seemed that very little happened during those years. There were three lines of development, all leading apparently nowhere. But in retrospect these developments are very interesting, because they foreshadow later events.

The first of these three lines of development was the attempt of the Swedish Lieutenant-Colonel Baron von Unge to create a new kind of war rocket, an air-to-air missile. This particular phase is described in Chapter 7.

The second was an attempt to influence the weather, or rather one particular phase of it: to avert hailstorms. That this attempt was a serious piece of work and not just somebody's wild idea is indicated by the occasion at which it was reported for the first time: the annual meeting of the Society of German Scientists and Physicians in 1906. The man who made the report, one R. Baur, a former Instructor of the Artillery School of the Turkish Army, had also performed the experiments. His idea had been that clouds and in particular hail clouds might be dispersed by explosions. Obviously it was desirable to have the explosion take place inside the cloud and since one could not very well equip townships with modern artillery for such a purpose, Baur had developed a rocket which was easy to handle, could reach a height of 3000 feet or more, and carried a powerful explosive charge. He stated that he had been able to disperse heavy rain clouds—no hail clouds happened to appear during 1905 in the locality where he experimented—repeatedly and added that some test firings in late fall had

had the curious result of causing local snow flurries which surprised all observers.[1]

Baur's work was discussed for a long time, and especially in Switzerland many experiments were made along the lines he indicated. Of course there was no agreement about the results, since nobody could prove whether a cloud which under rocket fire merely shed rain would have dropped hail if left undisturbed. Baur and his successors may have done some early cloud seeding; in any event rockets are in use for such a purpose now. (See Appendix 3.)

The third of the new lines of development was introduced at the same 1906 meeting by one A. Bujard, in a paper entitled "Rockets in the Service of Photography." It dealt mostly with the work of the engineer Alfred Maul who wanted to use camera-carrying rockets for military reconnaissance. But Maul was then just beginning and had only completed the first of several models. That first model, tested in 1904, carried a camera with plates measuring 40 by 40 millimeters and could reach an altitude of 900 to 1000 feet. But it was a case of pure good luck if a picture could be recovered. Either the shutter did not work or else the parachute failed to open.

Maul decided that everything was too small for ruggedness and reliability and constructed a much larger model. Its guiding stick was 13 feet long and carried at its lower end four square fins in cruciform arrangement. The camera case and parachute were attached to the other end of the guiding stick, the plate size of the camera was 120 by 120 millimeters (4¾ by 4¾ inches). A large blackpowder rocket—or more than one, depending on the altitude desired—was attached to the stick, which was the type used at the seashore for carrying lines. One interesting and useful trick was the insertion of a 30-foot line between camera and guiding stick. The heavy guiding stick with the empty rocket cases struck the ground first on descent, so that he camera was still 30 feet above ground and was lowered over this remaining distance by a parachute which did not have anything else to carry. This model reached altitudes of 1500 to 2000 feet. The next model had a 15-foot guiding stick, a camera with a plate size of 180 by 180 millimeters (7 by 7 inches) and was powered by two 80-millimeter blackpowder rockets. Take-off weight of the whole was 25 kilograms (55 pounds) and it reached a 1600-foot altitude in 8 seconds.

The largest model, made in 1912, was equipped with a stabilizing gyroscope, reached an altitude of 2600 feet, had a take-off weight of 42 kilograms (92½ pounds), and carried a camera with a plate size of 200 by

[1] *Verhandlungen der Gesellschaft deutscher Naturforscher und Aerzte* (78th Meeting, Stuttgart, 1906); published in Leipzig, 1907. Baur's report, read during the third session of the Section of Applied Chemistry on September 19, 1906, appears on pp. 167–68 of that volume.

250 millimeters, or about 8 by 10 inches. The rocket was ignited electrically from a distance and was quite reliable, but by the time this model was ready it was also possible to aim a camera at the ground from an airplane.[2]

And this is virtually all that there is to tell about rockets for the time interval from 1900 to 1914 when the First World War began in Europe. During the war there were some sporadic uses of rockets, in addition to the obvious and customary one of signaling. The probable reason why rockets appeared only sporadically is discussed in Chapter 7, "The Return of the War Rocket."

The "new era" began to dawn almost with the last shot of the war. On May 26, 1919, half a year after the Armistice, one Dr. Robert Hutchins Goddard, professor at Clark College, Worcester, Mass., finished the manuscript of a small pamphlet which contained only 69 printed pages. The manuscript consisted mostly of a study he had written in 1916 and sent to the Smithsonian Institution as a plea for financial support; the additions were some information about tests made and conclusions based on these tests. It was printed by the Smithsonian Institution as Publication no. 2540 and, although dated 1919, was released in January 1920. The title was *A Method of Reaching Extreme Altitudes.*

The main reason behind Goddard's thinking and the main purpose he had in mind appear in the very first sentence of the book. Reinforcing the title, it reads: "A search for methods of raising recording apparatus beyond the range for sounding balloons (above 20 miles) led the writer to develop a theory of rocket action." Neither the subject of the book nor the treatment should have interested the general public at all. The subject was an investigation of whether rockets could be used as instrument carriers for the exploration of the upper atmosphere. The treatment was that of a professional study of a special phase of applied physics, couched largely in mathematical terms, encumbered with many tables, tersely written. But toward the end there was something intriguing. The report, more or less by way of example, discussed the possibility of a rocket reaching the moon and exploding a load of flash powder there to mark its arrival.

That was something newspapers could write about. A rocket carrying strange instruments of unknown purpose is one thing. A rocket striking the moon and exploding into a huge blinding flash, visible from earth, is something else. And, even more important from the newspaperman's point of view, this was not a wild idea thought up by a poor and presumably half-mad inventor. It originated from a man who was teaching physics,

[2] Photographs of these rockets and by these rockets can be found in the monthly magazine *Kosmos* (Stuttgart), vol. 10, 1914, pp. 524–27.

who had earlier earned his Ph.D. in his own field[3] and who had held a Navy commission, his technical assignment having been the improvement of Navy signal rockets. Furthermore, his work had been published by an institution of semiofficial character.

From what I have been told, the idea seems to have created some stir in the daily press. But it does not seem to have caused much debate in scientific circles; at least I have not been able to discover controversial articles in professional magazines. And in continental Europe the book remained virtually unknown for a number of years. Some of the large libraries which receive all the Smithsonian publications naturally got Professor Goddard's book along with the others; by the end of 1920 there must have been a score or so of the total of 1750 copies on the shelves of the European libraries. But the importance, or even the contents, of the book was not immediately noticed. The noncommittal title may have created the impression that the book was simply a discussion of a new technique in meteorological research work, of no interest to anybody except a professional meteorologist. Librarians, when asked, were not even sure whether they had it or not. Besides, Europe, in 1920 and for a few years later, had other worries, worries about war and civil war, about food and clothing and unstable currency, about political uprisings and street battles. A book published in happy and prosperous America was only too likely to be overlooked.

By the end of 1923 the greater part of the political troubles had run their course. The immediate worries about daily livelihood had receded to something approaching their normal level. Those who were so inclined could begin to take an interest in cultural matters again. In short, it was again possible to think.

And it was precisely at that point that the great old problem of space travel was revived. It began in what looked at first glance like a particularly ineffective manner. The publishing firm of R. Oldenbourg in Munich brought out a paper-covered pamphlet of less than a hundred pages. (I learned later that its author had to pay the larger part of the printer's bill out of his own pocket.) The pamphlet was called *Die Rakete zu den Planetenräumen* (*The Rocket into Interplanetary Space*) and it was by H. Oberth.[4] Nobody knew who "H. Oberth" was and the book itself shed no light on that question.

[3] The following outline, taken from *Who's Who,* provides the main data of Dr. Goddard's life prior to the time when he devoted himself completely to rocket work: "Born Worcester, Mass. Oct. 5, 1882 . . . B.Sc. Worcester Polytechnic Institute, 1908; M.A. Clark University, 1910; Ph.D. Clark University, 1911. . . . Instructor in physics, Worcester Polytechnic Inst., 1909–11; student at Princeton Univ., 1912 and 1913; at Clark University, instructor and honorary fellow in physics, 1914–15; assistant professor, 1915–19, professor since 1919."

[4] The pronunciation of the name is *O-bert,* with the stress on the *O.* Hermann Oberth was born in Hermannstadt, Transylvania, on June 25, 1894.

The introduction began as follows:

> (1) The present state of science and of technological knowledge permits the building of machines that can rise beyond the limits of the atmosphere of the earth.
> (2) After further development these machines will be capable of attaining such velocities that they—left undisturbed in the void of the ether space—will not fall back to earth; furthermore, they will even be able to leave the zone of terrestrial attraction.
> (3) Such machines can be built in such a way that they will be able to carry men (probably without endangering their health).
> (4) Under certain conditions the manufacture of such machines might be profitable. Such conditions might develop within a few decades.
> In this book I wish to prove these four assertions. . . . (Oberth)

It is almost superfluous to say in so many words that proof of these assertions (except, perhaps, the somewhat "unprovable" fourth) was established. But the way in which it was established was meaningless except to mathematicians, astronomers, and engineers. As far as the general, even the interested, public was concerned, the book might just as well have been printed in Sumerian characters.

There is a good reason for writing of this type. It is that the originator of a new thought tries to convince the professionals first, that he seeks professional criticism of his thesis. It happens every day in science and the style fulfills its purpose with only rare exceptions. Oberth's book, for some reason, turned out to be such an exception. It became, incredible as that may seem, a limited popular success; the first edition was sold out in a short time and the orders that piled in at the publisher's office almost exhausted the second edition before it was even printed. But while the small book was successful in that respect, it did not do so well in the other. It sought the professional criticism of the professionals—but, of course, there was no such profession. Many of those who should have read it carefully did not do it at first because they felt that this was something beyond their ken, that it was not really a problem in their own profession.

Since Oberth's book became the sole cornerstone of all later space-travel ideas, a short outline of its contents may be useful. It was divided into three parts. The first dealt with more or less general questions of rocket motion and contained a lot of things that people—meaning physicists, etc.—should have known but didn't. It was a great surprise to many of them that a rocket should be able to surpass the velocity of its own exhaust. It was also surprising to many how important the so-called "free ascent" was. Of course one has to remember that in those days the word "rocket" meant only "fireworks rocket" to practically everybody. And a fireworks rocket does not burn long enough to attain even a sizable fraction of its own exhaust velocity. In a fireworks rocket, too, the moment propulsion stops is very nearly the same as the peak of the trajectory, just because it is so slow. Oberth pointed out that a fast rocket at the instant propulsion stops—

what is now called *Brennschluss*, or "end of burning"—has a high velocity and then behaves like a projectile. It may be compared with that of a cannon which has the muzzle located at the Brennschluss point. Naturally the shell will travel on from there. And if it is a fast shell, or rocket, the altitude attained *after* Brennschluss may amount to a multiple of the altitude attained *at* Brennschluss.

To say more about Part I of Oberth's book now would only result in a long chapter full of technicalities; the detail is explained as it appears in the course of the later chapters. Part II consisted of a description of an assumed instrument-carrying rocket for upper-atmosphere research, which was called Model B. This description was not supposed to be taken as literally as many critics took it. One, I remember, complained bitterly that no dimensions had been given in the diagram and that Oberth had neglected to specify the alloy and had not even given the wall thickness! Well, Model B was not a workshop drawing and probably would not even function if built in strict accordance with the book. Model B had the purpose of discussing the whole problem in its applications to one particular instance.

While the second part is a study of the same type as Goddard's Smithsonian report, the third part of Oberth's book went further, much further. This third part—incidentally the only portion of the book which can be read without continually getting caught in the barbed wire of interlocking integral and square-root signs—dealt with general prophecies about the probable achievements. Oberth described a theoretical spaceship, called Model E, discussed some problems of space travel and a number of possible objections, and developed the first sketchy theory of a station in space.

Just enough of the book was "readable" to intrigue a comparatively large number of people. But the professional critics, the people for whom the book had really been written, did not do so well. Some just announced it; a few others wrote articles which were a condensation of Oberth's statements. Some, finally, criticized it, as they were expected to do. But what they said was, for the most part, surprising—in one of the more uncomplimentary shades of meaning of that word.

Two or three well-reputed astronomers, men who certainly should have known better, simply "killed" the whole idea—or so they thought—by stating that all these things are very nice and interesting but lacking in foundation since everybody knows that there can be no recoil in empty space. Another critic, physician by profession, added some more devastation. Not only was there no recoil in empty space, the idea of manned rockets was preposterous for all time to come because people, as soon as they left the atmosphere of the earth (provided it could be done), would be subjected to the gravity of the sun which is powerful enough to squash their bodies.

The good doctor had taken his figures from tables in an astronomical book. Unfortunately the figure for the sun's attraction in that table referred to *the surface of the sun itself* and not to a distant place such as the orbit of the earth or a point near that orbit.

Another critic, an aviation expert, stated in a puzzled manner that he was inclined to believe Oberth, but that he could not understand, try as he might, why the exhaust gases should follow the rocket if the latter, after some time, surpassed its own exhaust velocity. That the exhaust velocity, *seen from the rocket,* never changes, apparently did not occur to him. The whole puzzle which he could not solve was caused only by the (unconscious) mental change of viewpoint: by looking at the exhaust once from the rocket and then from the ground.

Another, by his profession as a physicist more given to precise statements, said that the rocket, of necessity, could not surpass the velocity of its own exhaust because the efficiency would surpass 100 per cent in that case. And an efficiency of more than 100 per cent is, obviously, not possible. This was a strange mixture of correct and incorrect conclusions. It is true that the efficiency of a rocket which moves, say, twice as fast as its own exhaust velocity seems to be far above 100 per cent, if you consider only the time interval when that condition takes place. But it is wrong to do that. The apparently impossible efficiency occurs, in a manner of speaking, at the expense of the very low efficiency at the beginning of the ascent when the rocket moves very slowly. If you look at the whole performance, the efficiency never even approaches 100 per cent, which is what should be expected in the first place.

What could happen if you went off on a tangent somewhere in the middle of the whole problem was nicely demonstrated by another mathematician and physicist who calculated with gusto that the most powerful explosive known could not even lift its own weight to a greater height than about 400 kilometers (250 miles). He forgot that the fuel is not carried along, but only its energy.[5] The great calculator who burst into print with this discovery had even meant to be considerate. He had picked nitrogelatin, then the most powerful of the generally known explosives, for his example. Oberth, he mentioned in passing, had mere liquid fuels like gasoline in mind. I don't know whether anybody told him that a mere liquid fuel like gasoline will produce a higher exhaust velocity than any

[5] Dr. Otto Steinitz, who later worked with Oberth on some patents, nicely explained this mistake by the use of analogy. When shooting an arrow muscular energy is stored in the wood of the bow. It amounts to about 20 kilogram/meters for every kilogram the bow weighs. Twenty kilogram/meters would be just enough to lift the bow 20 meters or about 66 ft and the conclusion would be than an arrow cannot be shot higher than 66 ft, no matter how large the bow, since you cannot store more than 20 kilogram/meters of energy per kilogram of weight. But you don't shoot the bow, you shoot the much lighter arrow, weighing, say, one-tenth of the weight of the bow. Consequently it can rise ten times as high as the bow could lift itself. In the case of a rocket, the rocket itself represents the arrow and the fuel the bow, and the altitude is not limited by the weight (mass) of the exhaust gases—the product of the fuel—since these are left behind.

high explosive. I feel sure, however, that if somebody had told him, the critic would have refused to believe it.

Oberth, the preceding paragraph suggests, placed great emphasis on liquid fuels for rockets. As is mentioned in Chapter 3, the velocity of a rocket can be increased either by increasing the amount of fuel consumed per second (hence by increasing the volume of the reaction-mass) or by increasing the exhaust velocity. It so happens that even the most ordinary of liquid fuels, ordinary automotive gasoline, will produce an exhaust velocity about twice that of a blackpowder rocket. This fact alone had been sufficient for Oberth to decide in favor of liquid fuels; he also mentioned that mechanical considerations of storage and handling also favored a liquid fuel. Now, more than 25 years later, one can add that the latter considerations are more valid. Some modern solid rocket fuels produce exhaust velocities which almost compete with those of liquid-fuel motors.

Oberth could not know that; he chose liquid fuels for their higher exhaust velocities. Twenty years earlier Ziolkovsky had done the same for the same reason. Oberth did not know that either; not only was it virtually impossible to find files of the old Russian aviation journals that had published Ziolkovsky's articles, but Oberth could not read Russian. And Professor Goddard (who couldn't read Russian either) was actually experimenting with liquid fuels at the time Oberth's book appeared. However, he kept this a secret until 1936.

But in addition to confusing the critics, Oberth's book did something else. It encouraged others who had similar ideas to present them to a surprised public. One of the most important publications, following close on the heels of Oberth's book, was a volume of similar external appearance, written in a similar manner. As a matter of fact, it looked so forbidding to prospective buyers who leafed through it in the bookstores that the first edition of only fifteen hundred copies was never completely sold.

Its author was Dr. Walter Hohmann, city architect of the City of Essen-on-the-Ruhr, and the title was *Die Erreichbarkeit der Himmelskörper* (*The Attainability of the Celestial Bodies*). The title does not sound quite as clumsy in German as it does in English, but it is severe enough. The table of contents read: "I. Departure from the Earth; II. Return to Earth; III. Free Coasting in Space; IV. Circular Orbits Around Other Celestial Bodies; V. Landing on Other Celestial Bodies."

If that sounded intriguing, a reader was subjected to a cold plunge when he opened the book, say, at Chapter II and began to read: "In order to reduce the velocity of a vehicle of the type described in the preceding chapter from v_1 to zero, if that vehicle is falling toward the earth from a very great distance, we require the same time of operation t_1 as calculated by means of equation (10) but the stream of the exhaust dm/dt has to point in the direction of the movement." The whole book read like that. Years later the editor of a large newspaper came to my office, pulled a

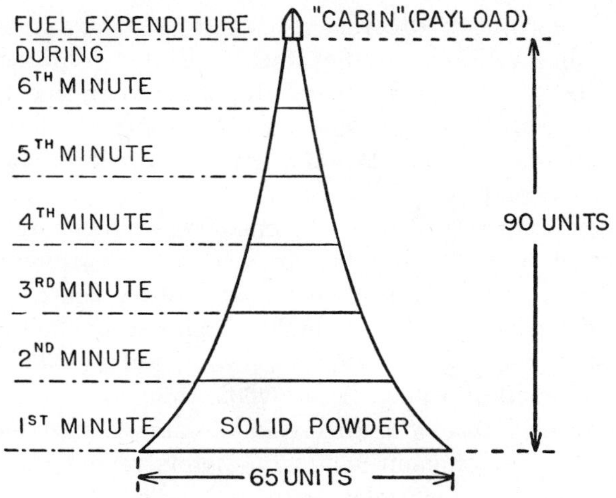

Fig. 23. The Hohmann powder tower.

copy of Hohmann's book from his briefcase, and said: "I want to write a feature article about you people, and you, Mr. Ley, have written that Hohmann's contribution is important. All right, I believe you and I want to say something nice about him . . . *but what does it all mean?*"

I can understand why Hohmann's book was even more obscure to people like that newspaper editor than Oberth's book. Oberth, at least, had spoken about a rocket called Model B and about another one called Model E. He had calculated their performance and stated what they ought to be able to do. To the engineer these Models B and E were only examples, sketches supplied for the purpose of making a definite calculation. To the layman, and that term included quite a number of engineers too in this particular case, Models B and E were not examples but projects, something that could somehow, even though with difficulties, be visualized.

Hohmann, as he wrote me later, had thought about actual constructions but had not come to presentable conclusions. Consequently he had restricted himself to a mathematical investigation of the quantities of a given or assumed fuel required for an assumed operation. It was all highly interesting and of great scientific value, but it read very much like a prescription. For illustration, Hohmann had drawn a powder tower which confused the issue a little more. It might not have done so if his powder tower had looked like Fig. 23 with its explanatory inscriptions, but Hohmann's original showed the "tower" only. It was not meant as a "design," as many people thought; it was just one more of those abstract examples.

Fig. 23 shows how it was meant. If you imagine that a certain operation would require 6 minutes and if you assumed that the tower burned at the base only you could draw six parallel lines through it. That would produce six disks of powder plus the payload to be moved. The six disks would be of equal thickness, but of different diameters; also of different weight, of course. Each slice represented the amount of fuel required per minute: the biggest slice at the bottom being the amount for the first minute of operation and so on. If you made a sufficiently large and careful drawing you could subdivide each slice into sixty parts and find the amount required from second to second.

While German scientists, preconditioned by Lasswitz to taking space-travel theory seriously, battled merrily and sometimes mistakenly about details, there was clearly a need for some writing which let the general public in on the concepts about which a high-tension battle of equations was going on. It was not just a question of popularization, it also had a very important practical aspect. If the whole thing was to be of any value, the matter must not rest with theoretical findings. Something had to be built or Oberth and Hohmann would be merely repetitions of Ziolkovsky, possibly to be revived again another twenty years later in another repetition of the same thing. The idea had to leave the study to enter the laboratory and later the workshop. But people who could unlock laboratory doors had to be interested first, and that required nontechnical language, not only for those who could not understand the technical language, but also for those who could but did not know of the existence of the formal treatises, or who needed an interest-rouser.

It was reasoning of this kind which prompted a man named Max Valier, known as a popular writer on scientific subjects, to approach Oberth with the proposal of emulating Dr. Perelman. Of course he did not put it that way, since neither of them knew either of Ziolkovsky or of Perelman, but the idea was the same.

Oberth accepted the proposal. He had had the idea of a popular book himself, as he told me later, but he was overburdened with work. His own studies for a teaching job were not quite completed. He worked on extensions of his own rocket theory and he also spent quite some time in discussions with financiers and in working out simple demonstrations for them.

The discussions about financial support came to nothing. From a few hints which Oberth dropped later and especially because of my own subsequent experiences in similar talks, I can well imagine what went on there. The financier or financiers had apparently held the customary conception of professors who invent things, the conception you can find in print in countless bad stories. They had expected an old or at least elderly man who had labored over one idea all his life and who carried battered manuscripts around with him, manuscripts wrapped up in newspapers and tied around by strings with many knots in them. They probably had had secret hopes concerning complete models which would fly around in the conference room. They had quite definitely expected that the investments they were—perhaps—going to make would bear fruit in not more than two years.

Then they saw that the professor was a slender man who looked taller than he was because of his gangling build, a young man who had just rounded the third decade of his life. The long matted hair and big beard they had visualized turned out to be a short crop of black hair and a small moustache. The only thing that was "professorial" about him was his manner of speaking—whenever Oberth opened his mouth you got the feeling that you were in a classroom and there was no way of leaving, not even on that famous urgent errand, until the lesson was over.

They learned that Oberth had not kept any great secrets to himself, as they had expected, but that everything could be found in his book (which they couldn't read), that the program of carrying out what Oberth had visualized was a program of decades and not of years. They learned that the demonstration models were rigged up at home and were not even patented! Like many people they did not know what a patent was, but they labored under the belief that a patent is an expression of official approval of an invention. Actually it only certifies that the idea was new on the day it was submitted and carries with it a license to sue imitators.

After having wasted a good deal of time and effort, Oberth returned to the country of his birth, Transylvania, an old German-speaking colony of German Saxons wedged in between Hungarian and Rumanian territories and belonging politically at that time to Rumania. He went to Schässburg in January 1925, and in February became professor of mathematics at the high school of Mediash, also in Transylvania.

Max Valier's book appeared soon afterward. But it failed to be what the author had promised. It was full of well-meant but ridiculous illustrations. When it came to difficult points, Valier resorted to flippancy which he took to be humor. A great deal of the book was in fine print ("to be skipped by the lay reader"), full of calculations, most of them made by Oberth. In spite of all these faults it sold at a fair and steady rate, a reprint became necessary roughly once a year, and some of the most glaring faults were remedied in these reprints. Some others were added later by Valier in an effort to display independent thinking.

As a matter of fact it was the poor quality of Valier's book which made me enter into competition with him, although I was not quite twenty years old at that time. With the enthusiasm peculiar to that age I decided that I could do better than Valier. I sat down and wrote a small and formula-free book on the same subject. It was printed in 1926 and sold some six thousand copies during the following six years. To my surprise many people, including Oberth, said that it was actually better than Valier's—at any event it did what Valier had failed to do, it told the whole story understandably and in as few words as possible.

It was then that the problem began to assume international dimensions. In Russia, as has been told in the preceding chapter, Ziolkovsky's old articles were reprinted, along with new editions of Dr. Perelman's book. In Moscow even a student society was formed, with the avowed purpose of advancing space travel. It was short-lived, which, in view of its name of "World-Center of All Inventors and Scientists," was probably just as well. It was not until November 1929 that a more serious society was formed in Russia. It called itself the "Group for Investigation of Reaction Motion"—in Russian, Gruppa Isutcheniya Reaktivnovo Dvisheniya, hence GIRD. There were two groups, one in Moscow, called Mosgird, and one in Leningrad, called Lengird, the latter under the leadership of Dr Perelman and Professor Nikolai A. Rynin.

The same Professor Rynin had then just begun to publish the first parts of his enormous nine-volume work, *Interplanetary Travel.* And in France, Robert Esnault-Pelterie delivered a lecture on the problems of space travel, addressing the Société Astronomique created originally by Camille Flammarion. The lecture, published in 1928 in book form, was given on June 8, 1927—precisely three days after several gentlemen, living in and near the comparatively small provincial town of Breslau in Germany, had met in the back room of a restaurant to found a society for the purpose of spreading the thought that the planets were within reach of humanity if humanity was only willing to struggle a bit for that goal.

The name of that society was Verein für Raumschiffahrt (Society for Space Travel). But people later referred to it as the VfR, while the common appellation in English-speaking countries became the German Rocket Society.

If I remember correctly, German law provided that the minimum number of people who can legally found a society is seven; I don't think the assembly exceeded that number by more than two. One of those present at the meeting, a man by the name of Johannes Winkler, agreed to accept the presidency of the society—Valier had refused because of his almost continuous lecture tours--and to publish a small monthly magazine which was to be the mouthpiece of the society. This monthly magazine, called *Die Rakete* (*The Rocket*), was actually published immediately afterward and appeared regularly until December 1929. Winkler invited Oberth and Hohmann to join and they did. Winkler also undertook the task of registering the society with the courts in Breslau. This was required for business reasons: only a society registered with a court of law existed legally and that fact was indicated by adding the letters *e.V.* (*eingetragener Verein,* or "registered society," implying about the same as "Inc."). There was a little trouble with this particular incorporation; the court found that the word *Raumschiffahrt* (space travel) was not defined in any dictionary and that the public, therefore, would not be able to judge the purpose of the society. A change of name was suggested. However, the court finally relented and, since new inventions require new words, it accepted the registration of the society under the condition that the document of registration itself define the name in an unmistakable manner.

The growth of the VfR was rapid. Within a year it acquired almost five hundred members, among them everybody who had ever written about the problem in Germany and in neighboring countries. Oberth and Hohmann, Dr. von Hoefft and Guido von Pirquet in Vienna, Professor Rynin and Robert Esnault-Pelterie—all had joined. The program was, in outline, to interest as many people as possible, to collect membership dues and extra contributions, and to create a fund for experimental work. And since existing literature in the field was already obsolescent to a certain extent, I began to plan another book, to be written in collaboration by all the leading men of the society. My publisher promised to publish it and to advertise the VfR in the process, and the VfR, in turn, promised to advertise the book. It appeared under the title *Die Möglichkeit der Weltraumfahrt* (*The Possibility of Space Travel*) in the spring of 1928.

But meanwhile several other developments had taken place, some of them good, others less so. One of the more important ones was a series of articles attacking Oberth in a rather vehement manner. They were published in the very important journal of the VDI (Verein Deutscher Ingenieure, Society of German Engineers) and they were signed by Privy Councilor Professor Dr. Lorenz of Danzig. Privy Councilor Lorenz did not make any elementary mistakes; he simply proved that Oberth's spaceship could not acquire the "velocity of escape" of about 7 miles per second. His arguments and calculations boiled down to the statement that a rocket fueled by known fuels, if it were to acquire that velocity, would have to weigh thirty-four times as much when fueled as it weighed when empty. The conclusion drawn from these calculations read: therefore it cannot be done.

Oberth, naturally, wrote a reply. It was rejected. Dr. Hohmann, being a member of the VDI as well as the VfR, also wrote a reply. It was rejected. The excuse given was lack of space. The real reason, which I learned through a personal conversation years later, was: "We cannot permit people half his age to contradict the Privy Councilor!"[6]

But another society, the WGL (Wissenschaftliche Gesellschaft für Luftschiffahrt, Scientific Society for Aeronautics), was not so narrow-minded. That society held annual meetings and the meeting for the year 1928 had been scheduled to be held in Zoppot near Danzig. The WGL invited Lorenz to attack Oberth and it invited Oberth to defend himself. Lorenz spoke at length; Oberth answered with a very short speech. He pointed out that he had followed the Privy Councilor's arguments and that one could arrive at the ratio of 34:1. Personally, by knowing that one factor

[6] The VDI finally did give in, in the winter of 1950!

was more advantageous than the Privy Councilor had assumed, he had arrived at the ratio of 20:1, as the Privy Councilor would have seen if he had finished reading Oberth's book. But in any case Oberth could not help it if the Privy Councilor refused to believe that it is possible to build an aluminum pot into which one could pour enough water so that the full pot would weigh twenty times as much as the empty pot.

I may add that Lorenz never wrote about rockets again.

After the meeting, Oberth, Mrs. Oberth, and I sat together and complained bitterly about fate. The big book, the *Möglichkeit,* had been published just in time for the meeting in Zoppot. But something else had happened too. The headlines of every newspaper in Europe told about "the successful trial run of the world's first rocket-propelled vehicle." They referred, of course, to Opel's rocket car which had just been demonstrated to the public in a big, carefully staged show. And that was, to put it mildly, exasperating. These rocket cars and the stories around them embodied the greatest possible misunderstanding and stupidity.

The instigator of all this colossal nonsense had been Max Valier.

For some time Valier had hinted in various articles and lectures that great developments could be expected in the very near future. In January 1928 we had learned just what this great development was going to be—it was such that I had sent the manuscript of the *Möglichkeit* to my publisher without waiting for the chapter Valier had promised to contribute to it.

Max Valier, as he put it himself, had "succeeded in interesting Fritz von Opel in rockets." This, of course, was a very euphemistic way of putting things. It was one of those statements that are correct and truthful as far as the wording goes, but which are, in fact, untruthful because the implications are different from those that may reasonably be expected.

The fact was that Max Valier had one day had an opportunity to see Fritz von Opel, who was the owner of a large factory manufacturing cheap cars. Although he always denied it, nothing flattered Opel so much as hearing people refer to him as the German Henry Ford. When listening to whatever it was Valier told him, Fritz von Opel had a brilliant idea. He saw an opportunity of purchasing unlimited amounts of publicity with what was for him small change. He and Valier were to build a rocket automobile. How long would it take to develop a rocket motor? Valier assured him that there was no need to wait for anything; such experiments would be just as valuable from the scientific point of view if they were made with large powder rockets, and large powder rockets could be purchased on short notice. Of course the experiments had no value either way, but Opel did not probe that question. That was not what he was interested in. Being a shrewd businessman, he answered reluctantly that he might consider it if Valier knew where to get the powder rockets.

Valier did—there was a factory making powder rockets in Wesermünde

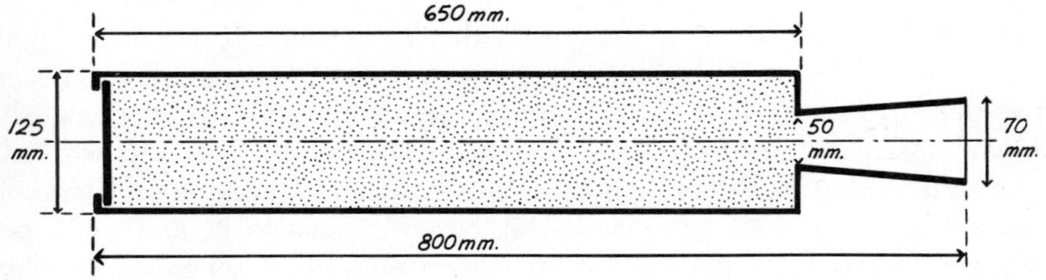

Fig. 24. Cross section of a steel-case solid-powder Sander rocket. This was the type used by Fritz von Opel for his cars and railroad cars. (125 mm is not quite 5 in.; 650 mm not quite 26 in.)

near Bremen, owned and run by an engineer named Friedrich Wilhelm Sander. Sander did not manufacture ordinary fireworks rockets for amusement; he specialized in practical types. He made line-throwing rockets for the coastal rescue societies, signal rockets for the merchant marine and the German naval units, and other pyrotechnic devices connected with navigation. Sander's rockets were acclaimed in seafaring circles for their good performance and that good performance was due to a special process of manufacture which Sander had developed. By means of this process Sander was able to compress the powder mixture to a much higher degree than "possible"—which is to say "customary"—and since he succeeded in doing that without blowing up both the machinery and the laboratory, it was quite a feat.

Valier and Sander, when talking about the proposed use of the rockets, decided to employ "mixed batteries," consisting of both bored and solid rockets. The largest bored rockets delivered a thrust of close to 400 pounds for not quite 3 seconds, while specially made large solid *branders,* as they were called, delivered a thrust of about 45 pounds for 30 seconds. The rockets were to be used to give the initial impulse and accelerate the car to a fair speed; the branders were to maintain that speed (Fig. 24).

Valier wanted to make a private test at Wesermünde before going to Fritz von Opel's race track at Rüsselsheim. But Sander refused to lend his car for the experiment and Valier did not own a car. Their arguments resulted in a draw and they compromised by going to Rüsselsheim without a preliminary test. The rockets were shipped by truck—the railroad refused to transport them—and the first secret test was made on March 15, 1928. It consisted of attaching one bored rocket and one solid brander to a small Opel car. Opel's test driver Kurt Volckhart took the wheel, released the brakes, braced himself for a violent start, and pressed the ignition button. The car began to move very slowly and rolled to a stop after about half a minute—450 feet in 35 seconds!

Both Volckhart and Opel were greatly disappointed and Opel wanted to give up then and there. He said something on the order of being too important and too busy to be annoyed with silly schemes. If he had stopped at that point the whole episode might have been over and might never have become known. But he added a remark about rockets that "obviously did not work," which insulted Sander. Sander knew that his rockets worked. He stared at Opel with his light blue eyes, growled, "Wait," found himself a long and heavy pole, tied one of his bored rockets to it, and set it off for a vertical ascent. Opel was visibly impressed and consented to another test. This time Volckhart started the car by means of its own engine. When the car was moving at the rate of about 30 miles per hour, he released the clutch, killed the ignition, and simultaneously ignited the rockets. They accelerated the car to 45 miles per hour. Opel, somewhat pacified, ordered a special car built.

The new car was not so very special; it was an ordinary racing model without an engine and with a mixed battery of Sander rockets attached to the back of the car. The first run (April 11, 1928) was made with six rockets. One of them failed to ignite, the other five propelled the car for about 2000 feet. The next run was made with eight rockets. Again one of them failed to ignite, two others exploded without doing any harm, but the car ran some 3000 feet with an average speed of 55 miles per hour.

The next test, officially called the First Test Run, was made the following day. Five of the battery of twelve rockets did not ignite, but the speed exceeded 70 miles per hour and the newspaper reporters present wrote enthusiastic stories keyed to a patriotic note: German science and German industry had created the first rocket vehicle in the world. Opel saw that his publicity scheme had worked after all, and while his advertising department placed full-page ads in the better magazines, his engineering department designed another rocket car.

It was a long streamlined car with stubby wings mounted upside down so that they would press the car against the road. After a secret test run, Fritz von Opel himself demonstrated the car—called the Opel Rak II—on the Avus Speedway near Berlin on May 23, 1928. This run was a complete success, all twenty-four rockets ignited without any trouble, none of them exploded, and the speed of the car came close to 125 miles per hour. Opel, photographed from all sides, delivered a speech over the radio, promising an even more astounding Opel Rak III.

This was the run that had taken place near Berlin while Oberth defended himself against Professor Lorenz in Zoppot. Small wonder that this victory did not make us very happy. After the VfR had worked itself to exhaustion explaining the difference between liquid and solid fuels, after pointing out scores upon scores of times that rockets could never be efficient unless they moved at a velocity which was a large percentage of their exhaust velocity, say, 800 miles per hour or higher—after all this, one of the founders of the VfR had gone and played around with solid-fuel rockets of commercial manufacture to pave the way for what we considered a cheap publicity stunt. Valier came close to being publicly expelled from the VfR. It did not happen mostly because it would have been a tedious procedure involving a lot of correspondence. I believe Oberth also found some kind words for Valier, who had already been pushed into the background.

Opel, naturally, went on.

He produced the Opel Rak III, which he had promised. This was a track-bound railroad car, built along the general lines of Opel Rak II but with a small battery of rockets in its nose, generally referred to as brake rockets. They were to stop the car at the end of its test run, being ignited automatically. The German government had granted permission to use the track from Burgwedel to Celle (near Hanover) for this experiment. That particular track had been chosen because it was perfectly straight and level. The first run was made on June 23, 1928, with a battery of ten rockets. The ignition was timed by a clockwork device, as were the brake rockets, and the car was not manned. It attained a top speed of about 180 miles per hour, the brake rockets failed to perform properly, they shot up into the air when ignited, and the car coasted for several miles.

Retrieved and towed back to the starting point, it was charged with a battery of thirty rockets—all speed records were to be broken. But the acceleration was too high; the car derailed almost immediately and was destroyed. The fate of Opel Rak IV was similar. One rocket of the first series exploded and a splinter short-circuited the ignition system, causing all the remaining rockets to go off at once. The car jumped off the tracks and was totally destroyed. Opel Rak V was ready to replace it, but the railroad officials had had enough; they did not permit further experiments which might ruin their best track.

Some time later Fritz von Opel switched to rocket airplanes but gave up after one fairly successful flight near Frankfort-on-the-Main on September 30, 1929.

Nor were Opel's rocket vehicles the only specimens of their kind. His former test driver, Kurt Volckhart, had made himself independent with a rocket car of his own design. Valier had found another manufacturer of powder rockets who lent his support for the construction of railroad cars that lost all four wheels under too violent acceleration. Finally Valier even built a rocket sled, and a manufacturer of small airplanes tried his hand at a rocket glider.

What had started out as a kind of scientific movement was almost smothered under a series of publicity stunts. Almost, but not quite. One of the serious attempts at furthering the progress of the idea was made in Paris with the creation of the Annual Award for Astronautics, somewhat reminiscent of the old *Prix Guzman*. But unlike the Guzman Prize, this one did not involve a large fortune nor were the requirements as high as stipulated by the far too optimistic founder of that old award. Robert Esnault-

Pelterie and the Paris banker André Hirsch set aside the sum of 5000 francs each year, to be given to the author or experimenter who had done most to further the idea of space travel during that year.

Oberth, meanwhile, had begun to rewrite his book for a badly needed third edition. But it was not just to be a new and somewhat enlarged edition. It was planned on a grand scale and was to consist of two volumes. Only the first, entitled *Wege zur Raumschiffahrt* (*Road to Space Travel*), appeared in Munich in 1929.[7] In spite of numerous small flaws, like poor proofreading of the formulae, somewhat clumsy arrangement, and the completely unnecessary refutation of absolutely unimportant newspaper articles, it is still the most important theoretical work on the subject. It won for Oberth the first REP-Hirsch Prize which, as a special compliment, was doubled.[8] When the wire containing this news arrived Oberth did not believe his eyes and called me up to hear my opinion on whether it might not be a dubious joke.

That was in the autumn of 1929. Oberth was in Berlin then, having arrived in the fall of 1928. But this time he had not come for personal reasons, or for a scientific debate, or for more or less vague discussions about financial support. He had been called by Fritz Lang to serve as scientific adviser for a motion picture. It was the film *Frau im Mond,* which appeared in English-speaking countries under the titles *By Rocket to the Moon* and *The Girl in the Moon,* the latter being the translation of the German title. The script for the film had been written by Thea von Harbou (then Mrs. Fritz Lang), and it credited Oberth's first publication and my second book as scientific source material.

The news that Fritz Lang was going to make a film on space travel was very good news indeed. It is almost impossible to convey what magic that name had in Germany at that time. The first showing of a Fritz Lang film was something for which there was no equivalent anywhere as a social event. The audience—it was an unwritten but rigid rule that one had to wear full evening dress, not just a dinner jacket—comprised literally everybody of importance in the realm of arts and letters, with a heavy sprinkling of high government officials. It is not an exaggeration to say that a sudden collapse of the theater building during a Fritz Lang premiere would have deprived Germany of much of its intellectual leadership at one blow, leaving mostly those who for one reason or another had been unable to attend.

[7] I mention the planned second volume only because the first contains some references to it. The second volume never was finished.

[8] The prize was not awarded in 1929, 1931, and 1932. In 1930 it went to the French engineer Pierre Montagne for a paper on "Gaseous Mixtures Utilizable in the Propulsion of Rockets." Montagne received a renewal in 1933 for a similar paper, but the real prize was not awarded in that year nor in 1934 when the French engineer Louis Damblanc received a "prize of encouragement" of 2000 francs for an exhaustive treatise on the tests and performance of powder rockets. The 1936 award, which was the last given, was presented jointly to the American Rocket Society and to Alfred Africano who was then its president. The prize-winning paper was a design for a high-altitude rocket.

A Fritz Lang film on space travel, consequently, meant a means of spreading the idea which could hardly be surpassed in mass appeal and in effectiveness. More than that, this connection might also mean funds, sizable funds, for experimental work on liquid-fuel rockets. I remember that I began needling Oberth about this during our very first meeting after his arrival. Oberth smiled and said that he had thought of that too. But apparently it took him a long time to get up enough courage to discuss the matter. When I met Fritz Lang myself a few months later I began to talk along similar lines, but then I was told that discussions about funds for rocket research were already under way.

They were. Needless to say there were some difficulties. Management felt that science was a fine thing, but it shouldn't cost their money. *They* were movie makers. Accounting wanted to know where the money was supposed to come from. In the end it was mainly Fritz Lang who brought the matter to pass, dragging the reluctant management of the Ufa Film Company at least part of the way along with him. There would be, he said (and I believe that was his own idea), commercial value in scientific experiments if they were used for advertising the film. If an Oberth rocket rose into the air on about the day of the premiere it would be quite defensible at the stockholders' meeting to have paid for this rocket out of the advertising budget. There was still some opposition, which claimed that the same unspecified amount of money—unfortunately nobody, not even Oberth, could name a sum, which did not help the talks any—spent for advertising in the normal manner would bring better results. Fritz Lang then declared that he would contribute a rather large sum out of his own pocket, if the company would divert the same amount from the advertising budget.

The result was a kind of confused agreement (which the management, in the end, did not sign, so that it was actually Lang's money and Oberth's savings which were spent) that Oberth should begin with whatever preliminary experiments he deemed necessary. The goal was a rocket (more or less of the type which he had described as Model B in his book) to be fired from a suitable locality on the morning of the day of the film's premiere. This actual rocket was to be represented as the first step toward the solution of the problem shown in the film. This, of course, would have been a truthful statement if everything had worked out. After that, the most difficult problem would probably have been to get the money for the *second* rocket.

It was not in itself a bad scheme. That it ended in utter confusion was due to a number of factors, none of which could be judged very well then. To begin with, nobody had any experience with liquid-fuel rockets. Secondly, Oberth had no experience as an engineer and had never organized anything like this. He had no idea how to go about it. He was—a point I wish to emphasize—the greatest authority on rocket propulsion at that time (he still is, in certain respects, or rather his work is still the most authoritative of all), but he was a theorist, not an engineer. He lacked the ability of an experienced engineer to say in advance what can be done and what probably cannot be done.

On top of all that he was greatly confused by his surroundings. He had grown up in the small towns of Transylvania and he had studied in the leisurely atmosphere of Heidelberg and of Munich. Now he was suddenly plunged into the strange atmosphere of fast-moving, efficient, flippant, and sophisticated Berlin. Not only was he a stranger in that city of four million inhabitants, but he found himself in the very spot where the apparent turmoil of a big city appears wildest. People spoke to him in a dialect which was strange and, to him, ultrarapid. Instead of the company of unhurried small-town intellectuals in semiretirement, he found himself on a movie lot in the company of film stars and directors. He ate his lunch in a canteen where Russian and English and French (none of which he knew how to use himself) were as common as German. He sat next to internationally known people; he had to speak to chairmen of huge enterprises of all kinds; he was beset by newspaper columnists and magazine writers with imposing names.

When a man has to adapt himself to a completely new set of surroundings he usually goes through three phases. The first is a kind of astonished disbelief that everything can be so different. The second is more or less violent rejection of the differences, or rather of the fact that there are differences. There shouldn't be; everything should be as it is at home. The final phase is a more or less thorough adaptation to the different surroundings.

By the time experimentation was to begin, Oberth had reached the second phase, the phase of refusal to adapt. He was generally angry with me because I tried to advise him. But he did realize that he was a stranger to engineering practice and, more important, that he did not know more about the city than the approximate location of a few of its railroad terminals. He needed an assistant who was a professional engineer, preferably with some specialization, preferably, too, a native, who did not need the telephone directory to find a place which would do a good job on aluminum welding but would know the phone number by heart and be on first-name terms with the foreman.

Oberth could have called any one of half a dozen specialized employment agencies or the VDI or another engineering society. But he had come to the point where he must show people that he did not need their advice. He put an advertisement in the classified section of one or several daily newspapers. The advertisement brought a number of applicants, some of them, no doubt, capable and experienced men.

One applicant was a small man with a strangely accented voice and a hard face, meticulously dressed and with precise military posture. He had introduced himself: "Name is Rudolf Nebel, engineer with diploma, member of oldest Bavarian student corps, World War combat pilot with rank of lieutenant and eleven enemy planes to my credit." And Oberth, who always stated his opposition to "Prussian militarism," hired Rudolf Nebel at once!

The "Prussian militarism" was, however, only a theoretical abstraction in this case, for Nebel was a Bavarian.

Oberth did not make certain whether Nebel had the qualifications he needed so desperately, whether he knew the town and its smaller and larger industrial establishments well, whether he had practice in working with aluminum and magnesium alloys or at least with liquefied gases. Nebel told me later, without regarding it as a personal secret, that he had been graduated in a hurry during the war because he had volunteered for the air arm, and that after the war he had never worked as a designing engineer but for some time as a salesman of mechanical kitchen gadgets instead. Since jobs were almost impossible to find, all this was probably not his fault—but he certainly was not the man Oberth needed.

But Nebel became Oberth's first assistant. The second assistant was Alexander Borissovitch Shershevsky, a Russian aviation student. Oberth had come across his name because it had appeared as a by-line under several articles in aviation magazines, articles which had met with Oberth's approval. Via the editor of one magazine, Oberth had contacted Shershevsky, who was simply destitute at the time. He had been sent to Germany to study gliders but had overstayed his *kommandirovka* and dared not go home again. Lazy by nature (and very proud of it), he earned just enough money for room and board by writing for professional journals. He could have done better if he had written about Russia, but he didn't. He knew that nothing much favorable could be said and he would not say anything unfavorable, certainly not in print. He was a frenetic communist who was a refugee by sheer accident.

The trio, consisting of a slightly bewildered theorist, a professed militarist, and a Bolshevist accidentally in disgrace, worked together, or tried to. Shershevsky would much rather have discussed the concept of infinity in mathematics, the importance of radicalism in politics, and the great work they would be able to do if they only were at the Central Aero-Hydrodynamical Institute in Moscow. Nebel was quite eager to work but, having no engineering experience worth mentioning and no theoretical grounding in rocket work at all, could not do so on his own initiative. And Oberth had to begin with some experimental research work because one of the critics had frightened him badly by stating that his whole theory was useless because of one established fact. No liquid-fuel rocket could ever be built because one could not bring liquid oxygen and a fuel, say gasoline, together for continuous rapid combustion. Such a mixture would never burn, it would always explode.

This was a very serious objection, especially since it originated with a man who had many years of experience in the manufacture and handling of liquefied gases. It was a point to be investigated, and Oberth was perfectly correct in making its investigation his first experiment. Liquid air was poured into an open bowl (liquid oxygen was considered too dangerous) and a very fine stream of gasoline was shot into the bowl and ignited at once. There may have been an ignition delay, and a minor explosion resulted. It did small damage; it also gently pushed out a windowpane. The experiment was repeated and the result was that the critics had been wrong. It did work or, rather, it could be made to work. As a matter of fact, Oberth was quick to see that something took place which had never been observed but which was very advantageous: the burning droplets tore themselves apart and were consumed much faster than had been assumed. This discovery meant that much larger amounts of fuel could be burned in a given space and during a given interval of time than had been believed possible. For one thing, it made the still theoretical rocket motors much smaller and lighter.

During one of these experiments another explosion occurred; this one was worse and come close to destroying Oberth's sight in one eye. It laid him up for several days, but there have been worse rocket explosions since. Oberth spent some time figuring out a theoretically ideal combustion chamber. In general shape it was conical (Fig. 25) and Oberth called it *Kegeldüse, Kegel* being the German word for cone. He had several specimens of it built somewhere—I don't recall whether in one batch or one after the other, but I suspect the latter.

While Oberth had stressed the advantages of gasoline as a rocket fuel in lectures, he did not want to use gasoline for his own experiments but looked around for methane (CH_4, known also as marsh gas) because of a slight theoretical advantage. It could not be had easily in Berlin, which disappointed him, because "home in Mediash we have it in commercially pure form from gas wells."

The rocket was to be torpedo-shaped and about 6 feet long, the construction material an aluminum alloy. Nebel took the plans to a factory where the parts were made, and after his return they began to work on a parachute release which they wanted to test by means of powder rockets. When they got in touch with a factory of powder rockets, they found that the mechanism developed by the factory for the ejection of complicated starlights was completely satisfactory for their purposes.

All this consumed time. And time was the one commodity which had been in short supply even before they started. It is now possible to say that about a year and four to six months would have been needed. They did not have even four months.

All of a sudden it turned out that they also had no place for firing the rocket. It had been taken for granted from the outset that the firing should

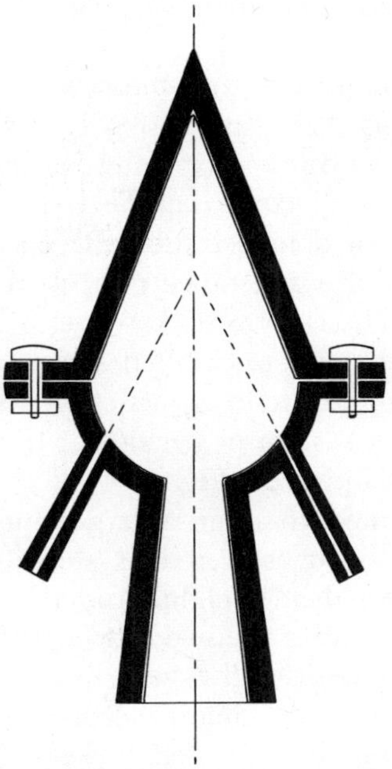

Fig. 25. Oberth's Kegeldüse. The material was steel, with a heavy copper lining on the inside. It was made in two parts and bolted together.

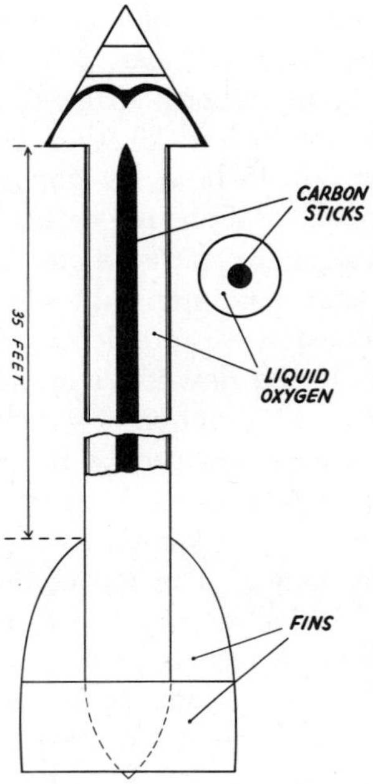

Fig. 26. Schematic diagram of the rocket for liquid oxygen and solid carbon which was to have been built for the Ufa Company and demonstrated near Horst.

take place at the seashore, for reasons of visibility. Better even than the seashore would have been a small island not too far from the coast. When this was mentioned, somebody suggested the Greifswalder Oie, a small and perfectly flat island in the Baltic a number of miles off the coast. The press department of the Ufa Film Company announced, accordingly, that the Oberth rocket would reach a maximum altitude of about 40 miles and that it would be fired from the Greifswalder Oie. But the authorities, when approached for permission in what was thought to be routine fashion, bristled and said no. They also advanced "reasons." On the Oie there is, they stated gravely, an important lighthouse. They could not endanger this lighthouse; it might be damaged by the returning rocket. It was a perfectly idiotic statement, considering the probable radius of descent from an altitude of 40 miles. Within that radius the whole Oie is a virtually invisible speck. And when I saw the lighthouse myself at a later occasion I could only wonder whether a direct hit from a field gun would be able to scratch it. And the dangerous rocket weighed about 30 pounds when empty and dangled from a parachute. But officialdom had spoken and another place, the seashore resort of Horst, was picked. That the Oie was still within the probable area of descent was something we neglected to mention at that time.

Suddenly Oberth realized, or was made to realize, that there were only a few weeks left. He then changed his plans. What he had planned first was meant to be a straight prototype of an instrument-carrying rocket for upper atmosphere research. Since that obviously could not be completed he designed a special model for the purpose of demonstration. It consisted of a long aluminum tube with several sticks of a substance rich in carbon in the center, surrounded by liquid oxygen. The carbon sticks were to burn from the top down. Theoretically things were dimensioned in such a way that the consumption of 5 inches of carbon would use up 5 inches of oxygen level. The gases were to be ejected through a system of nozzles at the top of the rocket, which were to pull the rocket upward (Fig. 26).

This system, known as "nose drive," seemed to offer a large number of advantages. The rocket did not have to be constructed as sturdily as a "tail drive" rocket; hence it would have less dead weight. And the idea of pulling it instead of pushing it seemed to obviate the need for a steering mechanism; as long as a vertical ascent was desired the very fact that a rocket had nose drive seemed to take care of everything. Oberth told me that he had always had that idea in mind but had wanted to keep it to himself. He had sacrificed it only because of the great need for hurry. (Actually, as is explained later, the nose drive does not do what one would expect it to do—at least not for what looks like the obvious reason.)

Oberth made more experiments, failed to find a "carbon carrier" which would burn with just the proper speed, and rapidly approached a nervous breakdown. He left town for a week or so—not announcing it beforehand, so that Shershevsky called me up to say laconically, "He ran away"—but came back to attend the premiere of the film (October 15, 1929). The Ufa Film Company released a statement that the rocket ascent had to be postponed indefinitely, hinting that the season was too far advanced for reliable weather. Nothing happened for some weeks after that, then Oberth suddenly left for home, releasing a statement that he would sue the film company.

Years later, in 1934, he explained in a letter to the new president of the VfR that he had not been accountable for his actions at that time. The liquid-fuel test explosion, he claimed, had given him all the symptoms of shell shock and he had never completely recovered.

6

Success, Failure, and Politics

History has been charged with the duty of judging the past and of educating the present for the benefit of the future. This work does not attempt to do anything as important as that; it only wants to state what really happened.

—Leopold von Ranke

To say that things looked bleak by the end of 1929 is to understate matters. The Oberth Rocket had failed to materialize. Winkler was forced to abandon publication of the monthly journal *Die Rakete*. Even the film was only moderately successful. By the time it was being distributed every town had at least one movie theater wired for sound, and while the artistry of the first talkies was doubtful, they had novelty on their side.

What saved the situation was the fact that early in the same year Johannes Winkler had resigned as president of the VfR for personal reasons. Professor Oberth had become president and I vice-president. And a patent attorney named Erich Wurm had offered the use of his own office in Berlin to the VfR. While the legal seat remained in Breslau, at least the work could be done from Berlin. Nor had the VfR done badly during that year: a tally in September 1929 showed a membership of 870 and new members were coming in every day. With the use of Wurm's office facilities it was possible to substitute mimeographed information bulletins and circular letters for the defunct journal and to maintain contact with the members in that way.

After the first shock of the lost battle was over, there began what military men call the regrouping. The leading members of the society recalled that the VfR had originally been founded to do experimental work. That was still a point of the program and money was not much shorter than it had always been. The film episode which had dazzled us into confusion had resulted in some equipment. If the society could get it the whole circumstance could still be considered a step forward which had merely been overshadowed by disappointed hopes. Things probably would have worked out in any event; the way it actually happened was that I met Nebel on the street by pure chance. Having no job, he had naturally been wondering whether his recently acquired experience could not be put to some use and he had toyed with the idea of founding a society for the advancement of rocket research. I was somewhat dumbfounded, because I could not see any reason why he should want to compete with the VfR, but, on the other hand, he did not sound as if he did want to. I asked him outright and learned that he did not know of the existence of the society. Oberth, although he was president of the VfR, had never mentioned it.

Naturally I brought Nebel and Wurm together. Oberth returned and a general meeting took place in Wurm's office. Oberth and Nebel glowered at each other across the conference table. Wurm busied himself as peacemaker and in general succeeded. I don't remember whether Winkler was present at that particular meeting; next to me sat a young engineer named Klaus Riedel (pronounced *Ree*-del) who had just joined the VfR and wanted to help. There was much useless talk, but one important decision was reached: the VfR would try to obtain the equipment that belonged to Ufa and was still on the movie lot. It would also obtain the Oberth Rocket which existed in parts in various factories, requiring "merely" the payment of a few bills. Then the Oberth Rocket was to be assembled.

From that point on all agreement consisted in the common conviction —shared by Oberth—that the Oberth Rocket would not work. What to do with it was the next point of discussion. Nebel revived a suggestion he previously had made to Oberth during the last days of the Ufa episode. He wanted to put one of Sander's most powerful powder rockets inside for a faked ascent—or for an explosion—just to show something to the public. When we all yelled at him in utter indignation, he was mainly surprised; he did not understand such a reaction. Nor did Oberth understand the equally violent reaction that greeted his counterproposal. He wanted to sell the rocket to a circus in Mediash and use the money to build a better one. I suggested that the rocket be given to the German Museum in Munich but that was voted down, because everybody present, and I include myself, still hoped that somebody would think of something to do with it. Nobody ever did; the Oberth Rocket just stood around for a few years.

Then Nebel revived another suggestion he had made earlier. He wanted to build a liquid-fuel rocket, not designed for any performance in particular, but merely designed to function. And that rocket was to be of the smallest possible size, for obvious financial reasons. He did not have a design at the moment, but he had a name. He wanted to call it *Minimumrakete,* or, abbreviated, *Mirak.* Oberth, who had vetoed the idea earlier, was still opposed. His reasons were that a liquid-fuel rocket should show its superiority by superior performance, in particular, high altitudes. And he said that it was, in his opinion, quite possible to build a liquid-fuel rocket with such small dimensions that its performance would be less than that of the larger blackpowder rockets then in existence. But in this case he was outvoted; the rest of us felt that a small rocket which worked was greatly to be preferred to a large one that didn't. There also was a strong desire to prove something to ourselves, for we were of the opinion that no liquid-fuel rocket had actually lifted itself off the ground.

I now know, of course, that this was a mistaken assumption because Professor Goddard had anticipated our earliest experiments by some six years. In what is usually called his second Smithsonian report[1] he stated:

> On November 1, 1923, a rocket motor operated in the testing frame, using liquid oxygen and gasoline, both supplied by pumps on the rocket. . . . The first flight of a liquid oxygen-gasoline rocket was obtained on March 16, 1926, in Auburn, Mass., and was reported to the Smithsonian Institution May 5, 1926. . . . The rocket traveled a distance of 184 feet in 2.5 seconds, as timed by stop watch, making the speed along the trajectory about 60 miles per hour.
>
> (R. H. Goddard)

In fact, while we were debating the merits of a Mirak, Goddard was working at Fort Devens, Mass., with liquid-fuel rockets. But even if we had known about it—that is, if our efforts to have correspondence with Goddard had not been curtly declined by him—we could hardly have followed a different line from the one we did follow.

Nebel was asked to draw up a preliminary design sketch of his Mirak, while the parts of the Oberth Rocket were brought together and assembled. After some hesitation Ufa also turned Oberth's equipment over to us, the biggest item being an iron launching rack that had been built for the rocket. Meanwhile, Oberth and Nebel spent a good deal of time trying to raise funds from science foundations of various kinds. The financial result was nil; either the coffers of these societies and foundations were empty or else what funds they did have were already earmarked for specific purposes. But the unsuccessful treasure hunt yielded an unexpected result.

There was a government-sponsored institute called Chemisch-Technische Reichsanstalt, the Reich Institute for Chemistry and Technology. This institute did some work of its own along the lines of that done by the Bureau of Standards. But it also tested industrial inventions and processes and testified as to their value. Originally the Reichsanstalt had been approached for money. It did not have any funds available but its director, Dr. Ritter, suggested that a rocket motor for liquid fuels be demonstrated to him. If the demonstration turned out well, he would issue a certificate and that should be a great help in approaching other institutions.

It was a good suggestion and we lost no time in accepting Dr. Ritter's offer. Oberth's Kegeldüse was made ready for the test. Nebel hoped to get his Mirak in too, but it was still incomplete when the day of the test came. I am tempted to say that it was just as well, because the day of the test was one of terrible weather. All through the day it poured, it had

[1] *Liquid-Propellant Rocket Development,* Publication 3381, dated March 16, 1936.

poured the day before, and it was still pouring the next day. The test was made in the open, in a clearing in the pine forest close to a small shack which offered some protection. But it was as wet inside the shack as out because everything was permeated with moisture; the clouds were hanging so low that they obscured the boughs of the trees.

The Kegeldüse had been rigged up on a registering scale and the whole assembly had been placed in a shallow slit-trench. Handling liquid oxygen under these conditions was pretty much a sleight-of-hand trick. Liquid oxygen, having a temperature of close to 200 degrees Centigrade below freezing, rapidly draws all moisture within reach out of the air and freezes it around pipelines, valves, and other points which should not be covered with ice. There was much moisture in the air on that day, clouds of ice crystals formed whenever liquid oxygen was poured from one container into another, and I saw one pipe collect a quarter-inch layer of ice in less than a minute. But, despite the great waste of liquid oxygen, Riedel, who did most of the actual handling of the equipment, finally got the motor going. He was assisted by another new member of the VfR, a young student named Wernher von Braun. While Oberth talked to Dr. Ritter and some other members of the institute and Nebel moaned that the Mirak was not yet ready for demonstration, the press photographers somehow succeeded in getting a few pictures in spite of the raindrops on their lenses.

Dr Ritter certified that the Kegeldüse "had performed without mishap on July 23, 1930, for 90 seconds, consuming 6 kilograms of liquid oxygen and 1 kilogram of gasoline, and delivering a constant thrust of about 7 kilograms." (One kilogram equals 2.2 pounds.)

Soon afterward Oberth had to leave Germany. Officially he was still a teacher of mathematics at the high school in Mediash and as far as that institution was concerned his whole year of work for the Ufa Film Company had been simply a leave of absence. While his endeavors and his spreading fame were appreciated by his superiors there was a limit to the time a man could be away and still retain his job. Pocketing the certificate issued by Dr. Ritter and thinking nasty thoughts about everybody he had met, Oberth went happily home to Mediash. Soon afterward Nebel and Riedel left Berlin to begin tests with the Mirak.

The test at the Reichsanstalt had been a test in another respect too. There had been voices demanding that rocket research be banned, chiefly because of a tragic occurrence earlier in the same year.

On April 11, 1930, the VfR had introduced itself to the German capital by way of a public lecture in the auditorium of the General Post Office. The completed Oberth Rocket was displayed for the first time, and a model of the Oberth Rocket, complete with parachute, was suspended from the ceiling. Representatives of various firms were present. So were a few well-known scientists, who, however, preferred to remain anonymous then. Winkler delivered the main lecture. Old Hermann Ganswindt had come; he and Max Valier sat behind me, loudly disagreeing about something.

That night Valier told me that he had "said good-by to powder rockets." Early that year he had come to terms with Dr. Heyland, the director of a firm with the clumsy title of Association for the Utilization of Industrial Gases. Among other things this firm manufactured liquid oxygen and Valier had built a car which was propelled by a liquid-fuel motor for liquid oxygen and gasoline. The large vehicle made a long and lumbering run on April 19; it was evident that the motor did not work well and the flame was reddish and smoky, certain signs of poor combustion. But Valier was sure that he could improve matters soon. He wanted to have the car ready for a demonstration during "Aviation Week" which had been scheduled in Berlin for the week May 25–31, 1930. The events of the Week consisted of public lectures, documentary films, some inexpensive air hops over the city, and an exhibit on Berlin's most important square, the Potsdamer Platz.

Valier and I met once more on May 14 or 15. He was confident that his improved car would draw a record crowd.

When I came home two days later I found a wire on my desk: Max Valier was dead! It was a Saturday and he had worked late in Dr. Heyland's empty factory. Apparently he had made several test runs of his motor—"idling" it without actually letting the car run—which had turned out well, and he had been observing another run at close range. Suddenly the motor exploded and a large steel splinter cut the aorta; he bled to death before anybody could do anything about it.[2]

Valier's death was especially tragic in view of the fact that nothing had ever happened to him during all his dangerous and useless experiments with powder rockets. He died while engaged in his first really useful experiment, although the idea of mounting his motor in a car was, of course, ridiculous.

The demands for the outlawing of rocket research were mainly prompted by Valier's death. It may also have been those demands which influenced Dr. Heyland to break off negotiations toward collaboration with the VfR. After the Reichsanstalt test had been publicized we watched for more comment of that type. Although none appeared we still decided to test the Mirak outside Berlin, so that, as Riedel phrased it, "we could keep further casualties secret." Testing ground was a farm near the small city of Bernstadt in Saxony, owned by Riedel's grandparents. In the course of the summer I received a succession of small progress reports, reading about as follows: "The Mirak burns, but the recoil is too small to be measured with our homemade thrust meter, probably a pound or so." Then: "The Mirak

[2] Valier was not the first victim. About a year earlier an adolescent boy had been killed by the explosion of a homemade powder rocket while trying to build a large model of the Opel rocket car.

now produces a recoil of 3 or 4 pounds." Some time later: "The recoil of the Mirak is slightly higher than its own weight; it would rise if we released it." And in September 1930: "The Mirak has exploded, no harm done; we'll come back and build a new one."

When we put these reports into the mimeographed bulletins by means of which we held the VfR together, two members saw fit to reveal that they were wealthy. One of them, an engineer named Dilthey, donated approximately $1000 in cash, in two installments. Another one, a manufacturer, Hugo A. Hückel, sent only $100, but promised the equivalent of about $150 every month, provided that none of it would be used for anything but experimental work. He absolutely refused to pay postage or carfare or delivery charges—nothing but bills for machinery, tools, gasoline, and oxygen could be charged to him.

It does (and did) sound somewhat funny, but it was a basis to build on. What was needed now was a permanent proving ground, an equivalent of the shack and the open field at Bernstadt. Nebel drove around the outskirts of Berlin looking for inaccessible places and he found several—an abandoned factory here, an unused tract of land there, a small island in a lake elsewhere. One of these places looked best and he checked on its ownership, etc. Then it looked ideal.

It was a tract of land about 2 square miles in area, difficult to find, situated in the district of Reinickendorf, a typical worker's suburb, near the northern periphery of the city. Two roads led into the vicinity, but one was so bad that a city ordinance forbade its use by motor vehicles. I still wonder why the use of the other one was permitted. At the point where these two roads met was an old army garrison, used as local police headquarters.

The tract we were interested in was across the street from the police garrison but did not border the road. A secondary dirt road branched off there, leading for 1000 feet or so through a conglomeration of small single-family houses, minor manufacturing establishments half of which were idle, truck garages, and shacks of uncertain purpose and use. After you had negotiated that dirt road you came to a wire fence which enclosed the 2-square-mile tract.

The City of Berlin owned the land, and to the municipal officials in charge it must have been something of a white elephant. To make it usable for manufacturing or settlement purposes would have required an enormous capital investment, not to mention a new road. The place itself was suitable for practically nothing. Half of it was hilly and covered with trees, and some of the depressions between the hills were swampy.

To make it worse from a businessman's point of view, the jurisdiction was somewhat doubtful. During the First World War, when the police garrison had been an army garrison, the place had been used to store ammunition and the War Ministry had erected storage buildings. These were massive concrete barracks with walls a foot thick, surrounded by blast guards in the form of earth walls, 40 feet high and about 60 feet thick at the base. The earth walls had slits to make the buildings accessible for vehicles, but the slits were a little too narrow for large motor trucks. The roads connecting the various buildings could hadly be seen any more and the buildings themselves had only a very few windows. While the ground itself belonged to the city, the concrete buildings and earth walls belonged to and were under the jurisdiction of the War Ministry. The city might have liked to rent the place but the War Ministry insisted that no damage of any kind must be done to its buildings and earth walls. That, of course, prevented any business deal.

We had to make an enormous number of promises. We were to use only one of the two gates, we were to occupy only two of the buildings and were not to enter any of the others (there were five or six more, all perfectly empty, as we quickly found out), we were not to make any changes in the two buildings we were permitted to use, and we were not to move in machinery and/or equipment which could not be moved out within forty-eight hours. We promised everything and then rented the place for a nominal sum—some $4 per year. We received the key to the smaller building on September 27, 1930, which date we proclaimed to be the "birthday" of our testing ground. Nebel called it the *Raketenflugplatz* (rocket airdrome) and it became so known to interested circles all over the world.

The smaller building next to the gate, not surrounded by an earth wall, had only one story. It was an oblong about 30 by 45 feet, divided into two sections by a massive partition, one section being about twice as large as the other. The smaller section was subdivided again so that it had three rooms, the large one being four times as large as either of the two small ones. It had obviously been the guardhouse, with a rest room and an office for the officer in charge and a room for the soldiers of the guard. A replica of this building could be found near the other gate. During the interval between the time the guard had moved out and we had moved in, somebody had used it to store lumber which was afterward forgotten. When we finally got the door open we found a solid layer of thoroughly rotted wood, a yard thick. It was a full day's work to drag this wood out into the open, to burn it, and to clean the house.

After that Nebel and Riedel set up a bachelor household in the two small rooms and used the larger room as temporary storage space for our equipment. This included the Oberth Rocket, a full-size wooden model of it, and the iron launching rack for the Oberth Rocket. The wooden model was later mounted on a pole near the road, pointing in the direction of the testing ground. Then we had the second Mirak, which had been completed in the meantime, and the stand in which the first Mirak had been tested at Bernstadt. The rest was odds and ends.

This room was later used as a combination office, reception, conference, and board of directors' meeting room; I called it the *chambre à tout faire.* But during the winter it became an incredible jungle of machinery and raw materials. We wrote hundreds of letters to firms manufacturing things we could use, explaining what we wanted to do and asking for contributions which would help us a lot and not hurt the contributor. We did not ask for cash; we asked for raw materials, unused machinery (unused, that is, by the man or firm who owned it), for tools and similar things. And we got them: boxes of hand tools, aluminum in sheets and rods, magnesium alloy in rods and seamless tubes, angle iron, screws, nuts, bolts, pots of paint, a drill press, and a small lathe—a whole machine shop came in piecemeal.

Somebody, in an hilarious moment, even wrote a letter to the Department of Internal Revenue asking for a waiver of the tax on the gasoline we would use for our experiments, transportation, etc. To our great surprise it worked, bringing the price of gasoline down from about 80 cents per gallon to something like 13 cents per gallon.

"Tell me," I have been asked since, "just what was your program?" This question is by no means easy to answer, even in retrospect. We did have a program of sorts, but while we knew precisely what we were *not* going to do, we could not formulate clearly what it was we were going to do. On the negative side we were certain that we would not touch solid fuels in any form. We also were not going to stick a rocket motor for liquid fuels on a car, railroad car, or glider. We were, in short, not going to do anything but build rockets. But how those rockets would look and what they might be used for was something we could not tell because we did not know. I wish I had known in those days the answers Michael Faraday is said to have given to visitors to his laboratory who asked him about the use of his experiments with electricity. Once, when a lady asked the question, Faraday is said to have replied, "Madam, what use is a newborn baby?" And another time, when a member of the House of Lords asked it, Faraday allegedly answered, "Some day you'll tax it." Both these answers would have come in handy if I had known them.

The basic theme of our program, in our opinion, did not need any elaboration because it was given in the very name of the society, which was "Society for Space Travel." It was exceedingly obvious, therefore, that one had to look upon a rocket as an embryonic spaceship. Everything that came before the first spaceship we had a strong tendency to judge simply in terms of whether it worked or not. But there was also a scheme which had been drawn up by Baron Guido von Pirquet of Vienna in his contribution to my book *Die Möglichkeit der Weltraumfahrt* a few years earlier. Looking at the problem of early and intermediate experimental stages both from the point of view of reasonable technological sequence and probable financial self-support, von Pirquet had outlined three successive steps, which he called "Registering Rocket," "Long-Distance Rocket," and "Space Rocket." It was understood that this did not mean three distinct models, but rather three types, each probably represented by dozens of models. It was also understood that the boundary lines between these types would be flexible and fluid, so that, for example, an especially large "Registering Rocket" (research rocket) might also be a small "Long-Distance Rocket." A point especially stressed by von Pirquet was that there are two kinds of engineering projects: one kind which is useless unless completed and one which is useful even if not completed. A bridge or a tunnel, for example, has to be completed to be useful, while a highway which is to connect five towns is useful even if the work is interrupted when it connects only two.

Future rockets, von Pirquet wrote, belong to the second category. Even if, for one reason or another, development work gets stuck when it has only reached the stage of the "Registering Rocket," such rockets will be useful and "pay their keep." The struggle was, of course, to progress far enough to reach the first useful type, to connect the first two cities of the projected highway.

This is what I told lecture audiences and what Riedel and Nebel repeated to visitors. When we were alone, however, our thoughts revolved around the next problem on hand. Every once in a while somebody looked thoughtfully at the plans of the Oberth Rocket and at the original, but that was all that ever happened to it. After a while we were content to polish the rocket nicely so that it stood like a big gleaming airplane bomb in the *chambre à tout faire* where visitors were received, causing these visitors to throw away their cigarettes in a hurry.

Our first job was to make the second Mirak work.

It was a replica of the first, except that it was slightly larger.[3] When Nebel had drawn up the design of the first Mirak he had followed the pattern of a powder rocket as closely as he could. Like a powder rocket it consisted of a "head" and a "guiding stick." The latter was a long thin aluminum tube which served as gasoline tank. The "head" was made of cast aluminum and machined to size; it looked like an artillery projectile. The nose could be removed to fill in the liquid oxygen and it contained a safety valve (Fig. 27A). The bottom of the head was made of copper and carried a combustion chamber in the center, a small replica of the

[3] In some English and American books and magazine articles it is stated that the second Mirak differed in various respects from the first, mainly in having a ceramic-lined combustion chamber. This resulted from a series of misunderstandings which occurred when G. Edward Pendray, one of the founders of the American Interplanetary Society (later the American Rocket Society), visited the Raketenflugplatz in April 1931. Pendray did not speak German and my English was very poor. What I meant to say, but apparently didn't, was that one specimen of the Kegeldüse had been tested with a ceramic lining. As was to be expected, the lining cracked and one piece of it blocked the exhaust nozzle, causing the combustion chamber to explode. That happened before the first Mirak existed even on paper.

Kegeldüse. The combustion chamber, therefore, formed the bottom of the liquid-oxygen tank. It was assumed that in this way it would serve two purposes: the liquid oxygen would cool the rocket motor, while the heat from the rocket motor would evaporate some of the liquid oxygen, thereby creating sufficient pressure to feed the liquid into the combustion chamber. The gasoline was to be fed into the combustion chamber by means of carbon-dioxide pressure, produced by a carbon-dioxide cartridge of the same type that is used for making soda water. That cartridge was carried in the end of the tail.

The launching rack of the Mirak was equipped with a simple gear which could be turned from a distance. By turning that gear the carbon-dioxide cartridge was discharged in the customary manner. But the launching rack was also provided with a clamp that held the Mirak tightly so that it did not rise when fired but only pulled that clamp. And the pull could be measured.

The second Mirak performed like the first: it would have been just powerful enough to lift itself off the ground. Finally, like the first, it exploded. The hot combustion chamber made the oxygen develop more pressure than the tank could stand or the safety valve could handle. We came to the conclusion—it sounds terribly obvious now—that a liquefied gas could not be used to cool the rocket motor. Water was needed for that. Possibly—the thought was then voiced for the first time—possibly one might be able to use the fuel for cooling purposes.

The second Mirak blew up in the spring of 1931.

After that several things had to be done more or less simultaneously. Of course a third Mirak had to be built, avoiding some of the sources of trouble that had shown up in the first two. The motor was to be below the bottom of the oxygen tank. And instead of having only one tubular tank it was to have two, fastened symmetrically to the oxygen tank. The second tail was to contain compressed nitrogen gas for pressure-feeding of both liquids to the motor. This avoided the carbon-dioxide cartridge. Most important, the third Mirak was to have a motor of a new type, no longer a Kegeldüse. But that new motor had to be developed first and in order to do that a test stand was needed.

The idea behind the new motor was one which could have sprung only from practical experience. And it is precisely for this reason that nobody who did not have such experience even believed it. Everybody learning that a rocket motor burned a liquid fuel, gasoline in those days, with an oxidizer, liquid oxygen, asked how hot the flame would be. Upon receiving an answer he would ask about the material for the motor, the "obvious" conclusion being that the melting point of the material of the motor had to be much higher than the temperature of the flame. The actual temperatures involved happen to be such that only a very few substances have a higher melting point. And these few are either unsuitable or so expensive that they could not be tried. Nor would it surprise me very much if somebody some day wrote a research paper reporting that these expensive substances are unsuitable too.

With the Kegeldüse, made of steel, Oberth had, possibly without realizing it, followed the example of gun designers. The burning temperature of all types of gunpowder is also above the melting point of the steel of which gun barrels are made, but the time is too short to do harm immediately. The principle is still useful for rockets of very short burning times, say 5 seconds or less, as is the case in all solid-fuel bombardment rockets. Unlike gun barrels, these rockets don't even have to be usable a second time. But for a liquid-fuel motor longer burning times are required—several minutes instead of several seconds. The problem, therefore, was to prevent the metal from heating up. Lining the motor, and especially the exhaust nozzle, with a heat-resisting substance such as a ceramic is difficult in practice, usually fails to perform as expected, and really is not a solution of the problem.

The real solution is to prevent the metal of the motor from heating up by cooling it. Having little to lose but time and effort we decided to look at the material for the motor from only one point of view. Disregard melting point, pay little attention to structural strength, forget about expansion factors temporarily, concentrate on one thing: *what is the factor of heat conductivity?* That factor probably determined whether the motor could be cooled effectively. The material chosen was aluminum, almost pure. The motor (Fig. 27B) was made in two sections which were welded together. When the development was finished it weighed about 3 ounces and a fine specimen swallowed 160 grams of liquid oxygen and gasoline per second, producing a thrust of 32 kilograms or about 70 pounds.[4] Because of these figures it was referred to in front of visitors as the "160/32 motor"; privately it was usually called "the egg" because it had approximately that size and shape.

I cannot say who invented it; it is almost impossible to tell who invented anything at the Raketenflugplatz. The Kegeldüse was Oberth's invention and the first Mirak was Nebel's, but after that almost any device or development was the result of informal talks and conferences between three or four or even five or six people. We never paid any attention to the question of who had thought of what, knowing that it was a long way from our experiments to definite shapes, and knowing also that our glory was a collective glory. If the question of who invented the standard motor were important, I'd be inclined to give most of the credit to Riedel.

The standard motor was not only powerful, it could also be regarded as safe after it had been through one or two test runs. We had just enough accidents to keep us from getting careless. Once a motor burned through

[4] This indicated an exhaust velocity of 2000 m per sec or about 6560 ft. per sec.

at a faulty place in the welding seam with the weird effect that there were two jets of fire, one vertical and one horizontal, the latter carrying a steaming spray of water with it. Once the man at the controls pulled the oxygen switch first by mistake and the whole motor blew up—a piece of it imbedded itself deeply into a spade handle which I kept for years as a souvenir. It would have penetrated two inches in living flesh. Another time the whole welding seam broke and the top of the motor shot skyward. The violent explosive combustion resulting from the jets of oxygen, gasoline, and water meeting in midair almost tore the test stand apart before the man at the controls could turn the pressure off.

But practically all of these accidents happened during first runs and even then we sometimes knew they were coming. If the motor roared like a waterfall with a steady and fairly small darkish flame everything was fine. When we suddenly got a bright flame and machine-gun staccato it was better to duck quickly and with a great disregard for curiosity.

This was the motor that was to go into the third Mirak. But the third Mirak was never built, or rather it was never assembled.

To develop this motor it had naturally been necessary to create testing facilities. When I described these in detail for the first edition of this book—reinforcing my memory with copies of articles I had written for German engineering journals while all this was going on—it was "news" to my readers. Now, when many of them may be working in beautiful bombproof concrete firing pits, with all the latest electronic controls around them, this may read like ancient history. But it all took place three and a half decades ago, and I think it deserves to be remembered, because this is how it began. It will never be done that way again.

The testing site was near one of the walled larger buildings which was not only bigger in area than the others but also had a second floor which may have been a later addition. Not far below the roof a ceiling of 2-inch planks, resting on 8- by 8-inch wooden beams, had been built in. This ceiling was at the same level as the crest of the earth wall around the building, and a door led to a narrow bridge which ended on the crest of the earth wall. Walking over the bridge and down the earth wall one came to a small depression in the ground and this was chosen as the site for the testing stand.

The testing stand was the old iron launching rack for the Oberth Rocket, fitted with a balance. The rocket motor was attached to one end of the balance and the tilt of the balance registered on a revolving drum. An insulated oxygen tank and a gasoline tank were buried in the ground at either side of the test stand, each with a nitrogen steel bottle for pressure. The valves were operated by thin wire ropes which led into the top floor of the machine shop. The ropes were connected to large railroad switches and operated by a man standing behind the doorway. The same man also ignited the rocket motor by closing an electric switch. The operator was perfectly safe, but he could not see the test stand; he followed orders shouted by the man standing on the crest of the earth wall and directing the test.

The procedure was as follows: the rocket motor was fitted into a metal container which, in turn, was attached to the balance of the test stand. The cooling water came from a large water barrel standing on the ground just outside that small depression in which the test stand was placed. The water was drained off through a pipe fitted into a hole near the bottom of the container; the pipe ended on the ground a few yards away. The ground crew filled the water barrel and the fuel tank and attached the motor to the balance. Then they attached the ignition device to the muzzle of the exhaust nozzle.

This device was actually a kind of small powder rocket, but with strange qualities. The powder had been compounded by the chief chemist of a large pyrotechnic factory near Berlin, and a specialized chemist's specialized knowledge had been required to produce a powder that did just what we wanted it to do. It had to produce a very hot flame, but it also had to burn for at least ten seconds and with little generation of exhaust gases; the ignition device was not supposed to have any recoil. In addition, the flame was not to be extinguished either by water or by a jet of cold compressed gas. A thermite cartridge approximately filled our needs but was too hard to ignite; we wanted electric ignition too. The chemist, Dr. Feistl, after hearing all these requirements, said they could be met, and he delivered a dozen capsules only a week later. They proved excellent.

After that device was attached, the clockwork of the registering drum was wound up, and after that the liquid oxygen was poured into the oxygen tank. Then the ground crew scrambled to safety, leaving only one man behind who turned the stopcock in the cooling water line. When he too was under cover the test really started.

There was a definite sequence in the orders shouted by the observer. It went: *"Feuer!—Benzin!—Sauerstoff!"* "Fire!—Gasoline!—Oxygen!" At "Fire" the electric switch was closed, igniting the small powder tube which jetted its flame horizontally across the mouth of the exhaust nozzle. When that flame was well established the shout for "Gasoline" was answered by a spout of yellow fire from the rocket motor. The call for "Oxygen" followed immediately. The flame grew first bright and then bluish, shortening rapidly at the same time. The sound produced by that short and sometimes hardly visible flame resembled the roar of an enormous waterfall and kept up until the switches were closed or the fuel gave out. Test runs were limited in time by the capacity of the oxygen tank; the longest we could produce was 90 seconds.

When G. Edward Pendray, of the American Interplanetary Society, and Mrs. Pendray visited the Raketenflugplatz in April 1931 the new motor was nearly finished; we could show them a fine test run. I may add here that such test runs not only furthered motor development but also added to our income. Nebel, when asking prospective donors of machinery and tools for donations, had had to promise on occasion that he would invite them for a demonstration. When the promise could be fulfilled, it occurred to him that admission could be charged. Of course we did not want to advertise rocket motor tests as a public diversion; we invited engineering societies, charging a flat fee of, say, $50, regardless of attendance. Then we could invite "donors" for the same time and test one or two new motors, holding a tested one in reserve in case something went wrong. This system was a very practical one, but it helped to cause delays.

The time spent on demonstrations also helped to deprive us of the honor of having sent the first liquid-fuel rocket in Europe into the air. That first European liquid-fuel rocket (since Goddard had not then published his second Smithsonian report we naturally took it to be the first liquid-fuel rocket anywhere) was not ours; it was Winkler's. The rocket stood about 2 feet high and weighed 11 pounds of which 3¾ pounds were fuels. It looked like a prism placed on end and consisted of three tubular tanks, partly covered with sheet aluminum much in the manner of the old box kites. One tube held liquefied marsh gas (methane, CH_4), the second liquid oxygen, and the third "an inert gas under pressure" which was Winkler's way of saying "compressed nitrogen." The motor consisted very simply of an 18-inch piece of seamless steel tubing, placed near the center line of the system. "The first test took place on the drill ground near Dessau on February 21, 1931, but because of a mal-functioning part the apparatus rose only about 10 feet. The first unassailable test took place in the presence of public witnesses on March 14, 1931, at 17 o'clock, also on the Dessau drill ground. During the ascent the apparatus left the vertical and for this reason did not attain the previously diagrammed altitude of 500 meters, but otherwise the experiment was undisturbed." (Winkler, in *Männer der Rakete*)

Hückel had paid for the experiment. While we felt somewhat foolish that Winkler's private tinkering had achieved a rocket before we did, we liked the news. A liquid-fuel rocket had performed. It could be done. That meant that we could do it too. Two other pieces of news which emerged during the next month pleased us less. They had to do with rocket automobiles and powder rockets, and it was faint consolation that at least the two did not appear in combination.

The rocket automobile in question was Max Valier's last experiment. After some hesitation, Dr. Heyland had permitted his chief engineer Pietsch to go ahead and to design and build a new rocket motor for this car. Somebody told me at that time that the motor was cooled by its own fuel; I don't know whether this was actually the case. The figures released then merely stated that the motor of the car weighed 18 kilograms (40 pounds) and that it produced a thrust of 160 kilograms (about 350 pounds) "for a period of several minutes." I take it that there were a number of private test runs; the two public runs took place on April 11 and on May 3, 1931. This was no doubt the most powerful liquid-fuel motor then in existence and the duration of the test was remarkable too. Apparently Pietsch had to pay for this by poor combustion and low exhaust velocity. The flame was red and smoky.

If this revival of a rocket automobile was annoying, the new powder rockets were, from our point of view, worse. They were the invention of Reinhold Tiling (pronounced *Tee*-link), an engineer who resided at Osnabrück. The first *public* demonstration took place April 15, 1931, at the shore of a lake near Osnabrück. Tiling, who had the financial backing of one Count von Ledebour, must have worked in secret for a long time because the rockets he displayed then were finished products.

They consisted of a streamlined aluminum body, housing the rocket proper, and four long fins. There was no parachute. The functioning of Tiling's landing mechanism is incorrectly reported with annoying consistency, mainly because Tiling patented a misleading description (U.S. patent No. 1,880,586, Oct. 4, 1932). The patent states, in essence, that the rocket's four fins will spread out at the peak of the ascent and act like the blades of an autogiro. Actually Tiling's rockets did not use that system at all. Two of the fins remained fins, while the two others snapped out like the blades of a pocket knife. The two fins transformed the rocket into a glider, not into an autogiro. The arrangement worked well, provided the air was calm. It was better looking but much less effective and much less versatile than a parachute. The Tiling rockets had an over-all length of about 6 feet, and they rose to altitudes varying from 1500 to 2500 feet. They never showed any improvement, just because Tiling had demonstrated a finished product to begin with.

Meanwhile we were still busy with demonstrations of our rocket motors and with the enlargement of our workshop. After quite a number of demonstration runs, Nebel felt that he could do with a vacation. There was an Aviation Week of some kind in Kiel and Nebel decided to go there, taking the Oberth Rocket along for exhibition. There were no demonstration runs for several weeks and Riedel and I had time to straighten out odds and ends and talk things over. It had been vaguely agreed upon that the third Mirak should be assembled, but that was no definite promise.

Riedel had another test in mind which we had planned together some time earlier but had not been able to perform, partly because of the pressure of other work, partly because Nebel would have frowned upon anything not directly connected with his third Mirak. The idea was to take two pieces of magnesium tubing of the same capacity as the Mirak tails, put one of the standard motors between them, and measure the thrust that would be developed. We talked for a while about the simplest possible arrangement for such a test and then just sat and looked over our Raketenflugplatz, which had grown very beautiful with the coming of spring. The hilly part was covered with the young green of pine shoots and new birch leaves, the depressions between the hills were full of young willows. Crickets sang in the high grass and frogs croaked somewhere in the distance.

A few days after that conversation Riedel telephoned me in the office to tell me that he had "taken the secret baby out for a test run" the day before. He did not expect anything to happen; it was well weighed down with oversized valves, a large cooling water-jacket, and heavy struts. "But the beast flew! Went up like an elevator, very slowly, to sixty feet. Then it fell down and broke a leg."

I inquired just what "a leg" meant. "A fuel line," chuckled Riedel. "Home in Mediash that wouldn't have happened." This accidental flight had taken place on the 10th of May; on the 14th the rocket was repaired, lightened, and ready for the first planned ascent. The only thing lacking was a parachute. We did not have one ready and we were much too eager to wait. The two tubular tanks of the rocket were inserted into two stovepipes which served in lieu of a launching rack. The spring-operated valves we had were too heavy so we used light turncocks which had to be opened by means of "keys" mounted on broomsticks. It was done through an open but sand-bagged window of one of the buildings.

In spite of all the makeshift the "flying test stand" took off with a wild roar. It hit the roof of the building and raced up slantwise at an angle of about 70 degrees. After two seconds or so it began to loop the loop, rose some more, spilled all the water out of the cooling jacket, and came down in a power dive. While it was diving the wall of the combustion chamber—being no longer cooled—gave way in one place, and with two jets twirling it the thing went completely crazy. It did not crash because the fuel happened to run out just as it pulled out of a power dive near the ground; actually it almost made a landing. Examination showed that it was intact save for the hole in the motor. We could not examine it at once, since we were dizzy from watching and jumping out of the way and had to sit down in the grass for a moment. In retrospect it seems as if all this took place very close to the ground, but my notes say that we estimated altitude as about 200 feet immediately afterward.

That was the beginning and end of Repulsor No. 1.

I took the term straight from Lasswitz' novel, because we wanted to avoid the word "rocket," which then still meant a powder rocket to virtually everybody. And of course it was not a Mirak either. I reminded Riedel that in the novel the "engineer Fru" had made a lot of money with *his* repulsor and wished him the same.

Work on Repulsor No. 2 began the same night. We used the same tanks but substituted another motor. The struts were eliminated and better valves were substituted. We decided to do once more without a parachute, but we connected the tanks by circular aluminum hoops. We cut a large circular disk of sheet aluminum into four quarters, which were attached to the lower hoop to serve as fins. The repulsor was to stand on these fins on the ground so that no launching rack would be needed.

No. 2 was ready on Saturday, May 23, 1931. It happened to be an exceptionally beautiful day. At first the repulsor was fueled with oxygen only, since we had to find out how long it would take to build up the necessary pressure of 300 pounds per square inch in that particular tank on that particular day. The time required was just short of four and one-half minutes. One of the mechanics then opened the valve which let the oxygen escape through the motor and the exhaust nozzle. He swore that he felt a noticeable recoil.

In this and in all later repulsors the oxygen was fed into the rocket motor by its own gas pressure; this necessitated a few minutes' wait only. The gasoline was fed into the motor by means of compressed nitrogen. Since a fuel tank was filled to only one-third of its full capacity at the most, there was enough empty space on top of the fuel for the compressed nitrogen.

After the oxygen pressure test we had some coffee and buttercake in the open before we got the repulsor ready in earnest. While Riedel waited, wristwatch in hand, for the oxygen to build up pressure, I climbed a low hill to have a vantage point from which I could not only see the repulsor but also look over most of the Raketenflugplatz. From down below I heard the shouts of "Ready!—Fire!" and saw a white flame shoot out of the nozzle. It shortened quickly, became less bright, and began to roar. At the same instant the repulsor rose off the ground, slowly at first but accelerating rapidly. It climbed to about 200 feet and then turned sideways, like a car taking a curve. In that position it raced across the whole place, still under full power. It slowly lost altitude and it also slowly turned around its longitudinal axis while in flight. The four aluminum fins reflected the red rays of the setting sun like mirrors.

We saw that the repulsor was going to land outside the Raketenflugplatz and started on a wild pursuit, by car, motorcycle, and bicycle. We knew that people often walked in the forest outside the fence and we had visions of horrible slaughter. Fortunately the repulsor had not done any worse harm than breaking off a 2-inch branch of a stout tree. When we found it, it was hanging in the smaller branches some 30 feet above the ground, totally smashed. The distance from take-off to tree was 600 meters, about 2000 feet.

That night we wrote two very triumphant letters, one to the American Rocket Society in New York and one to Nebel in Kiel. And then we quietly put the parts of the third Mirak into a big wooden box. After that it was repulsors.

No. 3 was built within a few days, and it was a greatly improved model.

The two fuel tanks were placed much closer to each other than previously. They were only about 4 inches apart and were held in place by two sets of aluminum braces that jutted out for an inch on either side, fitting into U-channels of aluminum screwed to a wooden launching rack. The bottom braces also carried the parachute container, a vaned can some 3 inches in diameter. It had a fairly loosely fitting lid with a hole in the center through which the main shroud of the parachute was threaded; with this arrangement the lid was not lost when the parachute was ejected from the container.

The actual ejection was produced by a thick disk of cork which carried a small charge of ordinary black gunpowder. The charge, in turn, was ignited by a timing device of the type used by photographers. The timing device was started by the rocket itself when it pulled out of the launching rack; it was set for the time interval that was calculated to elapse before the repulsor reached peak altitude. This solution, presenting as it did a mixture of mechanical and pyrotechnic principles, was not quite satisfactory from a theoretical point of view. It was, I am tempted to say, theoretically unclean. But it was reliable.

Repulsor No. 3 was tested on a beautiful afternoon early in June. It quickly climbed to 1500 feet. About then the fuel was exhausted and about then something went wrong: the parachute was ejected but much too early. Either there had been more fuel in the repulsor than we had thought or else the timing mechanism had been set wrong. At any event the rocket was still climbing at a fast rate when the parachute blossomed out. It was effortlessly torn off, the repulsor rose for at least another 600 feet, but because of the momentary pull of the parachute it rose at an angle of about 60 degrees while it had climbed almost vertically before. Describing an enormous arc, the repulsor landed outside the Raketenflugplatz like a howitzer shell—by coincidence in the same clump of trees where No. 2 had found its end. It buried itself more than a foot deep in the hard forest soil, completely smashed.

Three more models of No. 3 were built during the next few weeks and one or two more later. They all made fine ascents, even though there were a few more parachute accidents of the same kind. The next step was Repulsor No. 4, which turned out to be even more impressive. It was essentially type No. 3 taken apart and reassembled differently, consciously copying the pattern of the late Congreve rockets in which the guiding stick was placed along the center line of the rocket tube. The motor, enclosed in a small bullet-shaped cooling water-jacket (see Fig. 27B), was on top; two struts and two fuel lines, forming a four-pronged fork, held it in place. The handle of that fork was the oxygen tank. The gasoline tank was placed in line below the oxygen tank, and the vaned parachute container below the gasoline tank.

We called this model the "One-Stick Repulsor" and the older types, consequently, were referred to from then on as "Two-Stick Repulsors." The first of the one-stick models was tested in August 1931 and reached an altitude of about a kilometer, roughly 3300 feet. At that altitude it was only a small dot in the sky. Suddenly we saw a tiny white parachute against the blue background and the repulsor drifted gracefully back to the ground. Several more were built, two of them larger in size but with the same standard motor. The One-Stick Repulsors, not fueled to capacity (which we did not dare to do because it would have caused trouble with the authorities), reached altitudes of about 1 mile; one that by accident took off at an angle covered a distance of more than 3 miles. Often we released them half fueled because they were sent up for demonstrations and they presented a better spectacle when the ejection of the parachute was clearly visible from the ground.

All the time progress was steady with hardly any disappointments. We had only one partial failure—the test runs of a larger motor. It had been designed as early as April 1931 and since we called the small ones eggs I named the big model the "Aepyornis Egg," in reference to the gigantic eggs of the extinct "Madagascar ostrich" Aepyornis, which are still found in swamps on occasion.

The Aepyornis Egg was expected to yield a thrust of 64 kilograms or 140 pounds; actually it barely made 50 kilograms or about 110 pounds. This partial failure was not conclusive, however, since it might well have been a partial failure of that particular motor. It happened occasionally that one of the standard motors did not perform so well. The tentative designing of a still larger model seemed justified.

There was no further progress during the remainder of that year of success, but things still went well. The Ufa Film Company devoted a large portion of its sixtieth newsreel[5] to the first year of work in the Raketenflugplatz, showing a test run of the Aepyornis Egg and the ascent of a One-Stick Repulsor. By that time 87 ascents and 270 test-stand runs of rocket motors had been made, not counting the Miraks at all.

During the filming of the newsreel one repulsor tore off its parachute and landed on top of a shack, setting the roof afire with the last remaining drops of gasoline. It was an old shack and there was nothing of value stored in it, but it belonged to the police garrison across the road. The police descended on the Raketenflugplatz like an invading army and any further experimentation was forbidden then and there.

Long discussions and hearings followed, revolving around the question: "When is an accident just an accident and when is an accident a symptom that something is essentially dangerous?" We quoted airplane crashes and

[5] They were numbered and there was one every week. Number one had been the first Ufa newsreel with a soundtrack.

train derailments, bus collisions and fires from short circuits. Finally a demonstration was staged just for the police, after which the order was withdrawn, and we were allowed to experiment, subject to five provisions:

1. The weight of the rocket with fuel must not exceed 11 pounds.
2. The rocket motor used must have made three flawless test runs.
3. Heavier rockets require special permits.
4. Flights must be made only on workdays between 7 A.M. and 3 P.M.
5. No rocket flights are permitted on windy days.

Considering that we worked within the city limits, these provisions were sensible enough.

The newsreel and the victory over the police were our last triumphs. What followed afterward was a hopeless struggle against political tension and economic misery. It was a hard winter climatically. And it was the fatal winter under Chancellor Bruening when Adolf Hitler suddenly assumed prominence. It was a winter during which the roster of VfR members shrank to less than three hundred, most of them deprived of their livelihood. It was a winter during which there came many letters saying that no further dues would be forthcoming because "all money belongs to the Führer." Hückel wrote that he could not help us any more and Dilthey disappeared. Six months later his body was found in a narrow cliff-bounded valley. He had met with an accident while vacationing in the mountains.

In order to raise money I went on a lecture tour in East Prussia. At the Raketenflugplatz the men struggled along. Riedel planned a reorganization of the testing equipment, which was put into practice during the following year. The new arrangement consisted of putting a dugout into the earth wall, only 30 feet from the test stand. The pressure bottles were placed in that dugout, to be operated by the observers who were protected by thick wooden planks and peered through a slit less than an inch wide. (Plate VIII shows a picture taken from that dugout.) In addition to making the layout for this arrangement, Riedel also prepared a few motors of a new type and for a new fuel: alcohol.

This was the outgrowth of a few conversations we had had in private during the fall. We had discussed the fuel question and had admitted to each other that gasoline did not seem so flawlessly ideal as we always said when somebody was listening to us.

In his book Oberth had very often mentioned alcohol as a rocket fuel. He had even used the terms A.R. and H.R., standing for alcohol rocket and hydrogen rocket, respectively, as designations. But in private conversations he said that the reference to alcohol had been intended only as an example of a liquid fuel. When asked pointblank he had said that he considered gasoline the most practical fuel, partly because of its high energy content, partly because of its availability any place. When he got to actual design he wanted methane, as has been told; my telling Winkler of this preference was the direct reason for this fuel in the first Winkler rocket. Oberth, in short, had often written about alcohol as a rocket fuel but if he had been serious then, he had changed his mind later.

When Riedel and I spoke about alcohol we did mean alcohol. I had advocated switching to alcohol for one specific reason. In order to burn 1 pound of gasoline completely you need 3½ pounds of oxygen. In order to burn 1 pound of alcohol you need about 2 pounds and 1 ounce of oxygen. While alcohol had a slightly lower energy content, I saw an advantage in that 2-to-1 ratio as compared with the 3.5-to-1 ratio for gasoline. Riedel agreed with me, but for a different reason, though he accepted mine too. In spite of the cooling water-jacket the motors still burned out; only on the test stand where there was running water for cooling did they stand up well. Riedel said to himself, and to me, that cooling the motor from the outside was apparently not enough and he wondered whether one could cool a rocket motor by injecting some cooling water. But to do this would have required another injection nozzle, which presented practical difficulties. Riedel's ready acceptance of alcohol was based on the fact that the cooling water could be mixed into the fuel, obviating the need for a separate injection nozzle.

At this time I was about to leave for my lecture trip to East Prussia. Knowing that my father owned a small liquor factory there, Riedel said, "You know, there are some liquors which are served flaming. Ask your father how much alcohol they have to contain for this purpose." The next day I wrote him a postcard from Königsberg, saying: "Flaming liquors must contain 40 per cent by volume; 38 does not burn any more." While I was still in Königsberg, because my lectures were not spaced very closely, Riedel ran a few preliminary tests. He found that a rocket motor running on 40 per cent (80 proof) alcohol ejected a thick jet of steam "like a waiting locomotive." With 90 per cent alcohol the motor lasted hardly longer than with gasoline; an alcohol-water mixture with 40 per cent water seemed to be about right. Later in that year the next step was taken. The watered alcohol was no longer injected directly but was sent into the cooling water-jacket at one point, taken out at another point, and then injected into the combustion chamber (Plate VIII and Fig. 27 C). The invention of cooling a rocket motor with its fuel has been made at least three times, but this was the first.

This very important advance, however, was hardly noticed, because the general situation was deteriorating from day to day. Nerves were frayed; practically every meeting of the Board of Directors (in the meantime Professor Oberth had resigned and a Major von Dickhuth-Harrach, since deceased, had become president) led to violent clashes, caused, in the last analysis, by political differences. The deterioration was rapid and the events

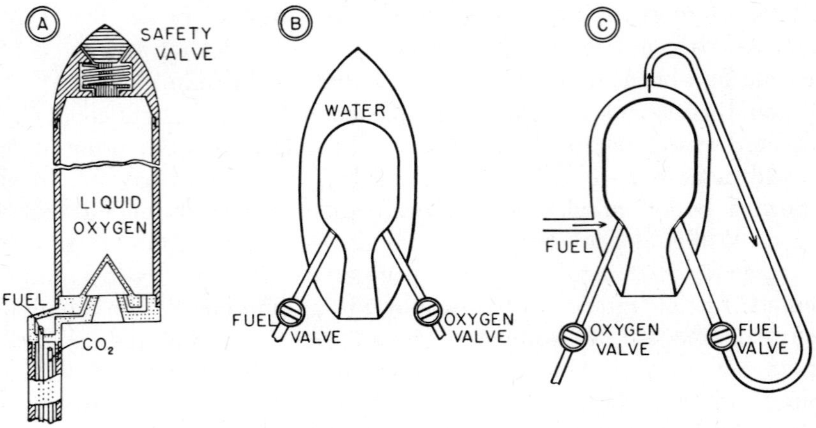

Fig. 27. Accomplishments of the Raketenflugplatz.
A. Cross section of the "head" of Mirak No. 2 with combustion chamber placed inside the liquid-oxygen tank. Dotted portions were made of copper, others of aluminum.
B. Cross section of the "head" of one of the later repulsors, showing Riedel's motor and cooling water sealed into the jacket.
C. Rocket motor for liquid oxygen and watered alcohol. In this design the fuel was circulated around the combustion chamber before injecting it, thus obviating the need for a separate cooling fluid. Most modern rocket motors are developments and elaborations of this principle. For photograph of this motor in action see Plate VIII.

of that last year of the VfR can be told only in a series of episodes which are not closely connected with one another, even though some influenced others.

To begin with, there was the near-comedy of Winkler's second rocket. One day Winkler had come to the Raketenflugplatz with a business proposition. If we could assure him absolute privacy he would rent the building farthest away from those we used, for a period of a few months. Nebel promised that he would not be disturbed and Winkler carried out a number of tests in great secrecy. Once in a while somebody heard the roar of a rocket motor—which was to be expected—and that was all we knew about his work. Winkler then built a rocket resembling the Oberth Rocket in size and general shape. It had three stabilizing fins around the motor's exhaust nozzle, a parachute in its rounded nose, and two spherical tanks for methane and liquid oxygen. The trouble with this rocket was that it, like the Oberth Rocket, was based on a large amount of computation with only a very slight seasoning of experience. Winkler had made just one full test run in the course of which most of the exhaust nozzle had burned away.

Like the Oberth Rocket, the HW-2 (Hückel-Winkler 2—Hückel having provided the cash) was to take off from the Greifswalder Oie. But the lighthouse was still there and the same officials were still in charge—at least the same kind of officials. After months of waiting, Winkler had to leave the unfriendly isle. He took his rocket to East Prussia where he finally got permission to test it on the beach not far from the harbor of Pillau. He made his first attempt on September 29, 1932. Nothing happened; icing around and probably inside the valves prevented the fuels from entering the combustion chamber. This turned out to be the lesser evil. During the next attempt on October 6, the valves apparently leaked and permitted a trickle of gases to accumulate somewhere between the tanks and the outer skin of the rocket. When the rocket was ignited this mixture caught fire too and the rocket, which had barely lifted itself 10 feet, was thrown to the ground. It was an extraordinary smash-up. Winkler complained afterward that not a single part was usable.

Just before Winkler began this work Nebel had tried to influence Hückel to increase his monthly donation to the VfR, even though Hückel had declared in a letter that he would have to discontinue his contributions. He had insisted that he was financially unable to do so. His sponsorship of the HW-2 was a blow aimed at Nebel.

That caused what in retrospect looks like pure comedy: Nebel's "collaboration" with the German Army. Casting around for additional money, Nebel sat down and wrote a technically wholly inadequate and in fact senseless "Confidential Memo on Long-Range Rocket Artillery." With a copy of this memo in his briefcase he went to see Professor Becker, co-author of *Cranz-Becker: Ballistics,* the most-used handbook on ballistics. The irony of the thing was not only that Professor Becker could have written a similar memo four times as good without any preparation at all, but that Professor Becker was already toying with the idea of bombardment rockets of some kind and had conceived the idea of what is now called a "missile." There was some political reasoning behind this. Heavy artillery had been forbidden to Germany in the Treaty of Versailles; rockets had not been mentioned.

The German Army consented to a demonstration on the Army's proving ground at Kummersdorf near Berlin. Nebel did not notify the Board of Directors, but he took Riedel and von Braun along. The Army had demanded that the rocket (it was one of the larger One-Stick Repulsors) eject a red flare at the peak of its trajectory. The Board of Directors learned about the experiment because one of the mechanics complained about the unnecessary work of adapting the repulsor to such a flare. Once the secret was out, Nebel began to complain too. The repulsor had worked reasonably well, but soon after take-off it had veered from the vertical, which may have annoyed Riedel but did not annoy anybody else. But Nebel did not get the large contract he had been hoping for; he claimed that he had been snubbed. I now know that a somewhat unusual kind of "collaboration" might have been accomplished if he had listened to reason and been frank with his co-workers. But he wasn't. A month or so later the Army hired Wernher von Braun, who disappeared from our view for a while.

Then there was the tragedy of Tiling's laboratory. Some time after the Kummersdorf demonstration—and here the sequence is purely chronological —Tiling came to Berlin and Nebel went to see him. I was not present at that meeting but Riedel was and produced a highly humorous account of it. He told me that Tiling greeted them with the words, "Gentlemen, I don't quite know why I received you, as you must know I am used to dealing with nobility only." In spite of this not particularly friendly reception they sat down and began to talk. Nebel offered to join hands with Tiling, the VfR to provide its liquid-fuel motors and Tiling to contribute his well-developed rocket body with the sailing-fin mechanism. Tiling coldly refused, saying that he did not see any sense in a collaboration with people who were still at the stage of building small vertical rockets while he, Tiling, was ready to span the English Channel.

Tiling had come to Berlin for a demonstration. But he did not do so well. The day happened to be gusty and rainy and the whole enormous Tempelhof airdrome, where he made his demonstration, had hardly a few hundred visitors. To make matters worse, one of his rockets almost smashed into the grandstand, so that the police intervened and declared the demonstration over for that day. Tiling returned to his home and set out to build rockets capable of crossing the English Channel.

This, at least, is what he announced. But whatever it was he had really in mind did not come to pass. On October 11, 1933, the newspapers reported that Tiling's laboratory had exploded the previous night. Tiling was dead, so was his laboratory assistant, Miss Angelika Buddenböhmer, while his mechanic, Friedrich Kuhr, lived for another day.

The accident had been the result of great carelessness, just like Valier's death. They had been working late the previous night, compressing not less than 40 pounds of powder into compact tablets. In the middle of that work the catastrophe occurred. It seems that the powder press exploded; pieces of the heavy cast-iron press were found all over the ruins. The flame of that explosion reached a large supply of processed and unprocessed powder which was *stored in the same room.* The whole laboratory caught fire instantly, but Tiling and his two helpers succeeded in leaving the burning ruins, their clothes on fire. They threw themselves into a nearby pond. Then Kuhr, who was the least injured, ran to Count von Ledebour's castle to give the alarm. Nothing could be salvaged and medical aid came too late for all concerned.

The next affair was something for which I cannot think of a literary designation. Its name was "Magdeburg Project" or "Pilot Rocket," and it started only a few months before Hitler became Chancellor. The chief provision of the project, Nebel said, was that it was to be entrusted to him personally, not to the VfR. Major von Dickhuth and I agreed readily. It looked like something in which we did not like to see the VfR involved. Much of our worry was groundless because Nebel had accidentally kept the society in the clear. When talking to Captain Becker and other army officers he had insisted that there was no connection between VfR and Raketenflugplatz other than the fact that he, Nebel, also was secretary of the VfR, or rather had been in the past. When talking to the representatives of the Magdeburg interests, he had said the same. He could say so with impunity because, if anybody thought of checking up with the Court of Registry, he would learn that the VfR, through its secretary Mr. Rudolf Nebel, had filed under oath a petition of voluntary bankruptcy in February 1930.

Nebel had not asked anybody about that, he had not even told anybody. When Major von Dickhuth and I learned about it, which was in the spring of 1933, we had ledgers and correspondence seized and found facts which had to be submitted to the District Attorney. To make the story short, nothing happened. The District Attorney, seeing that Nebel wore a swastika armlet, was afraid to act.

As for the Magdeburg Project, it began like a story by Jules Verne. A mentally decrepit "philosopher" had written a badly printed pamphlet about the true shape of the universe, in which he insisted that the earth *is* the universe, that we live *inside* a hollow globe of the dimensions of the earth, that there is nothing outside that globe, and that the universe of the astronomers is only an optical illusion. Since every crank can find some fellow cranks, the "hollow earth philosophy" had found some too, among them an engineer named Mengering who was somehow connected with the municipal government of the City of Magdeburg. Being an engineer, he conceived the idea of testing the hollow earth theory by means of a rocket. If a rocket going vertically upward crashed among the antipodes, proof would be established.

Now Magdeburg is a proud city. It was a burgomaster of Magdeburg, Otto von Guericke, who performed that famous experiment with the evacuated hemispheres which could not be pulled apart by several teams of horses. Magdeburg would welcome another great scientific achievement of that kind. Mengering offered his idea. The municipal officials did not believe any of the "hollow earth" trimmings, but they were interested in a spectacular rocket ascent, a man-carrying rocket, if possible.

Mengering talked to Nebel and Nebel at once agreed. He would be ready at Easter 1933 or, better, at Pentecost, or, if they did not want a holiday (Pentecost was an official holiday in Germany), the Sunday after Pentecost. They agreed on the latter date, which was the eleventh of June,

1933. It was agreed that the rocket could cost 25,000 marks and it was to be shown as the crowning feature of a kind of city-wide holiday which would require another 15,000 marks.

Mengering did not succeed in getting pledges for the whole amount, and since Nebel did not fulfill the contract he got only a part of what was collected. Mengering later spoke of 3200 marks.

Here is the story, as told by my friend Herbert Schaefer, who worked some sixty hours per week on that project:

> The man-carrying rocket, called the "Pilot Rocket," was to be a huge monster about 25 feet high with a rocket motor of 600 kilograms (1300 pounds) thrust. The passenger cabin and the fuel tanks were to be one unit, shaped like a huge artillery shell, while the other unit, comprising the rocket motor and the parachute, was a smaller "shell" topping the bigger one. The rocket was supposed to reach an altitude of 1 kilometer (1100 yards) at which point the parachute was to be ejected while the passenger—he was not a "pilot" since he did not do anything—was to jump out with his own parachute.
>
> A smaller rocket of the same shape was to be built first; except for size it differed from the large one only by having its parachute where the large one would have the passenger. This rocket was 15 feet high and equipped with a rocket motor of 200 kilograms (440 pounds) thrust.
>
> We all began to work feverishly although we knew that it would be impossible to get such rockets ready in the time interval agreed upon. But it meant an opportunity to build large rockets without being handicapped by lack of funds. The actual work began around Christmas 1932. Motors of 440-pound thrust were designed and built, and also a new test stand was designed to "take" 1000-kilogram (2200-pound) rockets. That stand was ready in March 1933, but one of the new motors was ready earlier and was tested on a provisional test stand on March 9.
>
> It could be heard for miles. The big new stand actually worked for the first time on March 22. It worked to our full satisfaction. Three days later one of the motors exploded at the instant of ignition; the concussion was so bad that the eyeballs of the observers pained considerably. We had expected some such mishap and had eight such motors scheduled. For a week I spent all my time in the welding shop supervising the construction of them. The second motor exploded, also on ignition, on April 3. This was the last real explosion, but on three successive tests the exhaust nozzles burned through at the throat (the narrowest part). Some twenty test runs were made in April; the thrusts obtained varied between 150 and 200 kilograms, depending on the fuel-oxygen ratios. Finally we considered that type of motor reliable.
>
> The City of Magdeburg hurried us and it was agreed to let the big rocket (but not the man-carrying) ascend on June 9. A large launching rack was built in a cow pasture at Wolmirstedt near Magdeburg. It was 30 feet high. Then a series of mishaps began. The rocket could not be attended properly so far away from the workshops. On the morning of June 9 it was fired, the rocket began to rise slowly, but before she reached the top of the rack she began to slide down again, very slowly. The thrust was insufficient and the reason could not be found. Another attempt two days later was spoiled by a leaky gasket. The motor got only one quarter of the fuel it was supposed to get, the rocket roared for 2 minutes instead of 30 seconds, and we approached to within 10 feet. Of course the rocket did not move. Another test on June 13 also ended prematurely. When

the rocket was 6 feet high a vent screw popped out and the rocket fell back, getting no more fuel.

We decided on a complete overhaul. After that new tests with all kinds of little mishaps; once a valve froze tight, another time the ignition capsule was blown out before it ignited the rocket, a diaphragm in the fuel line burst, etc. Heavy rains interfered and warped the wooden launching rack, not enough to be noticed, unfortunately. The city accountants had not granted the expenditure of a metal rack. Thus when the rocket finally did take off on June 29, one of the rollers derailed and became stuck. The rocket just stripped it but took off almost horizontally because of that. Losing altitude rapidly, the rocket made a belly landing 1000 feet from the rack, the motor still going full blast. It slithered for another 30 feet. It looked totally smashed, but the motor and the tanks were unhurt. Only the casing, fuel lines, etc., had been smashed.

(Schaefer)

This was the Magdeburg Project itself. It had an aftermath which is still of interest. Schaefer's shorthand notes describe it as follows:

We decided not to wait for another casing but converted the motor and the tanks into a four-stick repulsor, 7.5 feet high. This repulsor was tested on "Lover's Island" (official name on maps is Lindwerder) in Tegeler Lake near Berlin. She rose with terrific velocity to about 3000 feet, suddenly tilted over up there, made a few loops, and came down in a power dive, landing some 300 feet from the island in the water. The parachute was ejected at the last moment before striking, thus only minor damage was sustained. Another repulsor, with minor modifications, was tested from same island on July 21. The first try was unsuccessful; the valve burst, but after replacement and refueling the repulsor took off. One oxygen valve failed to open, and the repulsor rose slowly and off balance to about 200 feet. It landed in water with only minor damage.

The captain who owned "Lover's Island" objected—we scared his summer campers. We went to Schwielow Lake, using a motor launch instead of an island. Same type of repulsors. Test on Saturday, August 5, 1933. Exploded soon after take-off. Next test same type repulsor, same place, August 11. Seems that one valve did not open, horizontal flight, touching water. Sank in steamboat channel, the only dredged and deep section of the whole shallow lake. Rescue impossible. August 31, same type, same place. Flew away, could not be found again.

September 9 and 18. Two tests with new design, two-sticker with very long tanks and 200-kilogram motor. Same place. Both poor.

(Schaefer)

The end of the Raketenflugplatz looked quite dramatic to an observer on the spot. A large group of young men in beautiful powder-blue uniforms, who called themselves the *Deutsche Luftwacht,* came and told us that the place had been assigned to them as a drill ground. Whether this was true or not—I now doubt it, especially since that organization did not last very long—they did use it as a drill ground, making life difficult. On one occasion I was not permitted to enter, being told by one man with some insignia of rank on his collar tabs that the Gestapo was there to seize documents and equipment. Possibly a routine investigation took place that day, but if so, I, as vice-president of the VfR, should have been questioned.

The end of the Raketenflugplatz was far less dramatic if you knew the inside facts, which I did not learn until Wernher von Braun and I started reminiscing long after the war. At my request he put his recollections into writing in a long letter dated December 30, 1951. Here is what he had to say about this particular episode:

> Shortly after the test on the Lake (described by Herbert Schaefer) Nebel was visited by a representative of the city administration of Reinickendorf, who handed him an enormous water bill—if I remember correctly it was about 1600 marks. Nebel first thought it was all a mistake, but soon it was discovered that a number of leaky faucets in some of the buildings which had never been used had run up this exorbitant water bill, silently, during all the time from September 1930 to the summer of 1934. Having always worked on an almost cash-free basis, Nebel couldn't pay. The city boys became unsympathetic and canceled the famous lease; the area had to be cleared by a certain date. Klaus Riedel happened to know one of the directors of Siemens well; he was a man by the name of Altvater who had been a friend of Riedel's father, long deceased, and something like a foster father to Klaus. Altvater gave jobs in the Siemens branch to Klaus and several other of the Raketenflugplatz men
>
> The files, plus all the hardware (notably the old Oberth rocket, the Ufa spaceship model, rocket motors, testing equipment, and machine tools), were stored in a Siemens warehouse under an arrangement between Altvater and Riedel. The stuff lay there for several years.
>
> I do not know what happened to it later. Zoike (who is with me at Huntsville, Alabama) believes that Nebel sold the equipment after Riedel, Hueter, Hainisch, and Zoike had gone to Peenemünde.[6] In view of what Nebel had done for the Raketenflugplatz he may have felt that he was morally entitled to such a sale. I have not been able to find any evidence that the Gestapo ever had their fingers in the ultimate disposal of this material and I cannot imagine why they suddenly should have developed an interest in it after leaving it untouched for years in the Siemens warehouse. . . . One thing I know for sure: the Heereswaffenamt never tried to secure files or equipment of the Raketenflugplatz, nor was it ever turned over to Peenemünde or any other Army installation. (von Braun)

After reading this letter I dimly recalled having heard about an "unpaid bill" at the time, but I had not paid too much attention—there were always some unpaid bills.

By Christmas 1933 the picture looked bleak again, much bleaker even than it had looked by Christmas 1929. Then it had been mostly a matter of disappointed hopes which could be overcome by hopes for the future. But now it seemed as if there were no way out, with everybody's hands tightly tied by a ruthless totalitarian regime. During that winter the VfR collapsed during a stormy session, and while I pretended to try to reorganize it under another name, I quietly took the first steps toward leaving Germany. There seemed to be only one group left which was carrying on experimental research, if only intermittently and on a small scale: the American Rocket Society. Fired by the reports which Pendray had brought back from the Raketenflugplatz, this society had gone ahead along similar lines.

[6] See Chapter 8.

To an omniscient chronicler things would have looked quite different during those same winter months of 1933–34. He would have been aware of the existence of several places where rocket motors howled and thundered and, on occasion, blew up.

At the College of Engineering in Vienna, Dr. Eugen Sänger late in 1933 began a long series of rocket-motor tests which lasted till the end of 1934. They resulted in still another design for a fuel-cooled motor; since Dr. Sänger had rocket propulsion for aircraft in mind, the story is told in Appendix 1.

In France Robert Esnault-Pelterie was working quietly on a design of his own, investigating tetra-nitromethane. That was in 1931 and it was at about that time that our correspondence suddenly stopped. Now I know why; an explosion had occurred in which he lost the fingers of his left hand. Mostly because of this accident the French Air Ministry began to pay attention to the pioneer work of the French citizen Robert Esnault-Pelterie—and provided financial support. It was not much, and the condition was secrecy. As a result of this secrecy most detail of Esnault-Pelterie's work is now lost; it was destroyed when the Germans marched into Paris in 1940. The main accomplishment was a rocket motor, operating on gasoline and liquid oxygen, with a thrust of 300 kilograms (660 lbs.) for 55 seconds. The rocket contemplated for this motor was to weigh around 100 kilograms and would have reached an altitude of about 60 miles.

In America Professor Robert H. Goddard had quietly moved to Mescalero Ranch in New Mexico, where he was working in self-imposed secrecy on rockets which were meant to be instrument carriers for the exploration of the upper atmosphere.

Finally, in Kummersdorf near Berlin, where the 3-liter repulsor had been demonstrated, von Braun was at work on something called "A-1." An omniscient chronicler would have known how A-1 would grow.

7 The Return of the War Rocket

In Stahl gehüllt
vom Strahl umwittert
die Schar, die Reich um Reich zerbrach
sie treten auf,
die Erde schüttert
sie schreiten fort, es donnert nach!

—Goethe, *Faust*, II

Sixty years after the last of the Congreve rockets had been fired in the mountains at Glok-Tepé, the war rocket came back into history. It would be wrong to say that there had been no war rockets during that long time interval. There were some, but they appeared only sporadically; their use was somewhat half-hearted, often experimental, and sometimes even makeshift.

The earliest attempt to reintroduce war rockets after the gradual but thorough dissolution of all the older rockets corps had been made in Sweden. In about 1890 a Swedish inventor, Lieutenant-Colonel Baron von Unge, had approached Alfred Nobel with the project of an "aerial torpedo." It was actually a large rocket, patterned after the stickless war rockets of William Hale, with some changes and improvements.

Nobel was interested; he had worked with rockets himself on more than one occasion. One unfruitful contribution which he made was a novel type of signal rocket. The signal rockets then existing, like those of a much later day, released a white or colored flare at the peak of their trajectory. It was a visible signal. Nobel had reasoned that a signal should be audible too and had put a few ounces of nitroglycerin, along with his own patented fuze, into the head of a large blackpowder rocket. The nitroglycerin exploded with a white flash and a report that could be heard for 10 miles on a still day. Nothing came of the experiment, however, because nitroglycerin was so dangerous that Nobel himself stopped selling it even before some countries, including Great Britain, had prohibited its use.

Baron von Unge had several ideas about ways to make a rocket more useful for modern war. One was to provide the igniting spark, not through the exhaust nozzle which might become fouled by the igniting device, but

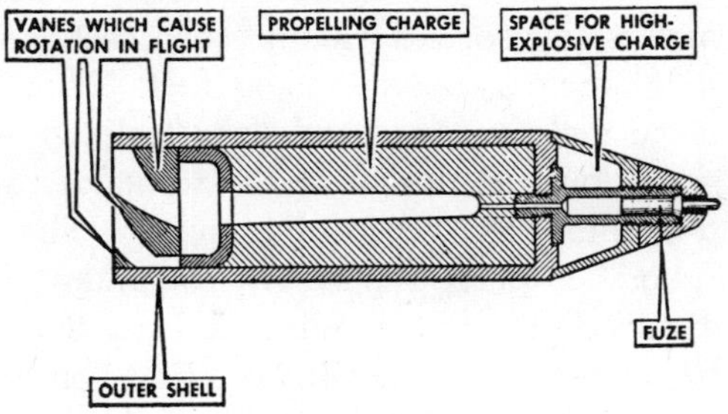

Fig. 28. Baron von Unge's "aerial torpedo." A cross section of a late model tested by Krupp in 1909. Over-all length was 30 inches.

from the front, through a thin hole drilled through the solid portion of the propelling charge. Another and more important innovation was to fire the rocket from a short mortar. The rocket would then take off with a certain velocity, say 300 feet per second, which would not only increase its range but also make it a more accurate weapon—accurate enough, von Unge hoped, to enable it to compete with artillery again. He could point out as an established fact that a rocket could be fired from something like a mortar. Fireworks makers had what they called a *pot-à-feu,* a mortarlike tube from which bundles of skyrockets were fired. This provided an especially nice display at night because, the "ballistic ascent" being naturally invisible, the rockets seemed to originate at a point in midair.

Nobel's interest in rockets was not merely academic. He put his compatriot to work, paying all the rapidly accumulating bills, which would have looked large to anybody of lesser means than Alfred Nobel. But in spite of the considerable expense nothing was accomplished that was good and finished enough to be shown to military men, who in those days had absolutely no taste for developments but preferred innovations in serviceable form. Then, in 1896, Nobel died and von Unge was apparently left at loose ends.

Five years later, in 1901, a corporation called Mars was founded in Stockholm to enable von Unge to complete what had been left unfinished because of Nobel's death. The results of these experiments were never published, but some facts became known later on in a roundabout manner. The propelling charge was the same in composition and method of manufacture as that of the line-throwing rocket used by coastal lifesaving units. It was a blackpowder mixture with an excess of coal, hammered into the rocket tube by hand. The shell was attached to the top of the rocket tube. Its charge consisted of dynamite, and the fuze was of the percussion type, set off by striking the target (Fig. 28). Nitrogelatin was also tried, as were time fuzes which were ignited by the propelling charge.

The weight of the high-explosive charge was 2 kilograms (4.4 pounds). The over-all length of the aerial torpedoes was 750 millimeters, the diameter 110 millimeters. The over-all weight of the early models was 35 kilograms, the highest velocity attained along the trajectory was 300 meters per second (about 1000 feet per second), and the extreme range 5 kilometers, about 16,400 feet. The mortar supplied an initial velocity of 50 meters per second (165 feet per second). This was found to be too small, but the design of the torpedoes apparently did not permit a higher propelling charge in the mortar. The accuracy of fire was admittedly "poor." Experts reportedly estimated that, to hit a given target over a distance of about 2 miles, at least five times as much ammunition would be needed as for howitzers firing comparable projectiles.

Lieutenant-Colonel von Unge then decided to drop the launching mortar completely and to use open guiding tubes instead. When the two Swedish dirigibles *Svenska I* and *Svenska II* attracted much attention in his homeland, von Unge began to advocate his aerial torpedoes as weapons for aircraft (meaning dirigibles) either against ground targets or for aerial combat. He emphasized that his torpedoes did not produce a recoil, which still is important for aircraft armament.

That was in July 1908.

In 1909 it became known that the firm of Friedrich Krupp, Aktiengesellschaft, in Essen had purchased the Unge patents and had also bought the existing supply of aerial torpedoes, about a hundred, as well as a launching tube and other equipment. The equipment was transported from Stockholm to Krupp's proving ground at Meppen, and extensive experiments were carried out.

Some information about the latest models was supplied later by Krupp's chief ballistician, Professor Dr. Otto von Eberhard,[1] in the course of a general discussion on the mathematics of shell trajectories. Dr. von Eberhard stated that the aerial torpedoes had an over-all weight of 50 kilograms (110 pounds) and that they attained a range of 4 to 5 kilometers, say from 13,000 to about 16,500 feet. Since von Unge's first models, in Nobel's time, had attained a maximum range of 4 kilometers, that was not much of an improvement.

In 1910, however, Krupp announced that experimentation with von Unge's invention had been discontinued because accuracy of fire could not be achieved. Naturally this announcement was disbelieved, if only because only a few months earlier Krupp's had applied for some patents relating to the invention. The patent applications may have been a matter of principle, or just part of the routine of a large industrial firm. In any case no weapon bearing even a faint resemblance to von Unge's aerial torpedoes made its appearance on the German side during World War I. In all probability Krupp's engineers had tried to convert von Unge's "torpedoes" into short-range heavy artillery, and failing to accomplish that

[1] In his *Freier Fall, Wurf und Schuss* (vol. II of the *Handbuch der physikalischen und technischen Mechanik*), Berlin, 1928.

had turned their attention to other means of throwing heavy shells over short distances.

The only country to use rockets operationally during the First World War was France. The facts—military secrets while the war was going on—became known only gradually since the technological accomplishments of the First World War were not written up and published as quickly and as enthusiastically as those of the Second World War. One account of French rocket weapons in the First World War can be found in the book *Zeppelin,* by Captain Ernst A. Lehmann, who later died in the disaster of the airship *Hindenburg* at Lakehurst.

> During the first months of 1916, I piloted the new LZ 90, one of the seven airships directly under the Highest Army Command. It was a beautiful ship of 32,000 cubic meters gas volume, with a tapered stern and more efficient control organs. . . . A few days earlier, after an attack on the railroad junction at Revigny, near Brabant-le-Roy, the LZ 77 had been shot down in flames. This was astonishing, for generally the shells went straight through the envelope, and even during shrapnel fire the damage was confined to a few gas cells so that the ship could still remain aloft. . . .
>
> Our goal was the railroad yards at Bar-le-Duc, which supplied France's key position of Verdun. The LZ 90 carried almost 7000 pounds of bombs. By throttling our motors and concealing ourselves in the clouds, we crossed the front at an altitude of 10,000 feet. I do not know whether we were detected or not, but, at any rate, Bar-le-Duc was not warned and at first greeted us only with a few ordinary shells. Though the railway station was an important one, it was small, and since it was well darkened, we had difficulty finding it. No sooner had the first load of bombs been released, than we were obliged to stop because the LZ 90 had overflown its mark. We circled around and were just making a second attack on the station, which was now marked by fire and smoke, when we saw a number of thick yellow rockets coming toward us. They moved fairly slowly, but rose steadily, climbing past our ship, which was then at an altitude of 10,600 feet, and then continuing on still higher!
>
> Incendiary rockets! The latest and surest means of igniting an airship filled with hydrogen. A single hit infallibly sufficed to destroy any airship. I ordered full speed ahead, brought the ship to its greatest possible altitude and fled in haste. The incendiary shells had come from a point near the station, and I hoped quickly to get beyond their range. But they continued to follow us, and a few came uncomfortably close to our stern. We discovered that the guns were mounted on motor cars which were following us along a highway running parallel to our course. Instantly we changed our course and escaped them. Passing over the lines we slipped into a cloud. . . .
>
> Now we knew why the LZ 77 had gone down in flames.[2]
>
> (Lehmann)

But the French not only created the first antiaircraft rockets, they also produced what von Unge had tried to achieve: the first air-to-air combat rockets. It is true that they were aided by having a fantastically vulnerable target. Following a tradition started by Count von Zeppelin during the

[2] Ernst A. Lehmann, *Zeppelin,* translated by Jay Dratler (New York: Longmans Green, 1937), pp. 103–104.

American Civil War the Germans used men in captive balloons as observers, especially to direct their own artillery fire. The balloons, which could not move, were filled with hydrogen gas—or sometimes with illuminating gas if a local gas company was in operating condition. The French attacked these helpless balloons with large Le Prieur rockets, rockets of the same type which served for throwing lines at the seashore. The rockets apparently did not even have special warheads; their incendiary action was sufficient.

The plane used was the French Nieuport, a biplane which had on each side of the fuselage a very sturdy V-shaped vertical strut connecting the two wings. Four Le Prieur rockets were attached to each strut, so that each plane carried eight missiles. After a few battlefield tests, several special squadrons of rocket-firing planes were established; they were called *Les Fuséens.* The Germans did the only thing possible: they stopped providing captive balloons as targets. Then the *Fuséens* were dissolved.

I have read somewhere that Russian aviators improvised the same weapon against the same type of target; but for obvious reasons there is very little source material about Russian operations during the First World War, especially as regards what we call "unit histories." These Russian aircraft rockets may well have been in the nature of private improvisations of a few pilots, as was the only German use of rockets during that war, not counting the traditional use of rockets for signaling. On the western front a few German units supplied themselves with the largest type of rocket they could find, attached a line to the stick and a small boat anchor to the head, and then fired the rocket from a first-line trench across the opposing barbed-wire entanglement. Then they pulled it back by means of a hand winch.

This completes the story of rockets during the First World War.

The scarcity of combat rockets during the First World War and their abundance during the Second is not due either to "accident," or "habits of thought," or even tactical considerations of some kind. Both can be explained as an industrial problem, which might be labeled "production problem," or "storage problem," or even "safety problem," but which hinged on the nature of the fuel employed.

When William Congreve defended himself against his critics he did so by comparing performances and by drawing up an expense sheet. His figures were perfectly truthful and looked impressive, but under present conditions they would represent only a very small portion of the story. As things stand now, a combat rocket has to conform to the standard demands made upon a war weapon. And the first of these standard demands—often overlooked because of its obviousness—is storability. A weapon might be manufactured, say, in Detroit. Then it has to be stored until it is shipped to some arsenal or military base where it is stored again. After an interval of time it might be shipped either to Africa or to Greenland and stored once more at an advance base. Then it would be brought up to the fighting lines to be held in readiness for a coming operation—but all the time, at least theoretically, it must be ready for instant use. All cartridge ammunition, from pistol to antiaircraft-gun cartridges, lives up to these expectations. The second important demand is that a weapon be mass-produced, if possible on automatic machinery which merely needs supervision.

When you consider these two main demands it becomes clear why a liquid-fuel rocket would be useful as a combat rocket only in some specific cases. Of course the parts of a liquid-fuel rocket, as well as the warhead, can be mass-produced and the rocket can be stored, either in parts or assembled. But it would be very hard to store a liquid-fuel rocket with fuel in its tanks, even if a fuel combination is used which avoids liquid oxygen. The fuels would have to be stored separately, to be filled into the rocket only when the time for its use is reasonably close. This is possible in fixed installations, such as antiaircraft establishments defending cities and harbors, and aboard ships designed for this purpose, and also for long-range weapons. But it is impossible in the front lines—one might as well expect a soldier of the Signal Corps to assemble his radio in a foxhole.

Logically, then, combat rockets have to be storable, solid-fuel rockets, of a type that can be mass-produced.

The latter requirement was not met by the large blackpowder rockets prior to, say, 1935. Their manufacture was handwork, piecework. Even Sander's far-advanced hydraulic presses only relieved the operator of the necessity of using his own muscles. It was still piecework. And it was highly dangerous. If one put only a few hundred such presses in a factory hall one could be virtually certain of a major accident every 24 hours. The fact that mass production of blackpowder rockets of large size is impossible may have been the main, if unmentioned, reason that prompted Krupp's to write off the money spent on von Unge's projectiles.

Storage of large blackpowder rockets was also impossible. Not that it was necessary to put the fuel charge in immediately prior to use; the fuel charge was there and in position. But that fuel charge did not stand storage or, to be precise, it could stand storage only under very carefully controlled conditions. And it could not stand much handling.

The reason was this: in order to make a powder rocket large and powerful, the powder mixture had to be compressed to a much greater extent than it is in small fireworks rockets. The specific gravity of the powder in fireworks rockets is about 1.25. The rockets built by Sander for the Opel experiments had a specific gravity of about 1.5, or even 1.7. While that high density improved the performance, it made the compressed powder mixture excessively brittle, much more brittle than customary. If such brittle rockets are subjected to changes of temperature in storage, or if they are transported, fine cracks which cannot possibly be detected are likely to form in the charge. When such a rocket is used, the performance will be normal until the flame reaches the crack. Then the surface of the crack will also become a burning surface, causing a sudden large increase in gas generation. In comparatively harmless cases unburned lumps of composition—those between crack and nozzle—will be thrown out. But usually the tube cannot stand the sudden increase in pressure, which may be aggravated by the blocking of the exhaust nozzle by a still unburned lump of powder.

Such cracks caused the explosions that were so frequently observed during the Opel experiments. One sudden drop of temperature, one minor act of carelessness during transportation and handling, and the rocket was unsafe. That all these things were not just academic worries is proved, in my opinion, by the refusal of the German railroads to transport those rockets. Even though the railroads were government-owned, their purpose was to provide revenue and they would not have passed up freight without good and sufficient reasons.

In addition to the possibility, or even likelihood, of such "hair cracks," as they were called, blackpowder rockets with wall-fitting charges were unsafe for still another reason. If the rocket was large, the walls, of necessity, had to be metal. And if the burning time was longer than a second or two the metal wall might transmit enough heat to ignite the powder farther forward, at a point the flame had not yet reached.

Any explosives expert who was acquainted with these problems must have thought that there should be a way out, consisting in the main of a switch from compressed blackpowder with its saltpeter-charcoal-sulfur base to a smokeless powder with a base of guncotton or nitroglycerin or both. Any ordnance officer or civilian expert was familiar with the spaghetti-like smokeless-powder tubes used as propulsive charges in artillery ammunition. These thin and fairly long tubes were not only not brittle, but actually flexible; they could be bent to a considerable extent before breaking. Powders of such type should stand rough handling and were known to be reasonably resistant to changes of temperature. Almost incidentally, such a charge would be far more powerful than even the most highly compressed blackpowder charge.

The problem was to find a smokeless-powder composition of just the proper rate of burning. That used in artillery cartridges burned much too fast; obviously the whole charge had to be consumed and converted to compressed gas before the projectile had time to leave the muzzle of the gun. For rockets there was no such hurry; in fact if the powder burned less violently the wall thickness of the tube holding the charge could be reduced.

Apparently the first man to experiment with smokeless powders was Professor Goddard. His first Smithsonian report shows that he did so. But he was interested primarily in measuring the exhaust velocity of smokeless powders to obtain a basis for calculations. Whether some arsenal worked on the same problem at the same time for the purpose of actually creating a smokeless-powder rocket, appears highly unlikely. One should expect the detailed reports on later experiments published after the Second World War to mention such early work if it had taken place. But they don't.

Therefore Friedrich Wilhelm Sander may have been the first to try his hand on such rockets. According to Max Valier, who was an eye-witness, Sander began experimentation with smokeless powders soon after the Opel experiments. But the results were discouraging; after a few seconds of smooth, if violent, burning, an explosion usually occurred. Of course I don't know what Sander did wrong, for the number of possible mistakes, each of which would lead to an explosion, is fairly large. He may have had the wrong kind of composition; he may have used wall-fitting charges which responded to heat transfer through the metal walls; he may simply have made his rockets too long, which can also cause an explosion. In any event the problem obviously was too complex for his facilities.

For as long as Sander's smokeless-powder rockets worked, they produced, according to Valier, "better than 1800 meters per second." This works out to 6000 feet per second, a figure which agrees well with the figures given for the bombardment rockets used operationally some fifteen years later.

Virtually all the military rockets of the Second World War were propelled by what are technically known as "double-base" powders. This term requires an explanation. Guncotton bears its name because it was at first expected to replace powder in guns. But whenever it was tried the gun barrel cracked. Guncotton evidently burned too fast and had to be slowed down. One method for doing that was to put finely chopped guncotton into a container with aceton. Aceton did not dissolve the guncotton, but made it swell up into a jelly. The jelly was then mixed with simple substances—charcoal was often used—then partly dried and rolled into sheets which were cut up into small squares or rhomboids. This was a "single-base" powder; the "double-base" variety was compounded first by Alfred Nobel. It received names like "cordite" and "Ballistite," and these names are still in use, although the compositions and the manufacturing processes have been changed several times since.

The two "bases" of the cordite and Ballistite powders are the two high explosives nitroglycerin and nitrocellulose (guncotton is one specific kind of nitrocellulose), and the main characteristic of the manufacturing process is the gelatinizing of the nitrocellulose by the nitroglycerin. Since nitroglycerin is not the most perfect gelatinizer imaginable this process is aided by the addition of other substances. The British explosives experts used diethyl-diphenyl-urea, which one is tempted to abbreviate as D.D.U., but the actual British trade name is Carbamite. It is not only a gelatinizer but also a fine stabilizer. Without it, the double-base powder would, after some time, become either unreliable or simply unsafe, but Carbamite neutralizes the decomposition products of nitric esters which otherwise would be annoying.

British rocket cordite consists of:

	PARTS BY WEIGHT
Nitroglycerin	**50**
Nitrocellulose	**41**
Carbamite	**9**

Its manufacture is called a "dry solventless process." Well, it is solventless, but not completely dry. The nitrocellulose pulp, which is moistened with water, is put into a tank with water where it is agitated while the necessary amount of nitroglycerin is run in. After some time the mixture is run into another tank where the Carbamite is added, and after having been agitated some more the wet pulp is pumped onto sheeting tables closely resembling those used by paper manufacturers. The result; called a "paste sheet," still contains 20 to 25 per cent of water, which is driven out by drying the sheets with hot air. The dried paste sheets, which are still just a mixture of the ingredients, are then passed through heated rollers. Gelatinization is caused by both the heat and the pressure. The gelatinized sheets are then rolled up tightly and put into heated cylinders from which they are extruded through dies under moderate heat and high pressure.

American explosives experts got their first call for a smokeless-powder rocket charge in 1940. The Ordnance Department wanted a rocket charge for accelerating a falling bomb. Airplane bombs, unless dropped from very high altitudes, do not strike the ground with as high an impact velocity as an artillery shell of the same weight. Hence, when dropped from a low altitude, they show less penetration; when dropped from a very high altitude, they are obviously less accurate. A rocket charge to push the bomb down seemed a logical method of combining the accuracy of low altitude and dive bombing with the high-impact velocity which ordinarily would require a high altitude. The rocket which was to do this was ready by late spring 1941, but the bomb for which it was designed was never used.[3]

The propelling charge in that rocket was a double-base powder consisting of about 60 per cent nitrocellulose and 40 per cent nitroglycerin, with a small amount of diphenylamine added as a stabilizer. This is similar

[3] The idea of a rocket-accelerated airplane bomb must have been "in the air" in those days, for it had occurred to me too. In early fall 1940 Schaefer and I set out to test the idea on a few examples on paper. The result was an article entitled "How about Penetration Bombs?" which made the rounds of various technical and semitechnical magazines without evoking the response we wanted. The editor of *The Scientific American* wrote curtly that it was "too far-fetched to be considered." It finally appeared in the December 1941 issue of the *U.S. Naval Institute Proceedings;* a few months later somebody in the National Inventors' Council wrote me that such an idea should have been kept secret. That was at about the time the Russians announced that rocket-accelerated bombs were used from their *Shturmovik* tank-fighter planes.

in composition to the British rocket cordite. But the American method of manufacture was entirely different, especially for Ballistite, which differs from cordite in being a "hotter" powder, with more nitroglycerin and a more potent version of nitrocellulose.

The American method, known as the solvent-extrusion method, worked as follows: The ingredients, after having been prepared separately, were compounded in the presence of a volatile solvent. The result was a thick blackish paste which was then rolled into sheets for gelatinizing. The sheets were then split lengthwise into narrower ribbons and these ribbons were extruded. Later in the war most of this was done by fully automatic machinery so that no human being was present during various stages of the manufacturing process. This solvent-extrusion process was considered much safer than the British method. It was admitted that the drying process afterward was time-consuming and even that somewhat thicker powder sticks did not dry too well in their inside portions. But the American experts felt that the British method was not the proper way to treat a high explosive. Squeezing a high explosive under heated rollers and forcing it through dies with heat and pressure did not seem exactly safe. However, the British could point to superior quality and to the fact that they had yet to blow up a factory, and in the end the dry method was adopted in the United States too.

The Germans also had a long and loving acquaintance with double-base powders and also knew that rocket charges of military usefulness would have to be composed of such powders. But when development work started in Germany it was decided not to use nitroglycerin. The reason for this decision may have been that glycerin is derived from fats and in case of a long war Germany might experience as severe a fat shortage as took place during the First World War. Whatever the reasoning, they substituted for nitroglycerin a liquid known to chemists as di-ethylene glycol dinitrate. They knew from the outset that this was less sensitive than nitroglycerin and therefore safer to handle. And when they began experimentation they discovered to their joy that it was also a better gelatinizer than nitroglycerin. The propelling charges which resulted from this different double-base powder did not differ much in performance from the Allied varieties.

In Germany as elsewhere there must have been a steady demand for bigger and bigger rocket charges, for larger projectile rockets as well as for assisted take-off for aircraft. In America this demand resulted, for the particular purpose of take-off help, in the so-called Galcit fuels which are described in Appendix 1. In Germany the so-called *Giessling Pulver* was invented. This is an interesting compound in many respects. It was manufactured by making a paste of nitrocellulose and di-ethylene glycol dinitrate, plus some diphenylamine and some of the substance the British call Carbamite. When finished and still moist this substance was crumbled and added gradually to molten TNT in a vat and the mixture stirred constantly.

The composition of the finished powder was:

	PARTS BY WEIGHT
TNT	**50–52**
Nitrocellulose	**28–30**
Di-ethylene glycol dinitrate	**17–18**
Carbamite	**½**
Diphenylamine	**½**

The whole was placed, hot, in a vacuum which removed air and water from the powder; that done it was poured into steel molds and subjected to a slow and controlled cooling lasting 24 to 48 hours. It produced charges which in sheer size far surpassed any other type; some experimental charges were 40 inches long and 21 inches in diameter.

Although the British were the first to use rocket weapons during the Second World War—small wire-trailing rockets fired from coastal vessels against dive bombers—the first war rocket to receive extensive publicity was a German model.

In 1942 the Russian news agencies began sending out pictures of a strange-looking German weapon which had been captured along the Russian front. The device had six stubby barrels which looked as if they were about 5 feet long and which were arranged in a circle very much in the manner of the cartridge chamber of the old Colt revolver. The undercarriage, as could easily be seen in the pictures, was that of the German 37-millimeter *Pak* (*Panzer Abwehr-Kanone,* anti-tank gun) with only minor modifications. The Russian caption on the back of the pictures merely read "*Novy nyemetski minenwerfer*" ("New German mine-thrower"), with date and place of capture. The news stories which augmented the pictures stated that the strange-looking object was a German rocket gun. The name given was *Nebelwerfer 41* which means "Fog (Smoke) Thrower Model 1941," which indicated that the weapon had been originally designed as a kind of chemical mortar for laying smoke screens. The Russians reported, however, that the weapon was used as a howitzer, firing high-explosive ammunition. Chemical shells for the weapon were captured later, indicating that the original purpose had not been completely forgotten. Over-all length of the shell was a little more than 40 inches, over-all weight about 80 pounds. The propelling charge, located in front of the projectile (Fig. 29, top), consisted of seven sticks of smokeless powder, each 16 inches long and 1⅝ inches in diameter with a ¼-inch hole through the center, the assembly weighing around 14 pounds. The caliber of the shells was 15 centimeters, or close to 6 inches. The Russians reported that all six firing tubes could be discharged in as many seconds, but German manuals, captured later, stated a much slower rate of fire; possibly the Russians, at the receiving end, mistook the impacts of a battery for the work of one rocket gun. The maximum range observed was somewhat above 18,000

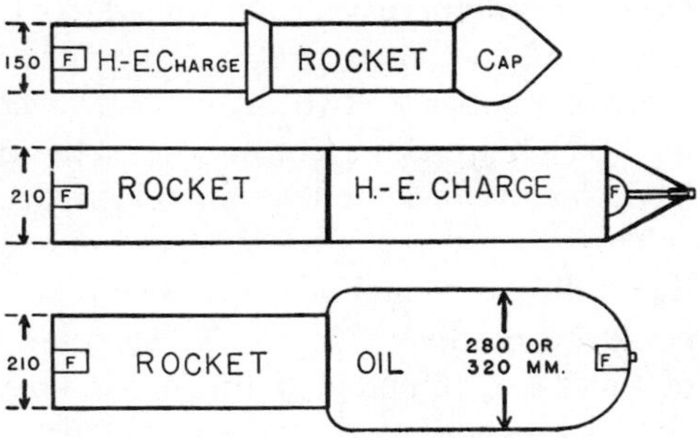

Fig. 29. German projectile rockets, Second World War (dimensions in millimeters).
Top: Projectile of Nebelwerfer 41.
Center: The larger Nebelwerfer.
Bottom: Projectile of the Wurfgerät.

feet and the accuracy of fire was fair but, of course, much inferior to artillery under similar conditions.

The main weakness of the Nebelwerfer was not the lack of accuracy, which could be compensated for by volume. The main weakness was that while the Nebelwerfer could be camouflaged while being transported and hidden when in firing position, it could not be fired without giving itself away. The flash of the rocket charge, leaping out through the open rear ends of the firing tubes, was some 40 feet long and brilliantly luminous. The rocket, in rising, burned for some 450 feet of its trajectory. And even in daytime, when the backflash was not so conspicuous for luminosity, it kicked up large dust clouds.

About a year after its first appearance the 15-centimeter Nebelwerfer was joined by a bigger variety with a caliber of 21 centimeters (about 8¼ inches) and a different design. In the larger version the propelling rocket formed the rear end of the whole projectile. Instead of having a propelling charge consisting of separate thick-walled tubes, the new type had a solid charge, which weighed 14½ pounds, was 16¼ inches long, and had a diameter of nearly 5 inches. There were eight longitudinal corrugations on the periphery of the charge and eight lengthwise holes through it, arranged in a circle, with a ninth, somewhat larger, hole through the center line. This charge was analyzed later and found to consist of:

	PER CENT BY WEIGHT
Nitrocellulose	**63**
Di-ethylene glycol dinitrate	**35**
Carbamite	**0.5–0.6**
Wax	**0.2**
Graphite by difference	**1.2**

The range of this heavier projectile was about 3000 feet more than that of the original type.

The projectors for the larger version came in several forms. One looked like the original Nebelwerfer but had only five firing tubes arranged in a circle. Then there was one in which the five tubes were placed side by side. And then there was a super type, mounted on railroad flatcars, with ten firing tubes in two layers of five each. By that time the two calibers of the Nebelwerfer were no longer the only German rocket weapons in use. They had been joined by another, named *Schweres Wurfgerät.*

The name was a revival of the old term for the medieval siege engines, the *ballistae,* which lobbed heavy stones into besieged towns. While it could be rendered simply as "launching device" the name preserves more of the flavor intended if translated literally as "heavy throwing engine."

This weapon used a 21-centimeter rocket, presumably the same rocket which propelled the projectile of the heavy Nebelwerfer in a different manner. A bulky container for a mixture of oil and gasoline was attached to the propelling rocket. That container measured 32 centimeters in diameter and held about 11 gallons of liquid. The whole looked like a huge war club of old and weighed over 200 pounds.

At first the Wurfgerät came singly, in individual packing crates which served as launching racks. The crates were stripped of their outer wrappings and propped up on anything that happened to be around, and the Wurfgerät were ready for use—huge incendiary bombs which traveled under their own power to a distance of about 2000 yards.

Some of these 32-centimeter Wurfgerät projectiles were found with a yellow cross painted over their noses. This is the German marking for mustard gas, but when the projectiles were opened by men of the Chemical Warfare detachments they were found to be oil-filled like the others.

The method of firing these rockets individually from their shipping crates had apparently proved reasonably accurate on the testing grounds, but the Germans soon found that it did not work too well under battlefield conditions. To increase the accuracy somewhat and to provide for a higher volume of fire at the same time, they then put six of these crates together, in two layers of three each, and mounted them on a gun carriage.

At about the same time they produced a smaller Wurfgerät by reducing the bulbous warhead to a diameter of 28 centimeters and filling it with high explosive.

In addition to Nebelwerfer and Wurfgerät the Germans had airborne rockets (caliber 8 centimeters) and several flare rockets (caliber 8.6 centimeters). But I'll mention only one more German type, which had a particularly interesting design. It was known as the 21-centimeter R-LG. The dimension given refers to the caliber, of course, but strangely enough the inside diameter had been chosen for the designation. The actual diameter of the projectile was 21.4 centimeters. It was developed jointly by the laboratories of the Oberkommando der Kriegsmarine (Naval Chief of Staff) and the firm of Rheinmetall-Borsig in Düsseldorf, which was also responsible for several other interesting developments. The "R" stands for "rocket"; "LG" stands for "*Leuchtgeschoss*" ("luminous projectile" in literal translation), meaning flare. The whole looked like a big artillery shell 3½ feet long. The propellant was a single thick-walled tube, with (in round figures) a length of 20 inches, an outside diameter of 8 inches, and an inside diameter of 4 inches. Inside this wide hole was a metal tube inside which the illuminant and its parachute were placed. Maximum vertical range was 16,000 feet; maximum horizontal range was determined as about 25,000 feet. Presumably its use as a high-explosive shell had been considered. The development was about finished when Germany capitulated, but too late to put into production.

When Nebelwerfer and Wurfgerät were fired against Russian positions, the Russians answered with rocket fire. But while we know everything about the German types, we know very little about the Russian war rockets except for one case, the rockets which were fired in enormous quantities against the encircled army of General von Paulus. The launchers used then were of two types: one, strongly reminiscent of the old Congreve launchers, looking like very wide stepladders placed on the ground; the other, truck-mounted projectors. The projectors were of Russian manufacture. The trucks were Studebakers. And the rockets had mostly been made in Tennessee.

A real Russian type was the boxlike projector, mounted on a gun carriage, which was often called the Stalin Organ. It consisted of 48 firing tubes holding rockets of 8.2 centimeters caliber (3¼ inches) which could be fired in a rapid "ripple." After the war the Russians went in for mass production of 13.2-centimeter and 30.0-centimeter rockets, but anybody who knows details about them is more tight-lipped about this knowledge than an average earthworm.

In Japan rocket development began as early as 1935 but proceeded in a slow and uncertain fashion. All the navy rocket work was under the direction of a Lieutenant-Commander Kumao Hino. The general impression one receives from published reports and interrogations is that the higher echelons did not precisely want to stifle rocket development but that their interest did not go beyond that. Appropriations were small, priorities low, and when in spite of these handicaps the researchers accomplished something, it was not put into production. The Japanese solid fuel consisted of

	PER CENT BY WEIGHT
Nitroglycerin	27
Mono-nitro naphthaline	7
Nitrocellulose	60
Carbamite	3
Potassium sulfate	3

The last-named had the purpose of slowing the rate of burning. By the time the war had already begun to take a turn for the worse, from the

Japanese point of view, somebody discovered that enormous numbers of 250-kilogram (550-pound) demolition airplane bombs had been manufactured, which had no aircraft to carry them. These surplus bombs were converted into rocket projectiles by attaching a pushing rocket unit at their tail end. The projectiles were launched from an inclined trough of wood or metal and attained a maximum range of about 16,000 feet. Similar "conversions" were then made with other airplane bombs, and even with artillery shells. Some of the latter seem to have been improvised "on the spot" on various islands. (See table of Japanese rocket ordnance in Appendix 2.)

The first country on the Allied side in the Second World War to go in for military rocket research was Great Britain. Most of the credit for what was accomplished there is due to Sir Alwyn D. Crow, Head of Technical Services, British Supply Office. Naturally very little definite information came out while the war was going on. It became known through scattered news releases and publications that the program, or part of it, was code-named "Development U.P.," the letters "U.P." standing for "unrotated projectile"; that some tests took place on Jamaica in the West Indies; and that rocket batteries defended London along with gun batteries. That latter was not always appreciated by Londoners. When G. Geoffrey Smith, the editor of *Flight* and of *Aircraft Production,* was in New York, he said to me jokingly across the dinner table that absolutely the first job of postwar rocket engineering must be to make rockets noiseless—or at least to make the noise one that did *not* sound like a falling bomb.

Much of what did take place was told by Sir Alwyn D. Crow in a lecture (November 21, 1947) to the Institution of Mechanical Engineers, a copy of which was furnished to me by the British Interplanetary Society. The following quotations are from that lecture.

> The reports that were received by our Government in 1934 about German activities led to the serious consideration by the War Office of the desirability of developing high-velocity rockets in this country [England]. The first meeting to discuss the subject was called in December 1934; and in April 1935 the Research Department at Woolwich Arsenal was asked to put forward a program. . . . (Crow)

Several possible uses were discussed, antiaircraft rockets, long-range rockets, air-to-air rockets, and take-off help for aircraft. The consensus seems to have been that one should try for antiaircraft rockets first and it was decided to strive for the equivalent in destructive power of the British 3-inch A.A. gun. A 2-inch rocket seemed to be about right for this purpose and development began; experimental models were built and tested.

> Early experiments during the spring and summer of 1937 were encouraging, and good and reliable performance was obtained in firing trials; but with the onset of colder weather in the winter of 1937 it became apparent that the development of a satisfactory material for the plastic surround had by no means been achieved. . . . (Crow)

The term "plastic surround" refers to a plastic lining of the rocket tube, intended to save weight.

> About a year after the start of the development of the 2-inch rocket a demand was received for a larger and more powerful weapon to have a performance approximating that of the new 3.7-inch antiaircraft gun, which was just on the point of coming into the service. This led to the investigation of a 3-inch caliber rocket. . . . A design that was regarded as generally satisfactory for extended trial was evolved by the end of the summer of 1938; and . . . trials were planned to take place the following spring and summer. The Munich crisis precipitated action, and it appeared inadvisable to delay the trials. Consequently, as winter weather conditions in England are unsuitable for work of this kind, a search was made for a suitable site in a more favorable climate. Such a site was found in Jamaica and some 2500 rounds were fired there during the winter of 1938–39 in an extensive series of ballistic trials.
>
> The results were unacceptable to the General Staff; since the performance fell short of that specified, and the accuracy of shooting compared unfavorably with the new 3.7-inch antiaircraft gun. Nevertheless, the development work was continued up to the outbreak of the war in the hope that the accuracy could be improved; but when the war started the priority accorded to the work was drastically reduced. (Crow)

Some four months later it was decided that even a not perfectly accurate weapon was welcome and word was given out to get the 3-inch rocket into production. A projector for the rockets was designed too. "Several thousands of these projectors were made during 1940–41 for use in the defense of vital targets such as key factories and rail heads. In November 1940 a twin projector was designed on the general lines of the single projector." Multiple projectors came later and the massed batteries of projectors fired salvos up to 128 rounds. Another later step was the development of a 5-inch rocket for ground use; the specification read that the rocket should be able to carry a warhead weighing 30 pounds over ranges varying from 2 to 4 miles.

In 1940, as has been mentioned, the United States entered into the picture of military rocket research. And while the American researchers worked independently, they were made acquainted with the British models and could avoid any false starts that had been eliminated at Woolwich. The story of American rocket research has already been told by the men best qualified to tell it—the ones who directed and did the work—and the interested reader can find it in *Rockets, Guns, and Targets* edited by John E. Burchard (see Bibliography). I shall restrict myself, therefore, to technological principles and their background.

Obviously the invention of a better powder charge is not the whole story. That powder charge also has to work properly. From the point of view of propulsion it was greatly desirable to get even thrust, something that the old blackpowder rocket never had. In a blackpowder rocket the thrust would climb almost instantly to a certain value, say 15 pounds for a middle-sized rocket, stay at that value for ¼ second or so, then fall off fast to, say, 1 pound, and stay at that value for a second or two. What designers wanted was a rocket which would reach a certain thrust (preferably the thrust in the specifications) quickly, maintain that thrust for a while, and then stop. Plotted as a curve (thrust versus time), the resulting figure looked like a profile of a long flat building with sloping sides, and it was consequently termed "flat-top curve."

Easy to say, but not quite so easy to do. Obviously the desired thrust curve could result only if the rocket exhaust stayed the same, both in exhaust velocity and in volume (mass), throughout the burning period. It was therefore necessary to get a piece of powder that would burn evenly. To understand this problem, imagine that your lump of powder has the shape of a ball and is burning at the surface only. But, as the ball is being consumed, its surface gets smaller all the time. Hence the amount of gas generated gets smaller too; hence the thrust curve would be a slope. In reality the problem is complicated by the fact that the burning takes place in a closed space with only one exit, the exhaust nozzle. And as the pressure rises in the container the rate of burning changes.

One way out, often used, was to give the powder the shape of a thick-walled tube. Such a tube would burn from the outside in, with lessening burning surface, and from the inside out, with increasing burning surface. One could hope that the reduction of the outside surface and the increase of the inside surface would counterbalance each other so that the total burning surface would remain more or less the same through the whole process. But in order to do that, one could not use a wall-fitting charge: the powder stick had to be suspended so that it could burn without any restriction (Fig. 30).

The British experimenters had learned that early in their career; they referred to it as a "loose charge." American researchers thought that sounded as if something were loose that shouldn't be; they called it an "unrestricted burning charge" instead. An explanation is also needed for the terms "grain," "web," and "trap." A "grain," to a rocket technician, is a piece of powder charge. Shape and size do not matter; there were "grains" which were 40 inches long and weighed 18 ounces per inch of their length. But the diameter of the grain does not tell all; with those thick-walled tubular grains, for example, the wall thickness of the tube is at least as important as its diameter. The wall thickness is called the "web"; the term used refers to the thickness of the powder at its thickest point. And a "trap" is what holds the grain in place.

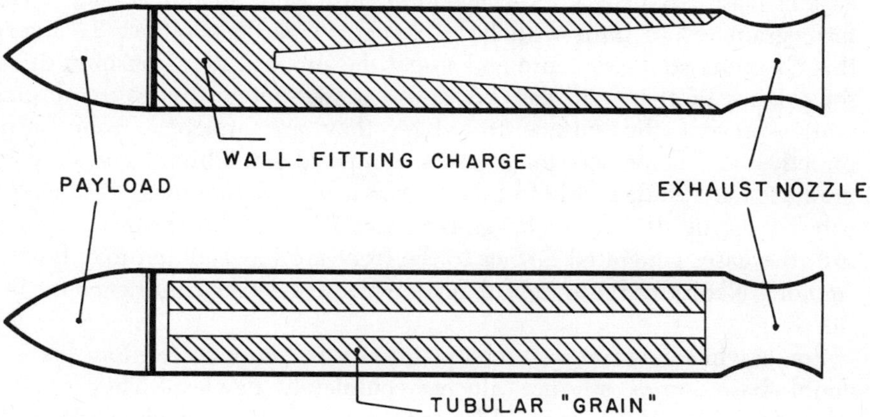

Fig. 30. Solid-fuel rockets. Top: So-called "restricted burning" or wall-fitting charge, customary for blackpowder rockets. Bottom: "Unrestricted burning" charge for smokeless powders. The latter has less variation of thrust while burning.

A fine example, in simplicity as well as in performance, of a modern solid-fuel war rocket is the 5-inch High Velocity Aircraft Rocket (HVAR), better known as "Holy Moses." The diagram (Fig. 31) shows the three main parts: warhead in front, then the rocket tube, called "rocket motor" (slightly confusing since it is not the same as the rocket motor of a liquid-fuel rocket, but the term is now established), and the tail assembly with stabilizing fins.

The powder grain in this rocket is what is known as the "cruciform grain," a shape which has the advantages of great web thickness and

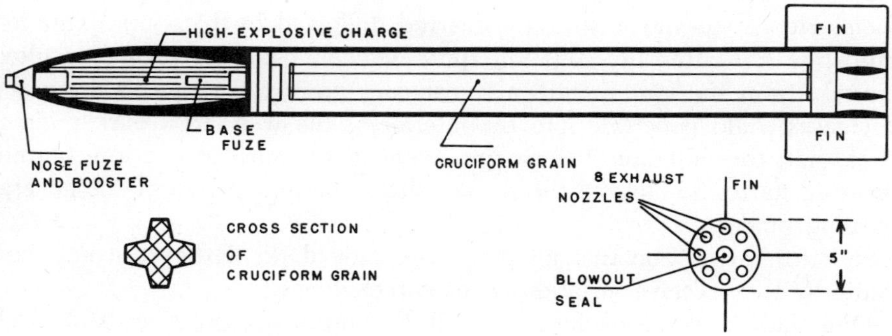

Fig. 31. Five-inch HVAR, called "Holy Moses." The "blow-out seal" is actually an additional exhaust nozzle which, however, functions as such only when the internal pressure rises beyond a critical point.

ease in mass production. Even if nothing else were done to it such a cruciform grain would burn with little change in gas generation. To improve this characteristic even more and to get the proper rate of burning desired, some areas of the grain can be, and are, "inhibited" by strips of plastic which prevent the surface to which they are attached from burning directly. The flame has to progress around the inhibitor. For very long grains it may be desirable to inhibit only a part of the grain, that near the exhaust nozzle. If too much gas is generated near the exhaust nozzle too fast, the gases generated farther to the front are blocked and may burst the "motor." Naturally the nozzle diameter has to be taken into account in all this.

For a while researchers were plagued by a curious trouble. Naturally double-base grains are not always completely flawless. They may, for example, have entrapped bubbles, which produce the same result as does a "hair crack" in a blackpowder charge. To inspect for bubbles was not easy, especially since the stabilizing compound that was used caused the powder to darken as it aged. Great enthusiasm, therefore, greeted the discovery that Carbamite (usually called "Centralite" in America) could produce translucent grains. These grains were easier to inspect—but as often as not they blew up the motor. The dark grains, which might contain the dreaded bubbles, caused fewer explosions than did the translucent grains which had been inspected. More careful investigation showed that something took place which was called "termite fissuring," because partially burned grains looked as if they had been eaten by termites.

A special research program had to be scheduled to find out just what did happen. It was this: When the grain burned, the gases generated were not only hot but also luminous. The radiation from these gases could penetrate the translucent grain and it would then be absorbed by tiny, in fact microscopic, specks of dirt which were dark in color. Absorbing the radiation, these specks grew hot, hot enough to ignite the powder near them. Hence "termite fissuring" occurred, followed by blow-ups. This improbable, but after investigation quite logical, chain of events resulted entirely from the fact that in a translucent grain the radiation from the hot gases could penetrate into the interior of the web. In a dark grain it could not; the radiation was stopped right at the surface and any termite fissuring started so close to the surface that it became a part of the general burning process.

It is for this reason that all grains are now dark; if they are not dark naturally they receive an addition of carbon black.

One additional headache, after all the problems of grain size, web thickness, nozzle diameter, and so forth had been solved was the problem of stability in flight. Earlier military practice had shown that stability could be accomplished in two ways. The old arrow demonstrated the one, the newer rifle bullet the other. With reference to rockets the two methods came to be called "fin-stabilized" and "spin-stabilized." Fin-stabilization was in itself simple. It required a heavy warhead at one end and fins at the other. And the fins could be foldable, if desired, for fitting into a launching tube. But fin-stabilization did not make for very accurate fire, except from aircraft. Aircraft rockets had the advantage that they were moving with speeds of 300 miles per hour and more even before firing, because they were carried externally. Fin-stabilized rockets used as airplane artillery against ground targets were quite accurate. But used as artillery from the ground against another ground target they left much to be desired and had to be considered as an area weapon. Apparently the fins could not serve efficiently during the early part of the burning time.

Spin-stabilization, first used by Hale during the nineteenth century, did not depend on the velocity of the rocket, since it was caused by the exhaust blast. For causing a spin, two methods suggested themselves. One was Hale's, which used vanes in the exhaust blast. The other was the use of several exhaust nozzles, arranged in a circle and placed at a slight slant (that is what the Germans did with their Nebelwerfer projectile). The latter was presumably the better method, since vanes in the exhaust would cause a loss of power. When discussion had progressed to that point, along came the questions: how much spin would be required for good stabilization? how many revolutions per minute? Slanted nozzles obviously involved a loss of power too; hence the rocket should be given only as much as needed for stability, not more. But how much was that?

Nobody knew the answer.

Accordingly Dr. C. N. Hickman, Chairman of Section H, Division A, of NDRC, and entrusted with the development of rocket ordnance, decided that the effect of various amounts of spin on rockets should be investigated. The method was suggested by R. F. Mallina, who was working on rocket developments in the Bell Telephone Laboratories at that time. His idea was to take a rocket without any stabilizing features and prepare to fire it from a tubular launcher, but, prior to firing, rotate the tube of the launcher, which made it possible to test the same rocket with varying amounts of spin. The suggestion was immediately accepted and a special launcher built. It consisted of a rocket-launching tube, mounted in large ball bearings which sat in a larger stationary tube. The whole could be elevated and traversed like a gun barrel. Power for spinning was furnished by a 1½-horsepower electric motor mounted below the outer tube at the rear end, driving the inner tube by means of a belt. There were three pulleys, so that the launching tube could be driven at 800, 1400 and 2400 rpm.

An old artilleryman might have felt that a gun with a wildly rotating barrel was mildly silly, but as a testing device it was a great success.

It was found that a considerable decrease in dispersion was obtained with very moderate spins, and that the speed was not critical—essentially the same improvement was obtained with a wide range of spin. Without spin, the dispersion for the standardized rocket was 39 mils, that is, the rocket would deviate 39 feet in 1000 feet of travel, while with spins of 800, 1400, and 2400 rpm the dispersion was reduced to 13, 11, and 9 mils, respectively. To test the effect of spin on rockets which normally have very bad dispersions, 25 such rockets were fired from the projector while rotating at a speed of 2400 rpm. The dispersion was 13.6 mils. When these rockets were fired from an 11-foot non-rotating tube, the dispersion was 78 mils. The standardized rocket fired from the 11-foot tube gave 16.6 mils dispersion. The inferior rocket would have had over 200 mils dispersion if it had been fired from the 4-foot tube.

(*Bell Laboratories Record,* vol. XXIV, no. 5, May 1946)

But most of the members of what was later called the "spinner family" of United States ordnance rockets came too late for active service; only a few of them were used operationally (see Appendix 3). The vast majority of American war rockets were fin-stabilized, among them the most publicized, the projectile of the Launcher, Rocket, M-1 of official registry, known the world over under its nickname "bazooka." The early bazooka rockets suffered the usual design troubles. At first the propelling charge ripped the rocket open when fired on hot days. The charge was reduced and worked fine in hot and warm weather, but failed to perform on cold days. When a charge that worked at all temperatures was finally found, there were complaints that the launching tube was too long and unwieldy in forests and difficult terrain. But the firing tube had to have that length because the whole propelling charge had to be burned before the rocket left the tube, else the gunner would get a backblast in his face. That particular problem was later solved very simply by a folding firing tube.

The first operational use of the bazooka was in North Africa.

When the existence of this weapon was revealed for the first time, early in 1943, by Major General L. H. Campbell, Jr., and it was explained that this comparatively small rocket weighing only a few pounds could destroy a tank, most people jumped to the conclusion that the bazooka was so effective because of the high velocity of the rocket projectile. Actually the bazooka rocket moves very slowly for a projectile; it can be seen all the way from launching tube to target if you are in a favorable position for watching. The secret of the high destructiveness had nothing at all to do with the fact that the bazooka was rocket-propelled. The secret was hidden in the pointed warhead of the rocket. It was not a special kind of explosive either, just a special application of known explosives in a form which is called "shaped charge," or sometimes "Munroe charge."

As the latter name indicates it was discovered by a man named Munroe—the American explosives expert Professor Charles E. Munroe. While experimenting with explosives in about 1887, Professor Munroe found something new and startling. One of the samples of high explosives which he used was a disk of guncotton, with the letters U.S.N. and the date 1884 indented on its face for identification of ownership and year of manufacture. Professor Munroe exploded the guncotton disk near a slab of heavy armor plate. The damage to the armor plate was, as expected, negligible. But "U.S.N.1884" was now indented into the metal! Nothing like that had ever been observed before. There was only one factor which could account for this strange phenomenon, namely, that the explosive charge had not fitted snugly against the metal in these particular spots, for the indentation in the guncotton naturally created a short air space between metal and explosive. But charges had been exploded at some distance before and nothing like that had happened.

Professor Munroe reasoned that the combination of a short air space and snugly fitting high explosive around it was probably responsible. To test this idea he took a bundle of dynamite sticks and tied them together tightly, with the center sticks pulled out for an inch or so. That charge promptly punched a hole through the wall of a burglar-proof bank safe. Professor Munroe saw no reason why he should keep this discovery to himself. He wrote a few articles about it, in 1888, and other explosives experts began to call it the "Munroe effect." They soon agreed that it was in reality a focusing effect. Seen from a distance the explosion of such a charge was the same as any other: the explosive force expended itself in all directions. But inside the cavity something special took place; *there* the gases liberated by the explosion impinged upon each other, resulting in a powerful punch in one direction (Fig. 32).

Since the "Munroe effect" was neither secret nor hard to produce one would expect that the armament experts of all countries would go after it with diligence and thoroughness. The fact is that nobody did. The only group that did spend time and money on investigating it was a German manufacturing concern specializing in high explosives, the well-known

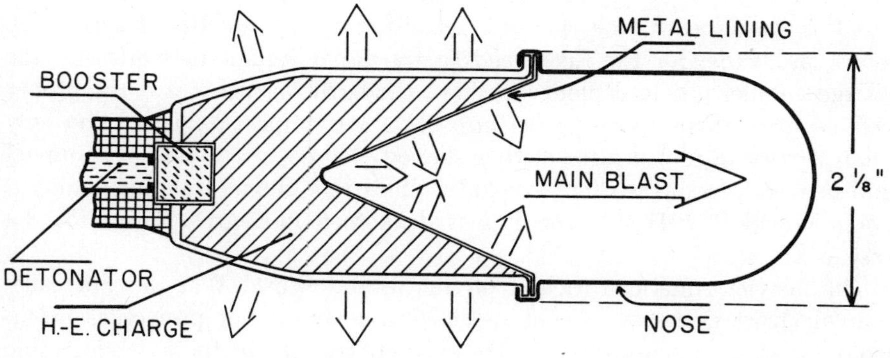

Fig. 32. Shaped charge or Munroe charge. The example of shaped charges shown is the grenade M9A1 (U.S.). The outline arrows show the directions of the explosive force. This force resulted in a blast directed ahead which carried the vaporized metal layer with it.

Westfälisch-Anhaltische Sprengstoff Aktiengesellschaft. This firm was interested because much of its business was the manufacture of safety explosives for mining. In 1912 it took out two patents for shaped charges to be used in ore mines. The charges were to punch holes into the rock preliminary to blasting, replacing, where feasible, the rock drill. Shaped charges are known to have been used underground in Germany and in Sweden all through World War I.

Military research on the shaped charge apparently did not get under way until just before or early in World War II. The most important innovation the military engineers turned out was a thin metal lining for the cavity. If the Munroe effect worked by shooting a high-intensity jet of hot gases in one direction, it was quite obvious that the penetrating power of that jet would be increased if it carried something more substantial along with it. Sheet metal lining the cavity, it was reasoned, would be torn into small fragments or vaporized, or probably both, by the explosion, and the metal particles carried along by the blast.

This reasoning was justified by experiment. The bazooka rocket, as a result of these findings, carried a shaped charge with sheet-steel liner. Meanwhile the laboratories tested other metals for liners. Lead, as was to be expected, produced a very shallow penetration into the target. Zinc, which has a higher melting point than lead, gave a much better result. But aluminum, which has a higher melting point than zinc, did not produce a correspondingly higher penetration. In fact it made shallower holes than zinc did.[4] Presumably there is a metal or alloy which has just the right combination of melting point, strength, toughness, and particle size for maximum effect. Naturally its name cannot be found in open literature.

Now the Munroe effect depends not only upon a cavity in the explosive, helped along by a metal liner, but also upon a certain distance between charge and target at the instant of explosion, a distance of a few inches. For this reason a Munroe charge is usually not very effective if carried by a fast projectile, for the fuze needs a fractional instant to work and the charge another one to explode. In short, the bazooka rocket was the proper vehicle for a shaped charge, *because* of its low impact velocity. The only other American rocket fitted with a shaped charge—not counting improved and heavier versions of the bazooka itself—is the "Ram." But the Ram was not a World War II weapon; it was developed in a great hurry for the war in Korea.

The heavier American rockets of the Second World War did not carry shaped charges; they were not specialized weapons but projectiles calculated for mass expenditure, a mass expenditure which, for a while, came to 120 million dollars per month. The 4.5-inch rocket was more versatile

[4] See "Explosives with Lined Cavities" by Garrett Birkhoff, Duncan P. MacDougall, Emerson M. Pugh, and Sir Geoffrey Taylor, *Journal of Applied Physics,* June 1948.

than the 7.2-inch type. The 4.5-inch, of uniform thickness along its whole length, weighed 38 pounds and had about the same destructive power as a 105-millimeter howitzer shell. Still, it could be used as a one-man weapon. In one version a single rocket came in a shipping tube which was also the launching tube, with camera-style tripod attached, the tube and tripod being considered expendable. The whole weighed 50 pounds, not too much for one man to carry.

Though it could be used singly and on the ground, the 4.5 inch rocket excelled in two other environments—in the air as flying artillery and in amphibious warfare as an area weapon. In spite of much ingenuity and a number of brilliant ideas, large-caliber airborne artillery had never really come to the fore. No matter how carefully the recoil of a gun was absorbed it was still there and required all kinds of modifications of the airplane. The true recoil-less guns were still in their infancy, and even if they had been well developed they would have created their own set of problems. The 4.5-inch rocket eliminated most of these problems. There was no recoil acting on the airplane. There was no heavy gun barrel to be carried. And even the drag of the launching tubes (mounted under the wings) could be eliminated after the rockets had been fired, by jettisoning the launching tubes and improving the "get-away speed" of the plane. The airborne 4.5-inch became the typical weapon for hit-and-run warfare. Single planes could penetrate into enemy territory and shoot up trains, ruin bridges, attack traffic on inland waterways, or spot specific targets.

This was one feature of the new weapon.

The other feature grew out of the fact that it is possible to launch very large numbers of rockets very nearly simultaneously. Both the 4.5-inch and the 7.2-inch rockets were stacked into massive batteries on the decks of special "rocket boats," rack after rack of self-propelled projectiles which could be fired electrically from a safe enclosure below deck. One such rocket boat, in the space of a very few minutes, could throw as much steel and high explosive as the turrets of three battleships. They produced a literal hail of projectiles which, while it lasted, made any counter-action impossible. Mass expenditure made breakthroughs and landings possible—the invasion of southern France, for example, took place under a hail of 40,000 rockets.

On the ground tanks substituted for the rocket boats of the amphibious forces. Sixty projector tubes for 4.5-inch rockets, in four tiers, were mounted over the turret of an M-4 General Sherman tank in an organ-pipe-like assembly that became known as "Calliope." The projector assembly turned with the turret on top of which it was mounted, and a hinged rod, connecting Calliope with the 75-millimeter turret gun, made it possible to elevate the projector tubes by elevating the tank gun. A special electrical firing arrangement, developed by Western Electric, enabled the gunner to fire the rockets in rapid sequence but without igniting two adjacent tubes simultaneously or in quick succession, so that the exhaust from one rocket would not disturb the trajectory of the other.

A secret device, secret all through the war, was the Anti-Submarine Projector, Mark 10, known as the "hedgehog." It was originally a British invention, but adopted and improved by the United States Navy. Its full charge was 24 heavy rockets, which could be fired in 2½ seconds. They fell in an elliptical pattern in the area which marked the approximate location of a submarine and sank immediately, heads down. But their charges were not conventional "depth charges." They exploded only on contact, not when they reached a specific depth. The sound of an underwater explosion, therefore, was definite proof that a submarine had been hit.

The largest American rocket happened to be airborne, however, being intended for use as siege artillery against targets out of artillery range. "Tiny Tim," as it was nicknamed, did not look like a rocket but rather like an airborne naval torpedo. Its over-all length was 10 feet 3 inches, its diameter 11.75 inches, and its over-all weight 1284 pounds. The propelling charge consisted of four grains of the largest cruciform type then available, with a combined weight of 146 pounds. The warhead of Tiny Tim weighed 590 pounds, holding a charge of 150 pounds of TNT, equivalent to the charge of the projectile of a very heavy siege gun.

When "Tiny Tim" was finished, there was the problem of launching it. If the rocket had been placed in a launching tube of proper size near the airplane's fuselage or wings, not only would the tube have caused a lot of drag while the rocket was in it (after firing the tube could have been jettisoned) but it also appeared likely that the backblast from a rocket that size would damage the plane. On a bomber the rocket could be carried internally in the bomb bay in an extensible launching gear which would produce drag only for the short time it was extended out of the bomb bay. The first experimental air firings of "Tiny Tim" were accomplished by such a device, but for fighter planes another method had to be evolved.

The method found was the lanyard drop; the externally carried rocket falls free of the plane, connected only by a lanyard which, when taut, fires the rocket.

In one of the earliest tests (late in August 1944) the event which the designers had feared actually occurred. Shortly after firing a Tiny Tim, the plane which had fired it nosed over into a steep dive and crashed, killing the pilot, Lieutenant John Murray Armitage, after whom Armitage Field at Inyokern rocket-testing station in California is now named. Investigation of the wreckage seemed to indicate that the tail structure of the plane had been damaged badly. The damage had been caused by a shock wave, apparently not from the propelling charge, but from the igniter of the propelling charge. The recommendations were to reduce the

I. Hermann Ganswindt's spaceship. The two cylindrical containers right and left of the bell-shaped explosion chamber were to hold the supply of explosion "pills." The long passenger cabin had a center well to permit passage of the exhaust.

II. Spaceship, as conceived by Hermann Oberth, as it is leaving the atmosphere. (From the Fritz Lang film *Frau im Mond*.)

III. Upper portion of the spaceship built for the film *Frau im Mond*. Its compartments housed parachute, pilot cabin, passenger cabin, storage room, and propulsion mechanism, including fuel pumps and exhaust nozzles. The model was later loaned to the VfR, disappeared during World War II, and must be considered lost.

IV. Spaceship leaving its hangar. (A scene from the Fritz Lang film *Frau im Mond*.)

V. Max Valier with his rocket sled on the ice of Starnberger Lake (1928).

VI. Reinhold Tiling, with his winged rockets, during a demonstration at Hannover, Germany (1932).

VII. A test run of the "Aepyornis Egg." Instead of producing a thrust of about 130 pounds, as expected, it did not quite reach 100 pounds.

VIII. Alcohol-oxygen rocket motor of the VfR, tested in 1932. The photograph was taken from the observation dugout, about 25 feet from the test-stand assembly. The thrust of the rocket blast was just above 130 pounds, and test runs were limited to 30 seconds because of the capacity of the oxygen tank. (See Fig. 27C.)

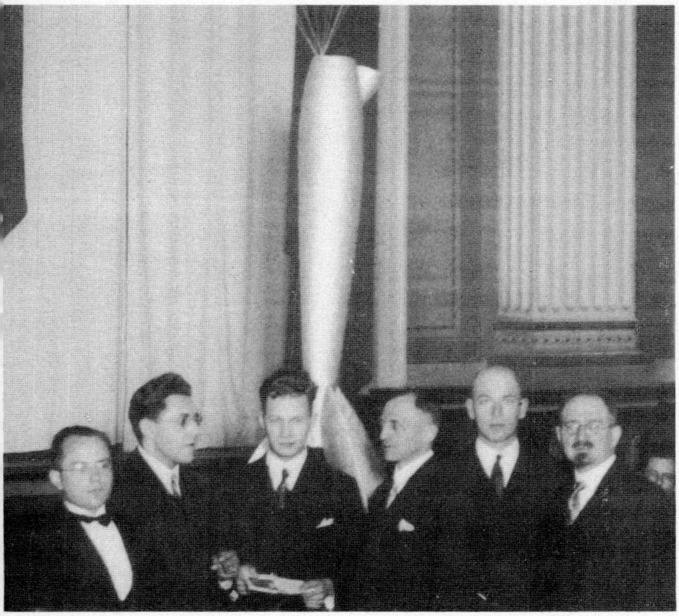

IX. First public photograph of Wernher von Braun (at meeting of Verein für Raumschiffahrt in Berlin, in the spring of 1931). Left to right: Johannes Winkler (then president of the society), Willy Ley, Wernher von Braun (holding books), Rudolf Nebel, Max Valier, Erich Wurm (patent attorney of the society).

X. Last experiment of the VfR: One of the tall repulsors built at the Raketenflugplatz ready for an early morning ascent from a raft floating on Schwielow Lake near Berlin, September 18, 1933. The motorboat is a patrol vessel of the river police.

XI. Dr. Robert H. Goddard in his laboratory, with one of the rockets he demonstrated in 1932, presumably the one which climbed to an altitude of 7500 feet. It was launched from a tower erected at Mescalero Ranch, near Roswell, N. M. (*Science Service*)

XII. One of the Navy's "rocket ships" bombarding Peleliu Island on the same morning that was D-Day in Europe. *(Official U.S. Navy Photo)*

XIII. Test missile Thiamat of the National Advisory Committee for Aeronautics during takeoff (1946). *(NACA Photo)*

XIV. V-2 rocket ready for road transport, resting on its Meillerwagen at White Sands, N.M., Proving Ground. (*U.S. Army Photo*)

XV. V-2 rocket being placed on the firing table. Note the blast-deflecting cone under the firing table. At this moment the rocket is still without fuel. The graphite vanes will be attached after it has been set up on the firing table. The Organ Mountains form the backdrop. (*U.S. Army Photo*)

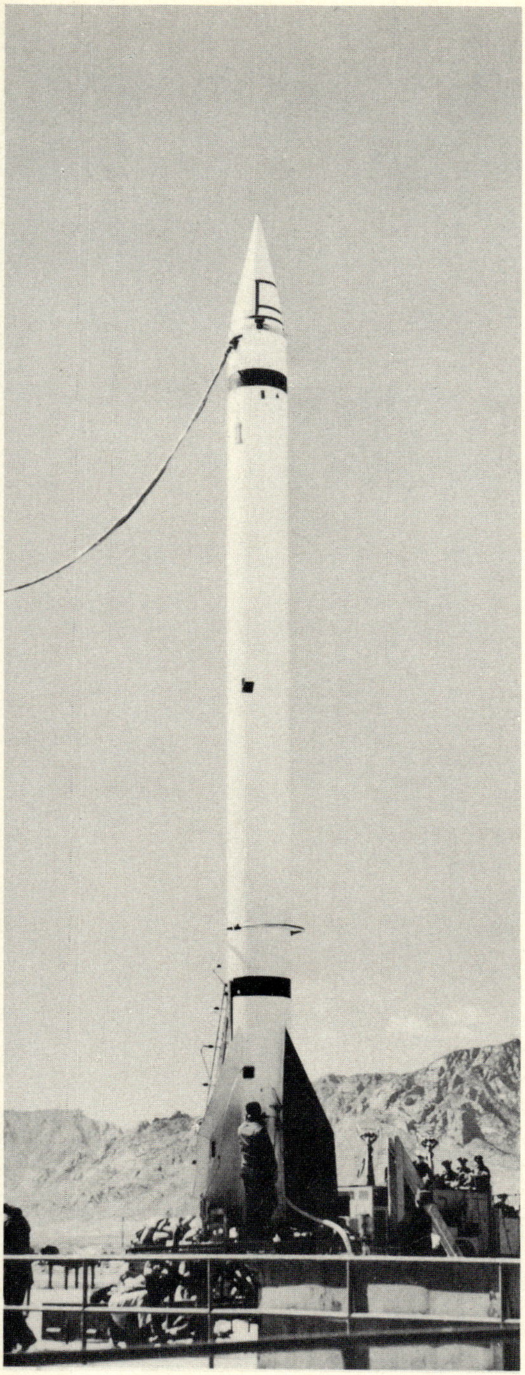

XVI. Viking 1, at the launching site of White Sands Proving Ground, at X – 30 minutes. The photograph was taken May 3, 1949. (*Courtesy Glenn L. Martin* Co.)

XVII. V-2 rocket carrying WAC-Corporal—the historic shot of February 24, 1949, which carried to 250 miles. Separation of the two rockets took place 20 miles up.

XVIII. The first American liquid-fuel rocket of the postwar period, the WAC-Corporal, ready for shipment to White Sands. (*Douglas Aircraft Photo*)

XIX. Aerobee rocket, with booster, on its transport cart. The rail on which the rocket rests is hinged in the back and serves to load the rocket into the launching tower. This, however, is not a hydraulic system as was the Meillerwagen. The loading is accomplished by attaching a wire rope to the front end of the rail and hoisting it from the tower. (*Aerojet General Corp. Photo*)

XX (above). The earth as seen through the camera of an Aerobee rocket, 57 miles up. (*Official U.S. Navy—Johns Hopkins APL Photo*)

XXI (left). Aerobee rocket rising from the deck of the U.S.S. *Norton Sound,* a converted seaplane tender, in the Pacific, near the magnetic equator. This is one of a sequence of firings made early in 1949; the rocket reached an altitude of 65 miles above sea level. Later a Viking rocket was fired from the same ship. (*Official Photo, N.M.E.*)

XXII. Take-off of a B-47 jet bomber with solid-fuel Aerojet JATO units. Each unit developed 1000 pounds thrust; thirty-three of them were used per take-off. (*Aerojet General Corp. Photo*)

XXIII. Take-off of a B-47 jet bomber, with the aid of liquid-fuel take-off unit, type YLR45AJ-1, an integral installation. Exhaust nozzles retract into fuselage after take-off, like the landing gear. (*Aerojet General Corp. Photo*)

XXIV. Hermes A-1 test missile no. 4., 1 ½ seconds after take-off, February 8, 1951. *(U.S. Army Ordnance Photo)*

XXV. Nike antiaircraft missile (operational) on its launcher, raised to firing position. *(U.S. Army Ordnance Photo)*

XXVI. Large solid-fuel rocket unit for missiles, called the Pusher, at its factory in McGregor, Texas.
(Phillips Petroleum Co. Photo)

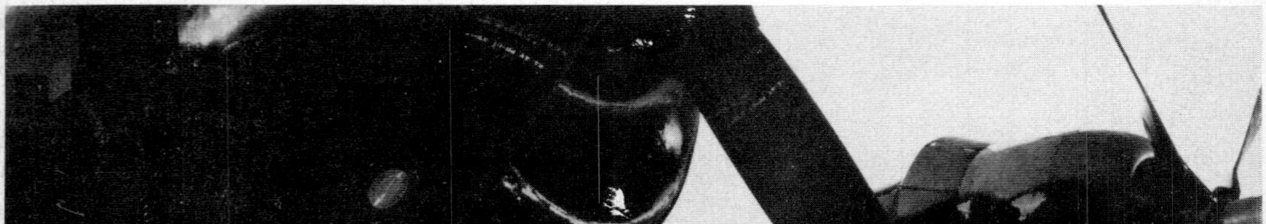

XXVII. The rocket airplane X-1 nestling under the wing of a B-29 bomber at Muroc Air Base, California. In order to prolong flying time, the X-1 was carried aloft by the B-29 so that no fuel would be needed for initial take-off.
(U.S. Air Force Photo)

XXVIII. Take-off of an early (1953) two-stage Nike antiaircraft missile. Picture was taken at White Sands Proving Ground. (*U.S. Army Photo*)

XXIX. Two-stage sounding rocket DAN prior to take-off. Booster is the same as that of the Nike. Upper stage reached an altitude of 67.4 miles (1955). (*NACA Photo*)

XXX. (above). Large liquid-fuel rocket motor being test-fired in Rocketdyne's testing area in the Santa Susana Mountains near Los Angeles. (*Rocketdyne Photo*)

XXXI. (left). Test firing of a large liquid-fuel rocket motor on a test stand at the White Sands Proving Ground. (*U.S. Army Ordnance Photo*) igniter considerably and also to increase the length of the lanyard. Since then nothing of the sort has happened.

During the Second World War, Tiny Tim was used on only one occasion, at Okinawa. Its effectiveness could not be ascertained then, because Okinawa was bombarded with such a multitude of weapons that the effect of any single kind disappeared in the general destruction. For that reason, Tiny Tim still had to be judged by testing-ground performance until it saw action in Korea.

By that time another development was just beginning to bear fruit. The reason the propelling charge of Tiny Tim consisted of several extruded "grains" was that single grains of the necessary size did not exist then. But everybody was in the market for single charges as large as would have been needed for Tiny Tim, or preferably larger. Antiaircraft rockets were under development; they needed a booster to provide them with an initial push for take-off; and that push should be as powerful as possible, which implied that the charge should be as large as possible. There were missiles under development which, once in flight, flew like airplanes by means of a jet engine. But to throw these missiles off the launcher and to start them on their way required a powerful rocket booster, or else it would be necessary to rely on catapults, which are expensive, clumsy, and not at all mobile. Moreover, the take-off rocket—its story is told in detail in Appendix 1—had become a standard item by then. But the available take-off rockets were comparatively small. Two, or sometimes four, of them could boost a fighter plane at take-off, but to speed a heavy bomber into the air would require scores of them. Logically, then, the aviation experts and manufacturers were asking for heavy and really powerful boosters as urgently as were the missile designers and the men who worried the antiaircraft rockets into shape.

The fuel chemists, of course, knew all the possibilities within reach. The fuel proper—that is, the substance which would be burned, which was charcoal and sulfur in the old blackpowder—was really the second question. The first one was the proper oxidizer, the substance providing the oxygen for the combustion. Once the question of the oxidizer had been settled the other things would be fairly easy, at least by comparison. The possible solid oxidizers fell into two groups, each with a number of members and each member with both advantages and drawbacks.

The first group was the nitrates, with potassium nitrate (KNO_3) the best known in pyrotechnical practice. Almost 40 per cent by weight is oxygen which becomes available in combustion. But the exhaust would be quite smoky, and just because this substance was so well known all the difficulties were known too. Next in that group was sodium nitrate ($NaNO_3$), with even more oxygen available—about 47 per cent—but also smoke-producing and with some other drawbacks. Third was ammonium nitrate (NH_4NO_3), which, one could foretell, would not have any solid substances in the exhaust; hence it might produce a virtually smokeless exhaust blast. Yes, but—the available oxygen is only around 20 per cent, since some of the oxygen would be used up by combining with the hydrogen contained in the same molecule. Moreover, at a comparatively low temperature, that of a really hot day when the thermometer creeps above 90 degrees Fahrenheit, ammonium nitrate undergoes a change in volume. This did not look safe, and for that reason most chemists passed it over.[5]

The other group were the perchlorates. At first glance they seemed superior to the nitrates, since some of them release more than 50 per cent of their weight in oxygen. The figure for the available oxygen is highest for magnesium perchlorate—$MgClO_4$—which releases 57.2 per cent, or 57.4 per cent according to another handbook. But chemists knew this substance too well to be fooled by this figure. They used it often in the laboratory as a drying agent. Anything that can be used as a drying agent is greedy for water—is hygroscopic, to put it more politely—and anything seriously hygroscopic is no good for military pyrotechnics, or even peaceful pyrotechnics, for that matter. The next highest figure for oxygen release, namely 52 per cent, was that of sodium perchlorate, $NaClO_4$.

Instead of having one drawback, like magnesium perchlorate, sodium perchlorate had two. One was the same: it is strongly hygroscopic. The other was that it would throw a solid substance, namely sodium chloride ($NaCl$), into the exhaust. The next in that group, potassium perchlorate ($KClO_4$) would release 46 per cent of oxygen and dump a white solid, KCl or potassium chloride, into the exhaust, but at least it was not noticeably hygroscopic. The last member, ammonium perchlorate (NH_4ClO_4), would release 34 per cent of oxygen, would not change volume like ammonium nitrate, and would not shoot anything solid into the exhaust. But it did yield an unpleasant combustion product, hydrogen chloride (HCl) which is toxic, corrosive, and fog-producing if the air is moist.

Of the whole list, only potassium perchlorate looked at all useful. In some cases it would not matter much whether the exhaust was smoky or not. And it was potassium perchlorate which was actually used in some of the GALCIT fuels.[6]

But there was one more group, the picrates. Their "base" is picric acid, which is a very strong yellow dye with a bitter taste that takes hours to wear off. It is also explosive. And toxic. The chemical name is 2:4:6 tri-

[5] Later it was found that the change either is unimportant or can be neutralized; the Rocket Fuels Division of Phillips Petroleum developed a solid fuel for boosters which consists of carbon black, synthetic rubber, and a few additives, with ammonium nitrate as the oxidizer. This fuel is amazingly stable over a very wide temperature range. It does produce black smoke when burning, but not much.

[6] See Appendix 1. They contained 70 to 78 per cent $KClO_4$ and 22 to 30 per cent asphalt, with small additions of asphalt oil.

nitrophenol, the formula $HO{\cdot}C_6H_2(NO_2)_3$, and chemists call it a "typical example of a nitro-compound of the aromatic series." Military men knew it as *lyddite* (after Lydd in Sussex) if they were English-speakers; if they spoke either German or French habitually they called it *melinite.* While picric acid by itself is reasonably safe, provided it is of high purity, it can be made to detonate by a sharp blow. But if the substance was permitted to form a salt—a picrate—with a metal, the result was something enormously sensitive to friction or heating. The picrates of heavy metals such as lead especially would detonate at the slightest excuse. Picrates of light metals were more manageable; in the past there were picrate powders, such as Brugères's powder and Designolles's powder, which were used both for blasting and for military purposes. The former consisted of 54 parts ammonium picrate, 45 parts potassium nitrate, and 1 part inert ingredients; the latter of potassium picrate, potassium nitrate, and charcoal.

None of this sounds exactly safe, but a modern rocket composition actually resembling Brugères's powder exists. It is one of the so-called "molded composites" and consists of ammonium picrate (40 to 75 per cent), potassium nitrate (20 to 50 per cent), and a solid fuel. But while all this was rather interesting and also to some extent promising, the best results were achieved by going back to Nobel's old double-base powders. There was one important variation on the theme, however. They were no longer extruded grains, but cast charges. In the extruded grains the composition had been 50 to 60 per cent nitrocellulose, 30 to 45 per cent nitroglycerin, and 1 to 10 per cent other substances. The cast charges contain 45 to 55 per cent nitrocellulose, 25 to 40 per cent nitroglycerin, 12 to 22 per cent of a plasticizer, and 1 to 2 per cent of miscellaneous substances with specific jobs, such as making a charge nontranslucent. The secret (not revealed) obviously lies somewhere in the manufacturing process.

But whatever it is, it enables the specialists to produce charges which are a foot or more in thickness and 6 or more feet in length, but which can still release all the energy stored in them safely within 2.5 to 3 seconds, exerting an enormous push. They burn with a loud roar and an orange-red flame, releasing a cloud of orange-colored smoke which looks luminous because of the flame in its center. The large cast charges are wall-fitting, with a plastic liner between them and the metal wall of the casting. Very often, when the charge has stopped thundering, a red and smoky flame emerges lazily from the exhaust nozzle. This is the plastic liner burning for a few seconds after the firing is all over.

A cross section through such a large booster unit is shown in Fig. 33. In this specific type the front plate is pushed against the charge by a powerful spring, which serves two purposes. One is to hold the charge in place; the other is to permit a little leeway for thermal expansion of the charge when it starts burning. Ignition is from the front, and the burning progresses from the center hole of the charge outward. The center holes

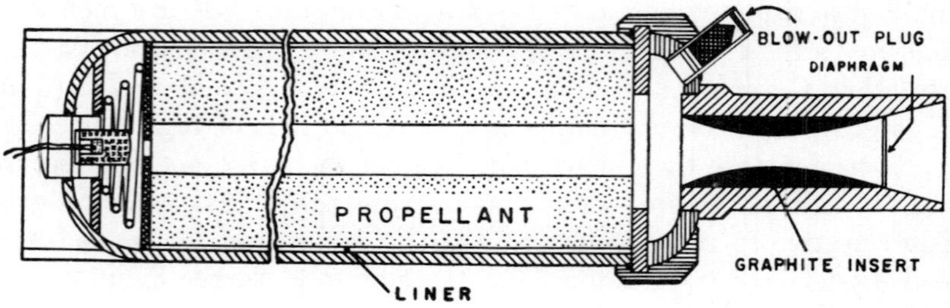

Fig. 33. Solid-fuel booster unit. The purpose of the diaphragm is to seal the finished unit against dust and moisture.

of such charges are not necessarily round but can have several different shapes that help to control the pressure inside the rocket while burning.

Evidently the internal pressure can do only three things. It can stay the same, within narrow limits, throughout the burning period; it can decrease during the burning; or it can do the opposite and increase as the burning progresses. In the case of a thick-walled tube that burns from the inside out and from the outside in, the pressure remains virtually the same; this is called "neutral burning." The cruciform grain mentioned earlier shows so-called "regressive burning;" that is, the pressure is highest immediately after ignition and drops steadily as the grain is being consumed. If the chamber pressure rises from the moment of ignition and keeps rising until the charge is used up, it is called "progressive burning." This "progressiveness" is most pronounced in the case of a rod with several lengthwise holes that also burns along its outer surface; it is less pronounced for a wall-fitting charge with only one center hole, provided that hole is round. If the center hole is not round but star-shaped the result is interesting: the burning is progressive, but not strongly so, soon after ignition, for the first quarter of a second. Then it becomes slightly regressive for the next two seconds and becomes progressive again for the last second. Obviously a star-shaped center hole will make the smallest demands on the strength of the casing and permit reduction of the casing weight.

Boosters of this type were used to boost flight-path missiles like the now obsolete Matador and the equally obsolete Snark. They have been tried experimentally with manned fighter aircraft. They were used to bring track-bound research "sleds" up to speed, for the purpose of testing the effect of very high acceleration of very short duration on humans—in reality such accelerations would happen if a fighter pilot were ejected from a fast-moving jet. They were used for throwing antiaircraft missiles into the air, fast. And these heavy cast charges also produced a new type of bombardment rockets of which Honest John (Fig. 34) was an early representative. Size, shape, and performance can all be seen in the dia-

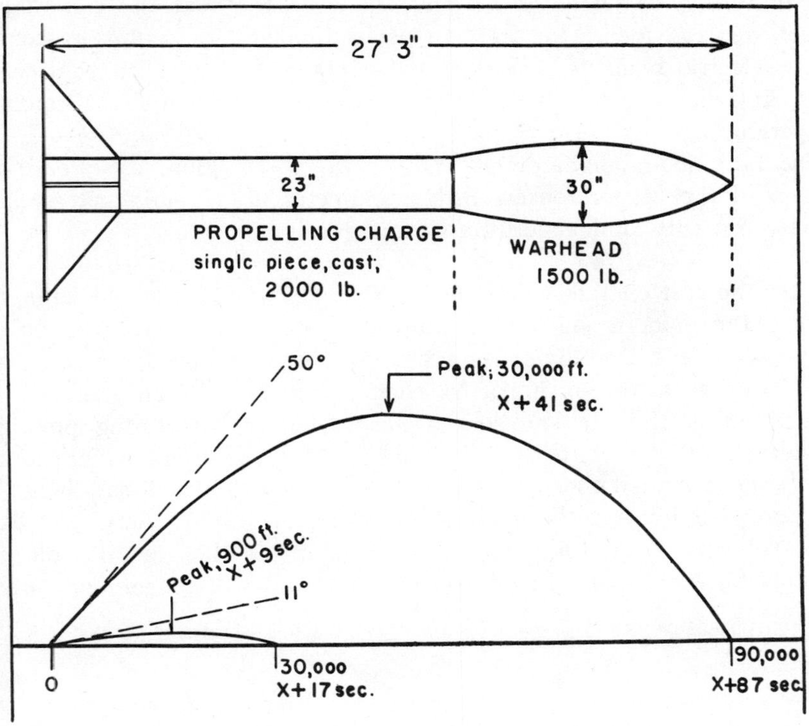

Fig. 34. Honest John and trajectories.

gram. What the diagram cannot convey, and even the printed word cannot possibly convey, is the earth-shaking roar with which the rocket departs from its launching rack.

Being a true artillery rocket, Honest John has no guidance at all. Its launcher is aimed like a gun, lining up with the target and the range is determined by the launcher's elevation at the moment of firing. But as all artillerymen know, a gun, firing with the same charge at the same elevation, will hit different points after a few rounds because the barrel has grown hot. This fact has to be taken into account; naval gunners called it the "hot gun correction." With rocket charges you have a similar effect, this time not caused by the gun barrel—since there is no barrel—but by the temperature of the atmosphere, that is, the temperature of the rocket itself. On a warm day a given charge will burn a little faster, giving more thrust for a shorter duration. On a cold day the burning period will be a little longer and the thrust a little lower. Although these two factors tend to compensate for a free-flying rocket, the compensation is not complete. Some difference in range would become apparent. Since the range table had to be computed for a temperature somewhere in the middle of the temperature spread likely to be encountered in the field, the Honest John rocket was provided with an item of equipment that sounds rather unlikely: electric blankets. For operations in the Arctic, or on cold winter days anywhere, these blankets kept the rocket charge at optimum temperature.

The next such bombardment rocket was Little John, a scaled down version of the bigger rocket, with a diameter of 318 millimeters (12½ inches) but with similar performance.

After the cast charges of double-base powders had grown to large size, a still more modern solid rocket fuel was introduced which has no size limit except the size limits imposed by the handling facilities. These fuels consist in the main of synthetic rubber with an oxidizer like ammonium perchlorate kneaded in during the manufacturing process.

These fuels burn with so little smoke that there seems to be none if the charge is a small one. When a small piece of such a fuel—say the size of a cigarette—is burning the flame looks white. In a large charge the flame looks yellowish. It is that type of fuel that made the gigantic solid-fuel boosters for the Titan III-C rocket possible, and still bigger ones are on their way.

8

Peenemünde

Wisdom is better than weapons of war; but one sinner destroyeth much good.—Ecclesiastes 8:18

Truth is the Daughter of Time—Francis Bacon

The German island of Usedom, according to legend, received its name in an unusual way. To understand the story, one has to know that the pronunciation of the name is *Oo-zay-domm,* with the accent on the first syllable. Well, it is said that some time in the past the local inhabitants, mostly fisherfolk, met to discuss a name for their home island. They spent most of a day discussing names, one more fanciful than the other, but everybody defended his own choice. When there still was no agreement late at night a disgusted visitor from the mainland exclaimed in good broad low German *"Oo, so domm!"* the last word being the local version for "dumb" or "stupid." That, legend claims, stuck, although I have never found the inhabitants of Usedom in any way "domm." There has always been something interesting going on on Usedom, but it took a war to make it world famous.

Usedom is one of two islands which separate the Bay of Stettin from the Baltic Sea. The one to the east is called Wollin; Usedom is the one to the west. Since there are two islands there are three channels connecting the bay with the sea. The easternmost of them is called the Dievenow. The one separating the two islands is the Swine, and at its seaward end there is Swinemünde, which later became a large seashore resort. The third channel is the Peene, and near its mouth there was a small village called Peenemünde, "Mouth of the Peene." If you walked to the northern shore of the island past a small lake (Lake Kölpin) you could see the Greifswalder Oie in the distance, if the weather was clear. The landscape looked the same as it did elsewhere along the shore of the Baltic: a sandy beach where, if you were lucky, you might find a "nest" of amber which had been washed up by a storm of the past, then sandy hills with stiff grass and a natural forest, largely pine, with a fair amount of underbrush. Red squirrels jumped from branch to branch and rabbits hid in the brambles and among the burning nettles. And there were wild ducks—after the summer vacationists had returned to the cities and to their jobs and left the island to Nature and to its regular inhabitants some gentlemen with more flexible work schedules would travel by rail to Wolgast, take the ferry to Usedom, and go duck hunting. One of them was the Baron von Braun.

World history is often influenced by odd factors. The first large rockets were built because a certain treaty did not mention rockets. And they were built where they were built because the chief engineer of the project knew about an out-of-the-way place where his father had gone duck hunting.

As has been mentioned in Chapter 6 the German Army, more specifically the *Waffenprüfamt* (Army Weapons Department) and still more specifically its Ballistics and Munitions Branch under Professor Becker, had been thinking about rockets. Solid-fuel rockets still offered the advantages which Congreve had defended so eloquently. They did not need guns; they needed only to be made safe and reliable, and some people had been put to work on this job. Liquid-fuel rockets offered, at least in theory, the possibility of shooting farther than any gun. And, again at least in theory, they could not be attacked en route, as an airplane could be. This was the reasoning behind the decision, made in 1929, to charge the Ballistics Branch with rocket development. The man who was "cognizant"—in view of the date involved this word has to be used quite loosely—was a Captain von Horstig.

It can't be much of an exaggeration to say that his job was comparable to an assignment to catch flies in a vacuum. He had nothing to go by. None of the engineering colleges in Germany did any work in the rocket field. Neither did industry. The only known rocket experimentation took place on a movie lot for publicity purposes. There were a number of individual inventors, most of them extremely secretive about their ideas, and there were a number of cranks. It must have been hard to tell the inventors from the cranks, especially since the attributes of the two can occur in combination. Captain von Horstig looked for a place where a judicious subsidy might have accomplished something. But he could not find one.

In 1930 Captain von Horstig greeted a man who had been appointed to the Ballistics Branch to assist him. He was a professional officer who had been with the heavy artillery during the First World War, just returned from an extended leave which he had taken to finish his engineering education and to acquire a degree of doctor of engineering. He was Captain Walter Dornberger. Dornberger helped von Horstig to look for a place where subsidies might accomplish results. He attended a test firing at the Raketenflugplatz. He talked Dr. Heyland (who after Valier's death had sworn not to permit any more rocket work in his factory) into developing a small liquid-fuel motor which could be used to test various fuel combinations. But as things progressed—or rather failed to progress—Dornberger began to realize that the Waffenprüfamt would have to do the job itself, with its own facilities, its own test stands, and on Army-owned territory. The idea was finally approved and on the artillery proving ground of Kummersdorf, some 17 miles to the south of Berlin, the new section was established. It bore the noncommittal name of *Versuchsstelle* (Experiment Station) *Kummersdorf-West,* Colonel Dornberger (now colonel) commanding.

The first civilian employee was Wernher von Braun, who had not been convinced by Nebel's insistence that, since nobody but the Raketenflugplatz knew anything about liquid-fuel rockets, the Army would sooner or later finance the Raketenflugplatz on his terms. (They had parted with full knowledge of the plans and hopes on both sides, accompanied by Nebel's prediction that von Braun would soon find out that he wouldn't get anywhere.) The second employee was an enthusiastic and capable mechanic by the name of Heinrich Grünow. In November 1932 Walter Riedel (no relation to Klaus Riedel), an engineer who had worked for Dr. Heyland's company, was added to the roster.[1] And, somewhat later, still another useful employee was acquired in a roundabout manner. Heyland's former chief engineer, Pietsch, approached the Waffenprüfamt with a proposal for a rocket motor for alcohol and liquid oxygen, with the motor located inside the alcohol tank. It was to produce 650 pounds of thrust for a duration of 60 seconds. He was handed advances for materials and labor—and then disappeared. Another man to whom he had referred as his "assistant," Arthur Rudolph, claimed to be the real inventor of this motor and proved it by finishing up the incomplete job and making it work to meet all specifications. He was hired.

The actual work of the Versuchsstelle Kummersdorf-West began with the building of a test stand, one which could produce some useful information. At the Raketenflugplatz the only measurements which were actually made were of the thrust and the duration of burning. The gas pressure in

[1] After the war, in one of my long discussions with Dr. Dornberger, the talk turned to the complaints of a number of people, Nebel among them, that they had been snubbed or suppressed. Dornberger was emphatic in saying that the policy had been just the opposite. The officers of the Waffenprüfamt felt that the publicity obtained by individuals would distract attention from their own work. The Weapons Office might even have strengthened such a "smoke screen" with money if it had felt certain that there would be no talk. This, of course, was the main difficulty as far as the Weapons Office was concerned, because if it had subsidized somebody and that somebody had talked, that would have been a definite tip-off that a real secret was hidden elsewhere. What the Army could hardly have prevented was the pressure brought by Nazi party organizations on all levels; Dornberger, for example, was unaware of the fact that Goebbels' Ministry of Propaganda, at one point, issued a directive to all newspaper editors that they were not to mention even the word "rocket" in print. The logical outcome of all these currents was what actually happened: individual experimentation ceased and all the better men were hired by the Army.

the fuel tanks was known—initial pressure, that is—and the fuel consumption was established by putting known quantities of fuels into the tanks. Then, by application of Oberth's formula $P = c \cdot dm/dt$ (the thrust P equals exhaust velocity c multiplied by dm/dt, which could be taken to mean the mass of the exhaust per second), the exhaust velocity could be calculated. The new test stand of the Ballistics Branch was designed to *measure* all the factors which had to be inferred at the Raketenflugplatz. There was to be a record of fuel consumption from instant to instant by recording the weight of the tanks; flow rates were to be measured; temperatures were to be taken; in short, everything needed for a mathematical analysis was to be recorded. The test stand was ready in December 1932 and the rocket motor to be investigated was installed.

It blew up on ignition.

A year of hard work followed. It was full of disappointments—rocket motors burning through in critical spots, the flame moving back all the way and setting the injection nozzles afire—not to mention some mechanical difficulties. But at intervals there were good test runs which proved that the rocket could be made to work; it was only necessary to learn a large number of lessons. A large number of lessons, no matter whether you use this term as a figure of speech or mean it literally, is expensive, and while von Braun, Riedel, and Rudolph thought about injection and exhaust nozzles, Dornberger had to think of money too. He had always thought that this experiment would be expensive, and what took place did not make him change his opinion; on the contrary. Of course money was potentially available. But to get it something had to be shown and even the best rocket-motor test run wasn't quite enough. To expand the program—as distinct from just keeping it going—some shooting was required. And for a good shot you needed, in addition to a reliable push, good stability.

What was then known about rocket stability could have been written on a postcard, leaving some blank space. But the stability of projectiles had been most thoroughly investigated, and Dornberger had the idea that the rocket should rotate. This is easy for a solid-fuel rocket, but with liquid fuels rotation would flatten the fuels against the walls and starve the motor. But this one could do: one could rotate the warhead, and if its shaft ran on roller bearings the fuel-tank portion of the rocket might not be affected. In any event the time for designing a complete rocket had come. The year was 1933 and the name chosen was Aggregate 1, or A-1.

> The rotating section, weighing 85 pounds, was placed at the nose of the missile, which was about 4.6 feet long and 1 foot in diameter. Approximately 85 pounds of propellants were to be forced by nitrogen pressure from the tanks into the combustion chamber, which developed a thrust of some 650 pounds and was built into the fuel tank at the rear of the rocket. The rotating section, made to form the rotor of a three-phase current motor, was to be brought up to its highest speed before launching. The A-1 would be fired vertically from a launching rack several yards high. (Dornberger, *V-2*)

This, unbeknownst to the designers, is a construction that had been tried some 60 years earlier in a naval torpedo, except for the rocket-propulsion part. Take-off weight of the A-1 was to be 330 pounds. The motor, to quote Dornberger, "was constructed and, after breaking down a few times, worked perfectly. But before the outward shape of the A-1 was finished we decided to go on at once to the next stage. . . . Shortly afterward we had ready the first design for a new motor, made of duraluminum, with a thrust of 2200 pounds."

This was fine, but it meant a new rocket, for a motor of this thrust would have swallowed the fuels in the A-1 tanks in a few seconds. It also meant a new test stand, since the old one was too small to accommodate a motor of such strength. Both meant money—and nothing much to show yet—which was somehow managed. And in December two rockets were ready; type: A-2; named Max and Moritz for fun—these are the German originals of the cartoon characters known to Americans as the Katzenjammer kids. They were taken to the island of Borkum in the North Sea and fired shortly before Christmas 1934. Both reached altitudes of about 6500 feet. But they did not use the new 2200-pound motor; their propulsion was furnished by the 300-kilogram (660-pound) motor developed earlier.

The main thing was: something had been shown.

The name of the next rocket was obvious, A-3, but nothing else was. One thing was sure—Kummersdorf did not provide enough room; they would have to move. Dornberger hardly dared to fire his solid-fuel rockets there, and von Braun said later that he simply could not concentrate on a calculation if bunches of soldiers were testing new machine guns behind his back. The Germans were the first to experience what every rocket engineer has experienced since: the need for wide open spaces. If you think this through, you'll find that only open space beyond the atmosphere is really large enough, but in the meantime rocket engineers are willing to settle for the open spaces provided by the seashore.

The large appropriations were still to come, but in the meantime there was no harm in looking at a few islands, and von Braun did not mind doing it in person. The North Sea islands were out—too close to foreign borders, Denmark on the one side and Holland on the other. Also too much traffic into and out of Hamburg. The Baltic offered better chances. There was the large island of Rügen, for example, with its high white chalk cliffs. Near Binz on Rügen there was a fine place, the requirements being mainly open sea and a forest in which military installations could be hidden. But the "Strength through Joy" men of the German Labor Front turned out to have seen it first. Then came Christmas and von Braun went to Anklam to spend the holidays with relatives. . . . "Wernher, speaking of out-of-the-way places, your father used to go duck hunting on Usedom, up near Peenemünde . . ."

Speaking of development work: a new motor with a thrust of 1500 kilograms (3300 pounds) was designed, built, tested, and made reliable. Speaking of appropriations: in March 1936 General von Fritsch finally came to Kummersdorf, was subjected to lectures, besieged with diagrams and calculations, and deafened by rocket motors of 300-, 1000-, and 1500-kilogram thrust and capitulated by saying, "How much money do you need?" Then the Air Ministry somehow got in and in April 1936 there was a conference in the office of General Kesselring. Kesselring smiled a bit when the explanations went beyond 200-mile rockets into the realm of transoceanic missiles, artificial satellites, space stations, and spaceships, but he agreed. In fact, as Dornberger wrote later, "High Authority virtually suffered an attack of acute generosity." A Ministerial Councilor of the Air Ministry drove to the City of Wolgast (which owned the land) the same day and reported the purchase by long-distance telephone the same night! Although the name of the new establishment was to be *Heeresversuchsstelle* (Army Experimental Station) *Peenemünde* it was a joint project of the Army and the Air Force. The Army was to take the forested section to the east of Lake Kölpin, and its portion was to be called Peenemünde East. The Air Force was to take the flat area of the island to the north of the lake where an airport could be built, and it was to be known as Peenemünde West. Peenemünde East was to report to Waffenprüfamt; Peenemünde West was to report to the Development Section of the Air Ministry. An army officer was supposed to be post command for the whole, being a mere administrator. But two years later, in 1939, this office was abolished, and there was no over-all head of all Pennemünde, administrative or otherwise, any more.

At that time the name was also changed to *Heeres-Anstalt* (Army Establishment) *Peenemünde* which was referred to as HAP. At a still later date the letters HAP were re-interpreted as Home Artillery Park, and for security reasons, since the very name Peenemünde began to acquire a meaning, it was officially located in Karlshagen, another nearby village.

But all this was in the future. So far only the land had been purchased and the work of developing rockets must have been delayed for a while by the good news. Plans of a different kind had to be drawn up first, and the expansion called for more men; von Braun could then remember his old friends of the VfR. As he wrote me:

> In 1937, after our initial success at Kummersdorf had led to the founding of Peenemünde, my up to then tight ceiling on personnel was off and I could employ more people. Naturally I was interested in the old Raketenflugplatz team, in particular Klaus Riedel, Hans Hueter, Kurt Hainisch, and Helmut Zoike. After the Raketenflugplatz had folded in 1934 these four had joined the Luftfahrtgerätewerk (aircraft instrument section) of the Siemens Company. I was definitely *not* interested in Nebel, whom I had always considered a successful if unscrupulous salesman with little technical and no scientific background.
>
> Klaus Riedel, who fully shared my view that Nebel would never fit into the Peenemünde framework, felt that it would put him into an embarrassing position

if he were to take service with Peenemünde and have to watch Nebel go empty-handed. While Klaus too considered Nebel a "dud" in technical matters, he pointed out that Nebel, through his tireless salesmanship, had given him the opportunity to work in liquid-fuel rocket development and thereby to acquire the experience which made him valuable for Peenemünde. Endeavoring to give Nebel a fair deal, I then looked for a way in which he could be remunerated for the effort and money he had spent on the Raketenflugplatz. Since Army money could be disbursed only for contracts and for transactions in which some "asset could be inventorized," Dornberger and I finally hit upon the idea of purchasing the old Nebel-Riedel patent DRP 633,677.[2] This patent covered a rocket motor made of aluminum with injection opposite to the main flow in the motor. [See Fig. 27B on page 142.] This system had been abandoned at Kummersdorf as early as 1934 and there was, consequently, neither a technical nor a legal reason for the Heereswaffenamt to acquire the patent. The purchase was nothing but a convenient bureaucratic vehicle to pay Nebel off. In order to prevent the shrewd Nebel from raising an additional claim against the Army for Riedel's and the others' joining of the Peenemünde organization, the purchasing contract contained a clause stating that Riedel, co-owner of the patent, would take service with the Heereswaffenamt as a civilian employee "in order to make their knowledge fully available." (von Braun)

While Peenemünde was being built up the work on the rocket A-3 neared completion. It would be ready for firing—or rather they would be, since there were more than one—in the fall or winter of 1937. A firing site was needed and Dornberger decided that the most suitable place was the Greifswalder Oie! Not the faintest objection was raised by anybody, even though they virtually rebuilt the island. If you dredge the harbor, transform a footpath into a road, pour a concrete apron for take-off, dig a dug-out, make four small concrete pyramids for phototheodolites, erect storage sheds and a large tent, string telephone wires all over the place—if you do all that on an island 1100 yards long and 300 broad at its widest point you can be said to have rebuilt the island.

The A-3 stood 6.5 meters (21 feet, 4 inches) tall and had a diameter of 70 centimeters (2 feet, 3½ inches). Its nose was filled with batteries; below the batteries was the instrument section, which also contained a barograph and a thermograph with a tiny motion-picture camera that filmed them in flight. There was also a device for emergency cut-off of the fuel flow by radio command. Below the instrument section there came the oxygen tank with a smaller tank for liquid nitrogen inside. Then came the parachute compartment, then the fuel tank and the rocket motor. The four tail fins were secured at their lower ends to a 10-inch-wide ring of plastic, to prevent flutter. The whole had a take-off weight of 1650 pounds.

The A-3 was equipped with a power plant of 1500-kilogram (3300 pounds) thrust, which was developed first at Kummersdorf and then completed at

[2] The purchase price was 75,000 marks (then $30,000) to be paid jointly to Nebel and Riedel. The contract was signed on July 2, 1937, but the payment was "due on the day Riedel reports for service with the Waffenprüfamt."

Peenemünde. Like the A-2, it burned alcohol and liquid oxygen, but it had one remarkable weight-saving device. The compressed nitrogen flask by means of which the fuel tanks of the A-2 had been pressurized had been replaced by a system using liquid nitrogen which was evaporated by an assortment of hot-wire resistors. Prior to take-off these resistors were heated by current from a source outside the rocket. They had sufficient heat capacity for maintaining a flow of evaporated oxygen for the 45 seconds of burning time.

The A-3 had a gyro-stabilized platform with "lateral accelerometers" for pitch and yaw flight-path control, with electric servo motors and molybdenum jet vanes. The guidance and control system was based upon ideas of one Mr. Boykow, who at that time was the German Navy's No. 1 expert on shipborne gyro-stabilized fire-control equipment. Boykow was a genius in more than one respect; his trouble was that he did not bother much with the engineering details of his very advanced ideas.

Three A-3's were test-fired in the fall of 1937. While the power plant behaved as expected, the guidance and control system was a flop in all three firings. At first the reason for these failures was not fully understood, since the system had worked well in laboratory tests and with complete gimbal-suspended A-3's fired on the static test pit. As a result of the failure of the A-3's we decided to develop a novel flight-control simulator technique which would enable us to subject the control system to *all* external forces acting upon the missile, including aerodynamical and wind forces. Investigations on this simulator then revealed that (a) the control moments (jet vanes) of the A-3 had been too small; (b) the response of the servo loop had been too slow; (c) the setting of the various "leads" was inadequate for the entire set of ever-changing conditions, such as velocity, stagnation pressure, center-of-pressure travel, center-of-gravity travel, etc., during the flight.

(von Braun)

The molybdenum jet vanes mentioned by von Braun were what later were more commonly referred to as "internal vanes." They had a long background in theory. It had been clear all along that the job of controlling the direction of a rocket's motion could not be entrusted to vanes in the airstream. The density of the air is sufficient for control surfaces for the first 10 miles, at the most. But since rockets were supposed to get out of the denser strata as fast as was practicable, something else was needed. For a purely vertical ascent it had been hoped for a while that the positioning of the motor ahead of the rocket would do. Goddard had built his first rockets for this so-called nose drive. Oberth had wanted to do it for his Ufa rocket. It was done all the time at the Raketenflugplatz. Now the rocket will of course follow the pull of the motor. But the laws of nature do not guarantee that the motor will pull vertically. If it should decide to pull at an angle of 45 degrees, the rocket will follow too. In the densest layers of the atmosphere the forces exerted by the air on the body of the rocket and on the stabilizing fins at the lower end do tend to keep the whole on a vertical course. But this was just another way of relying on the airstream, even though that was not clearly realized then.

But while the airstream was unreliable as to velocity and even existence, there was no question about the exhaust stream. Hence control surfaces should act against the exhaust stream rather than against the airstream. Ziolkovsky had thought of that first. Oberth had gone into the problem in considerable detail in his *Wege zur Raumschiffahrt* and had strongly advocated that the "exhaust vanes" should operate, not by dipping into the exhaust blast like rudders, but by pressing against it with their flat sides. Goddard used control vanes in precisely that manner in 1935. But for engineering reasons—and also probably for the sake of faster response —the vanes of the A-3 did dip into the exhaust.

Even while the A-3 was still in the planning stage in the summer of 1936, von Braun and Walter Riedel had been thinking about a much bigger rocket, to be called A-4 when the time came. Dornberger joined them; he had been thinking too and had a kind of goal in mind. Having been with the heavy artillery in the First World War, he was of course acquainted with a gun that had many names. Officially it was the *Kaiser Wilhelm Geschütz* but the range table called it "the long 222-millimeter cannon in the shooting cradle" (*die lange 22,2 Zentimeter Kanone im Schiessgerüst*). The gun crew called it *die Pariserin* (which translates, but not into English, as *la Parisienne*); the French counterbatteries referred to it as Big Bertha, mistaking it for an entirely different piece of ordnance; and the official name in English became Paris Gun.

This Paris Gun had shelled the city of Paris in 1918 from the forest of Crépy, 80 miles away. But although it had this then almost incredible range, the shell had not been very heavy because of the relatively small caliber of about 8 inches. And because the projectile had had to be much stronger than any other artillery shell most of its weight had been steel; it carried only about 23 pounds of high explosive. Still, the Paris Gun was responsible, in a devious way, for the design of the new rocket. The rocket should have twice the range of the Paris Gun. And the warhead should weigh a metric ton. From these two ideas everything else could be roughly calculated. The intended range of 160 miles meant that the rocket, at the moment of maximum velocity, should move at the rate of a mile a second. The payload weight meant that the structure of the rocket would weigh about three times as much. The intended maximum velocity, again, meant that the weight of the fuels would have to be about twice the weight of payload and structure together. This brought the take-off weight to 12 tons, and that meant that the thrust of the motor should be about 25 tons.

Even if every one of these figures was taken at face value, one could still design quite a number of different rockets. They could be very tall and thin, or short and squat. The external dimensions were determined by a different set of considerations. The rocket was to be a weapon which would have to be brought to the front lines, not too near of course (especially with that range), but somewhere near. And while nobody could tell where these front lines would be, the rocket would have to be transported considerable distances by road or rail. It could not be too long, or else it could not be carted around curves. It could not be too fat, or else it would not go through railway tunnels or under bridges.[3] The external dimensions—the maximum possible dimensions, that is—were dictated by the width of tunnels and the curvature of railroad tracks.

Now the A-4 had been discussed and to some extent calculated before the A-3 was even finished. Then the A-3 did not live up to expectations. What was needed, before the big job could be advanced any further, was an A-3 that worked. But this had to be a new rocket, a new type, even though the dimensions were to be the same. It had to be a new rocket for two reasons. One was that it had to be actually different in order to perform. The other was that the name A-3 had a bad reputation by then and was best not mentioned. Hence the next rocket was called A-5.

> The A-5 had the original power plant of the A-3. It had larger jet vanes made of graphite and an improved hull of almost exactly the aerodynamical configuration of the later A-4. Most important, it was equipped with an entirely new control system. In fact, we even had three different control systems for the A-5, and owing to the newly developed simulator technique, all three were successful. The first A-5 was launched, still without any control, in the fall of 1938. One year later, during the days when the war with Poland started, the first A-5 with all the trimmings took off and performed a flawless vertical flight to an altitude of 12 kilometers, say 7½ miles. We fired a total of twenty-five A-5's, first vertically, later along slanting trajectories. All A-5's were equipped with two parachutes, a small one which could be deployed at speeds up to sonic velocity, and a large one which was expelled 10 seconds later and would reduce the terminal rate of descent to about 14 meters per second (46 feet per second). All A-5's, like the A-3's, were fired from the Greifswalder Oie—the parachute-recovery system worked so reliably that a number of A-5's could be fired several times, since they suffered very little damage from the ditching in the Baltic Sea. (von Braun)

In one of the reports of the interrogations of Peenemünde researchers by Allied Intelligence Teams it is stated that the A-5 did not burn a fuel but derived its exhaust blast from the decomposition of high-strength hydrogen peroxide. This is wrong, but I now know, or think I know, how the mistake occurred. The interrogator probably ran several of his notes together. What actually happened is that there was a long delay in the development of the control mechanism and that the researchers still were not certain what shape the tail fins should have. To find out, a large number of scaled-down A-5's were made in the factory of Professor Hellmuth Walter in Kiel. These models were 20 centimeters in diameter (close to 8 inches), 160 centimeters long (63 inches), and weighed, empty, 27 kilograms (60 pounds). Their tanks held 20 kilograms (44 pounds) of hydrogen peroxide, which provided a thrust of 120 kilograms (265 pounds)

[3] The 200-inch mirror of the Palomar Mountain Observatory offered an analogous peacetime problem when it was finished and had to be brought from Corning, New York, to Palomar, California. It took weeks to figure out a route on which the mirror would not get stuck.

for 15 seconds. They were used to test tail fins of different shapes. It was the model of the A-5 that ran on hydrogen peroxide, not the full-scale A-5.

Hydrogen peroxide, the H_2O_2 of chemistry, had, of course, been known for decades. Nor had it escaped the notice of some rocket experimenters who, going entirely by its chemical formula, had occasionally viewed it as a possible substitute for liquid oxygen. It had never gone farther than some wishful thinking, for the rather decisive reason that useful hydrogen peroxide could not be bought. What was for sale in a pharmacy was a watery solution, normally of a 4-per-cent strength. Upon specifying "extra strong" one might get an 8-per-cent solution. And if one happened to know the chief chemist in a chemical factory one might, in time, get 15-per-cent hydrogen peroxide. Only a very few factories went to twice that strength, but even a 30-per-cent solution was completely useless as a substitute for oxygen.

The pure substance does contain 94.2 per cent oxygen, but when it decomposes two molecules of hydrogen peroxide rearrange themselves into two molecules of water and one of oxygen ($2\ H_2O_2 \rightarrow 2\ H_2O + O_2$). This simply means that half of the oxygen that looked "available" at first glance is tied up in the water molecules. Applied to even a 30-per-cent solution it means a yield of some 14 per cent of free oxygen. Obviously not a very good substitute for pure oxygen. But there was another factor to consider with peroxide. The results of the decomposition were not just water and oxygen; they were water, oxygen, and heat. If the peroxide of an 8-per-cent solution was made to decompose the solution grew noticeably warm. At a concentration of 13.5 per cent one would theoretically get boiling water; in reality, some of the heat would be lost. At a concentration of 64.5 per cent all the water, both the 35.5 per cent of the solution and that formed from the peroxide, would be converted into steam. For a 100-per-cent pure substance the steam's temperature would be 940 degrees centigrade (1725 degrees Fahrenheit), for an 80-per-cent solution it would be 465 degrees centigrade (869 degrees Fahrenheit), and that is steam that can do some work. And since each percentage of concentration would obviously produce steam of a certain temperature the peroxide reaction could theoretically be used to generate constant-temperature steam, something which is almost impossible by other means.

The drawback to all this nice reasoning was that nobody had made high concentrations commercially, partly because it was exceedingly difficult to do, partly because concentrated solutions were in the habit of blowing up suddenly. Sometimes it was even possible to ascertain afterward the reason for the explosion.

A new method of manufacture made 80-per-cent solutions available in about 1936 and it had also been learned that the earlier accidents had been caused either by impurities or by substances which acted as catalysts although they were not suspected of such action. For example, copper and brass and other copper alloys, containing as little as 2 per cent copper. By keeping strong solutions pure within a few parts per million (not counting the water, of course) and by using containers of copperfree nickel alloys, pure aluminum, etc., unwanted decomposition could be eliminated. But rapid decomposition could be introduced by spray-mixing the peroxide with a watery solution of permanganate; either potassium or calcium permanganate was used. If that was done in a rocket motor, a steam jet was the result.[4]

By that time every thought uttered, every line drawn, and every movement on the slide rule pertained directly or indirectly to the "big rocket," the one which had been somewhat prematurely called Aggregate 4, A-4. It was the one which later was to be publicized as V-2, and which all Allied, or at least all English-written newspapers called "Hitler's rocket" with beautiful unanimity. Of course, nobody, except some people in Peenemünde on the one hand, and Hitler's closest advisers on the other hand, could know what a misnomer this was. Actually, Hitler wasn't even interested.

He looked at rocket development once, just once. In March 1939 he came to Kummersdorf. He was shown diagrams and drawings, and Colonel Dornberger gave a report on the situation. Then he was taken out to the test range, and various rockets (with part of their skins removed) were shown to him. Dr. von Braun gave the technical lecture. Then various units were fired. Hitler had not said anything during the explanations—much to the surprise of the researchers, who knew that when he was shown a new gun or new tank, he spent hours asking questions about every smallest detail—and he watched the firings without speaking a word. I know what reactions Dornberger and von Braun expected; I know what reaction I would have expected. People who see—and especially hear—a large rocket motor for the first time either turn white and would cry "Stop, stop!" if only they could make themselves heard, or else they bubble over with enthusiasm as soon as their ears have adjusted to relative silence again. Nobody fails to be profoundly impressed—but Hitler stood and stared and did not say a word.

Only later, at lunch, he bestowed one sentence on the firing tests, the sentence: "*Es war doch gewaltig.*" It is a sentence which is hard to translate. The translator of Dornberger's book *V-2* rendered it as: "Well, it was grand," which is about the best one can do. But the word *gewaltig*, in

[4] For aircraft applications see Appendix I on rocket aircraft, also NACA Technical Memorandum no. 1170 by Professor Hellmuth Walter, listed in the Bibliography. The term *T-Stoff* ("substance T") was the code name for peroxide; the permanganates were called *Z-Stoff*, calcium permanganate *Z-Stoff C*. There are several government reports, on T-Stoff manufacture, summarized in the *Journal of the British Interplanetary Society*, no. XXIII, July 1948. That *Journal* will be referred to from now on as *JBIS*.

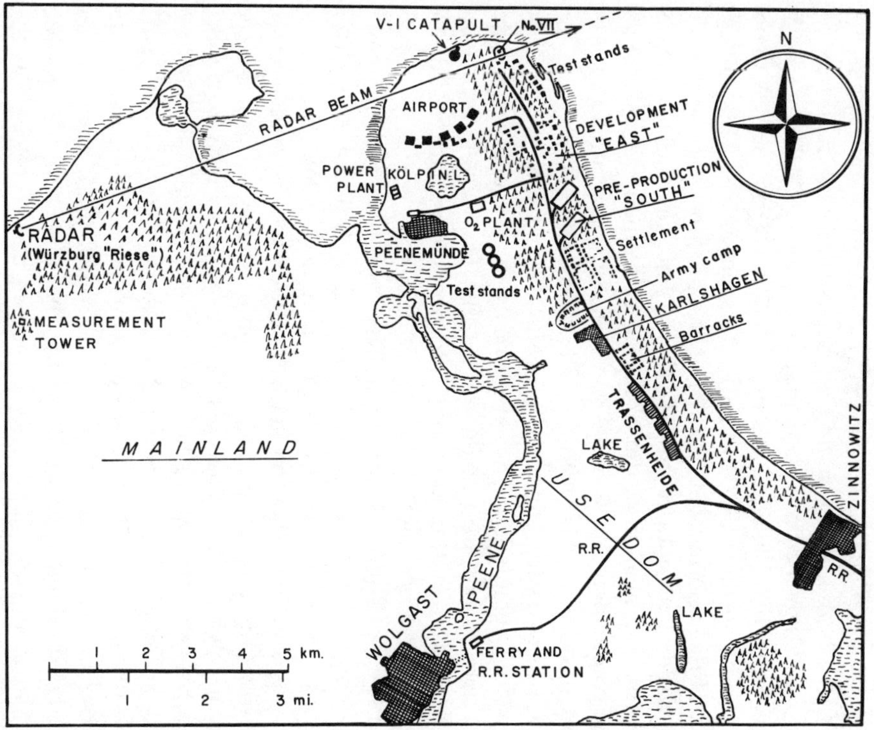

Fig. 35. The Peenemünde area.

Fig. 36. The Peenemünde rocket research station. The rockets were fired from the tip of the island east of test stand No. VII under the radar beam.

addition to "grand," can mean "enormous," "impressive," or "imposing"; in any event it was praise of a sort. It turned out then that Hitler, in his Munich period, had known Max Valier quite well, but while I don't know whether Hitler converted Valier to national-socialism (somehow I doubt it), Valier certainly had not converted Hitler to space travel. Hitler called him a dreamer. Hitler also did not like Zeppelin's airships.

After lunch Hitler left, with a formal "thank you" for the demonstration.[5] As far as the continuation of the work was concerned the rocket men at least had the consolation that General von Brauchitsch, who had been in Hitler's entourage, declared his pleasure and satisfaction.

It was another three years, counting from the date of that visit, before the first A-4's neared completion. They were ready in the summer of 1942. As a historical note it may be put down that the first seven A-4's were almost a ton heavier than the A-4's put into production later. And the finished product looked as shown in Fig. 37.

The A-4 consisted of four main sections. The nose was a conical warhead, weighing approximately 1 ton, made of mild steel about ¼ inch (6 millimeters) thick and filled with the high explosive Amatol. The reason for this choice was that the designers knew that the rocket would heat up considerably when falling back from its maximum altitude. Naturally the warhead was not supposed to explode before the rocket struck the ground. Amatol is remarkably insensitive to heat and shock. It will not blow up merely because somebody hits it very hard or heats it in a frying pan. Below the warhead there was the instrument compartment which also contained a few steel cylinders with compressed nitrogen, used, among other things, for putting a slight pressurization on the fuel tank. Below the instrument compartment came the fuel-tank section, the largest part of the rocket both in bulk and in weight. When fully fueled, that fuel-tank section accounted for three-quarters of the weight of the rocket. The fuel (alcohol) tank was placed on top; the fuel line to the motor assembly ran through the center of the oxygen tank. The space between the fuel tanks and the outer skin of the rocket, and also the space between the two tanks, was packed with glass wool for insulating purposes. Because of the low boiling point of liquid oxygen even a rocket of the size of A-4 cannot be insulated perfectly; hence the liquid oxygen had to be put in last. Evaporation loss amounted to 4.4 pounds (2 kilograms) for every minute of waiting time between fueling and firing. A 20-minute delay, therefore, caused an evaporation loss of about 90 pounds of liquid oxygen. This was (and is) considered bearable, but if there is for some reason a longer delay, the oxygen tank has to be filled up again.

The most important innovation was the presence of a pumping aggregate

[5] Goering, who received the same tour about a week later, was completely enchanted and prophesied rocket propulsion for airplanes, airships, ocean liners, railroad and passenger cars with complete disregard of scientific or engineering feasibility.

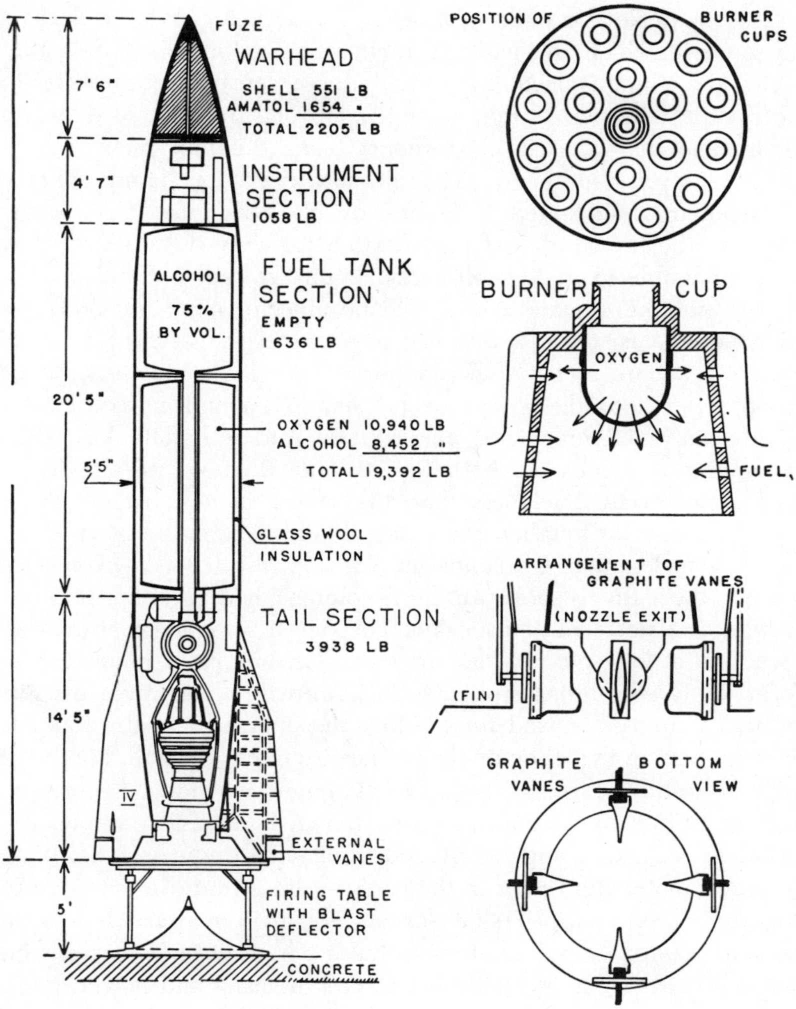

Fig. 37. A-4 (V-2). Left: Cross section of the whole rocket on its firing table. Right, top to bottom: Position of the eighteen burner cups in the top of the motor; cross section of a single burner cup; side view and bottom view of the exhaust nozzle, showing arrangement of the graphite vanes.

for the two liquids. In smaller rockets, up to and including A-3, the problem of forcing the liquids into the rocket motor could still be solved by pressurizing the whole fuel tank. Since the needed pressurization amounts to only slightly more than 300 pounds per square inch this does not increase the dead weight of the whole rocket too much as long as it is quite small. But when the rocket as a whole grows large, tanks which are to stand an internal pressure of that order of magnitude become heavy. Then it is advisable to make the tanks just strong enough to hold the fuel, in the same sense in which the gasoline tank of an automobile holds the fuel. But then something is needed to provide the necessary feed pressure for the rocket motor. The pumps took over the job of supplying feed pressure.

Again, like the vanes in the exhaust blast, the fuel pump for rockets was not a novelty in theory. The ultimate need for pumps was an old story. Goddard had stated it in one of his early patents, Oberth kept harping on the theme of fuel pumps from the very outset. But it seemed almost impossible to build such a fuel pump. It had to fulfill a whole set of rather exacting demands. It should be able to pump the fuels (one of them a liquid gas) with a delivery pressure in the neighborhood of 300 pounds per square inch. It should pump very large amounts, 50 gallons and more per second. It had to be simple enough in construction to eliminate malfunctioning if at all possible. And it had to be light, very light, since the pumps for both liquids and whatever mechanism drove them had to weigh much less than the saving in tank weight. To make the assignment a bit tougher, the pump should need only a very short time —in any event less than 6 seconds—to work up to full speed from standstill. Moreover, the delivery pressure (and volume) had to be constant within a fraction of 1 per cent of its value. The only redeeming feature was that the whole did not have to work for much longer than 1 minute.

To anybody who didn't specialize in pumps these requirements sounded very nearly impossible, and for a while the engineers at the Raketenflugplatz were hoping that they might not need pumps, after all. Maybe, it was thought wishfully, a really large rocket motor, when we finally get one, will draw its fuel by suction, provided that it is already running. At least an accident seemed to suggest this possibility. During a test run of a then large rocket motor the valves in the pressure lines from the nitrogen supply to the fuel tanks snapped shut for an unknown reason. But the motor continued to burn. It was obvious to those who saw it (I was not present at that test) that it burned with diminished intensity and power, but it did keep going. Unfortunately one could not tell just how the thrust fell off, for the thrust-registering device had gone out of action at the same moment that the pressure valves closed. This happened during the preparations for the Magdeburg experiment and there was no time to investigate the incident carefully.

Well, nature does not happen to help along in such a manner. But something else helped. As Werner von Braun told me later, he described the requirements for the pumps to the staff of a pump factory with a little trepidation. He expected protestations that these demands were impossible. Instead everybody listened in silence and when the pump experts finally opened their mouths they said that this was a fire-fighter's pump. The requirements were about the same. Simple construction, quick action, delivery of large volume of water, and constant delivery pressure, else the men would not be able to hold the nozzles steady. A design for such pumps could be based on the existing centrifugal pumps for fire-fighting equipment.

However, even the best pump still needed a power source; it has to be driven by something. It was at this point that the existence of high-strength hydrogen peroxide became important. By bringing concentrated peroxide and a permanganate solution together one could generate steam quickly, constant temperature steam which could be counted on to deliver a designated amount of work. Hence the motor compartment of the A-4—in England it is called the power bay—contained not only the rocket motor proper, but the turbine-pump assembly, a steam generator for the turbine, and two small tanks for the "auxiliary fuels," hydrogen peroxide and potassium permanganate. Of course the steam which had done its work in the turbine was still hot and could do an additional job. It was sent through a "heat exchanger" where it warmed up some of the liquid oxygen from the oxygen tank. That oxygen then went back, pressurizing the tank slightly. This not only made the job of the pumping unit somewhat easier, it also prevented the oxygen tank from collapsing when it was nearly empty. On the fuel side nitrogen gas performed the same task.

After the two liquids, now under pressure, emerged from the pumps, they were sent into the rocket motor. The oxygen went directly to the "burner cups" in the top of the tank, eighteen cup-shaped depressions from the center of which rose a brass stud, perforated all around to produce an oxygen spray. The alcohol emerged, in a similar manner, from the walls of the "cup." But the alcohol had traveled the round-about way through the cooling jacket. And some alcohol was used for cooling in a different manner, also first suggested by Oberth. Rocket experimenters had always had to learn by way of disappointments and misfirings that the most difficult part of a rocket motor was the "throat," the narrowest section of the exhaust nozzle.[6] When a rocket motor burned through, it usually was at the throat. Peenemünde East must have experienced that too. The solution, however, was astonishingly simple.

There was cold alcohol on the other side of the nozzle wall. The trouble was "merely" that it could not cool fast enough through the metal of the wall. So why not let it in by drilling a few small holes through the wall at that danger spot? If one drilled such holes the cold alcohol would come in, since the pressure in the cooling jacket was higher than the pressure inside the nozzle. As soon as it emerged, the alcohol would, of course, be caught by the moving flow of gas. Hence it would stay close to the nozzle wall, but on the inside. And since there was no oxygen provided

[6] Sometimes this exhaust nozzle is miscalled a "venturi," presumably because it sounds more impressive. Actually a "venturi" is a specific type of nozzle, but the nozzle of a rocket motor is *not* a venturi.

for the combustion of this fuel it could not burn. The over-all effect would be to produce a film of relatively cool alcohol vapor between the hot exhaust blast and the nozzle wall. Logically, the method is called film cooling.

The A-4 motor had four rows of such small holes in the nozzle wall, the first row slightly above the throat, the other three at suitable intervals below the throat. In the course of doing its job of cooling, this alcohol was heated enough so that it would have burned if oxygen had been available. But it did not find any oxygen until it reached the open air beyond the nozzle. For this reason the exhaust flame of an A-4 rocket looks about 50 feet long. If the motor could run without these film-cooling holes the flame would probably be only 20 feet long, maybe even less.

For the purpose of firing the A-4 was placed on a firing table, consisting essentially of a massive steel ring on which the rocket rested, standing on four "legs" built into the fins. The ring, of course, could be adjusted to be perfectly horizontal, so that the rocket would stand perfectly vertical. Below the steel ring there was the blast deflector, a pyramid of sheet steel which parted the exhaust blast of the rocket for the short time it stood there, burning. At first, that blast deflector was invariably ruined; the problem of making it last so that it could be used again was solved very simply by filling it with water which would absorb some of the heat.

Fueling began when the rocket was in position on the firing table, standing freely. After fueling it was heavy enough so that it would not be disturbed even by a 60-mile wind. (Of course one doesn't fire in such a high wind.) All this time the electrical equipment inside the rocket was running on an external power source, fed in through a cable which was held in place by a magnetic clamp. This cable was disconnected at the very last instant. The actual ignition of the rocket was accomplished by a pyrotechnic device, simply a pinwheel, running horizontally inside the rocket motor. Because it was cross-shaped it was called *Zündkreuz* ("ignition cross"). When the rocket motor caught fire, the Zündkreuz was destroyed by the exhaust blast.

Firing an A-4 then proceeded in three stages. The first was the electrical ignition of the pinwheel. When it burned, announcing the fact by the shower of sparks that fell out of the exhaust nozzle, the valves were opened, so that the alcohol and oxygen could enter the motor, at first by gravity alone, since the tanks were located above the motor. The Germans referred to this as the "little stage" or "preliminary stage"; the latter term became more customary in English.

During the preliminary stage the motor burned with the typical deafening cataract-like noise of a large rocket motor. Flames, parted by the pyramid-shaped blast deflector, splashed for many yards in all directions. The thrust was around 7 tons, which of course did not lift the rocket, which weighed nearly twice as much. The purpose of the preliminary stage was mainly to see whether the motor would burn properly. If it did—and only if it did—the auxiliary fuels were brought together in the steam generator. The turbine came to life and the pumps began to develop feeding pressure. It took some three seconds to build up to "main stage." The ground splash then increased in a frightening manner, the noise piled up to beyond what one can hear, and the thrust jumped from 7 to 27 tons, forcing the rocket into the air. It rose slowly, moving less than its own length during the first second, a little more during the next second, and then building up. At the end of every second the rocket moved 35 feet per second faster than at the end of the preceding second. Since the rocket lost 280 pounds of weight every second—its fuel consumption—the acceleration naturally built up, and as it did so this increase in acceleration was abetted by another factor: because of lesser air pressure the exhaust velocity increased, which, of course, made the thrust increase. Above a height of 10 miles that factor alone added 4.5 tons of thrust. When the tanks were nearly empty every second produced an increase in speed of about 150 feet per second.

The most critical period was the first few seconds when the speed was low and the rocket was highly unstable. Then the internal vanes had to do the whole job of balancing the rocket. Later, when the speed was higher, the external fins could help them. but soon the rocket had climbed into altitudes where the external air was too thin so that the job was again up to the internal vanes. For a vertical shot the vanes had only to balance the rocket and keep it vertical, but for a shot that was to cover distance the rocket also had to be tilted in the direction of the target. In such a shot the rocket stayed perfectly vertical for the first 4 seconds only, then the tilt began. The speed of sound was passed after 25 seconds, with the tilt uncompleted. It was completed during the 54th second; then rocket continued to accelerate along a straight, though slanted, line for another 8 to 10 seconds.

By the summer of 1942 the first small batch of A-4 rockets was very nearly ready for firing. In the meantime Peenemünde had grown to be a large establishment, so large, in fact, that Peenemünde East had split into two sections. The one in the vicinity of Lake Kölpin was then called Peenemünde North; it was the Development Section. The other, about halfway between the Development Section and the village of Karlshagen, was known as the Pre-Production Works of Peenemünde South. The Luftwaffe area kept its name of Peenemünde West.

Peenemünde had grown, but not without grief. The "acute attack of generosity" had not lasted very long after September 1939. Not quite a year later—after France surrendered in June 1940—Hitler ordered that Peenemünde be stricken from the priority list. Colonel Dornberger went to Berlin on "begging expeditions," as he called them later. Of course nobody could countermand an order from Hitler, who was commander-in-chief,[7] but Field Marshal von Brauchitsch could issue a few orders of his own: he ordered 4000 "soldiers" (mostly trained artisans and specialists) to Peenemünde, classifying them as "being on front-line duty," so that they could not be removed by anybody lower in rank than a field marshal.

The day of the first take-off was June 13, 1942. The rocket was ready, its motor had been thoroughly tested on a test stand, everything in, around, or connected with the rocket, however remotely, had been tested, tested, tested.

"Ignition! Preliminary stage! Main stage *on!*"

For the first time, amid the thunder of the main stage, an A-4 rose into the air. It did not look quite stable, however; the rocket rolled and made faintly pendulous motions as it climbed into a low overcast. For a while it could be heard rumbling noisily above the clouds, then there was a sudden silence. Soon afterward it fell out of the cloud layer, tumbling and without tail fins. It hit the sea, exploded, and sank.

The second rocket—with the production number 3—was ready August 16. It rose into the clear summer sky, and everything looked so fine that elation spread on the ground. Then its nose broke off. In altitude it did not match A-5, for it went only a little higher than 5 miles. But, well, a test is a test; if the outcome could be predicted with absolute certainty one would not need the test. Something could no doubt be done and the rocket with the production number 4 was pre-tested in every way conceivable—and a group of two dozen or so scientists and engineers can think up an enormous number of preliminary tests. The real test came on October 3, 1942. It was a clear day. The time was noon. The observers in the distance saw the cloud of dust and sand which the ground splash kicked into the air before the sound reached them. Then the rocket rose out of it, before anybody had time to think that the dust cloud meant an explosion on ignition. After 4.5 seconds of climbing vertically the tilt began. The trajectory of the rocket led east-northeast over the Baltic Sea, roughly paralleling the coast line at a safe distance. If the rocket functioned properly it would fall into the sea; if for any reason it did not, it would still fall into the sea. It carried bags of dye which would spread a large green stain when the rocket hit the water, and an airplane was aloft, looking for it.

A voice coming from loudspeakers counted out the seconds from take-off: "Eighteen, nineteen, twenty"—just figures. But there was an exception; soon after twenty-one and twenty-two had been counted, the voice

[7] Constitutionally the President of the German republic was commander-in-chief, not the prime minister (whom the Germans call chancellor), but when old President von Hindenburg died, Hitler had simply decreed that "in order to honor his memory," the office of president would not be filled again, and that the work of the president's office would be done by the chancellor. Thus Hitler became commander-in-chief—and also avoided a general election.

said, "Speed of sound." This was the first rocket to pass the speed of sound. It was clearly visible against the blue sky, and everybody watched, with or without optical aids. When the time keeper reached "Forty," a white trail suddenly appeared, and sixty people made seventy wrong guesses. It was simply a vapor trail, the condensation of the rocket's exhaust blast, rich in water vapor, in the freezing air of the high altitude it had then reached. There were more wrong guesses when that vapor trail began to zigzag; it was not the rocket which was moving erratically but air currents moving in different directions at different altitudes which pulled the vapor trail out of shape. As the trail seemed to stand there in the air, weirdly shaped and white against a blue background, somebody thought of a good name for it: "frozen lightning."

Fifty-eight seconds after take-off the last fuel valves in the rocket were closed by radio command. Brennschluss.[8] So far everything had gone splendidly. But it was not yet the end. The rocket was going to climb to a height of at least 30 miles along its trajectory. At 30 miles air resistance is negligible. But on the downward leg of its trajectory the rocket would enter denser atmospheric air again, moving faster than sound to begin with and being accelerated by the earth's gravitational attraction. The current name for this headache is "re-entry problem"—it did not have a name then but it existed. Calculations and wind-tunnel measurements indicated that the skin of the rocket might heat up to around 1200 degrees Fahrenheit—if this was correct and if it happened, would the rocket hold together? But the signal that came electronically from the speeding rocket continued. It was still coming 250 seconds after take-off; the rocket must still exist as a whole. The electronic signal still came at 280 seconds; by that time the rocket *had* to be back in dense air. Impact took place when the loudspeaker said 296, and as far as the people in Peenemünde could tell it had been an impact in one piece—only much later did they begin to wonder about that.

A few hours after the event they received reports—originating with fishermen who had been at sea—about a strange plane crash into the Baltic. The location of these fishing boats checked with the range: 118 miles.

It had been a marvelous success.

The next rocket did not do so well; it only made 91 miles horizontally. And of the next ten rockets fired not one was really perfect. Production number 12—the tenth shot—reached nearly 122 miles, but its trajectory was

[8] Literally, "end of burning." This German word has been retained in English rocket literature because its meaning is unmistakable. The word "burn-out" does not serve too well as a substitute because it may also mean that the rocket motor has burned out somewhere. But "burn-out" is often used for solid-fuel rockets where the associations are correct, for a solid-fuel rocket is actually burned out. In England the term "all burnt" is used for Brennschluss, but this is rather misleading, because at Brennschluss there may be still considerable quantities of either the fuel or the oxidizer in the tanks. Of course, if Brennschluss has been brought about by radio command, as it was in this case, the term "cut-off" will do nicely.

too shallow. The fifteenth shot was excellent again as far as performance was concerned, but contained a major flaw from the military point of view: it had somehow managed to follow a slightly different direction from the standard trajectory and fell on land.

Professor Oberth got to Peenemünde at about that time. The way he got there, and left again, is a story full of personal complications. After the experiments at the Chemisch-Technische Reichsanstalt near Berlin, Professor Oberth had returned to Mediash where he took no active part in experimental work and very little in organizational activities. A few years later he granted an interview to a correspondent of the influential Swiss daily *Neue Zürcher Zeitung*, but that interview only disclosed the rather obvious fact that he was continuing his investigations of rocket theory. What happened afterward is known from two letters written to the British Interplanetary Society (published in *JBIS*, xxx and xxxiii, September 1949 and March 1950, respectively), one by a Mr. Otto Hester, a relative of Professor Oberth, the other by Dr. Wernher von Braun.

In 1938 Professor Oberth was called to Vienna to the Technische Hochschule (College of Engineering) to work on rockets. Oberth began to feel after a while that this assignment—incidentally well paid—had no other purpose than to "put him on ice," to prevent his working for any other nation. At least one of Oberth's nominal superiors agreed with this point of view. But it probably was only a secondary result just the same. Von Braun advanced the more likely explanation that Oberth's call to Vienna was a case of service rivalry. Peenemünde was Army. Oberth was called under Luftwaffe contract. Peenemünde was not asked about this; von Braun did not even know Oberth's whereabouts. In any event the Luftwaffe did not give Oberth either proper facilities or sufficient help; "one miserable test pit, one mechanic," as von Braun, who visited Vienna on Luftwaffe invitation in 1940 or 1941, describes it.

Almost needless to say, the Luftwaffe was disappointed that no miracles had come out of that solitary test pit. It was willing to "relinquish" Oberth to the Army, meaning, of course, to Peenemünde. Oberth, of course, was ignorant of these decisions; all he knew was that he was transferred to the Technische Hochschule in Dresden, where he could be supervised more closely. He was given the job of developing a fuel pump for a large rocket; later he found out that the actual fuel-pump system of A-4 was already finished when the assignment was handed to him with great secrecy. When he learned this he felt himself "on ice" once more and told his superior, Professor Dr. Georg Beck, that he was tired of all this and that he was going to go back to Mediash.

Professor Beck felt that the time had come for some open words; that he had to tell Oberth something that Oberth should have realized when the call to Vienna came: namely that he, once he had even one foot on the soil of the Third Reich, was condemned to unconditional collaboration. "You are still a Rumanian citizen," said Beck, "and you have seen too much of German military matters to be permitted to leave. Once in a foreign country we could not prevent you from talking. The only thing for you to do is to acquire German citizenship; then you'll get a job commensurate with your abilities. If you continue the way you have and refuse to become a German citizen you'll have to expect concentration camp." Although Oberth may still have thought that a concentration camp was merely a place of confinement, he decided to apply for citizenship.

The bureaucratic machinery which deals with such applications is of necessity ponderous and slow, even in a democracy. In a dictatorship it also has a number of by-pass circuits producing unforeseeable complications. And although these circuits did not go into operation in Oberth's case—possibly because Peenemünde kept pushing, since naturally they wanted the "father of it all" on their home grounds—Oberth's citizenship took some time. When Oberth entered the top-secret area of the HAP for the first time, the rocket A-4 was finished and about ready to go into mass-production. Oberth, after his first surprise had worn off, pointed out that he would have done many things differently. But even if he had had a complete set of alternative and superior solutions it would have made no difference. The A-4 had progressed so far that any major change would have meant an entirely new development.

Oberth, apparently, was disappointed. He looked for other things he could do. Quite naturally there were a number of minor incidents, such as anybody experiences who has to dig a place for himself in an established organization. After some theoretical work, Oberth decided on an antiaircraft rocket as his next project. And he decided, with many good reasons, that it should be a solid-fuel rocket. But Peenemünde had been founded for the express purpose of doing liquid-fuel work. Peenemünde was not only not interested in solid fuels, it could not have devoted its time to solid fuels if it had wanted to. Hence Oberth's proposal was passed on to a firm dealing with solid fuels, the famous WASAG (Westfälisch-Anhaltische Sprengstoff A.-G.), well known all over Europe for the manufacture of commercial explosives. To Oberth it must have looked as if his former pupils just wanted to push him out, and the whole had the strange result that the outstanding pioneer of liquid fuels worked with solid fuels until the end of the war.[9]

[9] Johannes Winkler, the first president of the VfR, never saw Peenemünde. After the explosion of his HW-2 he went back to Junkers Aircraft where he had worked before. In 1939 he was transferred to the Luftfahrtforschungsanstalt (Aeronautical Research Institute) where he worked until the end of the war. At Junkers he had been mainly busy with small liquid-fuel units for aircraft take-off. Later he worked on the design of an instrument-carrying rocket. As far as I am able to determine, none of his developments was ever put into production. He died of a stroke on December 27, 1947, at Braunschweig-Querum.

After the war Oberth was temporarily detained by the English and released after interrogation. He then lived temporarily in Italy and in Switzerland, settled in West Germany near Nuremberg, and finally came to the United States to join the staff of Redstone Arsenal, Huntsville, Alabama. His arrival had been scheduled for July 4, 1955, and I had plans for showing him New York City. But the military transport made better time than expected and he was already in Huntsville on the date he was supposed to arrive in New York. I saw him there a year later, shortly before he left the United States and returned to West Germany. The reason for his leaving was that he had reached retirement age and had decided that retirement would be more comfortable for him if he lived in a country he knew better.

This story has a habit of getting ahead of itself and we now have to resume chronological narration a few weeks after the successful shot of October 3, 1942. Colonel Dornberger felt that now, having proved his case, he should receive the support due an establishment which, in time of war, had succeeded in creating an entirely new weapon. But very little could be done in a simple manner in Hitler's Third Reich; there always were political complications. Nor were these either accidental or unavoidable; they had been planned. The Nazi party, even though it had, after the "voluntary" dissolution of all other political parties, acquired totalitarian political power, still found that there were a few old established groups which followed their own traditions and did not change as much, or as fast, as desired. The Catholic Church was a prime example, and the German Army also remained something apart in the body politic, simply by failing to develop political attitudes, not to mention enthusiasm.[10] Most of the Army men still felt that their job was to be soldiers.

Early in its career the Nazi party had discovered that the establishment of parallel offices was a fine means of confusing everybody and everything. And the more time went by, the more this device was exploited. First the party set up parallel armed organizations like the Waffen SS. Then it went into detail. If the Army had a Weapons Office, the party, under some fancy name, created an equivalent. The War Ministry could be annoyed by a Munitions Ministry. In addition to whole new offices to produce overlapping jurisdictions, thereby causing confusion and, as the outcome, political supervision, there was still another means of getting political fingers into military pies: the Special Representative, the Commissioner Extraordinary, the Plenipotentiary for this and that. These individuals, in addition to doing whatever they were supposed to do, also had the very

[10] The Luftwaffe, built up by Goering, was considered politically reliable. As for the Navy, it was somewhere in between but developed a paradox on its own: it harbored a noticeable amount of nostalgic reverence for Kaiser Wilhelm II, who had created the German Navy originally.

natural desire to make their posts appear as important as could be managed for their own personal good.

Colonel Dornberger went to Berlin to inquire about the support Peenemünde should have received merely as the result of the many long reports and memoranda—in octuplicate—which it had sent through the proper channels. The only result had been an order to continue with development, to plan production on paper—and to recall all memoranda and destroy them! The War Ministry could not allot the necessary quantities of materials to Peenemünde—that was under the jurisdiction of Albert Speer's Munitions Ministry. So Dornberger went to the Ministry, and he even succeeded in seeing the Minister himself. But Speer regretted—no, Hitler was not convinced. But Speer would do something for Dornberger. He knew a man who had made a name for himself as a very successful expediter in the field of production. He was the type who could accomplish things even with low priorities because of his energy and his reputation. And this man had completed his last assignment; he was available. His name was Gerhard Degenkolb. Thank you, Herr Reichsminister, but who is Herr Degenkolb? "The chairman of the Locomotives Special Committee."

Speer, when pressed, refused to do more, pointing out that he had already stuck his neck out for Peenemünde. As head of the *Organisation Todt*—the fortification-building organization—he had agreed to start the building of a colossal bunker near Watten on the Channel coast, from which the rockets were to be fired. He had not yet told Hitler this and quite obviously did not like the position he was in. But the bunker would take a year to build, and it might be used for something else.

This was in January 1943.

In February something else happened that was interesting. Dornberger was asked to go to the Munitions Ministry to see Professor Hettlage, chief of its financial department. There he was told that Peenemünde was to be converted into a private stock company. After all, it would soon produce something and should therefore be converted into an industrial establishment, with the state temporarily owning all the shares, and under the management of a large industrial firm. When Dornberger objected, a Mr. Mackels, the Stettin representative of the Munitions Ministry, started to make charges of poor management and other things. Dornberger remained firm and for a time triumphed, though he had to put up with a few visits to Peenemünde by men who openly stated that they came to see whether anything was wrong. The party's Technical Office, hiding behind industrial firms, was determined to take the establishment away from the Army—but meanwhile Degenkolb, whose idea the whole scheme was, began to prepare for production.

In March 1943 the big bunker at Watten was under construction, and Speer, prodded by Dornberger, promised to see Hitler once more about the long-range rockets. The result was negative for a reason nobody could have foreseen. A few days later Dornberger was informed: "The Führer has dreamed that no A-4 will ever reach England."

On May 26, 1943, Peenemünde saw a large group of visitors, the Commission for Long-Range Weapons. It had come to see demonstrations and to make a decision. Beginning in 1942 the Luftwaffe station, Peenemünde West, had done development work on another long-range weapon, then called the Fieseler Fi-103, later to be called V-1,[11] or buzz-bomb, or flying bomb.

From the engineering point of view the V-1 missile was a perfect counterpart of the naval torpedo. Once launched, its course could no longer be influenced by the launching crew; the missile was kept on course (and in the case of V-1 at the predetermined altitude) by an automatic pilot. As a naval torpedo is actually a self-propelled, automatically steered, unmanned miniature submarine, so the V-1 was a self-propelled, automatically steered, explosive-laden, pilotless airplane. The V-1 happened to be large enough to carry a pilot, which fact was occasionally utilized during the testing and development period. But in action it was a pilotless aircraft and the first of its kind to see action.[12]

The flying bomb V-1 had a fuselage 25.4 feet long, the nose of which was a warhead housing 1000 kilograms (2200 pounds) of high explosive. Behind the warhead there was a fuel tank holding 80-octane gasoline. Then came two wire-wound spherical steel containers holding compressed air for the operation of the vanes, etc. The tail end was occupied by a simplified robot pilot capable of holding the flying bomb on a straight course and at a given altitude. The wings marked the position of the fuel tank; they spanned 17.67 feet. The most interesting innovation was the "propulsive duct," mounted on back of the fuselage and looking somewhat like an old-fashioned cannon barrel.

This propulsive duct, now classified as a "pulse-jet" was the invention of an engineer named Paul Schmidt who had been financially supported by the Army Weapons Office for some time. After the war he complained that his *Strahlrohr* ("jet pipe"), when it was finished, had been taken away

[11] The letter V stands for *Vergeltung,* meaning vengeance. That designation was coined in Dr. Goebbels' Propaganda Ministry at a later date and had nothing to do with the nature of the weapon. The Fi-103 became V-1, the A-4 became V-2, and the designation V-3 was reserved for the antiaircraft rocket *Schmetterling* (butterfly), which was never put into production.

[12] It was not the first to be built. The idea had been tried in America during the First World War, when the Army Air Corps, working with the Sperry Gyroscope Company and Charles Kettering's "Delco" had built a flying bomb which then was called the "Bug." By the middle of 1918, it could have been ready for production, even though it was not perfect. It was not produced because it was clear by that time that the war would probably be over before it could go into action. Development went on for a while even after the First World War and was finally dropped, for financial reasons, in about 1925. The Bug, of course, was propelled by an internal-combustion engine and a propeller. See General H. H. Arnold's book *Global Mission* (New York: Harper, 1949), pp. 74–76

from him by the Luftwaffe with all kinds of threats and had been handed over to the firm of Argus for final development and production. This Strahlrohr was in the main a steel tube, open in the rear and closed in front by spring-mounted shutters which were forced open by sufficient air pressure on the frontal end. When the air forced these shutters open and entered the tube it provided the necessary oxygen for the combustion of a fine fuel spray. The resulting explosion forced the shutters tight and produced one impulse of thrust. When the combustion gases had been exhausted through the rear nozzle the air forced the shutters open again and a second impulse resulted. The duct worked intermittently, and at night could be clearly seen from the ground. In night photographs the series of explosions looks like a string of luminous beads. Fuel consumption amounted to 1 gallon per mile, and the tank held about 150 gallons.

While such a duct can work on the stand, it will not deliver any useful thrust when at rest, and the flying bombs had therefore to be brought up to a speed of at least 150 miles per hour. The German method consisted of having a long slitted tube in the launching ramp. A piston, moving in the tube, was provided with a projection which hooked into the flying bomb and accelerated it. The piston was propelled by the decomposition of hydrogen peroxide. Once the propulsive duct took over, it quickly accelerated the flying bomb to a speed of about 360 miles per hour. A clockwork mechanism, set to operate at about the time the fuel supply was exhausted, locked the control surfaces so that the bomb dived into the ground, exploding on impact.

The decision which the Commission for Long-Range Weapons had to make was one which has been fought over again and again in innumerable conference rooms ever since. The two weapons represented the two fundamental types of long-range weapons which are possible. The Fi-103 relied on the atmosphere both for aerodynamical support and for the combustion of the fuel in its air-breathing engine. It was a "flight-path missile" or "cruising missile," which differs from an airplane by the one fact that it is not piloted. The A-4, on the other hand, was a "ballistic missile," which followed a trajectory like an artillery shell. The arguments are always the same, of course. The cruising missile tends to be cheaper than the ballistic missile, but it can be shot down by antiaircraft guns, antiaircraft missiles, and piloted fighter planes.

As for the decision between Fi-103 and A-4 the figures attached to these arguments were as follows: The warhead weight was almost the same. So was the range; both were expected to have a range of about 200 miles—later it turned out that the average range of the V-1 was around 150 miles (longest observed range 175 miles), while that of the V-2 was 190 miles on the average. But for a given sum of money one could build ten times as many V-1's as V-2's. Moreover, V-2 needed unusual fuels, V-1 did not.

Before the Commission started debating they enjoyed the spectacle of a demonstration. The weather happened to be just right for it. Two A-4's thundered into the deep blue sky which the atmosphere over the Baltic can produce if it feels so inclined, each ranging about 160 miles. One Fi-103 took off neatly but crashed after a short flight. Another one, in the afternoon, did not work at all. Air Marshal Milch made the best of it. Munitions Minister Speer protested that he personally had believed in the large rockets all along, but that his technical experts had told him every day that they could not work. The Commission as a whole decided to recommend the development and production of both weapons and to recommend also that they should be used in conjunction. Munitions Minister Speer promised, after the official verdict, that he would arrange for an audience with Hitler.

Two days later Speer called Peenemünde, asking for General Dornberger.

The audience with Hitler took place on July 7, 1943, at his headquarters in Rastenburg, East Prussia. He was shown the film of the shot of October 3, 1942; he was shown a model of the large bunker being prepared at Watten; he was shown models of the rocket and its means of transportation, the *Vidalwagen* which just transported it and the *Meillerwagen* with its hydraulic boom that erected the rocket and put it on the firing table. Hitler actually apologized in the end for his skeptical attitude. He also decreed that Peenemünde should get the highest priorities in existence. But there were a number of difficult moments. All of a sudden he demanded that the warhead should weigh 10 tons because he needed "annihilating destruction." And the question of how to fire the rockets also proved ticklish. There had been a recognized disagreement about that in Peenemünde itself. The engineers generally liked the bunker idea. To have a big underground workshop, with hundreds of rockets lined up, with all test equipment and spare parts on hand, and a liquid-oxygen plant too seemed to them the neatest way of getting things done.

The military men, especially General Dornberger himself, did not consider such a bunker a workshop. To them it was a target which would become known, and any target, no matter how strong and how strongly defended, can be destroyed if the other side is willing to pay the price. The military men had evolved the idea of "shoot-and-run" mobile batteries —which were actually used later. But Hitler wanted the bunker. Of course the Allies would find it, "but every bomb that falls on the bunker will not fall on a German city." He ordered that there should be more than one bunker. He also ordered that its concrete roof should be 23 feet thick. Dornberger and von Braun managed to get out of the audience without ruffling Hitler's feathers too much and with official blessings.

A few days later von Braun was informed that he had been granted a titular professorship.[13]

[13] A titular professorship is a very high civilian honor in Germany. It can be granted only by the head of state, a holdover from Imperial days.

As for the much-discussed bunker at Watten, it was bombed by Allied aircraft with great precision and exactly at the time when the concrete had been poured but had not yet set. The second bunker, at Wizernes, was never finished because of steady bombing raids.

The Allies had watched the once lonely but suddenly teeming spot on the Baltic shore for some time. Beginning with the spring of 1942 British Intelligence knew that Peenemünde was in some way important. The RAF sent photographic reconnaissance planes over quite frequently but in order not to tip the Germans off that they were after a specific installation they photographed the whole Baltic coast from Kiel to Rostock. And to make it look as if their photographic activities had a different purpose the RAF Bomber Command sent minelaying planes over every once in a while. RAF pilots reported that their presence was more or less accepted after some time; that the people on the ground did not seek cover any more but looked curiously at the British planes. One day one of the RAF pilots came back with a photograph showing something like a small airplane on a launching ramp. It was an early version of the V-1.

At the same time the Allies began to receive stories from fishermen from southern Sweden and the Danish islands. They told that they had seen devices flying through the air at very high speeds, making strange shattering noises. Somewhat later both American and RAF pilots reported stepped-up building activities along the Channel coast. When the things took shape they looked like skis and were dubbed "ski sites." They all seemed to point in the direction of London. Then more building activities and "big sites" were reported. Actually the "big sites," except the bunkers at Watten and Wizernes, had nothing to do with long-range weapons. Whether the building activities at the Channel coast and the activities on Usedom were directly connected was not then known to the Allies. The story which made the rounds during the last year of the war that British Intelligence had succeeded in placing two agents right into Peenemünde was a hoax. General Dornberger, when interrogated by high English officers after the war, did a little hedging in the other direction and learned that there had been two agents all right, but they had not been able to get into Peenemünde. The farthest they got was into the SS troops which guarded Peenemünde from Allied agents. This, I think, is the most frustrating fate that could possibly befall an agent.

In the late evening hours of August 17, 1943, the Germans learned that there was a concentration of RAF bombers in the north. Their approach made it look as if they were to raid Berlin and Berlin air defense was alerted. But over Rügen the RAF, instead of flying due south for Berlin, veered to the southeast and Peenemünde was subjected to a raid from more than three hundred heavy bombers. They dropped over 1500 tons of high-explosive bombs and enormous numbers of incendiaries. There was little counter-action, but the RAF lost forty-seven bombers just the same because German night fighter squadrons caught the bombers on the return flight.

The targets had been the test stands, the construction shops, and the settlement strung out along the shore of the island. The Luftwaffe establishment of Peenemünde West had been ignored, but the harbor sector with its power station and oxygen plant had been bombed. The loss of life amounted to 735 persons, the most important among the dead being Dr. Walter Thiel who had been in charge of motor development (and his family) and chief engineer Walther. The damage to the installations was not precisely minor, but less than the expenditure of bombs would lead one to expect; apparently many of the bombs were smothered by the soft sand of the dunes. Actually Peenemünde kept working and both the V-1 and V-2 were near production, the latter in the subterranean *Mittelwerk* at Niedersachswerfen near Nordhausen in the Harz Mountains. (The spelling "Hartz" is wrong, but the name is pronounced as if it were spelled that way.)

The Mittelwerk produced rockets, but they did not behave properly. After rockets became available the next step was obviously to train soldiers to use them. Because of the intricate criss-cross organization of the Third Reich it somehow fell to Heinrich Himmler, head of the secret police and the SS, either to select or to approve an area where rockets could be fired for troop-training purposes. The choice fell on a Polish artillery proving ground near Blizna, about 20 miles to the east of Mielec in the county of Kraków in southern Poland. The target area was a spot in the Pripet Marshes 200 miles to the northeast. The headquarters of the impact search organization was established in the precise target point, the theory being that at a range of 200 miles the rockets would fall close enough to the target to be watched easily but would not hit the target directly. Usually they fell from 1 to 3 miles away, but one landed only 100 yards from Dr. von Braun, who thought the firing was over for the day and was leisurely strolling to the place where his small plane was waiting.

The reason von Braun was there was that the V-2's misbehaved. A number broke apart and usually exploded while the motor was still burn-

Fig. 38. Long-range trajectories (diagram on facing page).
Top: Trajectories of the German Paris Gun of 1918, the rocket Rheinbote, and a V-2 fired for maximum range on the so-called "47-degree program." The angle of 47 degrees is measured from the vertical at take-off point.
Bottom: Three V-2 trajectories of the 41-degree program which was actually used. Mount Everest is drawn in to scale for comparison. Ranges and altitudes in kilometers as given in the original Peenemünde graph. In all cases T refers to the time elapsed since take-off in seconds, while V is the velocity at that point in meters per second. To obtain feet per second multiply figures for meters by 3.28; to obtain miles, multiply kilometers by 0.62.

(V-2 data courtesy *Antiaircraft Journal*)

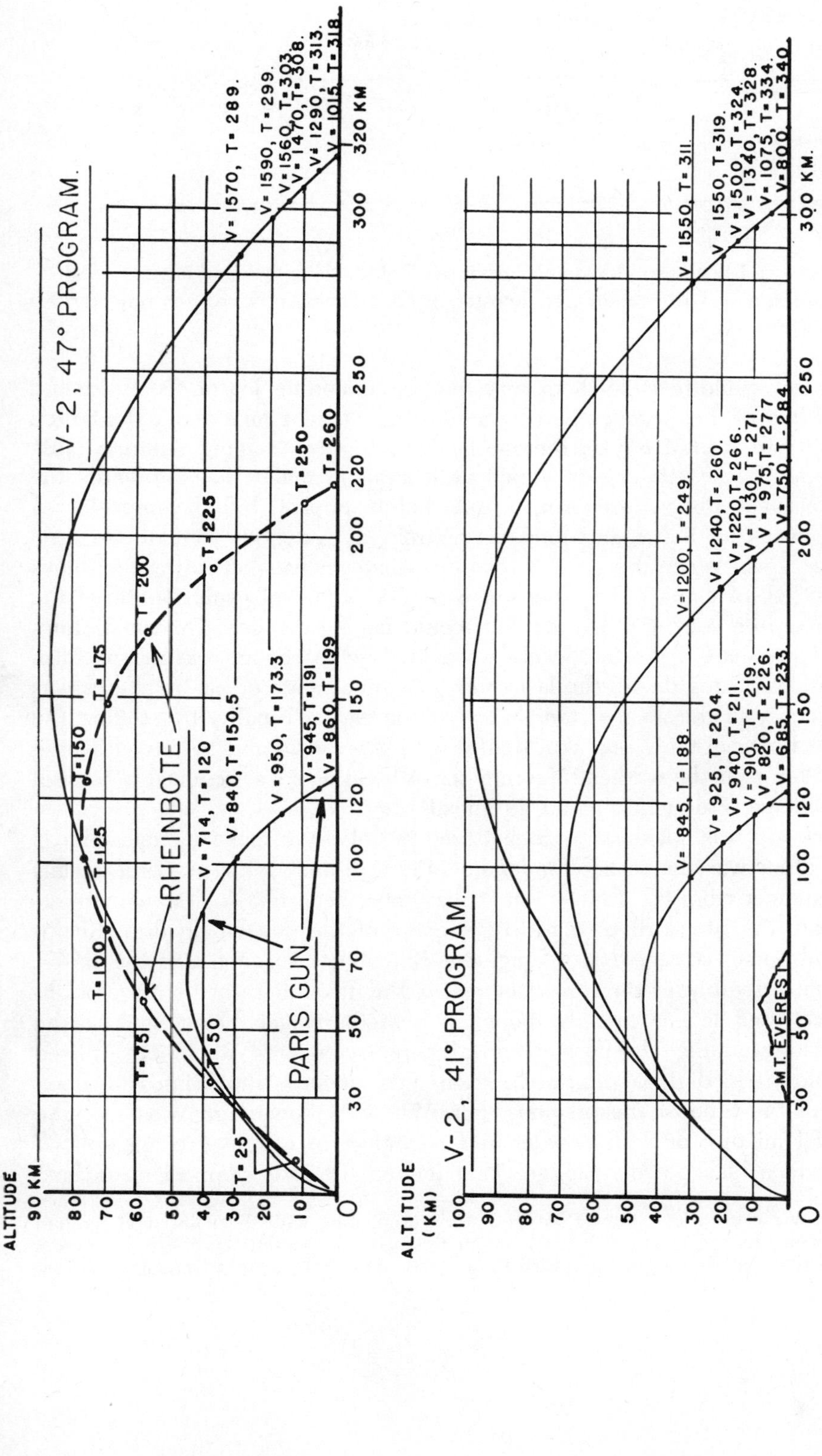

Fig. 38. Long-range trajectories.

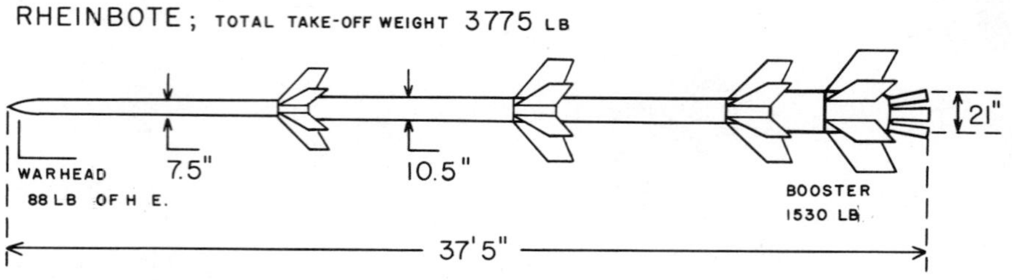

Fig. 39. Rheinbote (Messenger from the Rhine). The first multistage rocket, consisting of three steps plus booster; used against Antwerp. For trajectory see Fig. 38.

ing. But while most of them survived the ascending leg of the trajectory, fully half of them broke apart before hitting, about 1 mile above the target. What had started out as the routine operation of making a range table became a research project aimed at locating a major flaw. Probably the rockets had always broken up a mile before impact, but the impact area had been the Baltic and only very rarely did anybody actually see one come down. When the rocket broke up, whatever was left in the fuel tanks exploded, of course. The question was: does the fuel explosion cause the break-up or does it follow it? This could not be decided. On von Braun's insistence half a dozen rockets were fired with enough oxygen to burn the alcohol tank dry to the last drop. The percentage of air break-ups was the same, so the alcohol tank was not the cause. Finally, by stuffing the space between tanks and skin tightly with glass wool and by strengthening a section near the warhead the number of break-ups was reduced to a small percentage. Production could go ahead.

The V-2 was not the only rocket fired in Poland to make a range table.[14] The other was the rocket *Rheinbote* (Figs. 39 and 40). The name means "Messenger from the Rhine" but it may have been chosen less with reference to the Rhine River than to the name of the developing firm, Rheinmetall-Borsig. Over 37 feet long, the Rheinbote was not a single rocket but an assembly of three rockets, assisted at take-off by a booster rocket. It was fired like a gun, the boom of a Meillerwagen being used as the launching cradle. The booster and all three sections of Rheinbote utilized a solid fuel (diglycol-dinitrate); each unit, or step, fitted into the open end of the tube of the previous unit. When the lower unit was spent, a special mixture of flash powder and nitroglycerin caught fire and passed it on to a blackpowder charge. That ignited the next step, which pushed

[14] A few rockets which strayed led to a story, originating with the Polish underground movement, that a small Polish town was under rocket fire to test the effect; von Braun recalls that the whole operation caused 10 to 15 casualties, but mostly of German personnel.

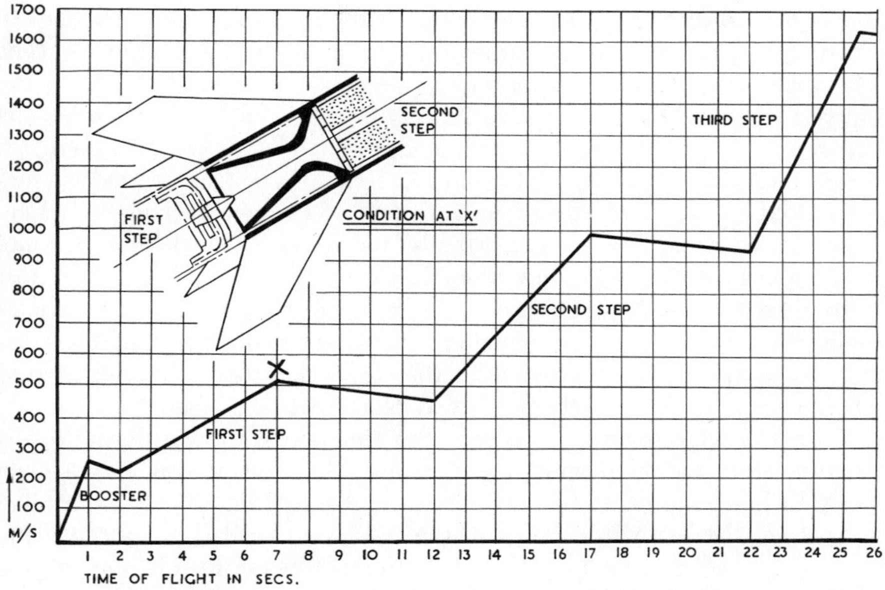

Fig. 40. Take-off performance of Rheinbote. Velocities are in meters per second. Notice how the velocity falls off in the intervals between the ignition of the various steps. Brennschluss at 25½ seconds. (Courtesy K. Gatland, B.I.S.)

itself out of the spent step. The last unit, a slender rocket slightly more than 13 feet long and only 7.5 inches in diameter, attained a velocity of 1 mile per second 25½ seconds after take-off. But because of its lesser bulk, which did not overcome residual air resistance as well as the massive V-2, it had a range of only about 137 miles.

This was still a remarkable range and as a piece of solid-fuel rocket design—considering that the year was 1944—the Rheinbote was a remarkable job. What made it mildly ridiculous as a weapon was that the warhead weighed 40 kilograms, holding a high-explosive charge of 20 kilograms (88 and 44 pounds, respectively). During the firings in Poland, one of the fins of a particular Rheinbote must have caught somewhere, so that the rocket rose vertically. The men present, Dornberger among them, ducked in their trench; the chance that the various pieces would fall on top of them was high. Moreover, the rocket had a live warhead. When the warhead finally did come down there was a muffled minor explosion, and after some searching an impact crater, shallow and 4 feet across, was found. "If this had been a real weapon," Dornberger said to me later, "we would all have been dead." He said about the same thing then, but Hitler and Special Commissioner (in charge of long-range weapons) Hans Kammler insisted that Rheinbote be used. Finally twenty rounds of

Rheinbote were fired from the Dutch city of Zwolle into Antwerp in November 1944. With all the other things that were going on at the time, Rheinbote went literally unnoticed.

Very early on March 15, 1944, General Dornberger received a telephone call from General Buhle in Berchtesgaden. He was to come to Berchtesgaden immediately to see Field Marshal Keitel. When he arrived General Buhle told him that Dr. von Braun and the two engineers Klaus Riedel and Hellmuth Gröttrup had been arrested by the Gestapo. Field Marshal Keitel added the next day that they were likely to lose their lives, since they were charged with sabotage of the A-4 project. They had been overheard saying that the rocket was something they had to do, but that their real object was space travel. Obviously, therefore, they had not applied their whole energy to the rocket, which was tantamount to sabotage.

The real reason for the arrest was that Heinrich Himmler, during one of his two visits to Peenemünde during 1943, had taken von Braun aside and told him how much better off he would be if the project were taken away from the Army and given to the SS. Himmler had not forgiven the setback he had experienced when von Braun refused to join the scheme, and he still wanted Peenemünde for his SS, so he had personally ordered the arrest. The conference with Field Marshal Keitel led to an inconclusive, if stormy, discussion between Dornberger and SS General Müller. The arrested men were finally freed, officially "provisionally," partly because of Dornberger's declaration under oath that they were needed for the conclusion of the project, partly because of the technicality that sabotage among employees of the Army had to be dealt with by Army Counterintelligence and not by the Gestapo.

During the early days of June 1944, an Intelligence report reached London, telling of a trainload of German missiles which had arrived at the Channel coast. The RAF reported that there was much activity around two of the "ski sites." An attack seemed imminent. In the evening of June 12 the German long-range guns began firing across the Channel, apparently in an attempt to cause a diversion. Main target for the guns was Maidstone, several miles inland from the Channel coast. Otham and Folkestone were also shelled. One Messerschmitt plane which flew into the London area while this was going on was shot down by antiaircraft fire.

> At 0400 hours (4 A.M.) the shelling stopped. A few minutes later an observer on duty at a Royal Observer Corps post in Kent was passed by an aircraft which made "a swishing sound" and emitted a bright glow from the rear. In common with all his colleagues, he had been briefed to recognize pilotless aircraft; and in accordance with his instructions he shouted "Diver." The missile continued over the North Downs "making a noise like a Model T Ford going up a hill" and fell to earth with a loud explosion at Swanscombe, near Gravesend, at 0418 hours. During the next hour three more of the missiles came down at Cuckfield, Bethnal Green, and Platt (near Sevenoaks) respectively. No casualties were suffered

except at Bethnal Green, where six people were killed and nine injured; in addition a railway bridge was demolished. (Hill Report)[15]

This was the beginning of the so-called "Robot Blitz," but the month of June 1944 was a lucky month for the Allies in so far as it brought them specimens of both German V-weapons. In the case of the V-1 the specimens were perfect. As has been described, there was a clockwork mechanism which, shortly before exhaustion of the fuel supply, locked the control surfaces to make the bomb dive into the ground. But in a few cases the fuel supply ran out before the clockwork mechanism went into action and in some others it may not have functioned; in any event these bombs went into a neat glide and landed without exploding. They needed only to be de-fuzed for study.

And in June 1944 a V-2 rocket, fired from Peenemünde, strayed to Sweden and broke apart over Kalmar. The Swedes, being both neutral and angry, protested; the British requested that they be handed the "specimen." The Swedes agreed and the job of salvaging the rocket fell to the American Transport Command. Lieutenant Colonel Keith N. Allen (later killed in action) flew a C-47 "Dakota" cargo plane to Sweden to get the wreckage which had been crated in the meantime.

> This particular V-2 was used as a test carrier for a radio remote-control system developed for the antiaircraft missile Wasserfall. The system enabled an operator on the ground to control the rocket in pitch and yaw by means of a joy stick. The man assigned to the job of maneuvering this V-2 was an engineer who had previously tested this remote-control system in simulator tests only. When the V-2 had reached an altitude of about 6000 feet he lost sight of it because a cumulus cloud drifted into the line of sight. In order to prevent the rocket from falling on land (south of the firing site), he deliberately maneuvered the flight path into a more northerly course. Now the Baltic is a small pond when you start maneuvering around with a missile of 200 miles' range. Twenty-four hours after the firing our Guidance and Control Branch informed me that preliminary evaluation of the Doppler tracking records indicated that the missile might have landed on Swedish territory. I told them to make absolutely sure before I reported their findings to my commanding officer, General Zanssen. Twelve hours later they confirmed their conjecture and gave a very accurate indication of the point of impact. I had just informed Zanssen when General Dornberger called from Berlin; he wanted to know whether we had anything to do with a mysterious missile which had landed in Sweden. They had just received a teletype from the German embassy in Stockholm. . . . The British intelligence team did a remarkable job in piecing together an accurate picture of the V-2 and all its components. The fact, however, that this particular V-2 was equipped with a radio remote-control system led them to the erroneous conclusion that all V-2's were radio-controlled.
> (von Braun)

[15] From the Supplement to the (official) *London Gazette* of Tuesday, October 19th, 1948; consisting of the report on "Air Operations by Air Defence of Great Britain and Fighter Command in Connection with the German Flying Bomb and Rocket Offensives, 1944–1945," submitted to the Secretary of State for Air by Air Chief Marshal Sir Roderic Hill, K.C.B., M.C. This report is subsequently referred to as the "Hill Report."

Meanwhile the V-1's kept coming. The total number launched against London—all launching ramps which were captured pointed at London; only one unfinished group pointed at Bristol—was 8070. Of these, according to the Hill Report, 7488 were reported by the Observer Corps and 2420 reached the target area. Fighters accounted for 1847 bombs brought down (some without a shot being fired—the Spitfire pilots raced alongside the bomb, put one wing tip under the bomb's wing tip, and flipped it over, causing it to crash), guns and rockets for 1878, and barrage balloons for 232. In the over-all picture, nearly 53 per cent of the flying bombs reported were brought down, while about 32 per cent of those reported reached the target area.[16]

Even these did damage enough: they destroyed a total of 24,491 dwellings, rendered 52,293 more uninhabitable, and caused minor damage to many more; 5864 persons were killed; 17,197 suffered major, and 23,174 minor, injuries.

But the over-all figures for bombs intercepted are misleading in a way, because they are over-all figures. The defense during the first phase of the "Robot Blitz" was lacking both in practice and in proper equipment and organization for this purpose. Guns and fighters had to be careful not to interfere with each other. Finally the guns were given the outer defense circle and the fighters the inner. Moreover the guns were directed by American M-9 gun directors that were exceptionally well suited for this purpose. Prior to this rearrangement, guns had accounted for 261 bombs and fighters for 925; after the rearrangement guns accounted for 1199 bombs and fighters for 847. Even more telling are the percentage figures: The British divide the whole Robot Blitz into three phases. During the first phase about 50 per cent of the bombs reported were brought down, during the second phase 63 per cent, and during the third 73 per cent, which makes a rather strong case in favor of ballistic missiles.

By September 1944 the V-2 was ready for action, with mobile road batteries. Each battery consisted of three Meillerwagens, carrying one rocket each. Each Meillerwagen was pulled by a prime mover, usually a half-track which also carried the firing crew. Then there were three tank cars, one with liquid oxygen for three rockets, one with alcohol for three rockets, and one for the auxiliary fuels and other equipment. Then there was one generator truck for electric power, and a truck which was the equivalent of a gun director, plus staff cars for the officers. When the firing site had been chosen, the "line of sight" to London was established first. Then the three rockets were placed on their firing tables in such a manner that stabilizing fin no. III was in the line of sight or parallel to it.

[16] These figures add up to a total of 3957 flying bombs shot down, the total given in the Hill Report. But 3957 plus 2420 does not make 7488, the number reported. The Hill Report does not account for the discrepancy of 1111 flying bombs; presumably they struck the ground outside the London target area.

The first operational use of the V-2 was not against London, as everybody seems to think, but against Paris. On September 6, 1944, two rounds were fired. One fell far short; the other hit the city. It was not reported then. The next two rounds were fired against London from road crossings in the outskirts of the Dutch city of The Hague.

> At approximately 20 minutes to 7 on the 8th September [1944] Londoners on their way home from work or preparing for their evening meal were startled by a sharp report which sounded almost, but not quite, like a peal of thunder. At 1843 hours (6:43 P.M.) a rocket fell at Chiswick, killing 3 people and seriously injuring another 10. 16 seconds later another one fell near Epping, demolishing some wooden huts but doing no other damage. During the next 10 days rockets continued to arrive intermittently at the rate of rather more than two a day. On the 17th September the Allied airborne operation against the lower Rhine at Arnhem was launched. Thereupon the German High Command ordered the rocket firing troops to move eastwards, and on the following day attacks on London ceased for the time being. Up to that time 26 rockets had fallen in this country or close enough to its shores to be observed; 13 of them had landed within the London Civil Defence Region. The higher figure does not represent the total fired during the period, which was certainly not less than 29 and probably well over 30.
>
> (Hill Report)

The firings showed that the protestations of the scientists—mainly Dr. Thiel—that the rocket was not ready for mass production had been correct. Two Dutch scientists, Professor Uytenbogaart and Dr. Kooy, tabulated the number of V-2's fired from The Hague and its suburbs and the failures which could be observed from the area of the firing site.

MONTH	V-2'S FIRED	FAILURES
September 1944	24	0
October 1944	81	4
November 1944	142	12
December 1944	132	17
January 1945	229	15
February 1945	207	12
March 1945	212	19
	1027	79 (7.7 per cent)

And of the nine hundred and forty-eight which seemed all right, a goodly percentage did not fall on London, possibly because the tilt was not carried far enough, and the rocket rose high into the stratosphere and fell far short of the target. In one of the shots from Utrecht the tilting mechanism did not go into action at all; the rocket rose to a height of 162 kilometers (100.5 miles) and fell down inside the city limits.

> The German offensive came to an end at 1645 hours (4:45 P.M.) on the 27th March (1945) when the one thousand, one hundred and fifteenth rocket to fall in this country or within sight of the shore fell to earth at Orpington, in Kent. The campaign had lasted seven months. During that time the Germans had fired at least 1300 rockets at London and some 40 or more at Norwich. Of these 518 had

fallen within the London Civil Defence Region and none at all within the boundaries of the latter city. Altogether 2511 people had been killed and 5869 seriously injured in London, and 213 killed and 598 seriously injured elsewhere. These figures would have been substantially smaller but for a number of unlucky incidents, in which rockets chanced to hit crowded buildings. (Hill Report)

The last operational use of the V-2 was during the battle of Antwerp.

The other German missiles of the Second World War, half a dozen of them, had in common the fact that they were antiaircraft missiles. Peenemünde contributed two of them: the large antiaircraft missile *Wasserfall* (waterfall or cataract) and the small antiaircraft rocket *Taifun,* a name which means and is pronounced "typhoon." Wasserfall was a rocket resembling V-2 but scaled down to half size. The outstanding feature, if one thinks of V-2 as the normal rocket shape, was that Wasserfall had four short wings. Obviously it was intended, after having climbed to the height of the bombers, to turn on its side and either meet the bombers head-on or chase after them. But there had been more thought about rockets with wings in Peenemünde, and that thought was reflected in the Wasserfall design (see Fig. 42).

The list of "A-projects" had not stopped with A-5, which had been needed to design A-4. A design designated A-6 was made, but no models were built. Then came A-7 (which was an A-5 with short wings), a test missile for a large, V-2-sized rocket called A-9 (A-8 was not built) which was to have wings. The idea was that the high speed with which a large rocket enters the denser layers of the atmosphere on returning from its peak elevation above the ground might be utilized to stretch the range. If by means of wings the empty and therefore relatively light rocket could be changed into an aerodynamically supported missile (a high-speed glider), the range would be increased correspondingly. Calculation indicated that the addition of short wings would add 100 miles to the range, bringing it to 300 miles. The rocket A-9 was to use a fuel combination different from that of A-4, because there was also a plan to have the A-9 carried by the rocket A-10, with a take-off weight of around 85 metric tons. That would make A-9 transatlantic. But A-10 was never built and of A-9 apparently only a few prototypes were finished. Meanwhile short wings were also tried on the A-4, resulting in the type A-4b.

But all these remained more or less in the planning stage. The rocket Wasserfall was finished, even though it was not put into production. Because antiaircraft rockets obviously have to wait until enemy planes show in the sky or at least on the radar screen, the oxidizer could not very well be liquid oxygen. Hence Wasserfall operated on a fuel combination which was cryptically called *Salbei* plus *Visol.* Salbei was nitric acid, used as an oxidizer; the code name was probably a joke originally. At the Raketenflugplatz it had become a habit for a while to call the oxygen (German *Sauerstoff*) *Sauerkraut.* Salbei, which, taken literally, is the name of the kitchen herb *Salvia officinalis,* may well have originated from the German name of nitric acid, which is *Salpetersäure.* As for Visol, that was a true code name. The Germans had gone in for chemical research on rocket fuels on a large scale and had come up with two groups of compounds. One group was based on the xylidines and was code-named Tonka; the specific mixture was designated by a figure following the name, for example, Tonka-250 which consisted of 50 per cent xylidine and 50 per cent triethylamine, by weight. The other group was based on vinyl compounds; they were the Visols. In the case of Wasserfall the fuel was vinyl isobutyl ether.

From the top down, Wasserfall was made up as follows: A radio fuze, exploded by radio from the ground, formed the nose; it was to be replaced later by a proximity fuze. Then came the warhead, filled with Amatol. The top compartment of the rocket proper, which had a diameter of 3 feet, was a spherical pressure bottle charged with compressed air which operated the control mechanisms. The valve compartment was placed immediately below that pressure bottle. Then came the Visol tank and the nitric-acid tank, and finally the power bay, housing the rocket motor and auxiliary devices. The stabilizers and vanes were attached to the power bay from the outside, the short wings to the outside of the compartments housing the pressure bottle and the Visol tank.

An early postwar report that Wasserfall was used operationally was mistaken. Records of some four dozen experimental launchings were found, but these records also stated that the rocket was a "full success" (*voller Erfolg*) in only twelve cases; only one out of four performed as expected. Because of its great weight, Wasserfall was launched like a V-2.

The rocket Taifun (Fig. 41) was a highly interesting, if unfinished, attempt to produce a small liquid-fuel rocket which could be mass-produced as easily as solid-fuel combat rockets and which was to be used in similar mass expenditure. Its body consisted essentially of a 50-inch length of seamless tubing with an outside diameter of not quite 4 inches. That was the fuel tank. The oxidizer tank was a length of smaller tubing, placed coaxially into the bigger one. Having coaxial tanks led to a considerable saving in weight; when pressure was put on the liquids to force them into the combustion chamber, the inner tank was subjected to the same pressure inside and outside, which means that it did not have to stand any pressure at all. Pressure was generated by the burning of a small cordite charge, a rather simple solution. And Taifun was a practically valveless rocket. When the cordite charge burned, generating a pressure of about 700 pounds per square inch, that pressure was momentarily confined by so-called rupture disks, metal disks holding thin metal membranes which broke under pressure of around 70 pounds per square inch. (Such rupture disks had been used for the Magdeburg rocket.) When the membranes broke, the pressure was transmitted through the liquids to similar rupture disks at the bottom.

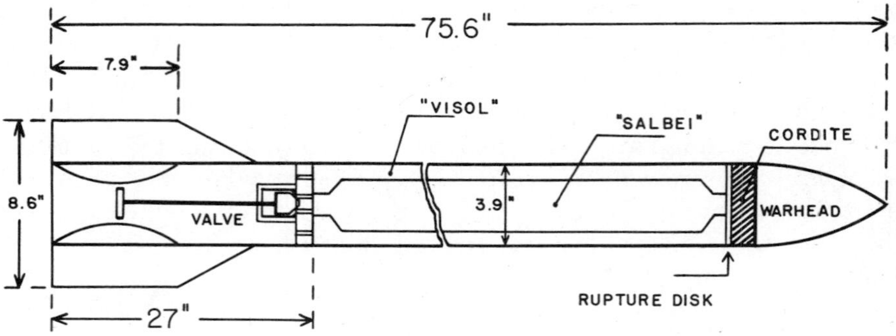

Fig. 41. Antiaircraft rocket Taifun (Typhoon). The smallest rocket developed at Peenemünde, holding the two liquid propellants in concentrically arranged tanks. Tank pressure was developed by burning a disk of cordite. This rocket was to be used for antiaircraft barrages but the development was not finished.

The fuel could then flow freely into the combustion chamber; the acid was still held back by a plug with a long stem which carried a second plug, closing the throat of the exhaust nozzle. The fuel flow pulled the stem down, so that the acid could flow too and start combustion. Then the stem burned through and the lower plug was simply thrown out. Taifun took off with a very high acceleration (31 *g*), moving with a velocity of over 1000 feet per second at the end of the first second of burning. With a burning time of about 3 seconds it reached an altitude of around 50,000 feet.

Two other German antiaircraft missiles resembled model airplanes, *Schmetterling* (butterfly) and *Enzian* (gentian). Both used solid-fuel boosters for take-off. Schmetterling, when the solid-fuel boosters were about spent, turned rapidly around its longitudinal axis a few times, presumably to throw off the booster units. Enzian just dropped them. The latter, incidentally, was of a mixed wood and metal construction which deceived Allied Intelligence Teams at first into thinking that the captured specimens were just so-called "mock-ups" (full-sized wooden models). They were not; they were the functioning missiles.

The same mistake was made with regard to the first specimen of *Rheintochter* (Rhinemaiden) developed by Rheinmetall-Borsig. It was, as may be observed from Fig. 42, a most fantastic-looking device, with four control surfaces in cruciform arrangement around its nose and six large swept-back fins near the end, the six rocket exhausts being placed between the fins. Because of the internal arrangement of Rheintochter the warhead formed the extreme tail end of the rocket. For take-off it was reinforced by a booster which sprouted four more fins. The operation visualized by the designers was that Rheintochter was to be fired as soon as a search radar caught a bombing plane. The rocket was to be followed by a second radar beam, and then the two beams were to be brought together.

None of these missiles saw operational use. One that did was the Hs-293, a development of the Henschel aircraft manufacturing firm. It was a winged bomb 11 feet 8 inches long, weighing about 1700 pounds. The wings had a spread of 9 feet 6 inches and the bomb carried a rocket in the extreme

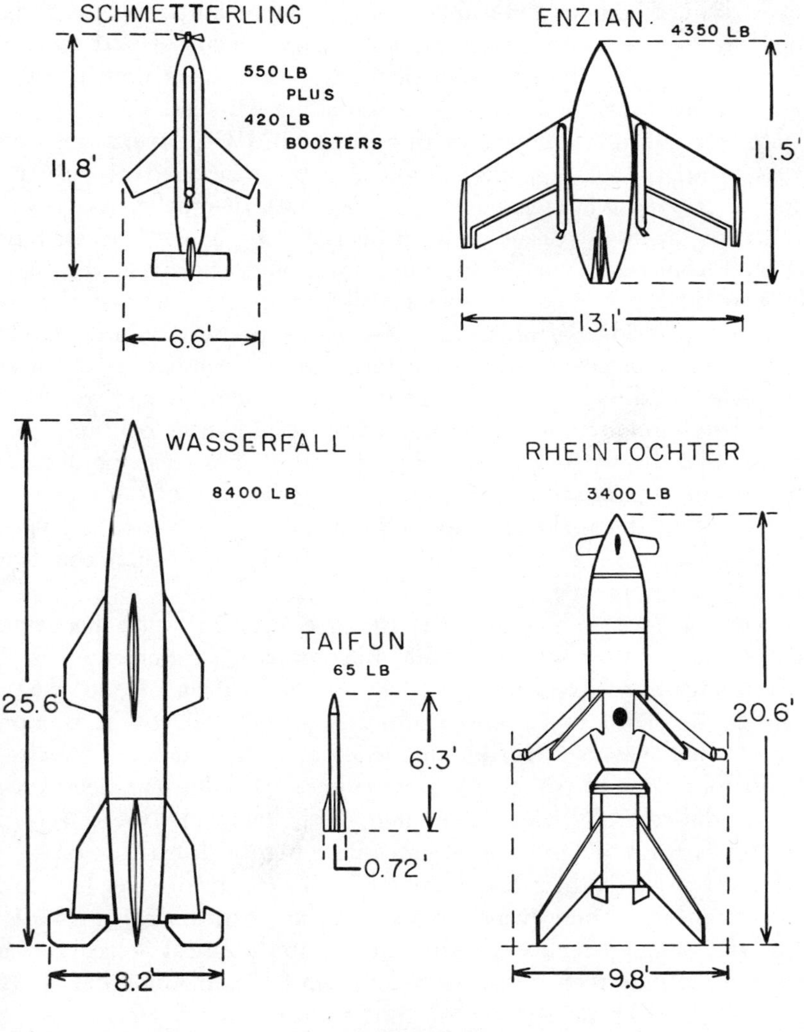

Fig. 42. German ground-to-air missiles of the Second World War. None of these types progressed beyond the experimental stage, though Wasserfall was test-fired in fair numbers to gather more information about performance under field conditions.

tail end of the fuselage. The Hs-293 was used late in 1943 and early in 1944 as a weapon against Allied convoys. The bomb was carried out to sea by German long-range bombers, Dornier Do-127, Heinkel He-177, Junkers Ju-290, and Focke-Wulf FW-200. Each bomber could carry only one of these bombs—of course they also carried some ordinary bombs inside. When the bomber was within sight of an Allied convey the winged bomb was released and its rocket began to work. The launching airplane influenced its flight path by radio. Although they missed as often as not, the Hs-293's did account for a number of Allied merchantmen. The designers blamed the misses on poor response and other difficulties and there were plans for "flying" the missiles by television. These were only plans then, but they would work now, especially for larger missiles.

A late wartime development of the firm of BMW (Bayerische Motoren Werke, Bavarian Motor Works) might also be mentioned because it had a number of very interesting features. This was the rocket X-4. It was a cigar-shaped air-to-air rocket with an over-all length of 6½ feet, burning Salbei and Tonka-250. One of the interesting points was that the two fuel tanks were just long tubes, coiled up like springs to fit into the rocket body, one coil inside the other. The X-4 had four rather small stabilizing fins at the tail end, arranged in cruciform manner, and four larger wings about in the middle of the body, arranged in the same way. Two opposing wings carried guide flares, the other two streamlined bobbins, looking somewhat like the wing-tip tanks of jet fighter planes, each holding about 4 miles of fine copper wire. When released by the pilot of the plane which carried the X-4, the rocket would forge ahead with more than twice the speed of the airplane, unreeling the wires, by which the pilot could guide the rocket by means of electric impulses (Fig. 43).

The port of Swinemünde and the island of Usedom with Peenemünde on it were taken by the Second White Russian Army, commanded by General Konstantin K. Rokossovsky, on May 5, 1945. Peenemünde itself was stormed by Russian infantry under Major Anatole Vavilov, who was put in charge of what was left. He declared later that there was not much—that they had found it "75 per cent wreckage" and that he had been ordered to destroy the rest. The German planning staff had gone to Bavaria prior to the Russian assault and had spent a few highly disturbed weeks. Not only did it look as if their lifework had been destroyed; they had to fear for their own lives. There were rumors that the SS or the SD—in short, one of the organizations which were part of the secret political police—had been ordered to liquidate them rather than let them fall into Allied hands. These rumors may well have been correct, but by that time indecision had pervaded even the political police. Finally, when it was clear that the area around them had been occupied by American troops, Wernher von Braun's younger brother Magnus was sent out to find somebody to whom the staff could officially surrender.

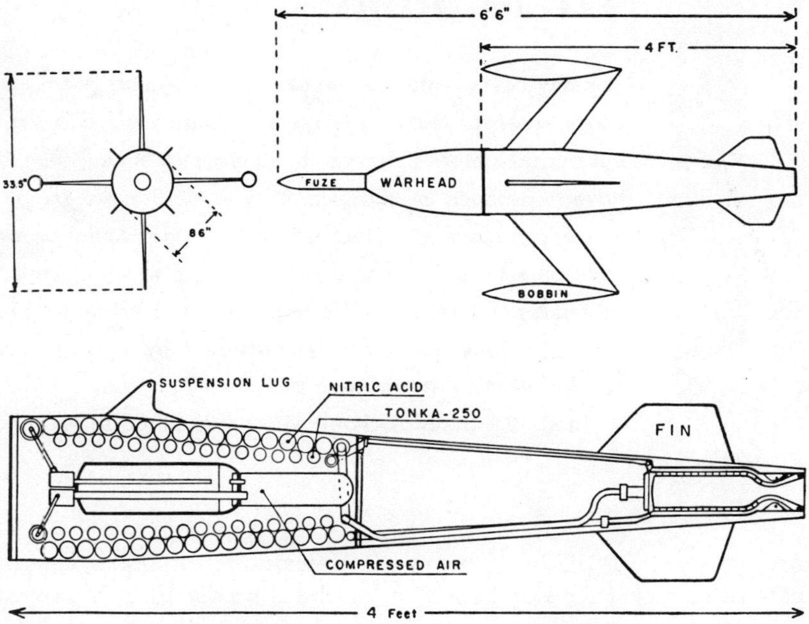

Fig. 43. The German air-to-air rocket X-4.

While the planning staff—among them all the old VfR men except Klaus Riedel who had died in a road accident a year earlier—surrendered to the Americans, the production engineers elsewhere were mostly taken prisoner by the Russians. Since production staffs are always larger than planning staffs, the Russians could release the story that "for every German rocket expert in American hands we got four or five." This story was dutifully printed all over the world—even in the United States.

Simultaneously, American forces had captured the underground factory near Niedersachswerfen. This was located in territory which, by agreement, was to be under Russian occupation. The underground factory could not be moved, of course. But by the time a few officers went through the necessary military formalities, three hundred boxcar loads of V-2 components and related equipment were on their way to the Western Hemisphere. And the researchers were taken care of by "Operation Paperclip," designed to bring specialists across the ocean as advisers and instructors. With only a very few individual exceptions they stayed on.

Peenemünde ceased to exist in 1945, except as a place name on out-of-date maps. But the rockets which had roared over the Peene continued to roar, with the Rio Grande underneath.

9 White Sands

In many cases Nature produces effects against our advantage; for Nature always acts consistently and simply, but our advantages change in many ways. When, then, we have to produce an effect contrary to Nature, we are at a loss and require skill [*techne*]. As the poet Antiphon wrote: "We by skill gain mastery over things in which we are bested by Nature. . . ." We term [these] mechanical problems. . . . Their method is demonstrated by mathematics, but the practical application belongs to physics.
—Anon. (Peripatetic School, wrongly ascribed to Aristotle)

History, in the shape of the Peenemünde Research Institute, slightly upset the pattern of rocket development as it had been predicted in the twenties. After the purely experimental types the high-altitude research rocket was expected to come first, the long-range rocket after that. But the predicted pattern was upset only temporarily. Although the long-range rocket V-2 came first, it turned into the predicted high-altitude research rocket as soon as the war was over. And it has been followed since by other research rockets soaring into the blue from the white sands of New Mexico.

The testing ground for large rockets was put into the area which also contains the White Sands National Monument—long rolling dunes of blinding white gypsum sands—for the same reasons which caused the scientists of the Manhattan Project to explode their first uranium bomb in that vicinity: large open spaces with very little commercial value and correspondingly low population density. More specific reasons were embodied in a set of instructions used by a group of officers from the Office of the Chief of Ordnance and the Engineer Corps when they looked for a site for missile research in 1944. The desirable characteristics for a proving ground were, first of all, large area, so that an experimental missile could be fired over scores of miles. Both the firing area and the impact area or areas should be flat and easily accessible. But there should be mountains nearby, on which radar and visual tracking stations could be set up. The whole should not be swampy; on the contrary, it should be dry, but enough water should be available. The testing ground should also be accessible by air (for speed) and by rail (for heavy loads) but should not be crossed by either a major air route or an important rail line. It should be noticeably above sea level and not too near the sea, to avoid frequent heavy overcasts.

It should have a climate which made it usable all year round. It should be uninhabited and the land if possible federal land, or at least the property of one of the states. It should be convenient to a permanent Army post.

After first studying detailed maps and then visiting several possible sites, the officers entrusted with the search for such a testing ground chose the present proving ground north of El Paso, Texas. It is conveniently near Fort Bliss of the Army and Alamogordo Air Base of the Air Force. No railroad or air line crosses it and only a single, not very heavily traveled highway. It is a flat, dry, and sandy valley between two mountain chains, sparsely covered with sagebrush and tumbleweed. The valley is some 4000 feet above sea level; the highest peaks of the mountain chains rise another 6000 feet into the thin air. Water can be obtained from deep wells. There is one small industrial enterprise (talc mining) near the northern end of the San Andres Mountains and some cattle-grazing elsewhere in the area. The climate has a slight tendency to extremes, but is suitable. The sky is clear almost all the time. In short, the White Sands Proving Ground, the establishment of which was approved by the Secretary of War on February 20, 1945, is almost ideal, except for its relatively small size. Beginning at the Texas-New Mexico state line, it stretches 125 miles north and is, on the average, 41 miles wide from west to east (Fig. 44).

As soon as the first preliminary "settling"—such as well drilling, barracks building, establishment of workshops, assembly halls, and communications equipment—was completed, a firing site was built. It consisted of a large patch of smooth concrete in the middle of the desert. At a distance of 350 feet from that firing site the Fortification Division of the Corps of Engineers constructed the world's most massive building. Locally referred to as "the block house," it is the nerve center of all the communications lines which are needed for the observation of a large "shoot"—as a launching is called on the Proving Ground. Not quite square in outline, the block house has concrete walls 10 feet thick. And the roof is a pyramid of solid reinforced concrete, with a maximum thickness of 27 feet. Inside there is high-intensity fluorescent lighting. All visual observation is indirect. There is air-purifying equipment for keeping noxious vapors out in case something goes wrong outside.

By the end of July 1945, 300 freight-car loads of V-2 components had been trucked to the Proving Ground. And a test stand for a whole V-2 rocket was built. It was put into the side of a cliff and was essentially a heavy concrete shaft, open at the bottom on the side away from the cliff. The inner corner was rounded in order to deflect the rocket's blast horizontally. The rocket itself was placed on top, held down by a massive steel construction which incorporated thrust-measuring devices.

But the first rocket to rise from the White Sands Proving Ground was not a captured German V-2 but an American rocket which preceded the first White Sands V-2 by half a year. And it was by only a comparatively narrow

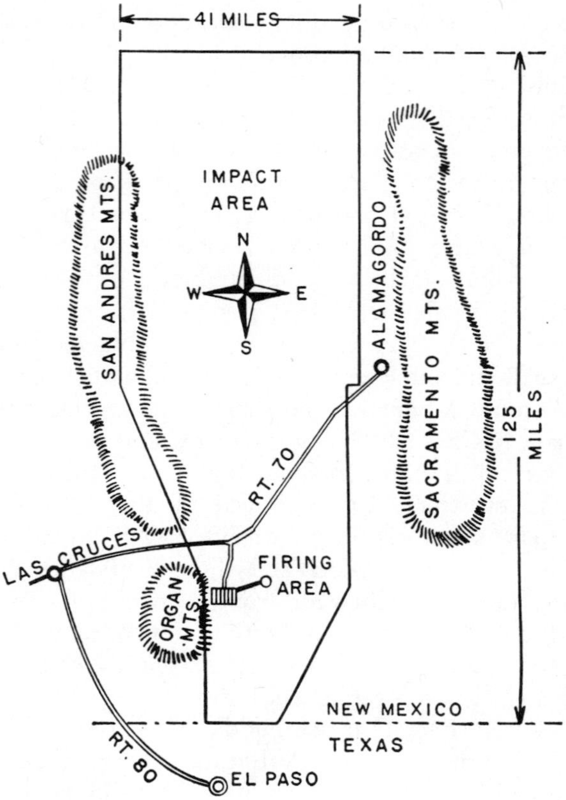

Fig. 44. White Sands Proving Ground.

margin in space that American-made rockets missed rising from the "proving ground" long before it was established. Fifteen years earlier, Professor Robert H. Goddard had tested some of his liquid-fuel rockets at the Mescalero Ranch near Roswell, New Mexico, just 100 miles east of the present impact area. Goddard had found the New England winter climate hardly suitable for rocket work; he also seems to have fought a constant battle with the fire marshal who, as was customary in 1930, did not think highly of rockets and had the layman's exaggerated ideas of the dangers of rocket research with liquid fuels. More specifically he considered them a means of starting forest fires.

Goddard, therefore, obtained a grant from the late Daniel Guggenheim (augmented by a grant from the Carnegie Foundation) to move to a more suitable climate. He too had picked New Mexico. The outline of his work there which follows is mainly quoted verbatim from his so-called "Second Smithsonian Report":

"A shop 30 by 55 feet was erected in September 1930 and the 60-foot tower previously used in Auburn and Fort Devens was erected about 15 miles away. A second tower, 20 feet high, was built near the shop for static tests, that is, those in which the rocket was prevented from rising by heavy weights, so that the lift and general performance could be studied. . . .

"The first flight obtained during this period was on December 30, 1930, with a rocket 11 feet long, weighing 33.5 pounds. The height obtained was 2000 feet and the maximum speed was about 500 m.p.h. . . .

"A beginning of the problem of automatically stabilized vertical flight was also made and the first flight with gyroscopically controlled vanes was obtained on April 19, 1932. . . . The method of stabilization consisted in forcing vanes into the blast of the rocket (U. S. Patent, *Mechanism for Directing Flight,* No. 1,879,187, September 27, 1932) by means of gas pressure, this pressure being controlled by a small gyroscope."

This series of experiments was discontinued in May 1932 "owing to the economic conditions then existing." In October 1934 a new Guggenheim grant permitted him to continue the work in New Mexico. The main objective of the experimental research from that date on was apparently not the rocket motor but the stabilizing mechanism. A pendulum stabilizer was used first, but it would function only at the beginning of an ascent, since a pendulum tends to point in the direction of the greatest acceleration. When the acceleration is high it will simply respond to that acceleration, which is not at all what it is supposed to do. Goddard had expected this performance and knew that he would have to use a gyroscope. The first test with gyroscope stabilizer was made on March 28, 1935. "The greatest height in this flight was 4800 feet, the horizontal distance 13,000 feet and the maximum speed 550 m.p.h."

What seem to have been the last two experiments of that series of tests attracted much public attention. Professor Goddard reported on these tests at a meeting of a scientific society held between Christmas and New Year's Eve, 1935. The newspapers carried the report of the meeting on New Year's Day, 1936, but some of them forgot to mention that Professor Goddard did not have the whole gathering of scientists out there in the desert near Roswell. He only showed films taken during the tests. From this misunderstanding it was generally concluded that the rocket ascents had just been made; actually they took place on May 31, 1935 (ceiling 7500 feet) and on October 14, 1935 (ceiling 4000 feet).

In the films the work of the stabilizer could clearly be seen, "the first few hundred feet of the flight reminded one of a fish swimming in a vertical direction," to use Goddard's own words. While the stabilizers worked satisfactorily the performance of the rocket motors was evidently poor. The rockets left a conspicuous smoke trail and there were even occasional flashes below the nozzle, the result of the explosions of gasoline vapor in the air. The weight of these rockets, the report states, "varied from 58 to 85 pounds." In these rockets the oxygen was fed into the rocket motor by pressurizing the tank, the fuel by means of a small centrifugal pump.

After Pearl Harbor, Professor Goddard offered his services to the United States Navy and worked at Annapolis, designing and building take-off rockets for Navy planes. He died, suddenly and unexpectedly, on August 10, 1945, after a throat operation.[1]

But while Goddard narrowly missed using the area of the present Proving Ground for his experiments, the first rocket that did rise from the Proving Ground is in a curious manner connected with him.

The story began in 1936 at the California Institute of Technology. There, encouraged by Dr. Theodore von Kármán, a group of researchers founded a kind of rocket society within the Institute. These original researchers were Frank J. Malina, Hsue-shen Tsien, A. M. O. Smith, John W. Parsons, Edward S. Forman, and Weld Arnold. Arnold financed the early part of the research work. The main goal was the design of a rocket for high-altitude research work.

As later developments show, that group did much more than just create a high-altitude rocket. It produced an entirely new group of rocket fuels, developed the first American take-off unit to be mass-produced (JATO), and did valuable research all around.[2]

As for the high-altitude project, it acquired concrete form when Doctors von Kármán, Malina, and Tsien sent a memorandum to the Ordnance Department in November 1943. In January 1944, Major General G. M. Barnes requested them to go ahead. The program received the name of ORDCIT Project, from Ordnance and California Institute of Technology. The first rocket was named Private A.

> Approximately 8 feet long, the Private A tapered to a sharp nose designed for supersonic flight, and it was guided at the aft end by four fins, each extending 12 inches from the motor body. Its gross weight was more than 500 pounds, including a pay load of 60 pounds. Driven by a solid-propellant rocket unit manufactured by the Aerojet Engineering Corporation, the missile delivered thrust of 1000 pounds for over 30 seconds.

[1] In his honor, the Daniel and Florence Guggenheim Foundation established two Robert H. Goddard professorships, one at the California Institute of Technology, one at Princeton University. The first Goddard Professor at Caltech was Dr. Hsue-shen Tsien. All the existing remains of Goddard's rockets have been placed on permanent exhibit at the "Goddard Wing" of the museum at Roswell, New Mexico, except for a number of smaller units and experimental devices that will be in the National Air and Space Museum in Washington, D.C. Books by and about Goddard are listed in the Bibliography; for a list of his numerous patents see *JARS*, no. 71, September 1947.

[2] Much of the work done was revealed in a report entitled "Research and Development at the Jet Propulsion Laboratory, GALCIT," released to the press on June 22, 1946. It was printed in *JBIS*, vol. 6, no. 2. (September 1946) and in *JARS* 66/67, June/Sept. 1946, and is usually referred to as the Galcit Report. See Appendix 1.

The booster unit, which supplied auxiliary thrust to initiate quick take-off, was a steel casing designed to mount four Ordnance aircraft armament rockets, each 4½ inches in diameter, and manifolded so as to insure simultaneous firing. Open at the centre to permit clearance of the jet blast from the Private, the booster was designed to deliver thrust of more than 21,500 pounds. Devices were installed to prevent rotation in the launcher both of the booster and the Private; and both vehicles were held in intimate contact by a shearing pin in order to prevent the destructive impact which otherwise would have occurred when the booster unit was fired, as it had to be, an instant after the main motor of the Private was fired.

The launcher was a rectangular steel boom of the truss type, with four guide rails inside its rigid structure. The boom was mounted on a steel base by means of a pivot joint so that it was adjustable both laterally and vertically. The length of the boom was 36 feet. The function of its length was twofold: to support the missile and guide it on its course until it attained velocity sufficient to gain aerodynamical stability, and to allow the booster unit to burn completely and to disconnect itself from the Private before the missile cleared the launcher.

Firing tests of the Private A were carried out at Leach Spring, Camp Irwin, near Barstow, California, 1–16th December, 1944. Twenty-four rounds were fired in all. The average range was approximately 18,000 yards; the maximum 20,000 yards (11.3 miles).

In the spring following the tests of the Private A, another experimental rocket was ready for testing. It was designed to explore the effect of lifting surfaces upon a guided missile. Called the Private F, it was essentially the same rocket as the Private A; but instead of four symmetrical guiding fins at the aft end it had one fin and two horizontal lifting surfaces with a total span of nearly 5 feet. At the forward end of the missile, to trim it in flight, were two stubby wings, their total span less than 3 feet.

Few changes were made in the booster for the Private F, but the spread of the wings and lifting surfaces on the missile dictated changes in the launcher. It had two rails above, rather than four inside, the steel framwork.

The firing tests were at the Hueco Range, Fort Bliss, Texas, 1–13th April, 1945. The range was equipped with radar for tracking the flight path of missiles, and with cameras for recording initial trajectories. Seventeen rounds were fired.

(Galcit Report)

Both the Private A and the Private F had been test missiles "pure and simple." They were not supposed to explore anything but missile design. The instruments they carried were to report on the behavior of the missile. But then a request came from the Ordnance Department to investigate the feasibility of a rocket which would carry a payload of 25 pounds for atmospheric exploration to an altitude of 100,000 feet, or about 19 miles.

Since the earlier rockets had been called "privates" this "higher" rocket obviously was a corporal, and it was named the WAC-Corporal. That name is always disbelieved at first. I remember a scene at White Sands where an old white-haired colonel was shown around by a young sergeant. "This, sir, is the V-2 which will be fired today after lunch. This building over there which looks like a pyramid, sir, is the block house, the control center. Colored flares will be fired from it to indicate the time just prior to firing. This is the meteorological tower and the tall one is the launching tower for the WAC-Corporal." "For *what?*" "The WAC-Corporal, sir. It's the name of a rocket we had here over a year ago." "WAC-Corporal—now who thought that up?" "I'm sorry, sir, I don't know." The sergeant looked puzzled himself all of a sudden. And the colonel looked as if he still wasn't sure whether he should believe it.

For years everybody had taken it for granted that the name WAC-Corporal was meant to be a kind of honor for the WACs, the Women's Army Corps. But at a later date, when the name had been nearly forgotten, the company explained primly that the three letters stood for "Without Attitude Control."

The WAC-Corporal was designed for liquid fuels, but neither for Goddard's favorite combination of gasoline and liquid oxygen nor for the German favorite, alcohol and liquid oxygen. During their earlier work the California group had made a thorough theoretical study of liquid oxidizers which could be substituted for liquid oxygen. They had decided on nitric acid (HNO_3) which will release 63.5 per cent of free oxygen for combustion. At first they worked with the "white acid" of commerce, which is relatively pure HNO_3, but found, after a while, that performance could be improved by dissolving from 6 to 14 per cent of nitrogen dioxide (NO_2) in the "white acid," thus changing it to what is commercially known as "red fuming." The fuel was still gasoline. And while the combination performed, the performance did not make everybody completely happy.

At Annapolis the rocket researchers experienced similar troubles. There were two groups active there at the time, and both had been given the same objective, the development of a liquid-fuel take-off unit for PBY flying boats. One of these groups was Goddard's, the other a Navy Development Project headed by Commander Robert Truax. The Truax group had been experimenting with a number of liquids and to their great surprise had discovered that some fuels burst into flame spontaneously when brought into contact with nitric acid. Turpentine was the first example they came across; then aniline was found to do the same. At that point Dr. Frank J. Malina visited Annapolis, saw what was going on, and telephoned Dr. Martin Summerfield in California. The WAC-Corporal was then designed for aniline with about 20 per cent of furfural alcohol mixed in to depress the freezing point. It was the first free-flight rocket to use this fuel.

The use of nitric acid as an oxidizer was less novel. In Germany its use had been pioneered by Helmut Zborowski and the Bavarian Motor Works group and it had been adopted by the Peenemünde group for Wasserfall and Taifun. But the first man to use nitric acid in a rocket motor, strange to say, was Friedrich Wilhelm Sander of Wesermünde, who demonstrated a liquid-fuel motor to the press and interested people in 1930. Unfortunately he did not name the fuel he used but referred to it noncommittally as a "liquid carbon compound," which description would hold true for at least six thousand different substances.

Before the design of the WAC-Corporal could be completed a few characteristics had to be established experimentally. For this purpose a WAC-Corporal of small size, scaled down to one-fifth of the original plans, was built. It was called the Baby-WAC and was test-fired at the Goldstone Range, July 3–5, 1945. These tests confirmed the choice of three tail fins, instead of the usual four, and the design principle of a solid-fuel booster for take-off. The WAC-Corporal, when finished, was a rocket of straight tubular shape, with a long conical nose section and three fins, an over-all length of 16 feet, and a diameter of 1 foot. Take-off weight without booster was 665 pounds. "Dry" weight, but with payload, was around 280 pounds. The thrust of the rocket motor was 1500 pounds for a duration of 45 seconds.

The pressure required to force the propellants into the combustion chamber was supplied by compressed air instead of nitrogen, conventionally used for the purpose. The substitution was made to simplify operation in the field.

Pictures of the WAC-Corporal show a blister running along part of its length. The blister is a fairing which covers the pipe lines running forward from the air tank, and aft from the propellant tanks.

The propulsive system was started by the operation of a device known as an inertia valve, incorporated in the compressed-air circuit. When the booster accelerated the missile out of the launcher, the force of inertia automatically opened the valve, which transmitted air pressure, at one and the same time, to the propellant tanks and to the actuating piston of the main propellant valve.

Fitted into the nose of the WAC-Corporal, in addition to meteorological instruments, were parachute and automatic devices for releasing both the entire nose cone and the parachute; an arrangement that recommended itself if the instruments installed were to be recovered intact.

The booster planned originally to accelerate the missile proved to be inadequate. Substituted for it was a modification of the Navy rocket known as Tiny Tim. Changes were made in the fins and nose and thrust was increased. Designed to deliver thrust of 30,000 pounds for one second, the rocket was modified to deliver 50,000-pound thrust for little more than half a second.

Calculations indicated, however, that in little more than half a second the booster and missile would rise some 216 feet, a prohibitive height for a launching tower. It was decided, therefore, to retain a tower height of 100 feet, the height agreed upon earlier to meet specifications as planned originally. Design had to allow, then, for part of the boost to take place in free flight, unguided by the launching tower.

Earlier experience with the Private A had taught much about the technique for coupling booster and missile in close contact for the duration of a boost, and about the automatic release at the end of the period. Experiments with the one-fifth-scale Baby-WAC had confirmed the design chosen for the contact-release mechanism, not to mention the fact that a 100-foot tower was high enough.

The launcher, fabricated of structural steel, was a triangular tower 102 feet high, with three launching rails set 120° apart, providing an effective length slightly more than 80 feet, after allowance for the height of the tower base. Piping was attached to the launcher for servicing the missile with propellants and compressed air.

Firing tests of the WAC-Corporal were carried out at the White Sands Proving

> Grounds, New Mexico, between 26th September and 25th October, 1945. Tracked by radar, the missile reached an altitude of about 43.5 miles in vertical flight. The great increase in altitude over that planned in preliminary estimates was the result primarily of reduction in weight achieved by changes and improvements made as the design matured, and of the added impulse provided by the Tiny Tim rocket adopted as a booster. (Galcit Report)

After several more WAC-Corporal rockets had been fired during the early months of 1946, the White Sands Proving Ground was ready for the "V-2 Program" which had been in preparation all along. The V-2 Program was a plan for systematic expenditure of V-2 rockets, at the planned rate of about two per month. The purposes of the Program were manifold: (A) to obtain experience in the handling and firing of large missiles, getting hints on the design of ground equipment in the course of the work; (B) to make experiments bearing directly on the design of future rockets; (C) to provide vehicles for operational tests of components of future rockets; (D) to obtain ballistics data, with the probability of gaining experience for the design of equipment to track trajectories and to measure missile speeds from the ground; and, finally (E) for upper atmosphere research. The latter was planned in such a way that as many different branches of knowledge as possible would benefit from the firings. (It may be remarked in passing that German scientists who had learned of the existence of the V-2 had asked for a few such rockets for research purposes. But this was near the end of the war with everything in such a turmoil that Dornberger had to refuse.) The organization of the V-2 Program looks a bit complicated to an outsider, but it probably was as simple as could be managed. The White Sands Proving Ground is under Ordnance, which did the actual firing. General Electric was responsible—as part of a larger contract called Project Hermes—for the readying of the rockets. Various scientific institutions, government agencies as well as private institutions of learning, were to furnish the instruments to be carried and the whole was coordinated by the Naval Research Laboratory which also issued the reports later.

Actually the British had had two V-2 rockets fired some time earlier. But the firings which were accomplished by German crews under British supervision from Cuxhaven were simply repetitions of wartime operational firings. The rockets were fired for maximum range into the North Sea. The first firing took place on October 15, 1945, but of the two rockets put together from captured parts one balked so that only one could be fired on that day. The one which was to have been the first seemed to have a leak somewhere and one of the Germans crawled into the motor through the exhaust nozzle to fix it, if possible. When, after some time, he failed to reappear, somebody else was sent to look for him and found him happily unconscious; the alcohol had slowly leaked into the motor and enough fumes had accumulated to do the blighter in, as the supervising officers probably phrased it. But that rocket also performed a few days later.

The American engineers in charge of the assembly of the rockets had a new problem on their hands. To make life more interesting, this problem was equally new to the German engineers who were around to help and to instruct, which still involved some linguistic difficulties anyway. The problem was this: the American troops had not captured complete rockets, ready to be fueled and fired—or at least only a very few of them. What they had done was to clean out factories and pack up whatever they found, using the same method in a few places where smaller quantities of rocket parts either had been stored or had been stranded in transit during the last weeks of the war. Only two rockets could be assembled from originally matched components. Otherwise Ordnance officers and General Electric engineers found themselves confronted with a large mess consisting of the following:

There were approximately 50 warheads, all in good condition but very nearly useless for the new purpose. They were too heavy and they had no access doors for the placing of instruments. The Naval Gun Factory made special warheads for the purpose, but until they were ready a few of the German warheads were used.

There were approximately 115 "control chambers"—the instrument section below the warhead. Some were all right but most of them were in very poor condition, requiring extensive repairs. Of the main section which houses the fuel tanks there were approximately 127 sets, mostly in good condition. There were roughly 100 thrust frames, mostly in good condition, as could be expected of such a sturdy component. There were 90 tail sections in usable condition, and a number that weren't. There were 180 oxygen tanks and an equal number of alcohol tanks, most of them in excellent condition. There were about 200 turbine-and-pump assemblies, most of them in excellent condition, and 215 rocket motors of which this could not be said. Among them there were obsolete earlier types; there were rejects; there were rocket motors that had never been tested, and even a number that had never been finished. There were 200 hydrogen-peroxide tanks and an equal number of permanganate tanks, most of them in excellent condition; since not all of them were needed not a single repair job was required. The same was true for steam generators, of which also 200 had been shipped. There were 100 heat exchangers, a large number of them in usable condition, others requiring extensive repairs. There were about 200 each of the various valves, many in good condition but not working well enough to satisfy peacetime standards. There were only 50 gyros—each rocket needs 2—and some were in poor condition; 140 American-made copies had to be ordered. And so it went down the list. There were 350 time switches, 500 servos, 600 inverters—it was necessary only to pick out the best of them. There was also enough cable for at least 100 rockets, but the American engineers felt that single-strand cables could not be trusted and threw the whole supply out after the first few flights.

Each rocket, then, was assembled from freshly tested components as close to the firing date as possible; the Germans warned that they had found out that performance reliability dropped sharply if a fully assembled rocket was stored for an extended period. In fact, they had made it a rule not to fire a rocket which had been fully assembled for more than 72 hours.

The first White Sands test was a ground test on the big test stand which had been put into the side of the cliff. It took place on March 15, 1946. The rocket thundered for over a minute and all went well. The first take-off was then scheduled for April and actually took place on April 16. But an additional precaution had been taken. Even if all components had been tested shortly before assembly, there was no denying that they were old. Failures of one kind or another had to be expected. If there had to be a failure, it was, of course, best for it to happen while the rocket was still standing on the firing table. The worst possible time for a failure to occur would be some time after take-off. Of course, once the rocket was moving there was no way of stopping it in mid-air. But the reasoning was that if at least the rocket motor could be stopped, a great deal was gained; in that case the rocket would not go very far, since its velocity was rather small. Hence an emergency cut-off was designed, a device for cutting off the fuel flow by radio order from the ground. As it turned out, that device was useful in the very first test.

Only 19 seconds after take-off the rocket suddenly turned through an angle of 90 degrees and headed east. Before the emergency cut-off could go into action the observers saw a fin tear off; it was fin no. IV. Analysis of the accident showed that the corresponding graphite vane had disintegrated very soon after take-off and that the trim tab on fin no. IV, in trying to do the whole job by itself, had weakened the fin to which it was attached. Of course nobody could tell afterward whether that particular graphite vane had been flawed or whether it was splintered by being hit with a piece of the Zündkreuz. To guard against both possibilities all graphite vanes were X-rayed and were then protected with cardboard covers which burned away quickly enough as soon as the main stage was on.

The test of V-2 no. 3, on May 10, 1946, was scheduled as a demonstration for the press and everybody else who had a reason to be there. It came off as nicely as a demonstration possibly can. So did the tests of V-2's nos. 4, 5, and 6. V-2 no. 7 tilted a bit too much, but that was noticeable only to the men servicing the tracking devices. But V-2 no. 8 went definitely wrong. It exploded 27 seconds after take-off, when it was just 18,000 feet up. It may sound incredible, but it was possible to establish the reason for that explosion. Because the turbine-and-pump assembly handles liquid oxygen on one side, the bearing on that side cannot be lubricated. An oily lubricant on the oxygen side would very likely lead to an explosion. The German designers, for that reason, had lubricated the oxygen pump with liquid oxygen. In this instance, however, that did not work and the bearing got stuck. With the pump standing still, the tanks ruptured. The German engineers present stated that they had ascribed some of the failures in Peenemünde to the same cause. At White Sands it has happened only once.

V-2 no. 9 performed very nicely, going somewhat higher than V-2 no. 5, which had held the record up to that time. With V-2 no. 10, however, everybody was again grateful for the emergency cut-off. At 13.5 seconds after take-off the rocket suddenly behaved erratically. Apparently something had gone wrong in the instrument which may be called the rocket's brain, the mixer computer (*Mischgerät*), and the computer had caused the servo unit to move one of the vanes into extreme position. Then all other vanes strained to counteract the nonsense produced by the first. At 20 seconds after take-off the ground observers were sure that the trouble would not remedy itself (it sometimes does; one rocket left at a decided angle and everybody expected it to crash, when it straightened out and made a beautiful climb) and activated the cut-off device.

Sudden deviations from the course also marred the performance of V-2's nos. 11 and 14. V-2 no. 11 turned east 4 seconds after take-off, traveling some 300 feet from the ground on a slowly rising course. V-2 no. 14 got off nicely, but some 4 or 5 seconds after take-off there was a sudden wiggling motion, something like a momentary dipping of the nose. The rocket then straightened and rose for another 2 or 3 seconds; then the same motion was repeated in a more pronounced manner. The rocket must have been about 600 feet up. From my position 2500 feet due south from the launching table, it looked as if the base of the rocket moved straight up, while the nose described a spiral. But before anybody could say just what it was doing, it turned its nose due south, stabilized perfectly but in a horizontal position, and roared over our heads toward the cantonment area and in the general direction of El Paso beyond the southern horizon. The man who operated the emergency cut-off crashed it neatly, beyond the cantonment area but still inside the proving ground.

What V-2 no. 14 threatened to do, namely to cross the Rio Grande and land somewhere in Mexico, actually happened in a later shot which was not a part of the V-2 high-altitude research program. A V-2, fired in the evening hours of May 29, 1947, suddenly shifted course at a high altitude and turned south. Apparently this happened so close to Brennschluss that the fuel cut-off signal failed to do much good. The rocket arced high over El Paso and struck the ground a mile south of the Mexican city of Juarez. There must have been a considerable amount of residual fuel left in the tanks because the impact was followed by a loud explosion. But it was just a hill that was struck and Mexican officials hastened to inform the commanding officer of the Proving Ground, Colonel Harold R. Turner, that there were neither casualties nor damage to property.

Another test shot outside the high-altitude program but a part of Project Hermes (under the name of Operation Sandy) was the firing of a V-2 from the flight deck of the aircraft carrier *Midway*. This test, which took place in the Atlantic Ocean near Bermuda on September 6, 1947, had more than one purpose. One was to see whether a missile of such a size could be successfully fueled and launched from shipboard. Another was to see whether such a launching could be performed while the carrying ship was underway. And still another was to find out whether such a ship, if it were an aircraft carrier, could fulfill its normal functions immediately after a missile shot, and, if not, how much time had to go by before normal operations could be resumed. The test answered all three questions in the affirmative. The rocket was made ready and fired in about the normal time required. The *Midway's* planes took off "immediately afterward." But the V-2 itself misbehaved. It lifted off at an angle and exploded after a 6-mile journey. Whether this also was caused by a stuck oxygen bearing or whether the motor set up vibrations which caused a leak in one of the fuel connections will remain an unanswerable question.

Operation Sandy had a counterpart, another phase of Project Hermes, which was called Operation Pushover. It was almost precisely what the name said: it consisted of the deliberate explosion of a fully fueled V-2 rocket under conditions simulating those of Operation Sandy.

Looking over the list of the V-2 firings[3] one might think that the performance figures mirror the results of slow aging of the equipment. The first seven shots did not reach 100 miles, probably largely because of a lack of practical experience on the part of the experimenters. But then as the men acquired more skill in matching components, greater heights were reached. No. 9 climbed to 104 miles; then, after two failures, no. 12 went to 102 miles. There followed two which didn't do too well and one outright failure (no. 14), but no. 16 went to 104 miles and no. 17 even beyond 110 miles. (There seems to be some uncertainty just how high it did go. The NRL Upper Atmosphere Research Report gives the figure of 113.7 miles, while General Electric's Final Report on Project Hermes retains the figure that was first announced, 116 miles.) In any event no. 17 proved to be the highest shot in the whole Program.

After that the record altitudes dropped. No. 21 still made 100 miles and no. 30 came close with 99 miles. No. 43 was the last to pass the 100-mile mark by going to 104 miles. But this gradual dropping of the peak altitudes reached was not a question of old age, although this factor may have entered to a slight extent. The V-2's were not fired just to see which one would go highest; they had work to do. Lots of people wanted to find out many different things. Twenty-four rockets underwent fairly important contour changes which presumably did not contribute to climbing ability.

[3] See Appendix 3; Sandy, Pushover, and a few other special shots are not included in the list.

Moreover, the take-off weight was increased all the time. The standard empty weight of the V-2, including the 2200-pound warhead, had been 8800 pounds. The shots of 1946 were overweight by 150 pounds. Those of 1947 weighed 400 pounds more than the original standard. In 1948 the added extra weight was 527 pounds, and in 1949 the original empty weight of 8800 pounds had grown to 9836 pounds. That the age question was minor was proved by one shot outside the Program (and chronologically near the end of the Program), which was fired for the purpose of seeing what altitude could be reached by a V-2 if altitude was the sole aim. The rocket climbed to 128 miles.

But I am getting ahead of the story once more.

With the coming of the V-2 to America a new chapter, or even a new volume, began for a fairly old science. This was the science which had more or less begun some three centuries ago when Monsieur Perier carried a barometer up a mountainside in France to see whether the air pressure would really drop. "Man," a famous biologist phrased it at a much later date, "is a creature which inhabits the bottom of the air ocean." True enough, but he should have added that Man, although he didn't actually dislike his environment, has always been possessed by the desire to leave the bottom of the air ocean, to get into the air ocean itself, and, if at all possible, above it, to see what space is like. The V-2 was Man's first chance to do this, even if, for a beginning, the rockets merely served as Man's messengers with their operators resigned to stay planet-bound at the bottom for a little while longer, until they were certain of the natural facts they would encounter higher up. To find out Man needed remote sensing organs —called instruments—which not only could be more sensitive than his natural organs but which also could detect and measure things like magnetism. In addition to such organs, Man needed a mechanical messenger which would carry them to places where Man could not or would not yet go himself.

The first modern concept of the atmosphere was proposed about 1730 by Dr. Edmond Halley, he of comet fame. In fact Dr. Halley's thoughts about comets and his thoughts on the atmosphere were connected because his chief contribution to the question of comets was to deny Aristotle's weird notion that they were atmospheric phenomena. It was relatively easy to show that comets were much farther away than the moon—even though numerical values for their distances could not be given—and that they, therefore, were far above the atmosphere. The next question, inevitably, was "And how high is the atmosphere?" Doctor Halley placed the limit at 45 miles above sea level and added that there were three distinct layers. The lowest extended from sea level to a height of about 9 miles and was characterized by a steady drop in temperature as one ascended in it. The next layer extended from 9 to 18 miles and had as its main characteristic a uniform and very low temperature. In the top layer, beginning at 18 miles, the temperature was supposed to fall off some more.

All this sounds surprisingly modern, except for the small figures. And Dr. Halley had nothing on which to base his conclusions except the indications of the still crude barometer and the then even cruder thermometer. Naturally the only way in which readings from these instruments could be obtained was by climbing mountains with them. Half a century later it became possible to ascend into the atmosphere even in places where no mountain happened to be available. The balloon had been invented.

The new means was utilized without the slightest hesitation. Literally the very first hydrogen balloon which took to the air (December 1, 1783) carried a barometer and a thermometer. One year later, almost to the day (November 30, 1784), came the first balloon ascent specifically devoted to meteorological research. That vertical expedition had been fathered by an American, Dr. John Jeffries of Boston, but took place in England because Dr. Jeffries resided in London at that time. He carried 6 bottles filled with distilled water along with him; they were emptied at various stages of the flight and sealed, thus producing air samples for later analysis. Also in the gondola were a mariner's compass, an electroscope, a thermometer, a barometer, and one of the relatively new hygrometers for measuring the humidity of the atmosphere.

During the following five decades quite a number of meteorological expeditions of this type took place, several of them to such altitudes that one wonders how the aeronauts survived in their open gondolas. During the same time-interval an important invention was made. The instruments no longer needed to be watched constantly, for they recorded their readings, either on a blackened drum or on paper tape. These instruments seem to have been misnamed from the very outset—they were always referred to as "self-recording instruments" which, of course, they are not. A recording barometer records the air pressure, not itself. But even though these instuments existed, their full possibilities were not realized at once. Something had to happen first.

In 1874 two scientists, Sivel and Crocé-Spinelli, had made a flight which carried them to 7400 meters (24,270 feet). During the following year they tried again; the name of their balloon was Zénith, and Gaston Tissandier joined them in the gondola. The balloon lived up to its name; it reached an altitude of 8840 meters (29,000 feet), just about the height of the peak of Mount Everest. But when the Zénith landed, only Tissandier was still alive. He barely survived and is said to have been sickly for the remainder of his life. It was this ill-fated flight which caused meteorologists to examine their premises of operation. The purpose of the vertical expeditions was to obtain information about conditions at high altitudes. But that information was not obtained by the meteorologists themselves, who quietly froze, had trouble breathing, and suffered occasional hallucinations; it was obtained by the instruments. So why not send up only instruments of the recording type?

A Frenchman named Brissonet seems to have been the first to make this suggestion (in 1879) and others took it up. No human lives would be endangered, they pointed out, adding that the expense would also be much less. All the instruments one could imagine then weighed less than a single aeronaut. And because of the lighter payload the balloons would probably go higher too. In addition to all the people who wrote about this idea there were a few who actually went to work. In France there was the team of Gustave Hermite and Georges Besançon. In Germany there were Professor Assmann and his pupils. Hermite and Besançon began by learning how to make small balloons; they intended to buy the instruments and just strip them of all unnecessary weight. When they tried that, however, they discovered it could not be done. Hermite said later that they had approached the problem as scientists, then had turned into apprentice mechanics, and had not been able to resume their character of scientists until after they had become expert instrument makers. But by the winter of 1892–93 they had run the full circle of this transformation and their first *ballon sonde* rose near Paris on March 21, 1893, and attained an altitude of 15,000 meters (49,215 feet). The German group under Assmann either dawdled a bit or indulged in extra thoroughness. Their first balloon was not ready until April 27, 1894. But as if to make up for the loss of time it rose to 21,800 meters (71,526 feet), higher than anything had ever gone before.

A few years later, in 1898, the French meteorologist Léon P. Teisserenc de Bort introduced some of the terms with which we are now so familiar. He had noticed, as had Professor Assmann and his group, that Dr. Halley's old idea about an atmospheric layer of constant temperature was actually correct. In high altitudes there was a uniform temperature of —55 degrees centigrade (—67 F) and apparently no vertical air currents. Logically, then, this layer would stratify according to density and since the main component gases of the atmosphere do not have the same density, the different strata might have different chemical composition. The whole, obviously, should be called "stratosphere." The layer below the stratosphere had constantly changing temperatures—if Teisserenc de Bort had been a biologist he might have used the biological term "poikilothermal"—and had both horizontal and vertical currents. Using a Greek word which means "to turn," but in this case rather "to churn," he called the bottom layer the "troposphere." The dividing shell between the two was called "tropopause."

Later it was found that the height of the tropopause depended on geographical latitude. The earth is not a perfect sphere and the tropopause exaggerates its deviation from the mathematical sphere. There are also seasonal variations, but the average figures for the height of the tropopause are: 6 kilometers over the poles, 11 kilometers at 50 degrees latitude and 18 kilometers over the equator, corresponding to approximately 19,700, 36,100, and 59,100 feet. Some 20 years later the existence of a very tenuous layer above the stratosphere, predicted by Dr. Halley, was recognized. Because that layer is subject to ionization it was named the "ionosphere;" the dividing shell between stratosphere and ionosphere is called "stratopause."

All this time there was an instrumentation problem.

True, the instruments were steadily improved in reliability and in accuracy. They responded faster to changes and grew lighter. The balloons improved too and went higher and higher before bursting. Once 30,000 feet had been "normal" and everything above that exceptional. Slowly the norm was pushed to 50,000 or 60,000 feet.[4] But the main problem remained: *the instruments had to be found afterward.* And the recovery rate is hardly ever as high as 50 per cent, even in flat country and far from the sea. It drops sharply in mountainous areas and near the seashore.

Since there was (and is) no means of improving the recovery rate the only way out was to make the instruments broadcast their readings. This is what is now called telemetering. While the V-2 Program would not have been impossible without telemetering it would have been far less efficient and useful. The principle of telemetering was invented quite early, by the Dutch instrument maker Olland of Utrecht in 1877. Of course, there was no radio then; Olland's method was one for telemetering ground stations over a wire. To understand the method, let us use as an example an ordinary factory clock, the kind which has three hands, one for hours, one for minutes, and a so-called sweep hand for seconds. Let's say now that we want to telemeter a seashore station which records the direction of the wind and the rise of the tide. We'll take the hour hand and connect it with a weather vane; due north is, say, on the "12," due south on the "6," while the other figures on the clock dial correspond to the intermediate points. The minute hand we'll gear to a float for the purpose of reporting the rise and fall of the tides. The sweep hand we leave alone to travel once round the dial every minute. The sweep hand, every time it passes the "12," makes a short contact, sending a short electric impulse through the wire. When the sweep hand passes the hour hand which reports the wind direction, it sends out an electric impulse of a slightly different duration. And the same happens when it passes the minute hand, reporting the tides. At the other end of the wire, then, you get a short impulse at regular intervals when the sweep hand passes the "12," which is called the "reference pulse." The other two kinds of impulse supply the actual readings; all one has to

[4] What is probably the record for small unmanned rubber balloons was obtained on September 8, 1930, near Hamburg. That balloon went to 35,900 m or 117,800 ft (22.4 mi). The large plastic balloons in use in the United States since about 1947 attain 100,000 ft as a norm; in exceptional cases they go 25,000 ft higher.

do is measure their distance from the reference pulse. If you substitute a transmitter for the carrying wire, you have wireless telemetering.

It was applied to balloon-borne instruments for the first time about 1925 by Professor Pyotr A. Moltchanoff of Slutsk in Siberia. He was then promptly hired by the German Zeppelinbau and a modification of his experimental type became known as the *Kammgerät,* in translation "comb apparatus," for the reason that the electric contacts were arranged like the teeth of a comb. While this system was the first, it could never be developed to perfection, even though hundreds of people tried their skill on the problem. The main difficulty was that just the "reference pulse" tended to be unreliable. Olland, for his ground stations, could assume that his driving clock would be accurate for long times. But when you lift a clock from the nicely warmed troposphere near sea level into the cold stratosphere it will cease to be accurate—at least if it is a clock small and simple and light enough to be carried by a small balloon.

Other methods were tried and succeeded. A Finnish scientist, Dr. Vilho Väisälä, transmitted readings by changing the wavelength of the carrier wave. The American system, developed by Diamond, Hinman, and Dunsmore of the Bureau of Standards, operates by modulating the carrier wave. One recent type of American "radiosonde," as such balloon-borne meteorological equipment with transmitter is usually called, uses a record with concentric grooves (not a spiral groove as in customary records) and sensing instruments with arms that reach into the proper grooves at regular time intervals. Each groove has a Morse code signal impressed on it, which is broadcast in the sequence: pressure, temperature, humidity, pause. The pause has the purpose of telling which is which: pressure comes first after the pause.

The histories of such divergent branches of engineering as instrument manufacture, electronics for telemetering, and rocket development happened to dovetail nicely. When the big rockets became available as carriers, both instruments and telemetering systems existed. They needed only to be attuned to one another.

In the V-2 Program not only the instruments carried as payload were telemetered, but a number of data about the rocket itself. For example, there was a telemetered gauge recording the pressure in the combustion chamber of the rocket motor, and another recording the pressure in the oxygen tank. There was a telemetered tachometer measuring the speed of the turbine-pump assembly—that is why we know that the pumps got stuck in test no. 8—and a set of instruments reporting on the positions of the graphite vanes, from which we learned how the vanes misbehaved when they did.

But, even with telemetering, certain things had to be recovered physically, such as the cameras and the film they had taken. Occasionally a bag of seeds or a box of fruit flies was carried up in order to test what the cosmic rays might do to their genes. Not all the instruments were put into the "warhead" of a V-2. Some were placed in the compartment below the warhead which houses the rocket's own instruments and incidentally has some space to spare. Cameras could be stuck into the main section between the fuel tanks, or fitted into the stabilizers. The telemetering antennae were strung from the stabilizers to a point higher up on the rocket, or else trailing antennae were used, which the rocket pulled out of their boxes as it rose.

But recovering things remained a problem, although in many ways a much easier one than it had ever been with balloons which drifted with the wind. The V-2 could be aimed to crash at a certain point in the desert. And even if it did deviate from the prescribed course it was tracked by radars and by telescopes and cameras on antiaircraft mounts. In every case one could tell where the rocket had crashed. But a V-2 rocket, falling back from an altitude of about 100 miles, tends to do as much damage with a warhead full of pressure gauges and cosmic-ray counters as it would with a warhead full of Amatol. That it makes a 60-foot hole in the desert of the Impact Area is in itself highly unimportant, but the rocket itself becomes part of the general débris and long and hard digging in the sands of New Mexico is required to recover even the things that hold together.

Parachutes were out. An empty V-2 still weighs 4 tons and while a parachute that will carry 4 tons can be built, it won't fit into a shoebox. Such a parachute would use up *all* the available space; one might wonder whether all the available space would be enough. Even a parachute just for the warhead was not too heavy, but too bulky. In a rocket especially designed for such usage one might put a special parachute compartment between instrument section and fuel tanks. But in a finished design like V-2 there was just no space available.

Somebody, I don't know who, found a way of preventing the high-velocity impact without adding extra bulk. The rocket traveled nose first under power and, in theory, maintained this attitude all the time. In reality it has happened that a rocket moved almost broadside on after Brennschluss or even tumbled slowly end over end, as a result of a small accidental off-center impulse received during the last fractional instant of burning. That did not influence the trajectory at all because there was no longer any power on and the air above 150,000 feet is too tenuous to offer measurable resistance. For the same reason the stabilizers were incapable of influencing the attitude. They did not become effective again until the rocket had descended to 120,000 feet or so. Then they did turn the rocket over, nose down, no matter what its attitude had been before. And then the rocket struck the ground nose first, like an airplane bomb, only several times as fast. If the rocket could be prevented from landing nose first, the recovery problem would be greatly simplified.

That could have been accomplished either by blowing off the stabilizers or by blowing off the warhead. The latter method was chosen, because it is simpler to blow off one warhead than it is to blow off four fins. And it had the extra advantage of reducing the falling speed to a greater extent. A V-2 without fins is still reasonably streamlined. A V-2 without warhead is certainly not. The separated warhead would still fall rather fast but in all probability rolling over and over, and it would strike the ground at an odd angle. As for the rocket's body, that should do much better than the warhead. The stabilizing fins would try to straighten it out while falling. But it would have a gaping and probably jagged hole in front in which the air would be caught, acting against the stabilizers. And the heaviest section of the rocket which was left, the motor and the turbine-pump assembly, would be located between the stabilizers. The whole would have the lowest possible degree of stability and would pancake through the air in such a way that it would hardly acquire enough speed to break through a pavement.

This method of aiding recovery by having the rocket arrive on the ground in pieces was dubbed "airburst" and it worked just as expected. But not at first. In the first try, with V-2 no. 5, 1 pound of TNT was tied to each of the four upright struts of the rocket body, just below the warhead. When the radar showed that the rocket was approaching the 100,000-foot level from above, a radio signal was sent out to explode these 4 pounds of TNT. The charges went off but an observer who watched through powerful glasses reported that the rocket was momentarily enveloped in a cloud of smoke—and then it came out of the smoke cloud in one piece and crashed as if nothing had happened. The telemetering record also proved conclusively that the instruments had kept on working after the explosion of the disrupting charge.

V-2 no. 6 repeated this strange performance in every detail. When the rocket for test no. 9 was to be equipped, the scientists of the Applied Physics Laboratory, in charge of that particular rocket, added 1 pound of nitrostarch to every pound of TNT as a backing. That was enough; the airburst did separate the warhead. Ironically that warhead was never found afterward; it must have buried itself in loose sand just deep enough to escape detection. The main body of the rocket was found in the desert, looking as if it had dropped only 100 feet or so—no longer usable and quite evidently banged up, but still in one piece. From then on the airburst became standard procedure, not only for V-2's but also for other large rockets. Customary altitude for the airburst is 100,000 feet so that the dense bottom layers of the atmosphere will retard the pieces. A higher airburst might lead to still lower impact velocities but would scatter the parts over too large an area.

How a fine ionosphere shot of a V-2 looks when plotted is shown in

Figs. 45 and 46. Figure 45 shows the actual shape of the trajectory; altitude and velocity for any moment can be read off Fig. 46. V-2 no. 21, which provided this performance, obtained, among other things, a set of readings of air pressure in various altitudes which proved that the earlier method of calculating them had been quite adequate to the task. It was in 1925 that the National Advisory Committee for Aeronautics (NACA) first published its Report No. 218, tabulating the characteristics of the atmosphere up to 20 kilometers (say, 65,600 feet). The following table shows the values of air pressure from that report, along with those actually measured by V-2 no. 21.

TIME AFTER TAKE-OFF	HEIGHT		VELOCITY ALONG TRAJECTORY		PRESSURE IN MM OF MERCURY	
(SEC)	(FT)	(KM)	(FT/SEC)	(M/SEC)	V-2	NACA
0	3,740	1.2	0	0	660	658
10	5,580	1.7	358	109	630	619
20	11,480	3.5	853	260	495	493
25	16,400	5.0	1141	348	410	405
30	22,970	7.0	1436	438	300	308
35	30,840	9.4	1738	530	210	217
40	40,350	12.3	2119	646	130	138
45	52,170	15.9	2657	810	76	79
47.5	59,060	18.0	2953	900	52	56
50	66,930	20.4	3265	995	33	40

Up to 20 kilometers the data could be compared with balloon data taken simultaneously, which agreed within 1 per cent of the values involved. Above that altitude they could be compared with those of rocket no. 12 (of Oct. 10, 1946) with which they agreed within 10 per cent. For these higher altitudes V-2 no. 21 gave the following results:

TIME AFTER TAKE-OFF	HEIGHT		VELOCITY ALONG TRAJECTORY		PRESSURE IN MM OF MERCURY
(SEC)	(FT)	(KM)	(FT/SEC)	(M/SEC)	
52.5	75,460	23.0	3576	1090	23
55	85,000	25.9	3937	1200	14.5
57.5	95,150	29.0	4265	1305	9.6
60	106,300	32.4	4658	1420	5.8
65	130,910	39.9	5050	1540	2.3
70	156,050	47.5	4887	1490	.95
75	179,800	54.8	4723	1440	.40
80	283,860	81.8	4560	1390	.18
100	287,300	87.5	3937	1200	.003

In addition to temperature and pressure measurements, rockets have taken pictures of the earth from high altitudes and, of less popular appeal but more valuable, pictures of the solar spectrum. Our atmosphere is remarkably opaque to all but the visible wavelengths and no spectrogram taken from the ground can be complete for this reason. Especially the short-

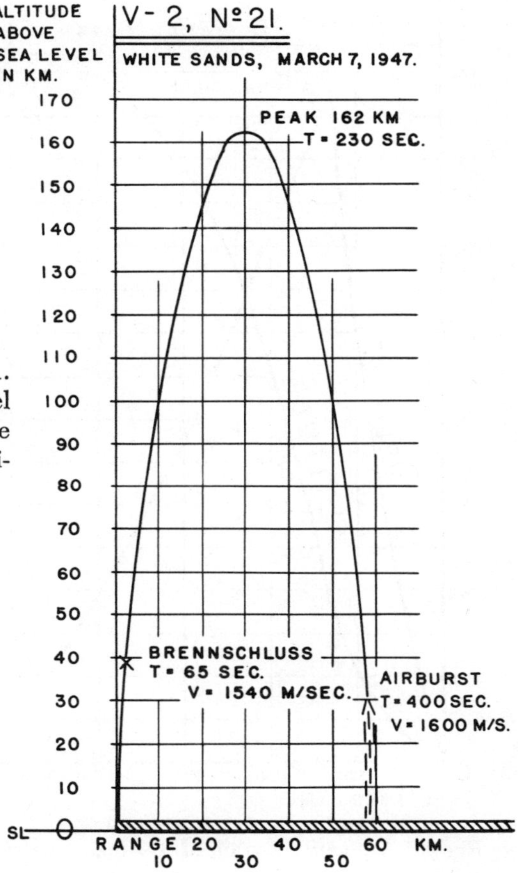

Fig. 45. Trajectory of V-2 no. 21. Take-off 4000 feet above sea level (SL), the altitude of the White Sands Proving Ground. Peak altitude slightly over 100 miles.

wave end of the solar spectrum is erased by a layer rich in ozone (caused by the radiation itself), the existence of which was indicated for the first time in 1935 by the stratosphere balloon Explorer II which ascended to 72,395 feet. V-2 rockets took pictures of the spectrum from below the ozone layer, from inside the ozone layer, and from points above the ozone layer. Finally, V-2 rockets have measured cosmic radiation in high altitudes and obtained air samples from as high as 45 miles.

One more detail may be mentioned. Astronomers are, of course, greatly interested in meteorites which fall from space. And even those meteorites which are too small to reach the ground provide information from which to draw conclusions about the atmosphere. But for such conclusions one has to make a few assumptions about which one cannot be perfectly certain. Now if there were some artificial meteorites high up, about which a few things, like particle size and velocity, were known, one could check the assumptions. The particle size of these "artificial meteorites" should be

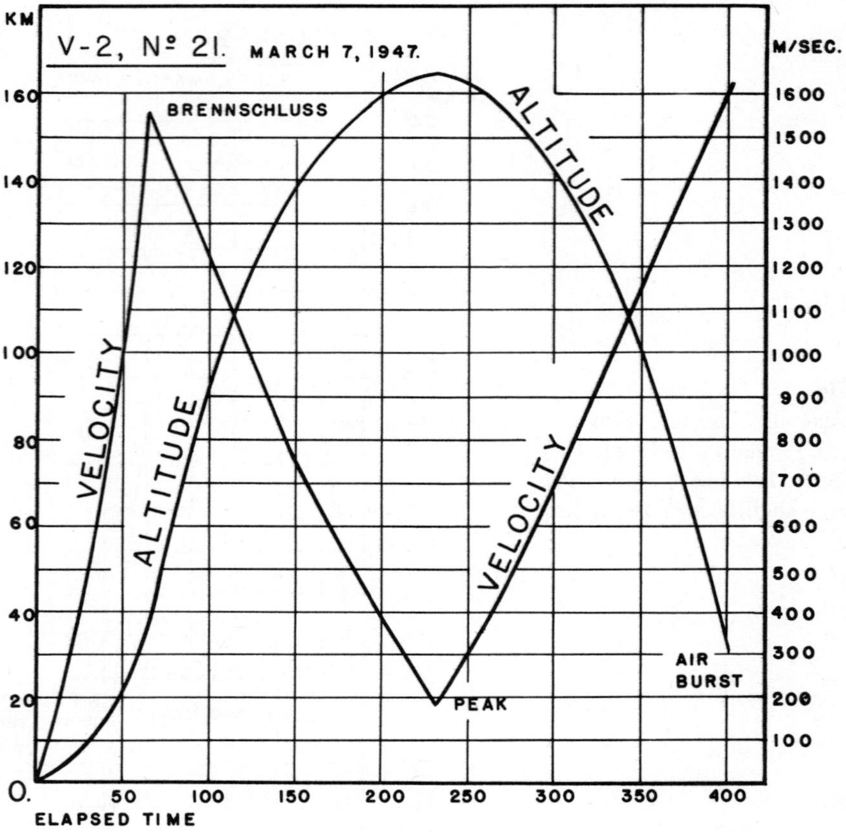

Fig. 46. Velocity diagram of V-2 no. 21. Altitude curve (km) at left; velocity (m per sec) at right.

small, incidentally; for example, the particle size of the smoke of gunpowder. Therefore, rocket no. 13 was equipped for what was labeled the "powderpuff experiment." It was equipped with blackpowder charges which could be exploded at the proper instant. As far as the *technē* was concerned, everything went well. But the blackpowder exploded so slowly that no "puffs" resulted, just streamers, which were of no use except for indicating that there was a high wind.

Since blackpowder did not explode fast enough, a shaped charge seemed the likely answer. Consequently no. 17, which happened to establish the altitude record, was "armed" with half a dozen M-9 rifle grenades (Fig. 32) for making artificial meteorites. The best available information seems to show that the rifle grenades did not go off; their firing was entrusted to the same device which produced the airburst and it functioned so sluggishly that the airburst was delayed for more than 100 seconds. Examination of the fragments indicated that the grenades had been discharged, but not at the right time. They may have gone off along with the airburst, or when the rocket struck the ground.

After something has been brought to a conclusion, the drawing up of a balance sheet is an inevitable procedure. Of course a formal balance sheet —called "evaluation"—of the V-2 Program has been drawn up, but to my mind it isn't a proper balance sheet. What kind of a "balance sheet" do you have when every factor is on the positive side of the ledger? Official reports go in for a careful listing of failures, but I feel that a properly understood failure is a positive item too; it points up what can go wrong and usually suggests what can be done to avoid a repetition.

But all right, let's list the failures. Officially thirty-two V-2 rockets were "failures," meaning that there was either a known malfunction or that the performance of the rocket clearly indicated that a malfunction existed. Of the thirty-two failures, an even dozen occurred at lift, the other twenty in flight. The earliest "flight failure" took place 8 seconds after lifting off the firing table, the latest 57.5 seconds after lifting. Eleven occurred within the first 30 seconds of flight, nine after 30 seconds of apparently proper functioning. This rather even distribution through the whole length of the burning time proves that there was no special cause which produced more failures than any other. Of the "in flight" failures, eight were due to malfunction of the steering system, the other twelve to malfunction of the propulsion system. Of the dozen failures that occurred "at lift," the steering system was blamed for six, the propulsion system for three, and miscellaneous causes for the other three.

While the V-2 Program was in full swing—and Robert P. Haviland was devoting his time to the Bumper Program, another portion of the Hermes Project, which was to create the next sensation—everybody realized that at one point there would have to be a big change. The supply of V-2 rockets would give out. The Program had started with getting the components for a series of twenty-five V-2 shots together and ready. Then a second series was started. Maybe a third series could be completed, but no matter how things were stretched it was inevitable that one day the Meillerwagen would carry the last V-2 to the firing site.

What then?

Of course more could be built, but this would have been very nearly equivalent to standing still. New rockets were needed—new types, new designs, not just new V-2's. The problem was: just what was wanted? The military men naturally wanted a ballistic missile, which is one thing. The scientists wanted to go on exploring the upper reaches of the air ocean and thought about a new sounding rocket, which is another thing. And even if the designers concentrated on a sounding rocket for the time being, the problem still existed—just what was wanted? The V-2 had been capable of carrying a ton of payload to around 100 miles. It could carry more, but it couldn't go quite as high then. But because of the way it had been originally designed it couldn't carry less than a ton.

One possible answer to the problem, then, was that a rocket was wanted which could carry as much as a ton of payload or as little as 200 pounds, and which would go as high as a V-2 with the 1-ton payload and correspondingly higher with lesser payloads. But others felt that this was too much energy expenditure. The military might like such high payloads, because they ultimately wanted to build ballistic missiles which would have to carry heavy warheads, but for scientific purposes a payload below 200 pounds seemed to be fully sufficient. Nor did every "mission" require a climb to above 100 miles; many research purposes did not need more than about 65 miles. Now, if one were satisfied with an altitude of 65 miles and a payload of a mere 100 pounds, one would, of course, get a much smaller rocket. This, in turn, meant that one did not need a large firing crew. Such a smaller rocket might be fired by just three men (not counting camera men and radar trackers, of course), and one would have a much more flexible research tool.

The outcome of two such trains of logic is obvious: two different rockets, one large, one small. The large one was *Neptune,* the small one was *Venus.* At least that is what they were called when they were just projects on paper. When they had become what in military circles is called "hardware," their names had been changed to *Viking* and *Aerobee,* respectively.

It is impossible, from here on, to maintain chronological sequence. Both rockets were ready and were actually fired while the V-2 Program was still running. Nor were they the only rockets that thundered upward from White Sands during the years from 1947 to 1951. So instead of following work from year to year we now have to follow it from rocket to rocket.

The Viking was announced for the first time—still under the name of Neptune—in the June 1947 issue of the monthly magazine *Aviation,* with an article written by Carl Harrison Smith, Milton W. Rosen, and James M. Bridger. I lived in Washington at the time, and friends in the Navy told me that there was some talk in the Department that the first firing should be attended by selected reporters, writers, and photographers. Naturally I wanted to be on the list and, having various schedules on my mind, I asked when the first Viking firing could be expected. Well, no date had been set and nobody could really tell, but February or March of 1948 seemed a likely time. The second half of 1947 passed without any news, and in January 1948 I decided to see to it that my name was put on the list before it was too late and called the Naval Research Laboratory. Whoever it was I talked to grew rather irritated at the very mention of the word Viking and finally exploded: "I don't know who told you that the firing could be expected in February or March but whoever it was, he is an awful optimist. It won't be for many months." Actually Viking I was not fired until May 1949, and visitors were not invited until the firing of II in September of the same year. Meanwhile I had moved away from Washington and never got to see a Viking take-off.

As could be expected, then, rocket shots have become very nearly routine performances, but at that time a shot needed, in addition to very fine weather and good workmanship, a reasonable amount of good luck too. And it became apparent, as short reports on the Viking rockets were published in the aeronautical journals, that the men who worked on and with those rockets had to do without any good luck for quite a long time. Not because the Vikings were so different from preceding rockets, especially the V-2, that malfunctions resulting from too many innovations were inevitable. The rocket used the fuels used in the V-2. The fuel was pumped into the motor by centrifugal pumps driven by a turbine which derived its power from decomposing hydrogen peroxide, just as in the V-2. The internal arrangement was the same too, with the exceptions to be mentioned.

The first Vikings were slim cylindrical rockets with a diameter of just a little over 32 inches; the reason for this particular diameter was the very simple fact that sheet aluminum comes from the factory with a width of 100 inches. The rocket stood almost as tall as the V-2, but the gross weight of the first Viking was less than the empty weight of the later White Sands V-2's—namely 9650 pounds. The V-2 had been designed for production with separate fuel tanks; in the Viking rockets, which were built by hand one after the other, the alcohol tank was an integral tank; that is, the tank was formed by the skin of the rocket. (In later Vikings the oxygen tank was made integral too.) This not only is a weight-saving device but also eliminates any space between the tank wall and the rocket skin, space in which, in case of a leak somewhere, alcohol vapors and oxygen gas can accumulate, mix, and cause an explosion.

There were certain innovations, the most interesting of which was the method of steering. The graphite vanes had been eliminated entirely, mostly because they exert a drag in the exhaust. In the Viking rockets the whole rocket motor was mounted in a gimbal ring so that the servo motors could swing the rocket motor to compensate for unwanted tilt. The trim tabs on the external fins had been retained, to correct for possible misalignment of a fin. The hydrogen-peroxide tank, in the early Vikings, took the shape of a pipe which was coiled around the turbine housing to save space. Unlike the V-2 the Viking was ignited with "main-stage thrust"; that is, with running turbine. Another innovation was the use of the exhaust steam from the turbine. Normally this escaped from vents in the side of the rocket in equal volume so that the reaction from the various vents would compensate. But if the rocket rolled, a valve arrangement throttled one or the other of the steam vents so that the "unevenly" escaping steam corrected the roll.

One more feature in which the Viking differed from the V-2 was the method of road transportation over short distances, as for example from the assembly hangar to the firing site. This was done by means of the so-called Barr car (after its designer Irwin Barr) which consisted of a square frame fitting over the tail-fin assembly and a nose "yoke." The nose yoke had one wheel at the bottom, the tail frame had two, so that the Viking was supplied with something resembling the tricycle landing gear of a modern airliner. It could be towed by a jeep; the erection of the rocket was accomplished by hoisting it from the Gantry crane. The earlier Vikings were towed as a whole, for the later Vikings the nose cone was transported separately.

In 1947, when the first disclosure was made, the rocket was under construction at the Glenn L. Martin plant in Baltimore, while the motor was built by Reaction Motors, Inc., in New Jersey. Delivery of the rocket was made in January 1949. The Glenn L. Martin men who traveled to White Sands with it still slipped occasionally and referred to the rocket as "Marco Polo," which had been a kind of nickname around the factory in the interim period between the abolishing of the original name of Neptune and the legalizing of the final name of Viking.

And then all those little things happened which can show up only on the proving ground. On March 7, 1949, a ground test (or "static firing") was attempted. It was called off at X – 15 (minutes) because it became evident that the disconnecting plug of the nose section failed to mate properly. On the following day that had been taken care of but then the vent valves of the oxygen tank did not close properly, draining off the supply of compressed nitrogen in the pressure bottles. Since the system had been planned for helium for actual flights, helium was substituted after the valves had been checked. Then a pressure line burst. Three days later, on March 11, this had been repaired and a static firing got underway for a period of 31 seconds, in which insulation burned off and steam leaked from the turbine.

On April 22, the control system misbehaved, which did not matter much for a static firing but would have been simply sad if a take-off had been made. On April 25 that defect was corrected, but the static firing was interrupted at 24 seconds because smoke began pouring out of the Viking. However, the cause was just new insulation on the steam lines baking and the rocket was declared ready for flight. The first take-off was scheduled for April 28 and then postponed for a day because of bad weather. The next day it was canceled because the oxygen vent valves misbehaved again. And the firing which was then scheduled for May 2 had to be postponed for another day because a broken part was discovered in time.

On May 3 the rocket rose, with a slight delay because the vent valves still did not function quite properly. But it made a beautiful take-off, just the same. The ascent, as far as it could be watched directly from the proving ground, looked fine too. But when the rocket was 17 miles high, at X + 54 seconds, the motor cut off. For this reason the peak altitude reached, at X + 160 seconds, was only about 51 miles. Maximum velocity had been 2250 miles per hour.

From all indications it was obvious that there had still been fuel in the rocket when the motor cut off. But why a motor which is going beautifully should cut off when there is still fuel was a mystery. The crew which had put in endless hours on the rocket was baffled. The crew chief, Pres Layton, could not answer their questions. The man in charge of the whole Viking Program, Milton W. Rosen, admitted to being mystified. And they were all disappointed. The official program had not said that they had to top the V-2 record, but that is what everybody had expected to happen. The payload had been less than 500 pounds (464, to be precise), and the tables published with that article in *Aviation* had predicted a 190-mile peak for a payload of 500 pounds. Of course that prediction had been based on 75 seconds burning time and the motor had stopped after 54 seconds. Why had it stopped? The talk and the questioning went around in circles, always coming back to that one question. Why, why, why should a rocket motor that has been going for nearly a minute stop burning when there was fuel for another half minute available? Why?

The whys were cut short by the arrival of Viking II which surely would clinch the high altitude record. Milton Rosen tried to anticipate every possible source of trouble and failure and probably worried quietly; Pres Layton saw to it that the crew worked with the greatest thoroughness and care possible—and probably checked everything personally five times over. But the general feeling was that whatever had happened to Viking I would not happen again, because everybody had so much more experience now. Viking II would climb to 150 miles at the very least—in fact it had to, because the press had been invited for this shot.

As soon as feasible a static test was run. For this test the rocket's fins were bolted to the firing table, and the firing table was placed over an open pit, water-cooled, so that the rocket motor could be fired for the full duration, if desired, and the functioning of all instruments could be checked while the motor was roaring. For the static test of Viking II a duration of 30 seconds had been decided on. The test went fine and the motor cut off neatly at X + 30. But for the last few of these 30 seconds black smoke had been coming out of the tail. That had happened more than once with Viking I—was this just some oil burning off or something more serious? As has been mentioned, the fuel-feeding system was like that of the V-2, hydrogen peroxide being catalyzed into steam which drove a turbine. The turbine housing had been made in two sections which were bolted together with a gasket in between. In Viking 1 that gasket had obviously ceased to be tight after a number of seconds of operation, and the escaping steam had caused the trouble. And because the peroxide tank had been coiled around the turbine to save space it had been impossible to get at the turbine bolts without taking the rocket back to the hangar and stripping it.

This time the men were prepared; they had special tools with which they could reach in and tighten the thirty-five bolts. They found that the turbine housing had actually leaked, which was all the more incredible since it had been tested before the static run and had been found to be tight. Even with the special tools it took two days and two nights to tighten all the bolts to the same degree of tension and to test the turbine housing several times to be certain that it was tight.

Firing was scheduled for 9 A.M. on August 26, 1949. Everything went well, if somewhat slowly—the press was not chased from the launching area until 11 A.M. At 11:29 Milton Rosen said "Fire!" Henry Hardin pushed the firing switch. The igniter burned and threw sparks, and the nose plug dropped, but nothing else happened. The motor did not catch and the needles of the instruments remained where they had been prior to the command. After 10 seconds of wonder mixed with faint hope, the cut-off switch was thrown. Then it soon became evident what had happened: liquid oxygen had leaked through a weak seal and flooded the turbine, freezing the valves shut. The fate averted after the static test—namely, that the rocket would have to go back to the hangar—could not be averted again.

The new firing date was September 6. A few minutes before 10 A.M. Milton Rosen said "Fire!" and the rocket took off. The reporters tried to coin new phrases in their minds as they watched the long thin rocket lift slowly and disappear in the sky with ever-increasing speed. But the men who watched the instruments felt uneasy: would something go wrong again? At X + 49 seconds the motor cut off. With the speed the rocket had at that instant it would coast to 33 miles at best. It did, or so it looked at first; later the peak altitude was revised downward to 32 miles.

Still, something good came of this flight. It was now fairly certain that the motor cut-off was somehow connected with the leaky turbine casing. The engineers at Reaction Motors, who were sure of the reliability of their motor because of the static tests they had run themselves, evolved a theory of the mishap. The two-piece turbine casing was tight at take-off. But after it had been subjected to hot steam under pressure for a while it became deformed, and the gasket could not hold the steam in any more because it no longer fitted. The steam escaped into the tail section, scorched the insulation off the wiring, and caused a short circuit which shut everything off. So far this was only a convincing theory, but one could find out whether it corresponded to reality. After the turbine had been tested in the factory there was no reason why anybody should want to get at the turbine wheel again. Hence the casing could just be welded together, making it impossible for any steam to escape. This was done, and the theory was proved correct. No more steam leaks, no more premature cut-offs.

But new worries promptly developed. Not only in connection with Viking III—incidentally the first of the Viking rockets to carry full upper-atmosphere instrumentation, but also about Viking IV, quite a bit ahead of time. Viking was a Navy project, Viking should go to sea—that is, it should be fired from shipboard. A ship, the U.S.S. *Norton Sound,* was being converted for this purpose. I believe that some of the more enthusiastic Navy men had wanted to shoot all Vikings from shipboard, beginning with the first one, but Rosen had wisely insisted on proving-ground firings. After all, even a ship is built on land, and at the beginning it was better not to add a heaving firing table to all the other possible difficulties.

But Viking IV was to be a shipboard firing, which brought a new problem into the picture. Something like a launching tower was needed. Since it takes hours to get a rocket ready for firing, even if everything is mere routine, this could not just stand freely on the firing table. On land the weight of the rocket is enough to hold it down in case there is a little wind —at White Sands, unfortunately, wind always carries sand and dust—and the fin bolts help make the rocket secure. But these bolts could not be expected to do their job if a movement of the ship caused the rocket to tilt. The device used for the firing of a V-2 from the carrier *Midway* did not seem suitable for Viking.

Irwin Barr at the Glenn L. Martin plant was given the job of designing the launcher. Not counting the supporting structure which held it rigid, the launcher consisted of two 20-foot rails sticking up vertically. Between the rails and the rocket were pairs of rollers, one set of them at the very tail end of the rocket. While the rocket was standing, the rails and rollers held it rigid so that it would not topple if the ship should list. On take-off the rocket would "roll" out of the rails. The question was whether it would clear the rails without touching them. If it did touch, it might still get away, but it might possibly fall over. Even if it fell overboard instead of onto the deck, the prospect was not pleasant to contemplate. But while contemplating it Barr had another idea. The take-off could be tested with a dummy Viking; a hollow shell which had the proper dimensions, with whatever weight it happened to have. A solid-fuel unit could be inserted into the shell and the thrust of that unit and the weight of the dummy balanced in such a manner that the dummy would rise at the same rate as the real Viking. Two such dummies were built.

In the meantime Viking III had to be fired.

It was the first with an integral oxygen tank. The turbine casing was welded; but just the same all the wiring, which had passed too close to the hot turbine before, was now on the outside, and each of the wires which had to enter the tail section was in a metal tube. Take-off was scheduled for February 7, 1950, and there was little leeway to this date, since Project Reach, as the shipboard firing was named, was coming up and could hardly be postponed. The first static test did not come off; it looked like a repetition of the first firing attempt of Viking II. But examination showed that the peroxide valve was not frozen, it was "merely" stuck. A second attempt was made late that night, but at X + 14 Layton called for cut-off. The control system was vibrating and wagging the rocket motor back and forth. One more static test was scheduled for February 6, and simultaneously the firing date was rescheduled for February 9.

The static firing came off well, but on February 9 there was a heavy overcast over the White Sands Proving Ground. The weather station said it was overcast all the way to the West Coast except for one large hole. The whole weather pattern was moving eastward, as it usually does, and the "hole" was expected over White Sands around noon. Alcohol fueling, which always comes first, was just finished as a narrow ribbon of blue sky appeared in the west over the Organ Mountains. The weather station cautioned that the open area was followed by an even heavier overcast. Shoot fast, or else. "X" was supposed to be 3 P.M.; it was moved to 2:45, and Viking III rose on the minute. At X + 34 the radar station reported that Viking III was heading too far west. The motor would have to be cut off, or else the rocket would fall outside the proving ground. But they would wait as long as possible. They had to cut at X + 59.6. Peak altitude, 50 miles.

The shipboard firing of Viking IV was to take place where the magnetic equator crosses the geographic equator in the Pacific. The place is usually designated by saying "near Christmas Island," but actually tiny Jarvis Island is closer to it. First, however, the dummy had to be fired, and the *Norton Sound* left Port Hueneme for the test. The preliminary calculations had been made for a 5-degree roll of the ship; if the dummy could be fired without mishap under this condition everything would be right for the firing of the real rocket.

There was a nice smooth sea.

A captain can make his ship roll even in smooth sea if he wants to. But the *Norton Sound* proved herself a stable ship. One degree of roll could be accomplished and sometimes almost two. But that was all. After three hours the captain returned to port. On the one hand it was nice to know that the *Norton Sound* was so stable. But the dummy had to be fired with a 5-degree roll. The next day another try was made. The weather was still fine. The captain allowed the ship to wallow with engines stopped. Not enough. He could produce a 5-degree list by steaming sharp turns, but what was needed was a 5-degree roll, not a 5-degree bank. Somebody suggested that they "sally" the ship—have the crew run from starboard to port and then back again. This shifting of weight should induce a roll. It did, but

3 degrees was the best that could be accomplished. The *Norton Sound* returned to port again.

Why didn't the weather get nasty? In the meantime there was still another possibility. For the actual firing the *Norton Sound* was to be escorted by the destroyer *Ozbourn*. The *Ozbourn* could be summoned from San Diego and made to run circles around the *Norton Sound* at full speed. That should make any ship roll. On the next try the *Ozbourn* joined the *Norton Sound* at sea and kicked up the biggest wake in her power. The *Norton Sound* rolled gently—not even 2 degrees. After returning to port the captain, John Quinn, decided to go next day to the Santa Barbara Channel, where he had always met rough seas.

Captain Quinn was right; the sea was rough. It did not produce 5 degrees of roll, but went a little past 4 degrees once in a while. There was a cycle to the roll, and after it had been learned, the dummy was fired when a big one was coming up. The dummy slid up the rails, cleared, and fell into the sea, tail first, 300 yards from the ship. It could be done. Meanwhile Viking IV had gone through a good 30-second static firing at White Sands. The *Norton Sound* sailed for the magnetic equator on April 26, escorted by the *Ozbourn*. Estimated Time of Arrival was May 5; the firing was tentatively scheduled for May 7. Of course there could be no static firing on shipboard.

The weather was bad on May 7. Moreover, a number of circuits in the rocket had gone bad in the high humidity and had to be replaced. The shot had to be called off—even though the weather had improved—at almost the last minute. New firing date: May 11. May 11 brought broken clouds, but the meteorologist felt certain that it would clear up at firing time. It did. Oxygen fueling began at 3 P.M. At 4 P.M. local time Viking IV thundered out of the rails and began to climb. It kept climbing, going to 105 miles in spite of an extra-heavy payload. It fell into the sea, 435 seconds after take-off, some 8 miles from the ship. In spite of the additional complications caused by shipboard operation and tropical humidity, it was the first fully successful Viking shot.

The second dummy was fired on May 22, during the return trip. It behaved like the first.

For Viking V the schedule was again tight, because Viking VI had to be fired on a specific date which had been designated as T-Day. This was December 11–12, running from noon to noon as does the (Julian) day of the astronomers who do not want to switch dates at midnight when they may be in the middle of an observation. Viking V was put through a 30-second static test in the morning of November 11, the day Viking VI arrived at White Sands. The test was unsatisfactory. Late that night (11:55 P.M.) another static test was run for 71 seconds. At 10:18 A.M. on November 21 Viking V took off. Its burning time, 79 seconds, was longer than that of any liquid-fuel rocket before it, but the peak altitude was only 108 miles. However, Viking v obtained many photographs of the earth from a high altitude.

I'll say more about T-Day later in this chapter. Viking vi was slated to take part in it and was to be fired in the middle of the period, which meant at midnight. The static firing was scheduled for December 1 and miscarried twice in a row because the igniter failed to ignite; there was a bad connection in the cable. Then it was found that the oxygen seal near the turbine leaked; this could have caused a repetition of frozen peroxide valves. Over the week-end the rocket was brought back to the hangar for replacement of the seal and returned to the firing site. T-Day came closer, and with it came the press, who, though present, missed a fantastic spectacle—partly because the firing was at night, partly because the sight took place too high up.

The firing button was pressed 4 minutes and 52 seconds after midnight; Viking vi rose in the glare of its own exhaust, plus floodlights. The trackers could not see the rocket, of course, but followed the exhaust blast. X + 15; the rocket could still be heard clearly and all was well. X + 30; all's well. X + 60; all's well, and hopes were soaring high, higher than the rocket, in fact. X + 62; the exhaust blast disappeared. Premature cut-off again? No, the blast became visible once more and stayed visible for another 4 or 5 seconds. The instruments reported cut-off at X + 70.3 but otherwise gave weird indications. Pressure gauges had jumped unaccountably; the radar had produced strange wiggles before its pen jammed. The impact computer predicted impact points inside the range to the west, inside the range to the east, outside the range . . .

Did the radar station order cut-off? inquired Rosen. Yes, it had, but only *after* something had happened which they could not understand. It was even likely that the rocket had cut itself off before the order could be carried out. The impact computer had predicted 70 miles altitude from the last information received, but this contradicted another reading which said much less. (Actually, peak altitude was 40 miles.) Rosen did not know what to make of all this, but he still had to face the press. It must have been a strange interview. Since he could not answer the main question of "What happened?" the press did not know what to ask. And he did not know what to say.

It took a number of days to establish what did happen, by comparing all measurements, especially the behavior of the pressure gauges and the radar readings, and investigating the wreckage. Dr. Ralph Havens was the first to say, "Viking vi looped the loop," and that was about it. The tail fins had been subjected to heating by air friction and one had buckled and given way. The control system, trying to compensate, could not and turned the rocket to a horizontal position (this was the moment that the exhaust flame seemed to go out), so that it moved up sidewise, losing speed at a fantastic rate as it did so. Then the rocket somehow straightened out, but the motor cut out a few seconds later, possibly obeying the radio command, possibly because of a short circuit caused by the heating that must have taken place.

In contrast to the unheard-of performance of Viking VI, the story of Viking VII sounds almost drab. Static firing on July 31, 1951; firing on August 7, with peak altitude of 136 miles, which was the record not only for Viking but for all liquid-fuel rockets.

Viking VIII looked different. Its diameter was 45 inches and it was shorter (see Fig. 47), with a better mass-ratio. Because the tail section was wider the hydrogen-peroxide tank no longer had to be coiled around the turbine; it was a separate tank like that in the V-2. And when ready for firing Viking VIII—and the others of its type which followed—no longer stood on its fins but on the base of the shell. For that reason only two tie-down bolts were needed. Viking VIII arrived at Oro Grande, the railway station serving the White Sands Proving Ground, on May 17, 1952. On June 6 everything was ready for a static test with tanks about half full. The motor caught well but after a few seconds the rocket looked as if it were wobbling. At X + 13 it tore itself free and rose. Of course no provision for tracking had been made, since this was to be a static test. Hence nobody could tell where the rocket was, and they hesitated to send the cut-off signal until X + 60. One minute later the rocket crashed, having broken apart at about one mile altitude, on the way down. Peak altitude of the runaway had been around 4 miles.

Viking IX began by giving trouble. The control system would not behave. Valve springs broke. Then the reservoir for the oil for the hydraulic lines cracked. When everything had finally been repaired, the first static test had to be cut at X + 3, because one of the meters went off scale. On December 15, 1952, Viking IX lifted off and climbed nicely. But it went to only 135 miles, mostly because at the end of the burning period there were over 500 pounds of liquid oxygen left in the tank with no fuel to burn.

Could anything else go wrong now?

Viking X arrived at Oro Grande on May 25, 1953, and went through 51 seconds of static firing on June 18. Firing date was then set for June 30, and preparations proceeded with no very important delays. But then, of all things, the radar gave trouble, and the rocket, fully fueled by then and pressurized, had to be "held." During the waiting period several hundred pounds of oxygen boiled away and had to be replaced. Meanwhile it had been decided to use another radar, not so accurate as the one that misbehaved, but more reliable. Finally, at 12:20 P.M., the firing switch was pushed. There was the customary puff of black smoke from the turbine exhaust, then the tail section disintegrated, and metal splinters flew around among steam and smoke and a ball of fire. The rocket motor had exploded and the rocket was on fire! There were four water nozzles around the rocket for just such an emergency; they were turned on. When the spray hit the burning rocket it disappeared in steam. Then all the oxygen was

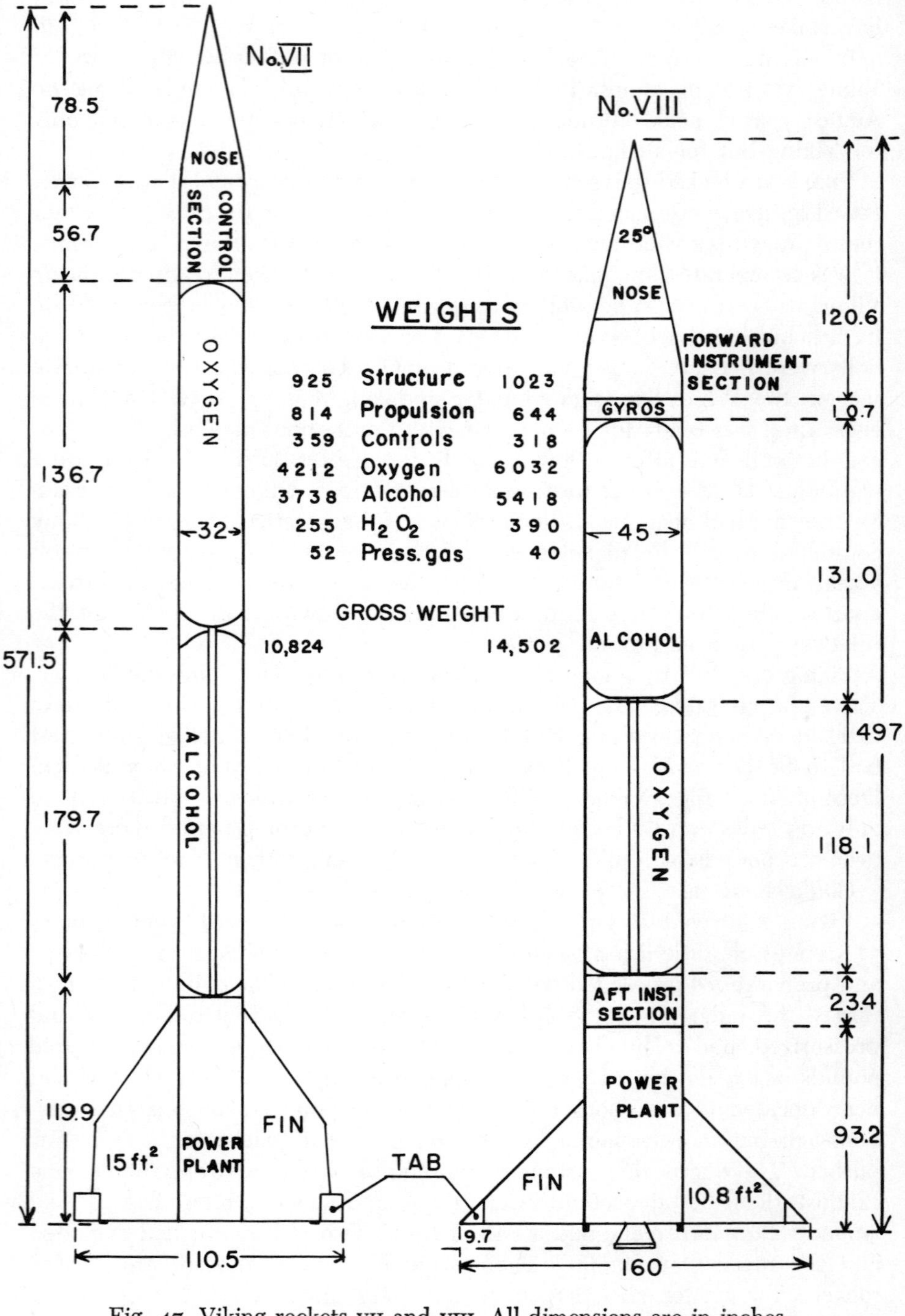

Fig. 47. Viking rockets VII and VIII. All dimensions are in inches.

dumped into the pit, forming more white clouds. Miraculously, after the clouds dispersed, the rocket was still standing, with a small fire in the tail section and three tons of alcohol in the upper tank. A man with a fire hose, taking the risk that the rocket might topple at any moment, approached it and put out the fire. But then the alcohol tank began to buckle. Some alcohol was draining away through a small pipe, and the main vent valves were stuck shut. Inside the tank a partial vacuum began to form, which still threatened to collapse the rocket. The only way to prevent this was to make a hole in the alcohol tank—by firing a bullet into it.

That the rocket did not collapse was one miracle. That there was enough left of it to rebuild it was the second one. A long time later, on May 7, 1954, Viking x climbed to 136 miles. But on May 24, 17 days later, Viking xi climbed to 158 miles. At last the Viking had reached record altitude. Viking xii followed on February 4, 1955, but reached only 144 miles, more than the highest V-2 but falling behind xi by over a dozen miles.

While the struggle with the Vikings was going on, other rockets roared up from the White Sands Proving Ground. There was the Consolidated Vultee 774, which looked like a somewhat smaller V-2 (32 feet tall) with three box-kite fins. It was fired for the first time in summer 1948. It was intended primarily for development work since it was a kind of preliminary model of the Atlas rocket. Only three of them were built; more detail on this rocket is given in Chapter 12. There was a smaller rocket called NATIV (North American Test Instrument Vehicle), built by North American Aviation. It was 18 inches in diameter, 14 feet, 6 inches tall, and had a take-off weight of 1237 pounds. Boosted by a solid-fuel unit through a launching tower it could climb to 50,000 feet.

Then there was, in Project Hermes, a rocket which was actually called Hermes, more specifically Hermes A-1 (see Plate XIII). Externally it was a copy of the German Wasserfall, but rather extensive changes had been made in the propulsion system. But the work on Hermes A-1 did not lead to a new design, and other projects took over soon afterward.

Most of the rockets counted down[5] at the White Sands Proving Ground in 1947 were still V-2's. But in the fall of that year a new one was added, the Aerobee.

[5] After the "Viking Hall" at the Hayden Planetarium in New York had been opened with a countdown Milton Rosen casually remarked to me that the countdown was a custom that had originated at the Raketenflugplatz. I assured him that we had not thought of it, but later I began to wonder who had. Thinking back, I realized to my own surprise that it had first been used in the film *Frau im Mond.* This was a silent movie, and at one point the words "10 seconds to go" flashed on the screen, followed by numbers, "6—5—4—3—2—1—0—FIRE." Knowing that Fritz Lang had been in the Austrian Army in the First World War, I asked him whether he had adapted some military practice which used a countdown. He replied that he had thought it up for dramatic purposes when working on the film; on a proving ground nobody would possibly *think* of that side effect!

The Aerobee (Fig. 48 and Plates XIX, XX, and XXI) had been initiated by the Applied Physics Laboratory of Johns Hopkins University and supported by the Navy Bureau of Ordnance; the actual construction was undertaken by Aerojet, Inc., and the Douglas Aircraft Company. The general scheme which was followed was that of the WAC-Corporal, a liquid-fuel rocket, fin-stabilized, without any guidance system, and boosted at take-off by a solid-fuel unit. The Aerobee was not quite 19 feet long and 15 inches in diameter. The fuels were the same as in the WAC-Corporal, aniline with an admixture of furfural alcohol and red fuming nitric acid. Feeding the fuels into the rocket motor is accomplished by pressurizing the tanks with helium. The rocket motor itself is of the fuel-cooled type. The 6-foot booster accelerates the rocket to about 1000 feet per second and then drops off. The Aerobee's liquid-fuel motor then takes over; the fuel supply lasts for 34 seconds. At Brennschluss the Aerobee has a velocity of 4100 feet per second and has attained an altitude of about 95,000 feet, if the trajectory is nearly vertical. Typical peak altitude with a payload of 150 pounds is 72 miles. The rocket can carry payloads of from 100 to 250 pounds; of course the peak altitude varies with the payload carried. Instrument readings are obtained partly by telemetering and partly by recovery of the instrumentation cone.

The Aerobee was ready for testing at White Sands in fall 1947. The beginning of the Aerobee program was summarized in an article in *Science* (December 31, 1948) by Doctors Van Allen, Frazer, and Floyd.

> Following preliminary firings of three dummy missiles, containing no sustaining motor, the first live Aerobee was fired on November 24, 1947. Due to an

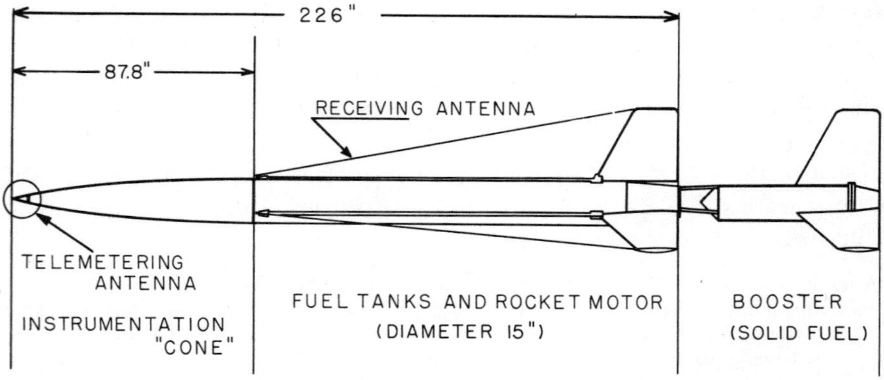

Fig. 48. Diagram of the high-altitude research rocket Aerobee. (*Courtesy U.S. Navy–Johns Hopkins APL*)

> excessive initial yaw, it was necessary to terminate the thrust of the motor by radio command at 35 sec. flight time in order to avoid impact outside of the military range. The resulting summit altitude was, therefore, only 190,000 ft. The second live firing was on March 5, 1948, and was thoroughly successful. Equipment for measuring the directional intensity and angular distribution of cosmic rays above the appreciable atmosphere was transported to 372,000 ft. and yielded valuable new data. On April 13 a flight of similar summit altitude occurred. Measurements of the earth's magnetic field were accomplished. . . .
>
> A fourth live round was equipped by this Laboratory with a set of aerial reconnaissance cameras and fired July 26, 1948. The flight was again normal, with a peak altitude over 70 miles. A large number of high-quality aerial photographs of large areas of the earth's surface and of cloud formations in the lower atmosphere was obtained by physical recovery of the film. (Van Allen, *et al.*)

Plate XX shows one of these pictures. Since these early tests had shown that the Aerobee was a reliable rocket, with the added advantages of being simple and relatively inexpensive, it quickly became the "work horse" of high-altitude research.

There had been slight misgivings in some quarters at first because the Aerobee has no control organs of any kind. It depends solely on the "arrow stability" provided by its three rigid stabilizer fins. But of twenty-four Aerobees fired at White Sands up to the end of 1949 only three misbehaved. And in only one case, the very first, was it necessary to use the emergency cut-off.

In March 1949 the Aerobee went to sea on the *Norton Sound,* preceding the Viking. Since a launching tower is needed one was built especially for the firing from shipboard. The Aerobee tower at White Sands is 140 feet tall, with a useful length of 90 feet. The one on board the U.S.S. *Norton Sound* was only half that height and probably could be lower. It was hinged so that it rested flat on deck when not in use. The rocket was fed into the tower in that position, like a shell into the breech of a gun. The tower was then erected with the rocket in place.

The first three Aerobees fired at sea were released a few hundred miles off the west coat of Peru under the magnetic equator. Some nine months later, in January 1950, two Aerobees were fired in the north Pacific and in the Gulf of Alaska from the same vessel. On one of these cruises an Aerobee misbehaved in a surprising manner. Apparently as a result of a leaking valve, the rocket suddenly took off 10 minutes ahead of schedule and climbed slowly and quietly by itself, leaving the booster, unignited, in the launching tower. Although everybody on board was struck speechless at first, the tracking crews recovered their wits quickly. The radar men followed it all the way up, to about 19 miles. And the camera men even got a fine color film of the rocket that took off by itself. The defective valve apparently never did open fully; calculations made as a result of this event indicated that an Aerobee fired without booster should go about twice as high as that one did. Of course no Aerobee has ever been fired without a booster on purpose.

You can deduce the temperature and density of the upper atmosphere in a number of ways. You can have balloons hovering with instruments which will radio their readings to the ground. Since the balloons hover the instruments do not need to respond rapidly. You can have rockets carry shaped charges and explode them at intervals, measuring the speed of the sound, from which the temperature can be calculated (see Appendix 1). You can also get some information from photographing meteors. All these things had been done, but the results did not quite agree. Scientists spoke jokingly of the "V-2 atmosphere," the "balloon atmosphere," the "shaped-charge atmosphere" and the "meteor-train atmosphere." The explanation of why there were several "atmospheres" was this: the V-2 shot had been made in, say, August and from White Sands. The meteors had been photographed in December and from California or Massachusetts. The balloons had been released from Ohio at any odd date. Moreover the meteor photographs had necessarily been taken at night, while the rocket shot might have been at noon.

To clear this up, all these different measurements needed to be made at the same time and from the same place. Because of the rockets involved —Viking VI and four Aerobees—the place had to be White Sands. The predicted arrival of a decent number of meteors fixed the date. The meteors in question were the Geminids which were due December 11–12, 1950. And the whole was called T-Day. The Signal Corps was going to launch balloons; a Harvard team under Dr. Fred L. Whipple set up their instruments at the Doña Ana and Soledad stations beyond the Organ Mountains. Stanford University, which had pioneered meteor detection by radar, went to Alamogordo. Viking VI carried numbers of pressure gauges (which were useful mainly in finding out what Viking VI did), and the Aerobees carried grenades.

The Aerobees were scheduled for 10 A.M., 9 P.M., 2 A.M., and noon the next day, the end of the T-Day period. The first Aerobee malfunctioned, rising only a few thousand feet; the other three were successful.

Because the Aerobee was so simple in operation and comparatively so cheap, it was built in larger numbers than any other research rocket. By May 1955 sixty of them had been fired, and in October of the same year an Aerobee was used for a spectacular experiment. It carried two steel cylinders of thermite with 2 pounds of sodium metal each. Thermite, consisting of powdered aluminum and iron oxide, creates great heat when ignited, but no combustion gases; hence the thermite would vaporize the sodium metal without upsetting the rocket by making an exhaust jet. Take-off was 20 minutes after sunset. When the radar said that the rocket was 40 miles up, the thermite was ignited by radio command. Sodium vapor jetted out through holes in the nose, and since the Aerobee rose to nearly 70 miles it put a 30-mile-high vertical streak of sodium vapor into the atmosphere. It was a glowing streak—the sodium atoms picked up the light of the sun, still above the horizon at that height, and reradiated it with the typical color of sodium. The light could be seen from Amarillo, Texas, 300 miles distant. Wind at high altitudes quickly distorted the streak into an enormous "C"—and that was the purpose of the experiment. The "frozen lightning" of the V-2 had been accidental but had betrayed the existence of winds at high altitudes. The sodium streak left by the Aerobee was a purposeful imitation of the same phenomenon.

An Aerobee which produced a flare 3 miles in diameter 60 miles above the White Sands basin on March 14, 1956, utilized a different scientific fact. At high altitudes the sun's radiation splits the oxygen molecule so that single oxygen atoms come into existence. They do recombine, and when they do they release a small amount of light, which is one of the factors causing the faint sky glow that intrigues geophysicists and occasionally annoys astronomers, especially when they try for an hours-long exposure of a very faint celestial object with a powerful instrument. It was also known that nitric oxide acts as a catalyst, causing single oxygen atoms to combine into oxygen molecules at a very rapid pace. The nitric oxide is not used up in the process; it causes the recombination simply by its presence. Normally at high altitudes there are only the faintest of faint traces of nitric oxide. But the Aerobee carried 20 pounds of nitric oxide to a 60-mile height. As it spread, all the oxygen atoms it touched combined into molecules. Seen from the ground, there was a patch of light in the sky, looking fifteen times as large as the full moon and giving off roughly half as much light. Since the sun causes these oxygen molecules to break up into single atoms again, this experiment could be repeated as often as desired without causing any permanent change.

In 1952 the Aerojet General Corporation (then still named Aerojet, Inc.) approached the Air Force and the Navy with a proposal for improving the performance of the Aerobee by redesigning the structure for magnesium alloys and stainless steel. The proposal was accepted and the new rocket was built. It was given the name Aerobee-Hi. The first tests were set for May and June of 1956, after the rocket had been completed early during that year. The tests did not go well at all at first; in fact, the first Aerobee-Hi did not perform at all. It was fired on May 2, 1956, and the booster accelerated the rocket to about 1000 feet per second as usual. This acceleration was supposed to open the fuel valves in the liquid-fuel system, but the valves stubbornly stayed closed; nobody knows why. After the booster had dropped off, the fully fueled but not operating rocket coasted to about 10,000 feet, fell back, and exploded on impact. The second Aerobee-Hi, fired on May 8, burned for 51 seconds, at the end of which time it had reached an altitude of 79,000 feet. It carried a payload of 120 pounds to that year. The tests did not go well at all at first; in fact, the first Aerobee-Hi could have gone under similar conditions. The third Aerobee-Hi, fired on June 4, also suffered from valve trouble: the valves did open but not all the way, so that the thrust was low. It reached only 195,000 feet. But the next shot, on June 29, was a complete success, reaching an altitude of 163 miles.

From then on Aerobee-Hi behaved as predicted. During the International Geophysical Year twenty-two of them were fired successfully from Fort Churchill in Canada. Others were fired from other launching areas, including some shipboard firings.

The purpose of the Aerobee-Hi was to carry 100 to 120 pounds of payload to about 165 miles, but there seemed to be a need for another version that would go only to 50 miles and would be smaller, cheaper, and easier to handle. The result of this thinking was the Aerobee-Jr. (officially Aerobee-100) which made its first flight to an altitude of 60 miles from the White Sands Proving Ground on February 18, 1958. In spite of this success there was not much demand for it, because a different kind of all-solid-fuel research rockets (described in the next chapter) began to appear. Aerojet General countered with another proposal of all-solid-fuel research rockets that were named Astrobee. Two types were proposed, Astrobee-200 for a payload of 120 pounds to be carried to 200 miles and Astrobee-500 which would be able to carry 20 pounds of payload to a peak altitude of 725 miles. Both rockets consisted of a solid-fuel booster and a solid-fuel upper stage (called a "sustainer"). The dimensions were as follows: Astrobee-200 had a booster 15 inches in diameter and 73 inches long, a sustainer 20 inches in diameter and 77 inches long, and a payload compartment 15 inches in diameter and 98 inches long. Take-off weight, including payload compartment but without payload, was calculated to be 1689 pounds. Astrobee-500 was the same as Astrobee-200 except for an extra top stage which brought the take-off weight to 1862 pounds and the overall length to 25.5 feet. Two Astrobee-500's were ordered, but after that this line of development was not pursued further.

For the sake of completeness, one special test, a firing from Fort Churchill, has to be mentioned. The rocket was designated as Aerobee-300 and in December 1958 it carried 65 pounds of payload to an altitude of 260 miles. Aerobee-300 consisted of the normal Aerobee solid-fuel booster, a liquid-fuel Aerobee-Hi, and a solid-fuel Sparrow I missile (an air-to-air missile) as the top stage.

It was a one-of-a-kind rocket, but other one-of-a-kind (or almost one-of-a-kind) research rockets were to proliferate soon afterward.

10 The Rocket into Cosmic Space

Du nimmst bei diesem kühnen Schritt
Nur keinen grossen Bündel mit.
Ein bisschen Feuerluft, die ich bereiten werde,
Hebt uns behend von dieser Erde.
Und sind wir leicht, so geht es schnell hinauf.[1]

—Goethe, *Faust*, I

In the early afternoon hours of February 24, 1949, Man took his first step into space.

The place was the White Sands Proving Ground.

The rocket was a modified V-2, carrying a WAC-Corporal.

The time was 3:14 P.M., Mountain Standard Time.

One minute later the V-2 had reached an altitude of about 20 miles and was moving with a velocity of not quite 1 mile per second. Then the WAC-Corporal took over, adding its own velocity to that provided by the V-2. Forty seconds later it had accelerated to 1.39 miles per second, expending all of its fuel in the process. The empty V-2, of course, continued to rise too, reaching an altitude of 80 miles before falling back. But the WAC-Corporal forged ahead with a higher velocity. When the V-2 crashed in the desert 5 minutes after take-off and 28 miles north of the firing site, the WAC-Corporal was still climbing. It kept climbing for another 90 seconds, reaching peak altitude 6½ minutes after take-off.

Peak altitude was 250 miles above sea level.

1 These lines about the "fiery air" are a reference to the hot-air balloons of the Montgolfiers; it is impossible to resist the temptation to "revise" them up-to-date and to make them read:

But only on so bold a flight
Be sure to have the payload light.
A hot exhaust which I shall soon prepare us
Above the earth will swiftly bear us,
Then after Brennschluss we'll be light
And travel fast. . . .

(With apologies to the spirits of both Johann Wolfgang von Goethe and Bayard Taylor.)

At that altitude there are fewer air molecules in a cubic inch of air than in the best vacuum we can produce in our laboratories here at the bottom of the air ocean. And one of these air molecules will have to travel a distance of 5 miles before it collides with another one.

The WAC-Corporal had reached empty space.

Naturally, it then began to fall back. It struck the ground not far from the northern edge of the impact area, 12 minutes after take-off and some 85 miles from the launching area. Since the WAC-Corporal was a small rocket and since it had struck the ground with a higher velocity than any other rocket had done before, it took a long time to find it, even though radar tracking gave a general idea of where to look. Nearly a year later, in January 1950, a badly smashed tail section was spotted and recovered. Parts of it were shipped to Robert P. Haviland, who had been the project engineer for this experiment, for identification.

Technically this was shot no. 5 of Project Bumper, a subproject of Project Hermes. A total of eight V-2 rockets had been assigned to Project Bumper and the number of failures was rather high, probably because so much modification was needed. Of the eight shots only three were fully successful, two have been labeled "partially successful," while three were failures, one of them suffering a tail explosion.

One of the problems that Project Bumper had to solve was the separation of two stages of a rocket while it was moving through the atmosphere at a very high velocity. This problem had never come up in this form before. In the case of the old fireworks displays involving staged rockets the velocities had been low; in the case of the Rheinbote missile the accelerations of the various solid-fuel stages had been high. But in the Bumper shots the upper stage, the WAC-Corporal, was a liquid-fuel rocket starting out with a low acceleration, while the V-2, on the other hand, moved with maximum acceleration at the end of its burning period. And, for reasons that will be explained, the time interval between the Brennschluss of the lower stage and the ignition of the upper stage should be as short as possible. Part of the solution consisted in throttling down the V-2 during the last few seconds of its burning time. Nowadays that problem might be solved by having a solid-fuel stage with a still higher acceleration and a burning time of only 1 or 2 seconds between the stages. The valves in the top stage would be adjusted in such a manner that they would respond to the very high acceleration of the intermediate solid-fuel stage.

Project Bumper's purpose was not merely to study the problems of the separation of two-stage liquid-fuel rockets and to establish a high-altitude record *en passant*. Rockets nos. 7 and 8 of that project were used for a special experiment. That experiment incidentally served for the "inauguration" of a new missile testing range in Florida. That the White Sands Proving Ground was somewhat "small" had been recognized all along; the

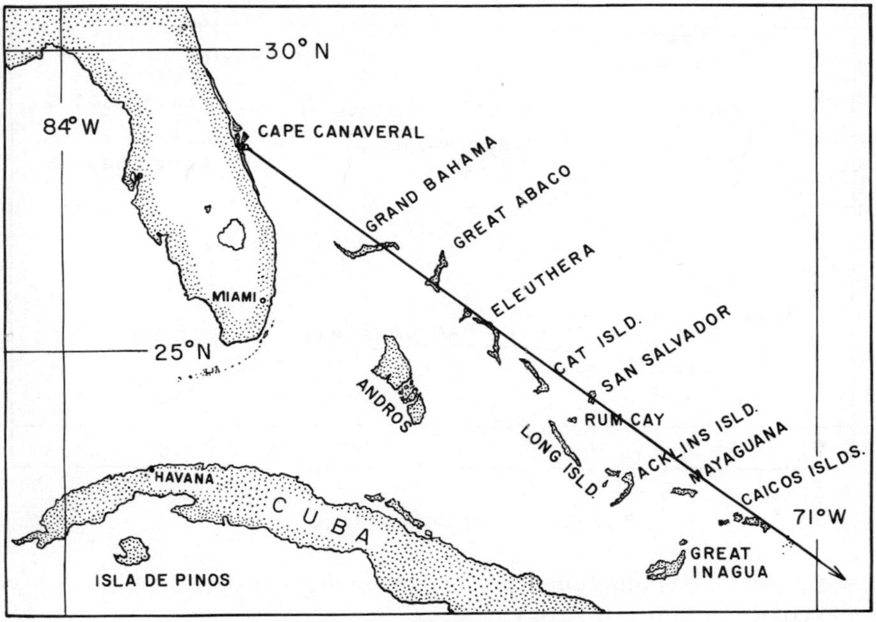

Fig. 49. The Florida Firing Range.
Tracking stations are located on the Bahama islands passed over by the firing track and on Puerto Rico farther to the southeast. A missile following this track does not touch a continental landmass for thousands of miles.

distance from the firing site to the extreme end of the Impact Area is only half the maximum range of a V-2. A firing range of greater extent could be found only at the seashore. In May 1949 negotiations were begun with the British Government to obtain permission for the establishment of observing and tracking stations on some islands of the Bahamas group. The firing site chosen was Cape Canaveral on the Banana River in Florida.[2] If one draws a straight line in a southeasterly direction from the Cape, it cuts across Grand Bahama (at about 200 miles from take-off point), Great Abaco (about 275 miles), Eleuthera (about 350 miles), misses Cat Island (about 400 miles), goes over San Salvador (about 500 miles), and continues across several thousand miles of open ocean; not counting the extreme

[2] During the Second World War the military installation inland from the Cape was known as the Banana River Naval Air Station. When it was made a rocket-firing base it was first referred to as the Florida Missile Test Center; the name was later changed to Atlantic Missile Range. All along, the geographical name Cape Canaveral was also in use. After the assassination of President Kennedy the name was changed to Cape Kennedy, and the whole aera acquired the name Kennedy Space Flight Center. Later, Merritt Island was added to the space-flight center; it is now referred to as MILA—Merritt Island Launch Area.

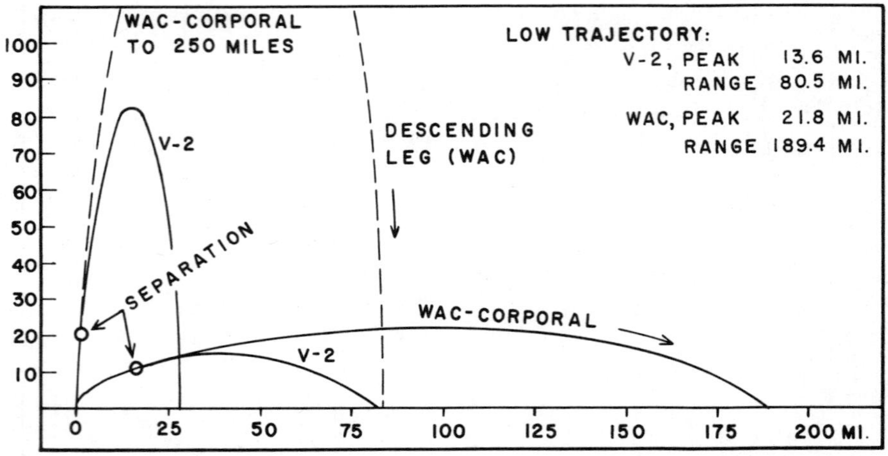

Fig. 50. Typical Bumper trajectories.

eastern point of South America, the nearest land in the direction of fire is the coast of southwest Africa (Fig. 49).

But the first tests made at Cape Canaveral with Bumper nos. 7 and 8 did not require any Bahama island observers; they were comparatively close-range work. The chief objective of these shots was to get the WAC-Corporal into a very flat trajectory at what is a low altitude for rockets (see Fig. 50).

The firing range was so new that operations which were the merest routine at White Sands turned into problems, such as the road transport of the rocket from the hangar to the firing site. And Haviland remarked later that a "barrier" of 1-inch pine boards does not inspire nearly as much confidence as does 8 feet of concrete.

The first test was scheduled for July 19, 1950. It happened to be one of those days, so well known to any rocket experimenter, on which everything goes wrong. While the rocket was being readied six airplanes were cruising over the sea to warn boats and ships which might stray into the danger area. Only a few minutes before firing time one of these planes had to make an emergency landing to replenish its fuel. As a consequence the firing button was not pressed, and since the whole schedule had been disrupted the test was postponed to the early afternoon hours. When everything was ready once more, a piece of electronic equipment did not work. The makeshift repair caused another delay and then everything seemed to be really ready. Somebody counted off the seconds, "X minus eight, seven, six, five. . . ." The pinwheel burst into flames on schedule. The preliminary stage was on. "Main stage, fire!" But the rocket just continued to burn.

Somebody shouted, "Missile does not lift." Colonel Harold R. Turner, who had come to Florida from White Sands, could see that for himself and reasoned that a valve was stuck. He ordered the preliminary stage cut. All that could be done that day was to drain the fuel and to remove the TNT bricks that were used for the airburst.

On July 24 the test was repeated with the other rocket. This was one of the days when everything goes smoothly. The rocket lifted on schedule and quickly disappeared in a thin veil of cirrus clouds. When it reached 10 miles the tilting began, through a full 90 degrees, so that it moved horizontally. Then the WAC was fired. The V-2 slowly lost altitude and was exploded at 3 miles up; the pieces fell into the sea some 50 miles from take-off point. The WAC, too small to carry instrumentation and a bursting charge, must have hit the sea in one piece, some 200 miles away. On July 29 the other rocket, which had not performed originally, was fired in a repetition of the first test. Everybody present probably steeled himself for all sorts of mishaps. But it was a smooth flight.

In this story of the prelude to actual space exploration two dominant themes vie for attention: one is the narrative of events as they took place; the other is the story of the theoretical reasoning behind these events. Since we have had several chapters of narrative we must now pay some attention to theory.

There is a host of questions to be answered, from the simple one of why in the Bumper shots only a very short delay between the burn-out of the first stage and the ignition of the second stage could be permitted to the more complex one of why several stages are desirable at all. The answers to these questions then lead to the question of how a satellite orbit comes into existence,[3] of why we don't concentrate on "anti-gravity" instead of building bigger and bigger rockets, and finally to the problem of what happens to a missile that has missed the moon.

Let's begin with Bumper no. 5 and the question of why the WAC-Corporal was not ignited when the V-2 was at the peak of its flight. This particular V-2 went to 80 miles; another one might have gone to 100 miles. No WAC-Corporal was ever fired without a booster, but if one had been it probably would have climbed to about 25 miles. If the WAC-Corporal had been ignited at the peak of the V-2 flight, it would have added 25 miles to the 80 or 100 miles of the V-2. But because it was ignited when the V-2 was fastest, not when it was highest, it went to 250 miles.

Why?

The answer can be derived from what has been known for centuries as Tartaglia's rule, after Niccolò Tartaglia, an Italian mathematician who was

[3] A magazine writer, informed that Dr. Wernher von Braun had in mind the "2-hour orbit at 1075 miles above sea level," asked me in all seriousness, "How does Dr. von Braun know that there *is* an orbit at 1075 miles?"

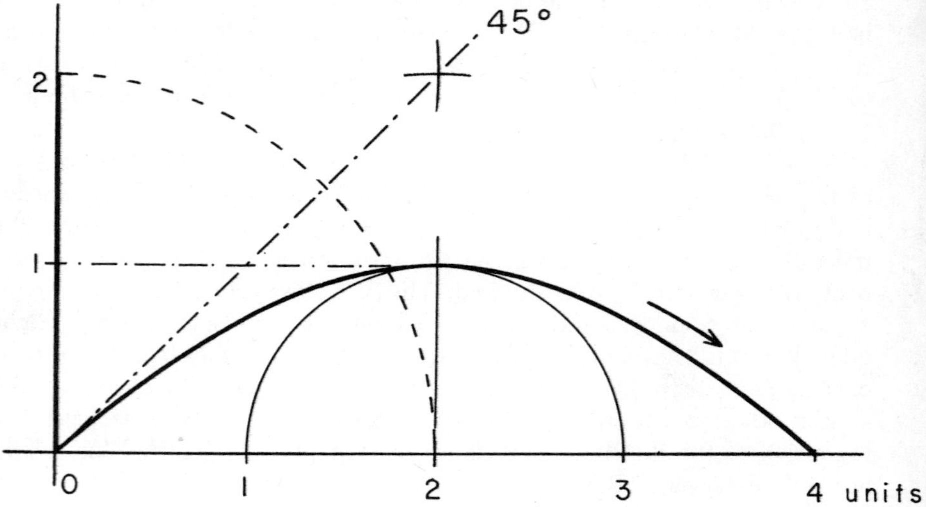

Fig. 51. Tartaglia's rule.

an expert on fortifications and probably the inventor of the gunner's quadrant. His rule, first stated in 1540, established a definite relationship between the range of a gun and the height to which it can fire. He said that a gun which can shoot to a height of 1000 units of length will attain a maximum range of 2000 units of length and that it must be fired at an angle of 45 degrees to throw its ball to maximum range. When it is so fired, Tartaglia added, the ball will ascend to a peak altitude of 500 units of length along its trajectory. (He either did not know, or at least did not say in so many words, that the peak altitude is reached at the halfway point between gun and target.) This simple relationship is somewhat distorted by air resistance when it comes to real shooting but holds very nearly true in two cases. One case occurs when the range is short and the projectile quite heavy, like the solid cannon balls of Tartaglia's time; the other when the range is so long that most of the trajectory is in what amounts to a vacuum. Remember the V-2 figures in Chapters 8 and 9: average vertical 100 miles, maximum horizontal about 200 miles, and peak along the trajectory for maximum range somewhat over 55 miles. (The V-2's were not fired at 45 degrees.)

Niccolò Tartaglia found his relationship by experimentation; he could not tell why it was so and he especially did not know why an angle of elevation of 45 degrees should result in maximum range. To us this is very simple indeed. The range R is determined by the formula $R = v_0^2/g \sin 2\beta$, where v_0 is the muzzle velocity of the projectile (or Brennschluss velocity of the rocket) and β the angle. Obviously the sine of 2β is largest when

$\beta = 45$ degrees. Similarly, the maximum height H is given by the formula $H = v_0^2/2\,g \sin^2\beta$, or, for a vertical shot, simply $v_0^2/2\,g$, but for a rocket, H must be counted from the Brennschluss altitude so that the total altitude $H^* = H + h$ if h is the Brennschluss altitude. Finally the formula for the peak altitude along the trajectory, h_p, is given by $v_0^2/4\,g$ for a 45-degree shot.

As a military tool Tartaglia's rule is on the same level as the electrician's insertion of his index finger into a socket to see whether it carries current. He can't measure either voltage or amperage that way, but it is a useful expedient for deciding "yes" or "no." Tartaglia's rule is still handy for a quick guess at performance but it does not explain anything.

Let's go on now and find out what determines the altitude reached by a projectile—to get rid of the extra value of the rocket's Brennschluss altitude we'll stick to cannon for the time being. As the simple formulas just given show, the problem is always that of velocity versus the gravitational pull of the earth. Obviously a projectile leaving the gun with a muzzle velocity of 1000 feet per second will go higher than one with a muzzle velocity of 500 feet per second. We are, at this moment, not much interested in the height to which the projectiles would go. We are merely watching their rise and fall and measuring their velocities at the instant of impact. Now, if there were no air resistance, we would find that the shell which left the gun with a muzzle velocity of 1000 feet per second in a vertical shot will strike the ground with 1000 feet per second while the other which left the muzzle with 500 feet per second will strike the ground with 500 feet per second. Both reached different heights, but if we dropped bombs from those two heights, the bombs also would strike the ground with velocities of 1000 feet per second and 500 feet per second, respectively.

The general rule is, then, that the velocity required to reach a certain altitude is the same that would be acquired by a fall from that altitude. Since we can calculate the impact velocity from any given altitude we therefore know automatically what velocity would be needed to go that high. A few figures of this kind look like this:

HEIGHT OF FALL		IMPACT VELOCITY (PER SEC)	
MI	KM	MI	KM
31.7	51	0.62	1
126.7	204	1.24	2
285.2	458	1.86	3
505.1	813	2.48	4
791.6	1274	3.11	5

The third line of this table resembles the actual performance of the shot of February 24, 1949. From these figures we can see at a glance that the altitudes grow much faster than the velocities involved. The altitude in the second line is four times that in the first, the velocity only twice that in the first. Therefore it was far more important for the WAC-Corporal to be fast than for it to be high.

While these five lines of figures explain this fact, the figures themselves should not be taken too literally. Not only was air resistance disregarded, but the formula used to derive them also fails to take into account that gravity itself grows weaker with high altitude (see Fig. 52). And if you have rockets in mind they neglect the Brennschluss altitude too. The following short table gives corrected figures for a rocket climbing under 3 gravities (3 *g*) of acceleration, taking the weakening of gravity into account, but still neglecting air resistance.

Both tables, with their fast growth of heights and slow growth of

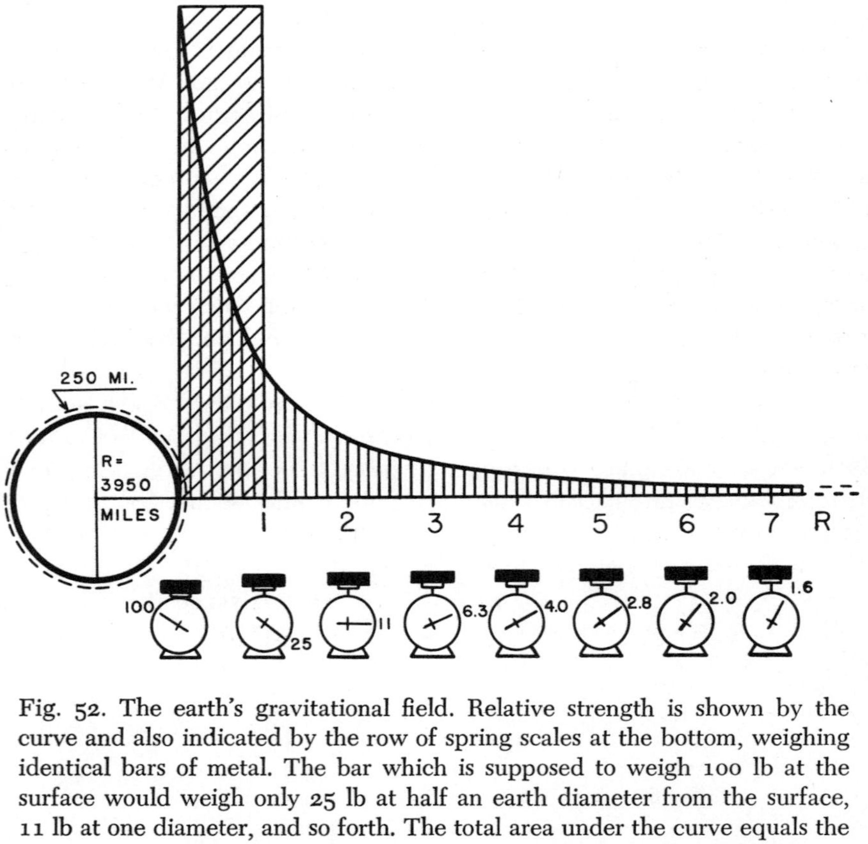

Fig. 52. The earth's gravitational field. Relative strength is shown by the curve and also indicated by the row of spring scales at the bottom, weighing identical bars of metal. The bar which is supposed to weigh 100 lb at the surface would weigh only 25 lb at half an earth diameter from the surface, 11 lb at one diameter, and so forth. The total area under the curve equals the rectangle; in other words the actual gravitational field is equal to a field of surface intensity extending to the height of one earth radius.

VELOCITY (MAXIMUM) PER SEC		HEIGHT	
MI	KM	MI	KM
0.62	1	42.2	68
1.24	2	172.1	277
1.86	3	397.7	640
2.48	4	814.0	1310
3.11	5	1224.1	1970

velocities, suggest the same thought: if something fell from an "infinite" height, its impact velocity would *not* be infinite. The impact velocity can be calculated; it would be 11.2 kilometers per second, which is almost precisely 7 miles per second.

Hence, if there were no air resistance, a gun with a muzzle velocity of 7 miles per second could shoot into "infinity." Its projectile would no longer be subject to the earth's gravity. The velocity is, therefore, called "velocity of liberation" or, more simply, "escape velocity." Such a gun could, for example, shoot to the moon, just as Jules Verne said.

Now I come to the discussion of this famous story which I promised in Chapter 2. Its basic idea is simple: an enormous gun fires a projectile vertically, with a muzzle velocity of a little over 36,000 feet, or about 7 miles, per second. As the projectile ascends, gravity, constantly at work, will reduce its velocity steadily. But the projectile would have the enormous velocity-wealth of 36,000 feet per second. Gravity would be able to demand a velocity-interest of 32 feet for every second at first. As the projectile recedes that figure would drop to 31 feet per second, then to 30 feet per second, etc., becoming less and less all the time.

In spite of this steadily decreasing rate of velocity attrition the projectile would have expended virtually all its velocity-wealth after 300,000 seconds. But by that time it would also have reached the distance where the gravitational fields of earth and moon balance each other. If the projectile lacked only a few inches per second of velocity at that point, it would still fall back to earth. But if it still had even a few inches per second of velocity left it would begin to fall toward the moon. After another 50,000 seconds it would crash on the moon, with an impact velocity of about 2 miles per second (a figure composed of the motion and the escape velocity of the moon) having needed 97 hours and 13 minutes or just about 4 days for the whole trip.

Because of that duration of 4 days Jules Verne pointed his gun at a point in the sky where the moon would be four days after the order "fire."

While this fundamental theme of the novel is very nearly correct, things began to go wrong when Jules Verne progressed to engineering detail.

The projectile to be fired at the moon is to be a ball 9 feet in diameter, under the erroneous assumption that a newly constructed telescope would be capable of following the path of a ball that size. Since a ball (later redesigned into a shell) of such size would be far too heavy if solid, it is decided to make it hollow, with walls 1 foot thick. The cannon required for such a projectile is, of course, too large to be built in the usual manner. The barrel, of the arbitrary length of 900 feet, is poured into a well dug for this purpose. Actually that barrel, which constitutes all of the cannon, is only the lining of a barrel with walls of indeterminate thickness, formed by the soil of Florida. An arbitrary amount of guncotton (400,000 pounds) is put into the barrel, filling about 200 feet of it. This is supposed to produce a muzzle velocity of 54,000 feet since the president of the Gun Club (or rather Jules Verne) expects to lose the difference to air resistance. All this, as may be expected, is somewhat vague. Elsewhere in the novel it is stated that air resistance will not be important because the projectile, with its high speed, will require only a few seconds to traverse the atmosphere.

The latter statement is like saying that a slab of armor plate 3 feet thick cannot very well be expected to stop a 16-inch shell since the shell traverses the distance of 1 yard in $1/1000$ of a second.

If the experiment had actually been made, the experimenters would have found, to their great surprise, that their ball would have landed 100 feet from the muzzle after rising not much higher. And the shell would have been flattened and partly volatilized. The reason is that Jules Verne, who did worry a little about air resistance in the open air, forgot to think of the resistance encountered in the 700 feet of gun barrel. That column of air, 700 feet high and 9 feet in diameter, could not possibly get out of the way of a projectile moving much faster than sound. It would have to be compressed, and the hollow ball or shell, as later redesigned in the novel, would have found itself between two very hot and enormously powerful pistons: the furiously expanding gases of the guncotton underneath and the column of air, heated by compression, above. The passengers, if everything had worked as Verne wanted it to work, would have been spread out into a thin film by the enormous acceleration of the projectile.

But even the shot itself would not have worked. As mentioned in Chapter 2, a modified moon gun was computed by Oberth and Valier for their own amusement. The results were as follows: The projectile must consist of a high-grade steel alloy—for example, tungsten steel—and must be practically solid. Its caliber must be 1200 millimeters (about 48 inches) and its length six times its caliber. The gun must be 3000 feet long and must be built into a mountain near the equator so that the muzzle is at least 16,000 feet above sea level. Before firing, the air has to be exhausted from the barrel and the muzzle has to be closed with a metal lid—padded with rubber around its rim—which is strong enough to resist the air pressure from above. Since the vacuum in the barrel cannot possibly be perfect, there will be a small amount of air left which is then compressed by the moving projectile and which will push the lid off just before the projectile itself reaches the muzzle.

Some years later Baron von Pirquet went over the problem again and came to the conclusion that even this improved moon gun would fail to perform fully as required. In the first place, von Pirquet made the mountain 3000 feet higher so that the muzzle of the gun would be at an elevation close to 20,000 feet. Furthermore, he found that the gases of the propelling charge would not expand rapidly enough if the whole charge were packed at the bottom. He rectified this by attaching a considerable portion of the charge to the bottom of the shell and by furnishing a number of extra firing chambers along the length of the barrel, their mouths leading into the barrel from the side. But even after all this we cannot say whether the gun itself could be built and whether the national budget for an average war would be sufficient to pay for it.

You cannot shoot into space through an atmosphere like the earth's and against a gravitational field like ours. Even a supergun could not overcome the set of conditions produced by the natural environment, but a sufficiently large rocket does not encounter any special difficulties. A rocket, especially a big one, climbs slowly at first and does not reach high velocities until it is a considerable height above sea level, in thin air. A projectile would have to fight the gravitational pull of the earth and an almost incredible air resistance, while a rocket, once it has passed the 100,000-foot level, has to cope only with gravity.

When I used to present this sequence of thoughts to a lecture audience before large rockets had actually attained open space, there was always somebody in the audience who would say that this was the wrong way to go about it. Now, when there is nobody left who hasn't watched a large rocket climb into space, such critics have become somewhat rarer but they are by no means extinct. They think, and say, that building rockets that can attain a high velocity is quite crude, something comparable to a battle ax made of flint. Why fight gravitational force with another crude force? Why not look for a way to eliminate gravity, maybe by discovering a substance that is unaffected by gravity or by generating "an opposing field"—in short, by "anti-gravity."

I suspect that all the talk about "anti-gravity" is ultimately based on H. G. Wells's classic science-fiction novel, *The First Men in the Moon,* in which a substance called "Cavorite" is used, a substance which not only is not affected by gravity itself but also creates a "gravitational shadow." When the novel first appeared, around the turn of the century, it was read by the aged Jules Verne who coined a classical phrase that is still quotable when "anti-gravity" is dragged into a discussion. He said, *"C'est très joli,* but let him show me this metal." Let me see your anti-gravity device in action. Or, at least, give a valid explanation of the nature of gravity; then we can see what can be done about it.

Look back at Fig. 52. This figure shows graphically the one thing we definitely know about gravity—namely, that it decreases with the square of the distance from the center of the body that exerts the "gravitational pull." Unfortunately this is not a key to the problem but merely a reflection of the geometrical fact that the area of a sphere is proportional to the square of its radius. Because the "inverse square law" is just a geometrical relationship expressed in figures, it is not restricted to gravity. Radio waves, visible light, and sound are all covered by the same law, which in itself shows that it is no clue to the nature of the phenomenon. All attempts to approach the secret of gravity experimentally have failed. Light or darkness, extreme heat or extreme cold, electric charges or magnetic fields make no difference. Screening gravitational attraction does not work. Small wonder that Dr. Paul R. Heyl of the National Bureau of Standards once called gravity "an intractable phenomenon."

As may be expected under these circumstances, all so-called explanations of gravity suffer from the common defect that they don't work. They all try to account for some of the observed facts by neglecting others or even by neglecting the very consequences arising from the explanation itself. Or else they try to explain the observed facts by explaining them away.

What may be called the "classic" explanation is the one Le Sage of Geneva proposed in 1750. According to him the whole universe was filled with "ultramundane corpuscles" moving at high speeds and exerting a steady pressure on the surfaces of all bodies, pressing human beings, for example, to the surface of the earth. The obvious present-day answer is to inquire about the heat generated by the impact of the corpuscles, but in 1750, when Le Sage advanced his hypothesis, the law of the conservation of energy was still to be discovered.

The general idea of the Le Sage hypothesis remained a favorite with gravity explainers for a large number of decades. Later elaborations stated that the corpuscles, of course, penetrate all solid bodies, but that they lose some speed while doing so. For this reason the push from below (at the surface of the earth) does not quite match the push from above, the difference being called "attraction." Again the law of the conservation of energy interferes with the hypothesis. And none of all these theories—by now the Le Sage hypothesis has been advanced about fifty times, each time as a "new" explanation of gravity—can be true, because the corpuscles would inevitably cause a screening effect, which should be measurable. If there were a screening effect, the moons of Jupiter would betray its presence; the actual behavior of the moons shows that no such effect exists.

When Albert Einstein became interested in this problem he looked for a similar intractable phenomenon and found it—inertia, especially in the form of centrifugal "effect." An observer in a rotating circular room, he said, would find himself in a kind of gravitational field turned inside out, forcing him away from the center and becoming more powerful with distance. In that case the observer would know what is happening to him. He would not look to the center or the rim of the room for the "origin" of the force but would know that it originated in the objects and that no screening of any kind can be expected.

Einstein proceeded to say that a "gravitational field is equivalent to an inertial field produced by a suitable change of coordinates," and then promptly disappeared from the layman's view in a forest of equations.

His idea is usually explained by comparing matter in space to stones forming cusps in the otherwise flat surface of a frozen lake. Another stone set in motion on that surface which is, by definition, frictionless, would follow a straight line in a given direction until it grazed one of these cusps. Then it would continue in an altered direction. If it got fairly deep into the cusp it would not leave again at all but would revolve around the larger stone. That would look like gravity, being the effect of a three-dimensional cusp in a two-dimensional world. Similarly, popularizers used to state sonorously, real gravity is the cusp made by matter in four-dimensional space—which was neither understandable nor a precise rendering of Einstein's equations.

Moreover, we have as little reason to believe in four-dimensional space as we have to believe in the two-dimensional world which had to be assumed for the purpose of explaining four-dimensional space. Higher and lower dimensions may be amusing as mental exercises on paper but that does not prove their reality.

What Einstein really suggested is that gravity may not be a separate "force" as the term is usually understood. If it is not, there can be no gravity screen.[4] If it *is* a force in the usual sense a gravity screen can at least be hypothesized, as H. G. Wells did. But then you get into an even weirder paradox:

A physicist refers to the points along that curve in Fig. 52 as the gravitational potential. It has a certain value at the surface and lesser values higher up. At an infinite distance the gravitational potential is zero. To move something from a higher to a lower potential requires energy; it takes a certain amount of energy to lift 1 pound 1 foot, and as a unit for measurement this is called a foot-pound. In the metric system it is a kilogram-meter, where the unit of measurement is the energy required to lift 1 kilogram (2.2 pounds) to a height of 1 meter (39.4 inches). To lift 1 kilogram from the surface potential to the zero potential requires 6,378,000 kilogram-meters and you can pack this energy into an actual kilogram of weight by accelerating it to escape velocity.

[4] See also "Nonexistence of Gravity Shields," D. C. Peaslee, in *Science*, December 28, 1956.

Now Cavorite is supposed to create the zero potential. Consequently a man who steps onto a sheet of Cavorite has to overcome the whole gravitational potential of earth. Supposing he weighs 165 pounds or 75 kilograms. His muscles have to produce a mere 6,378,000 times 75 kilogram-meters for this purpose. It does not matter that he walks only one step. Distance is unimportant; it is the difference in potential that counts. This puts the lucky inventor of Cavorite in a very curious predicament. Either his muscles are not powerful enough to enable him to enter his spaceship, in which case the marvelous invention is wasted because he cannot use it, or else his muscles are powerful enough for this task, in which case the invention is wasted too. For if he could overcome the whole potential he could jump to the moon directly.

One day, one may hope, the nature of gravity will be explained by somebody. And it is still possible that this explanation may lead to a method for manipulating gravity. But since nobody can say when this will happen—or even whether it will happen—it is useless to spend time in speculation.

We know what we have to do: we have to build rockets that can attain high velocities. That engineers have succeeded in doing so is attested by reports from Cape Kennedy in almost any morning newspaper, but how did they go about it?

Let us develop a specific example. Here we have a payload, a package of scientific instruments with telemetering transmitter and batteries. In spite of its complexity it weighs only 20 pounds. This payload is to be lifted to an altitude of about 170 miles and the table on page 273 tells us that to do this the rocket must attain a maximum velocity of about 1¼ miles per second (say, 6600 feet per second). So far we are on firm ground, but the next step is likely to be misleading. The misleading step would be to ask the seemingly natural question: how large would a rocket have to be in order to accelerate a payload of 20 pounds to 1¼ miles per second? The size of a rocket is not in itself an indication of what it can do, except that a larger rocket is likely to be more powerful than a smaller one.

The important concept here is one that has been hovering in the background for the last few chapters and has even been mentioned once before. It is that of *ratio*, the ratio between the rocket itself and the amount of fuel it can hold. If we call the mass of the rocket's payload M_P, the mass of the rocket itself M_R, and the mass of the fuel M_F, the take-off mass is clearly $M_P + M_R + M_F$. When all the fuel has been used up, the remaining mass is $M_P + M_R$, payload plus dry weight of the rocket. Dividing the take-off mass by the remaining mass results in the mass-ratio. This is usually written M_0/M_1, where M_0 is the take-off mass.

The next step would be to determine what the mass-ratio must be for a rocket that has to lift 20 pounds to 170 miles, but here we run into an interesting complication.

The answers could run all the way from 2:1 (for a modern high-energy rocket fuel) to 20:1 (for the original blackpowder charge), depending on the exhaust velocity of the fuel to be used. If we pick a high-energy fuel the figure for the mass-ratio will be low; if we pick a low-energy fuel it will be high. This looks as if we were faced with a dilemma: unless we agree on a specific exhaust velocity we can't get an answer. But if we do agree on a specific value for the exhaust velocity we get an answer for that value only. However, we were looking for a general answer.

The way out of this dilemma is quite easy; it consists in using the exhaust velocity itself as the yardstick. The general answer we want is found by looking for the mass-ratio needed to produce a rocket velocity equal to the exhaust velocity. Whether you take a low value for the exhaust velocity, or a high one, the mass-ratio required to reach exhaust velocity must evidently always be the same. Now if we call the rocket velocity v and the exhaust velocity c, our question can be sharpened to read: "What is the mass-ratio for $v = c$?" The answer is 2.72; thus, a rocket weighing 272 units (pounds, or ounces, or kilograms, or even tons) before burning and 100 units when empty will attain a velocity equal to its exhaust velocity.[5] This figure, which was mentioned in Chapter 3, is the figure mathematicians call e. Its value to 8 places is 2.71828183 . . ., but for everyday purposes 2.72 is close enough. Written as a formula, the statement that a mass-ratio of e to 1 will make v equal to c (regardless of the numerical value of the exhaust velocity c) reads

$$M_0/M_1 = e^{v/c}$$

Let us look at the expression $e^{v/c}$ for a moment. If v is supposed to be equal to c, the expression is $e^{1/1}$ or simply e. But suppose that v were to be equal to $2\,c$. Then the expression would read $e^{2/1}$ or e^2. The mass-ratio for attaining a velocity equal to twice the exhaust velocity must be e^2 (or 7.4). And for attaining three times the exhaust velocity the mass-ratio must be e^3, which is 20.1. If you are inclined at this point to wonder whether a mass-ratio of 20.1 to 1 can actually be built your doubts are justified, but at this moment we are not discussing what can be built; we are after principles.

How does this formula work for rockets that are not supposed even to reach $v = c$? It works the same way; if the rocket's velocity is supposed to be $c/2$, the expression reads $e^{1/2}$ which is the same as the square root of e or 1.64. If v were to be $c/4$, the mass-ratio would have to be 1.29. Such figures actually applied to some of the bombardment rockets of the Second World War. For a table of mass-ratios see Appendix 2.

Having found the principle for calculating mass-ratios, let us return to

[5] This statement is strictly true only if there is no external resistance (air resistance) and if no gravitational field has to be overcome. For actual take-offs from the earth's surface certain allowances have to be made.

our example. But first we have to learn about a few actual exhaust velocities. Here are a few samples.[6]

PROPELLANTS AND MIXTURE RATIO	COMBUSTION TEMPERATURE		EXHAUST VELOCITY			
			SEA LEVEL		"AVERAGE"	
	C	F	FT/SEC	M/SEC	FT/SEC	M/SEC
Nitric acid/aniline (3:1)	2800	5070	6,200	1890	7,300	2225
H_2O_2 (87%)/methyl alcohol (4:1)	2293	4160	6,300	1920	7,420	2262
O_2/ethyl alcohol (1.5:1)	2904	5260	6,870	2094	8,090	2466
O_2/H_2 (3:1)	2421	4390	10,150	3094	12,000	3657

Which of these fuels will be chosen in a given case depends on many factors which are usually unconnected with the layman's cherished belief that the most powerful fuel must always be the best. The first two examples of this short tabulation are somewhat weaker than the other two, but they have the advantage that they can be kept at normal temperatures. The third example involves one liquefied gas, oxygen, and the fourth even two liquefied gases, oxygen and hydrogen. These need special storage facilities, especially hydrogen, which has almost the lowest boiling point of all gases. Such storage facilities may not be available, on shipboard, for instance.

But for our purely theoretical example we can disregard such practical questions and choose oxygen/ethyl alcohol, with the justification that this combination has performed beautifully in literally thousands of rockets. It so happens that the sea-level exhaust velocity of this combination lies close to the rocket velocity we need; to take air resistance into account we will make the mass-ratio 3 to 1 instead of e to 1. Then we get the following weights:

	POUNDS
Payload (M_P)	20
Structure (M_R)	180
Sub-total (remaining mass)	200
Fuels (M_F)	400
Take-off	600

Students, when listening to such an exposition in the classroom, are apt to say: "Is this all that's to it?" Well, no; it isn't. Now you have to build a structure that will hold 400 pounds of fuels and be strong enough to carry the payload, under the accelerations that will occur, without being deformed. Nor must the structure weigh more than 180 pounds, which weight must include the rocket motor, the fuel pumps, and the steering devices. But what has been shown is the principle, or the "first approximation,"

[6] Adapted from "The Problem of Escape from Earth by Rocket" by Frank J. Malina and Martin Summerfield (*Journal of the Aeronautical Sciences*, vol. 14, no. 8; Aug. 1947). The calculations assumed a combustion-chamber pressure of 300 lb. per sq. in.; the figures obtained were then reduced by 12½ per cent to account for imperfections. These reduced "sea-level" values were increased by 18 per cent to give the "average" values valid above 100,000 ft. altitude.

giving the weight limits that must not be exceeded for such a mission.

Now we proceed one step further. The payload is the same but the mission is changed; the instruments must be carried to an altitude of 800 miles. This altitude calls for a velocity that is twice as high as the exhaust velocity of the fuel combination, meaning a mass-ratio of e^2 to 1, or 7.4 to 1. Rocket engineer *A* is a practical man who reasons as follows: "Nobody has ever built a rocket with such a high mass-ratio and this is a job, not a research project, so let's not start anything fancy now. Besides we have this 600-pound, 170-mile rocket, and we know it works. Therefore for the 800-mile job we want a two-stage rocket, with the existing rocket as the upper stage." This leads to the following "first approximation":

	POUNDS
Payload (the second stage)	600
Structure	4,400
Subtotal	5,000
Fuels	10,000
Take-off mass	15,000

What happens when that two-stage rocket is fired is that the first stage acquires a velocity of about 6600 feet per second. Then the second stage starts burning, adding another 6600 feet per second, so that the final velocity is 13,200 feet per second, enough to reach the 800-mile altitude. The two stages *add* their velocities, which is the same as if the mass-ratios of these stages were *multiplied* by each other; the "ideal mass-ratio" of a multistage rocket is *the product of the individual mass-ratios of its stages.*

The advantage of a two-stage rocket for the 800-mile shot can readily be seen if we imagine rocket engineer *B*, who once had a stage separation miscarry and therefore avoids stage separations if he possibly can. His "first approximation" will look like this:

	POUNDS
Payload	20
Structure	5,580
Subtotal	6,000
Fuels	39,200
Take-off mass	45,200

This leaves rocket engineer *B* with the small problem of packing nearly 40,000 pounds of fuels into not quite 6000 pounds of structure. Maybe it can be done, but there is no doubt that the two-stage rocket for the same mission will be cheaper by far. The reason that the one-stage rocket has this enormous disadvantage as compared to the two-stage rocket is simple: a remaining mass of 6000 pounds is needlessly carried to 800 miles. The two-stage rocket carries only a remaining mass of 200 pounds to that altitude, the 4400 pounds of structure of the first stage having been left behind as soon as they became superfluous.

If we take our oxygen/ethyl alcohol combination and consider that the exhaust velocity is 7000 feet per second in all stages (it is a little less than that in the first stage but around 8000 in upper stages) and give each stage a mass-ratio of 4 to 1, the velocity produced by each stage would be 10,000 feet per second, or a final velocity of 30,000 feet per second, or 5.7 miles per second. This is a velocity that is *higher* than is necessary for placing an artificial satellite into orbit around the earth at an altitude of 120 miles.

The father of the concept of an artificial earth satellite is Sir Isaac Newton. Of course he did not intend to launch an artificial satellite; he developed the concept for an entirely different purpose—namely, to explain the orbit of the moon around the earth. Newton's reasoning can be found in his *Philosophiae Naturalis Principia Mathematica*—now usually referred to as the *Principia*—which was written in Latin and printed in London in 1687. In 1729 it was translated into English by Andrew Motte.[7]

The pertinent section appears in the first book of the *Principia,* under the heading "Definition V." The definition itself deals with the use of the word "centripetal." It states that "a centripetal force is that by which bodies are drawn or impelled . . . towards a point as to a centre." Gravity is considered to be such a centripetal force:

> A projectile, if it was not for the force of gravity, would not deviate towards the earth, but would go off from it in a right [meaning "straight"] line, and that with a uniform motion, if the resistance of the air was taken away. It is by its gravity that it is drawn aside continually, and made to deviate towards the earth, more or less, according to the force of its gravity [meaning "weight"] and the velocity of its motion. . . . If a leaden ball, projected from the top of a mountain by the force of gunpowder, with a given velocity, and in a direction parallel to the horizon, is carried in a curved line to the distance of two miles before it falls to the ground; the same, if the resistance of the air were taken away, with a double or decuple velocity would fly twice or ten times as far. And by increasing the velocity, we may at pleasure increase the distance to which it may be projected, and diminish the curvature of the line which it might describe, till at last it should fall at the distance of 10, 30, or 90 degrees, or even might go quite around the whole earth before it falls; or lastly, so that it might never fall to the earth, but go forwards into the celestial spaces, and proceed in its motion *in infinitum.* And after the same manner that a projectile . . . may be made to revolve in an orbit, and go round the whole earth, the moon also . . . may be continually drawn aside towards the earth, out of the rectilinear way which by its innate force it would pursue. . . .

Here you have the whole theory of artificial satellites. The satellite must not be impeded by air resistance, that is, it must be far enough from the

[7] Motte's translation (mainly its mathematical terminology, which is no longer in use) was revised by Professor Florian Cajori and the new version published by the University of California Press in 1946. The quotation is from this revised edition.

earth's surface so that there is no atmosphere left where it orbits. It must be sufficiently fast so that the curvature of its orbit matches the curvature of the earth below. And its direction of motion must be "parallel to the horizon"—that is, horizontal (Fig. 53).

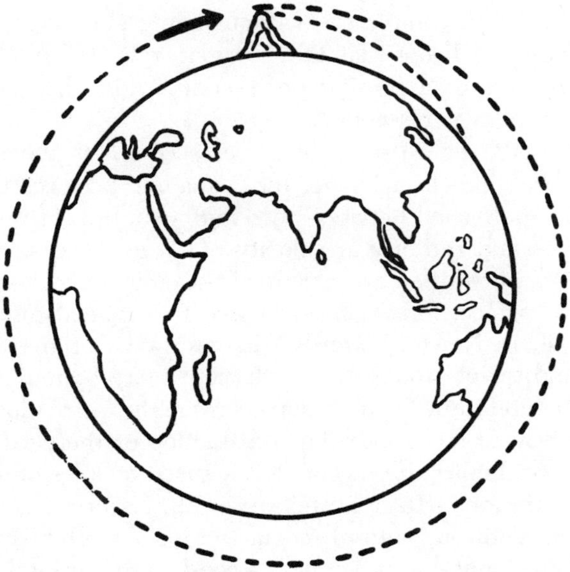

Fig. 53. Sir Isaac Newton's approach to the concept of circular velocity and the origin of an orbit.

The first of Newton's three conditions, the lack of resisting atmosphere, proved to be a tough nut to crack theoretically because it was so hard to get basic data for reasonable assumptions. Everybody was agreed that Newton's first condition would be met above 250 miles from sea level, and "probably" at a lesser altitude. It was that "probably" that used up reams of paper and hours of discussion. Some researchers had convinced themselves that 80 miles would be high enough if the artificial satellite had a reasonable mass and not too large a surface. Others felt very lenient indeed if they conceded that such a satellite might orbit for, say, half a year, at an altitude of 150 miles.

Because the earth's atmosphere, consisting of gases, cannot have a definite limit, it is still impossible to give a precise figure, but the practical experience that has accumulated from about five hundred successfully orbited artificial satellites points to 150 miles as a usable figure. Any satellite that does not come closer to the earth's surface than 150 miles is likely to orbit for a very long time.

Newton's last condition, that the direction of motion must be "parallel to the horizon" is rather easy to accomplish; the method will be discussed in Chapter 11.

As regards the second condition—a sufficiently high velocity—Newton himself would have liked to give an answer, but he was unable to do so because the size and mass of the earth were insufficiently known in his time. For the same reason Newton could visualize that a body might escape from the earth completely, without being able to say what velocity is required. Now we know that the "circular velocity" in the case of the earth is slightly less than 5 miles per second, while the "escape velocity" is just under 7 miles per second.[8]

Unfortunately the figure of 5 miles per second for the earth's circular velocity has been much misunderstood, mainly because the authors of popular articles published between 1950 and 1955 themselves believed that circular velocity and the orbital velocity of an artificial satellite were the same thing. Of course they are not, because circular velocity is calculated for the earth's surface, for sea level, where no artificial satellite can orbit. If one could—if, in Newton's words "the resistance of the air were taken away"—it would sprint around the earth once every 83 minutes.

An artificial satellite at some distance from the earth does not need to move that fast, since the gravitational attraction of the earth at a distance of several thousand miles can be counterbalanced by a lesser orbital velocity. Before there were any artificial satellites at all, Nature herself had demonstrated the case with our natural moon; the moon's orbital velocity is only 1 kilometer, or about ⅔ mile, per second. True orbital velocities for satellites between the earth and the moon therefore had to be higher than ⅔ mile and lower than 5 miles per second. And, of course, the farther a satellite is from the earth, the longer is its orbital period: Not only does it move more slowly, it also has a far longer orbit to complete.

Here are a few examples of circular orbits for artificial satellites that were calculated from astronomical data before the first artificial satellites actually made their appearance.

DISTANCE FROM SURFACE		ORBITAL PERIOD	ORBITAL VELOCITY (MI/SEC)
MI	KM		
346	557	96 minutes	4.69
470	756	1¾ hours	4.41
1,075	1,730	2 hours	4.39
3,200	5,150	3½ hours	3.58
4,000	6,430	4 hours	3.47
7,700	12,400	7 hours	2.90
22,300	35,900	24 hours	1.90

Every one of the orbits listed in this tabulation is a so-called permanent orbit; satellites in such orbits would suffer no interference from residual

[8] The relationship between these two figures is simple—circular velocity multiplied by $\sqrt{2}$ gives escape velocity.

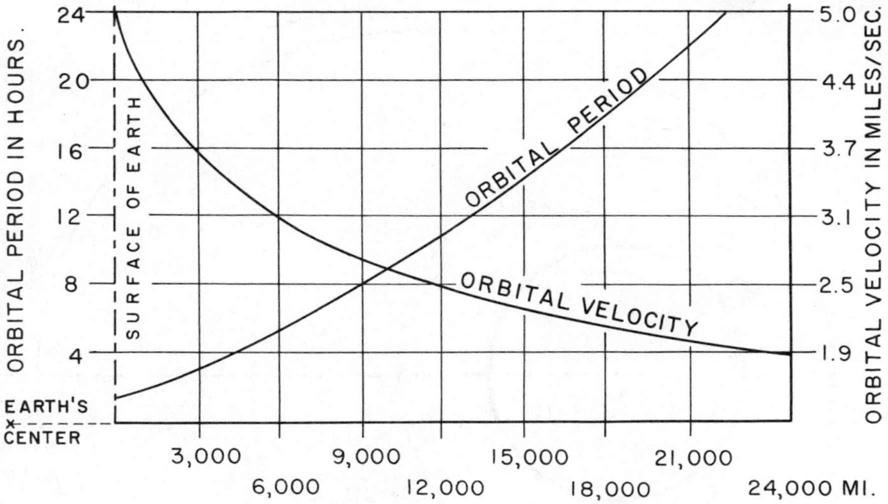

Fig. 54. Orbital periods versus orbital distances. Figures at left indicate periods; at right velocities; at bottom distances in miles. (*Adapted from a diagram by Arthur C. Clarke*)

air resistance and continue to orbit, unless subjected to artificial interception. The last orbit in the table, originally referred to as the 24-hour orbit, later received a special name. An artificial satellite in this orbit, if over the equator all the way round, would seem to stand motionlessly above one point of the equator because it would be synchronized with the diurnal rotation of the earth. For this reason it came to be known as the "synchronous orbit."

As can be readily seen from the tabulation, as well as from Fig. 54, each distance implies a specific orbital velocity. It is impossible to move in one of these orbits with a velocity that differs from the velocity peculiar to this orbit. If, by means of an additional impulse, the orbital velocity is changed, the orbit will be changed too. It will no longer be circular but will have become elliptical (Fig. 55).

We learned in Chapter 1 (pages 12–13) that Johannes Kepler revolutionized astronomy and disturbed a handful of philosophical systems by discovering that the orbits of the planets around the sun were not circles or combinations of circles but ellipses. This discovery compelled him to invent a number of new terms. If the earth, for example, described an ellipse around the sun, there had to be a point along that elliptical orbit where the earth was closest to the sun, and another one where it was farthest away. Kepler named the orbital point nearest the sun "perihelion," from Greek *peri,* meaning "around" and Helios, the Greek sun god. The farthest point he named "aphelion" (shortened from apohelion, *apo* being

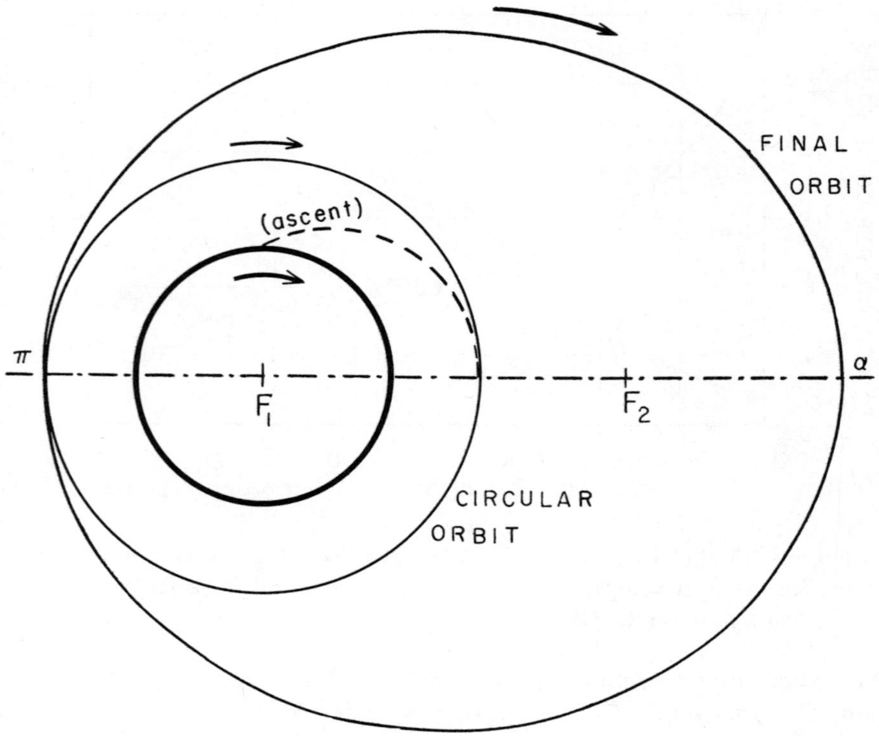

Fig. 55. Change from a circular to an elliptical orbit. The point marked F_1 is one of the two focal points of the ellipse; it is also the center of the circular orbit and the center of the earth. Perigee and apogee of the elliptical orbit are marked by the Greek letters pi (π) and alpha (α). The second focal point, F_2, is of geometrical interest only.

Greek for "to go away"). For orbits around the earth the corresponding terms are perigee and apogee (*Gaia* being Greek for "earth").[9]

Although Kepler could not doubt the observational evidence that showed that the orbits were actually ellipses, he was at a loss to explain why this had to be so. But with the knowledge that a circular orbit at a given distance must have a specific orbital velocity the origin of the ellipse can be quite easily visualized. Let us use the 4-hour orbit 4000 miles above sea level as an example. The orbital velocity in this orbit is 3.47 miles per second. If this velocity were increased to 3.5 miles per second the satellite would be too fast for this orbit; it would have more momentum than is needed to counterbalance the earth's gravitational field at that dis-

[9] For orbits around the moon the terms "perilune" and "apolune" have been suggested, but since *Luna* is Latin, the Greek *Selene* ("moon") is preferable, and the terms become periselenion and aposelenion.

tance. Consequently it would move farther away from the earth, not in a straight line but along a slanting curve, since it keeps orbiting the earth. In moving away it steadily loses momentum. When it has reached its farthest point on this new expanding orbit it does not have enough momentum left to settle into a new circular orbit at this greater distance. Logically it has to approach the earth again, again along a slanting line. After having moved through an angle of 180 degrees the satellite is back in its original circular orbit. But in approaching the earth it has gained speed and, along with speed, momentum, which means that it is again moving at the rate of 3.5 miles per second—too fast for this circular orbit—so that the same performance is repeated. Provided that the original circular orbit, and with it the perigee point of the new elliptical orbit, are well above the altitude where residual air resistance might interfere, this elliptical orbit is just as permanent as one of the circular orbits listed in the tabulation on page 284.

All this was astronomical knowledge long before the first rocket climbed to an altitude of even one mile, and since space travel can be called "applied astronomy" it is only logical that astronomers also knew the answer to the question: but what will happen if the perigee altitude is *not* outside the upper atmosphere? Astronomers had thought about this problem because they suspected the existence of "ephemeral satellites," meteorites that might enter the earth's atmosphere at a shallow angle and describe a few orbits around the earth before finally reaching the ground. The theory that applied to such an ephemeral satellite after its first grazing contact with the earth's atmosphere applied just as well to an artificial satellite with its perigee in the atmosphere.

Let us say that the perigee of such a satellite is at 90 miles, while the apogee is at 600 miles. The velocity of such a satellite (whether artificial or natural and ephemeral) at its apogee would be 4.4 miles per second. After having passed its apogee for the first time the satellite will approach its perigee altitude of 90 miles in its elliptical orbit. But for a few dozen miles before and after passing perigee the satellite will be inside the atmosphere, suffering resistance to its motion and thereby losing some of its momentum. The result is that the second apogee will be somewhat closer to earth than 600 miles, as the satellite now does not have enough momentum left to attain such distance. Returning from its closer apogee to perigee it suffers loss of momentum again. While the perigee altitude is hardly changed, the third apogee is now still closer than the second apogee. Thus, each succeeding apogee comes steadily closer to the earth, and as a result the ellipse becomes more circular in appearance. The technical term for this phenomenon is "orbital decay," because the ellipse "decays" into a circle, once considered the ideal shape for an orbit. By the time the ellipse has actually become a circle, all of the orbit is in the upper atmosphere, at the altitude which the perigee has assumed by that time—

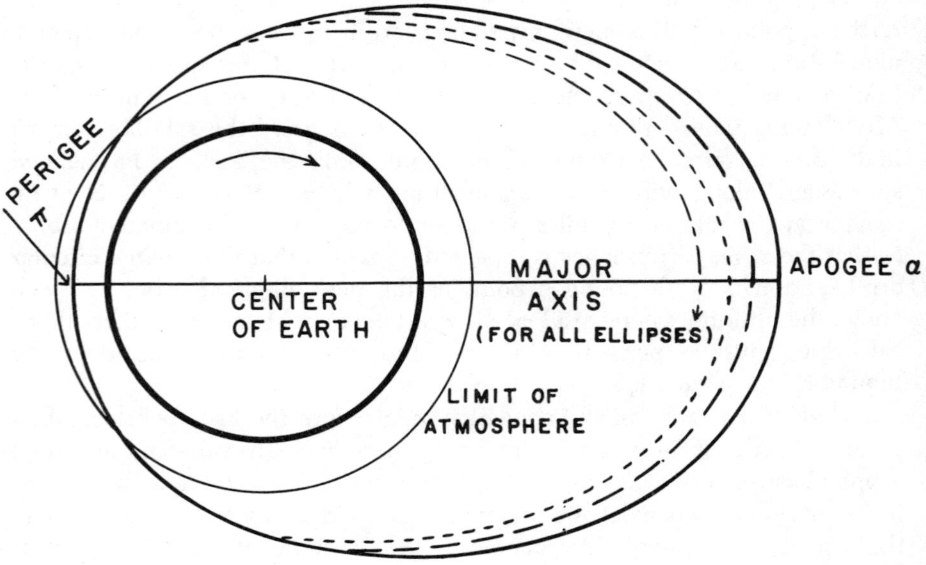

Fig. 56. Orbital decay of an artificial satellite in a temporary orbit. For clarity, the eccentricity of the orbit, the amount of shrinkage per revolution, and the arbitrary limit of the atmosphere have been exaggerated.

say, at 80 miles. Now the atmosphere is creating resistance all the time, the circle quickly becomes a tight spiral that gradually approaches the ground, and as soon as denser layers of the atmosphere are reached the satellite will be destroyed by aerodynamic heating (Fig. 56). This is now referred to as the "burn-up."

We now know that things progress precisely as theory predicted. The new information consists of figures for the altitude at which residual air resistance produces orbital decay and for the time required from first apogee to burn-up for a given orbit. These figures could not be calculated in advance because no information existed on the density—or lack of it—of our atmosphere at high altitudes. In fact, it was the tracking of the orbits of the early satellites that provided the needed figures.

The foregoing discussions have presented the theory that preceded the "things to come"—to use a Wellsian phrase—and now it is time to return to the story of rockets. During the years 1955 and 1956 new types of rockets were employed to study the upper atmosphere.

The study had started a few years earlier, in 1951, with the investigation of the so-called "cosmic rays." That a peculiar and penetrating kind of radiation came from outer space had been noticed during the years following the conclusion of the First World War by various researchers, (primarily, E. Kolhörster in Germany and Robert A. Millikan in the United States), but the nature of this radiation was to remain a major scientific puzzle for years to come. The uncertainty and the confusion were compounded by the finding that instead of the one type of radiation, several kinds seemed to be present. The true picture emerged slowly, though, in retrospect, it is a bit difficult to see why it took so long.

The actual situation is this: the "cosmic rays" that come from space are subatomic particles. Most of them are protons, the nuclei of hydrogen atoms, but occasionally a nucleus of a heavier element makes its appearance. The particles that come from space are called the "primaries," the majority of which probably originate in our sun, though the most energetic (fastest) particles are believed to have originated in deep space, far from our solar system.

A primary moving through space on a collision course with the earth may not enter the atmosphere at all, since many of these primaries are deflected by the earth's magnetic field. Of those that do enter the atmosphere only a very few reach the ground, because they collide with atoms or molecules of nitrogen and oxygen, producing more radiation of different types, called the "secondaries." Taking both primaries and secondaries together we find that the maximum density of all these cosmic rays lies at an altitude of about 25 miles; farther down, most of the secondaries have been absorbed by other atmospheric molecules.

Scientists interested in cosmic rays were especially interested in the primaries and conducted cosmic-ray counts in various localities. Although at each locality small fluctuations were detected depending on the time of day when the count was taken, it was doubtful how much meaning should be attached to these fluctuations. But stations located at various latitudes indicated very significant differences in the numbers of cosmic primaries counted. The number was smallest under the magnetic equator, where the earth's magnetic field deflected primaries most effectively. The number was greatest in higher latitudes near the magnetic poles, where the attrition of the primaries is mainly due to collision with gas molecules.

Of course it was important to find out what occurs a number of miles *above* the magnetic poles. The problem was to carry instruments cheaply to 10 or 12 miles above the northern pole, which is more easily accessible. Aerobee and Aerobee-Hi rockets had been used from Canadian soil, but they were still fairly expensive and could not be used too freely because they needed launching towers.

In 1950 a new 12-foot-long rocket called the Deacon had become available. It was one of the early large solid-fuel rockets, and the military who had sponsored its development apparently were not happy with it. Hence it was available for scientific-research projects.

The instrument package for this cosmic-ray research project weighed only 20 pounds, which was far less than the Deacon's carrying capacity. But the Deacon could not go very high when fired from the ground or from shipboard. The reason it couldn't was that it had a very high acceleration; hence it brought itself up to a high speed very fast and then expended most of its kinetic energy fighting air resistance at low levels. The solution was found by Dr. James A. Van Allen. If the Deacon rocket were first lifted to an altitude of, say, 20 kilometers (65,000 feet) where the pressure is only 41.5 millimeters of mercury instead of the sea-level value of 760 millimeters, the Deacon should perform nicely. The means of lifting it existed: the large plastic Skyhook balloons. The combination of rocket and balloon was termed a "rockoon," and the first ones were launched in 1952 from on board the Coast Guard Cutter *Eastwind* in the icy sea off Greenland. A pressure switch completed the ignition circuit for the rocket when the external air pressure had dropped to the desired low level. This took place about one hour after the balloon was released. The Deacon then climbed to over 50 miles.

Another research project involving rockoons took place in the Pacific Ocean 200 to 400 miles southwest of San Diego, California, during the last two weeks of July 1956. The launching ship was the U.S.S. *Colonial,* accompanied by the destroyer U.S.S. *Perkins,* which assisted in tracking with its radar gear. This time the problem was the detection of ultraviolet radiation and X rays from solar flares. A solar flare, as the name indicates, is an especially bright spot on the sun which appears quite rapidly and lasts for some time. On earth a solar flare makes itself felt by causing a radio fade-out in the medium and short wavelengths. If the launching of a rockoon had been delayed until a flare appeared the rocket would invariably have arrived too late, because the balloon needs at least an hour to float up to its peak altitude. Therefore a rockoon was launched every morning and permitted to float at 80,000 feet. Scientists in a land-based station in California watched for solar flares, using the radio fade-out as an indicator. This method had the advantage of working just as well under a cloudy sky. When they discovered a fade-out they radioed the ship, using, of course, a wavelength not affected by the solar flare, and the ship then sent a radio firing command to the floating rockoon. The rocket disposed of the balloon by literally tearing through it so that the torn plastic envelope floated down and disappeared in the Pacific. There was a total of ten flights.

The original rockoon proved to everybody's satisfaction—if to some people's amazement—that the larger solid-fuel units could be used for upper-atmosphere research. Suddenly a set of new names cropped up: Nike-Cajun, Nike-Deacon, Terrapin, HTV, and so on. All of them were two-step rockets, with both steps using solid fuel. The first step was almost always the booster of a Nike antiaircraft missile.

The Nike-Deacon combination—also called DAN—was ready for test firing in the summer of 1955. (See Figs. 57 and 58, Plate XXIX, and list of weights, etc., in Appendix 3). Two of them were fired from Wallops Island, alike except that no. 1 carried 34 pounds of payload, no. 2 39 pounds. Fired at an angle of 75 degrees, no. 1 reached a peak altitude of 356,000 feet (67.4 miles) while no. 2 went to 350,000 feet (66.3 miles). The second stage carried, at the base of its nose cone, an accelerometer sphere, mounted on tensed springs, one above and one below. Nose cone and sphere were held to the front end of the second-stage rocket by a duraluminum retainer ring. At 180,000 feet the retainer ring was exploded and the spring below the sphere (the stronger one) pushed both sphere and nose cone away from the rocket. Immediately afterward the weaker spring above the sphere separated it from the nose cone.

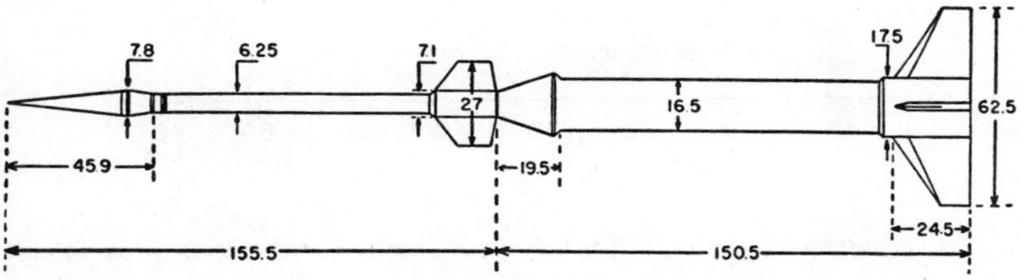

Fig. 57. The Nike-Deacon (DAN) rocket. All dimensions are in inches; for weights see Appendix 2.

If the procedure of the Bumper rockets had been followed in the Nike-Deacon rocket, the upper stage would have been faster than Oberth's optimal velocity. To avoid this a time interval was allowed to elapse between Brennschluss of step one and ignition of step two.

> Both second-stage rockets were allowed to decelerate after booster separation. The coast period permitted the vehicles to traverse the lower, denser atmosphere at lower velocities, thereby reducing the maximum outer-wall temperature of the nose cone and rocket case as well as lowering the aerodynamic drag. The optimum duration of the coasting period for attainment of maximum altitude was estimated to be 10 to 14 seconds, according to performance calculations. This variation in coast period was not critical. Since the burning time of the Nike booster first stage was approximately 3.5 seconds, a nominally 15.5-second time-delay powder squib, electrically ignited, was employed in the Deacon second stage in order to give a nominal coasting period of 12 seconds and permit ignition of the squib at the instant of launching. The accuracy of the time delay of the 15.5-second time-delay squib is about 1.0 second. In the flight test of the first DAN sounding rocket, the squib gave a delay of about 17 seconds. Accord-

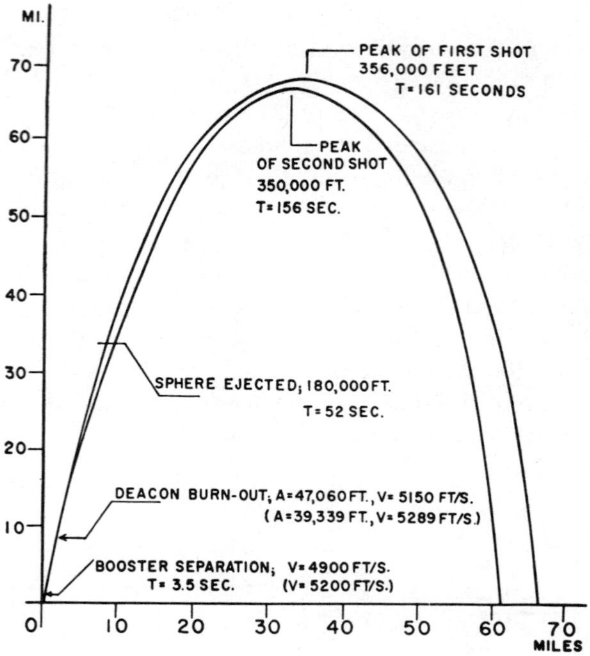

Fig. 58. Trajectories of two DAN firings. *A* stands for altitude, *V* for velocity, and *T* for time elapsed since take-off. Figures in parentheses refer to shot no. 2.

ingly, in the second DAN sounding-rocket test a nominally 13.5-second time-delay squib was employed which in the flight test gave a delay of about 12.8 seconds. (Heitkotter in *NACA T.N.* 3739)

The Nike-Cajun rocket, which looks very much like the Nike-Deacon, may be considered as a kind of successor to the latter. The lower step is an unmodified Nike booster, the upper step a Cajun rocket, which is essentially a Deacon with a more modern propellant. The first Nike-Cajun was launched from Wallops Island on July 6, 1956, at an angle of 75 degrees. Take-off weight was 1555 pounds, weight of the upper stage 257 pounds, weight of the payload 51.5 pounds. At booster burnout, 3.3 seconds after take-off, the rocket was 5300 feet high, moving at three times the speed of sound. Ignition of the Cajun took place at X + 12.3 seconds; meanwhile the velocity had dropped to twice the speed of sound but the rocket had climbed to 28,300 feet. At X + 15.3 seconds the Cajun stopped burning, having reached 39,000 feet and a velocity of 5.7 times the speed of sound. Peak was attained at X + 175 seconds with 425,000 feet (80.7 miles); the impact into the ocean took place at a (calculated) distance of 468,000 feet (88.6 miles).

After the information about the Nike-Deacon and Nike-Cajun rockets had been released, the NACA admitted somewhat reluctantly to having fired a four-stage HTV (hypersonic test vehicle) late in 1954. It consisted of two Nike boosters as the first two steps, then a 4.5-foot-long Thiokol T-40 solid-fuel rocket, with a 6-foot-long Thiokol T-55 rocket as the fourth stage. Over-all length of the HTV was 35 feet, 8 inches (somewhat longer than the lengths of the four rockets because of adapters and connections between the stages), and it had a take-off weight of 2800 pounds. The coasting periods were: 11 seconds between first and second stages, 5 seconds between second and third stages, and 2 seconds between the third and the fourth stages. Maximum velocity of the fourth stage was 10.4 times the speed of sound 84,000 feet up and the peak altitude somewhat over a million feet, or 190-plus miles.

While the NACA put superrockets together from available components other agencies did not remain idle. Under contract to the Navy's Bureau of Ships, the Cooper Development Corporation in Monrovia, California, developed a single-stage solid-fuel sounding rocket called the ASP (Fig. 59) with a propellant charge developed by the Grand Central Rocket Company of Redlands, California. Take-off weight of the Asp is 245 pounds, weight after Brennschluss 95 pounds. The Asp was test-fired both from Point Mugu and from White Sands, mostly at an angle of 30 degrees for ballistic data. Of the eighty rounds fired in 1956, thirty were used up in static tests at air temperatures varying from 30 to 130 degrees Fahrenheit.

Performance calculations based on these tests indicate the following possibilities for the Asp: when used by itself with a 25-pound payload it should reach 200,000 feet when fired from sea level and 240,000 feet when fired from White Sands (4000 feet above sea level). For a 50-pound payload the figures would be 170,000 and 200,000 feet, respectively. Used as a rockoon the Asp should reach 640,000 feet with a 25-pound payload and 500,00 feet with a 5-pound payload. And when boosted with a Nike booster it should reach 850,000 feet with a 25-pound payload with a suitable ignition delay (Fig. 59).

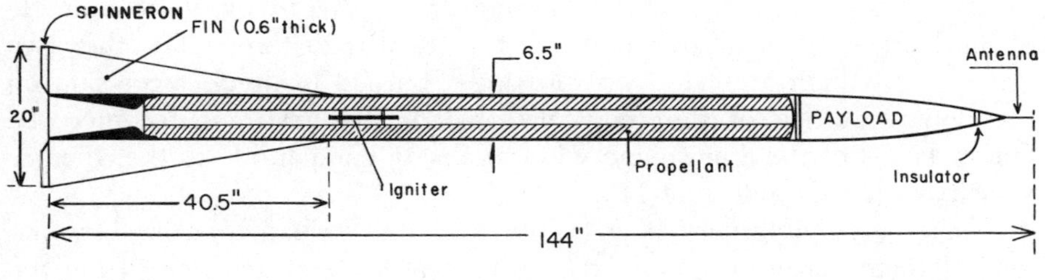

Fig. 59. The ASP (Atmosphere Sounding Projectile). A smaller version of the ASP is known as the WASP.

While the Asp streaked over the Pacific, the first Terrapin (designed jointly by the University of Maryand and Republic Aviation Corporation, with Dr. S. Fred Singer as head of the research program) roared over the Atlantic from Wallops Island on September 21, 1956. It, too, is a two-stage solid-fuel rocket and its upper stage climbed to 420,000 feet (79.5 miles), falling into the ocean 5.6 minutes after take-off. The almost fantastic things about the Terrapin are its small size and weight. The rocket is only 15 feet long, with its largest diameter 6¼ inches. And the take-off weight is a mere 224 pounds, of which the instrumentation weighs 6 pounds.

Another HTV was flown for the first time in November 1954. It had been built for the Air Force by Curtiss-Wright Corporation and was an innovation in so far as both its stages consisted of *clusters* of solid-fuel rockets. The bottom stage, 5 feet long, was a cluster of seven rockets, with an aggregate diameter of 9 inches. The top stage, with the same length but a diameter of 6 inches, was a cluster of four rockets. The whole was topped by a 2-foot cone housing the instruments. This HTV was flown without an ignition delay for the second stage in order to find out what would happen if a rocket moved very fast in dense air fairly close to the ground.

The idea of clustering solid-fuel units for special purposes led to an early triumph in Lockheed's X-17 rocket. This rocket was designed by rocket engineers who were anticipating ballistic missiles of intercontinental range which would pose special and novel problems. Though Dr. Vannevar Bush had gone on record that intercontinental missiles, if they could be built at all, were at least twenty years in the future, most practicing engineers, even though their opinions did not carry as much weight, knew that it was a question of only a very few years. One of the problems that such a missile would pose is what is now known as re-entry. But how do you observe re-entry of a 4000-mile shot if you don't have a 4000-mile missile?

The purpose of the X-17 was to simulate such a re-entry. The rocket had three stages; the first one was an XM-20 ("Sergeant") solid-fuel unit with a thrust of 50,000 pounds and a burning time of about 24 seconds. The second stage consisted of a cluster of three XM-19 ("Recruit") units, each with a thrust of 35,000 pounds and a burning time of 1.53 seconds. The third stage was a somewhat different "Recruit" unit with a 36,000-pound thrust and a burning time of 1.53 seconds.[10] The X-17 took off at an angle of 80 degrees. The purpose of the first stage was to throw the other two stages into a high arc. The two top stages burned in quick succession on the downward leg of the trajectory, making the instruments enter the denser layers of the atmosphere with a velocity comparable to the re-entry velocity of a 4000-mile shot (Fig. 60).

By the end of February 1957 twenty X-17 rockets had been fired, seventeen of them performing perfectly with re-entry about 200 miles from the

[10] A comprehensive article on the X-17 appeared in *JARS*, August 1959.

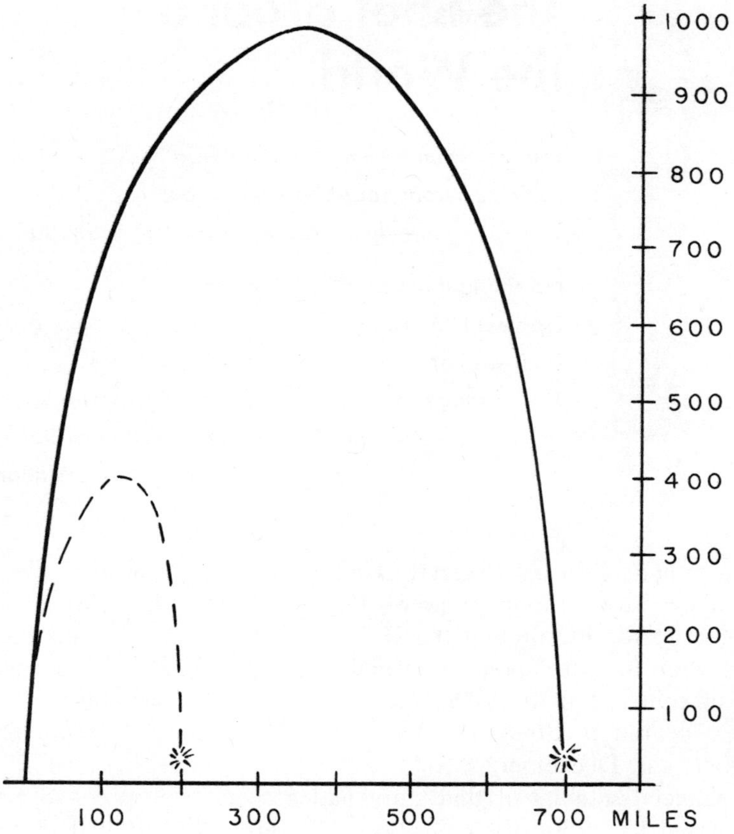

Fig. 60. Probable trajectory of the "run-away" X-17.

firing site. One of those that did not perform perfectly fired all three stages on the way up. The third stage re-entered about 700 miles from the firing site; the peak altitude is not known because the instrumentation was not geared for such an event. Calculation indicates that the peak must have been around 990 miles, over twice as high as that of the third stage of an X-17 which was later fired for altitude, carrying a useful payload.

Some readers may find it ironic that the first shot to go higher than Bumper no. 5 was an accident. The point is, however, that neither shot was for the purpose of establishing a high-altitude record—that was scientifically unimportant. The Bumper shots were made to test the separation of stages at high velocities and a few other matters. The X-17 shots were made to provide information about re-entry conditions and did provide this information. There were no shots for altitude because to reach a point beyond the earth's atmosphere was of no value; the goal was to *stay* at such altitudes.

11 The Shot around the World

Tout ce qu'un homme est capable d'imaginer, d'autres hommes seront capables de le réaliser.

—Jules Verne, in a letter to his father

No single thing abides; but all things flow.
Fragment to fragment clings—the things thus grow
Until we know and name them. By degrees
They change, and are no more the things we know.

—Titus Lucretius Carus, *De Rerum Natura* (Mallock translation)

The Reverend Edward Everett Hale, 1822–1909, Boston clergyman, writer, and editor, grandnephew of Nathan Hale, and, so far as I know, no relation to the inventor of the Hale rockets, was the first man after Sir Isaac Newton to write about an artificial satellite. What he wrote was only fiction, of course, but it was fiction at an early date; his story "The Brick Moon" appeared in three installments in the *Atlantic Monthly* October, November, and December, 1869.[1]

The artificial satellite of that story had a specific purpose—to aid navigators at sea in finding their position. For this reason it had to be quite large, with a diameter of 200 feet, so that it might be seen even with such portable spy glasses as were likely to be on shipboard. The satellite was made of brick, but hollow in order to save weight. For better visibility it also was supposed to be whitewashed before launching. The device that was to send it into orbit consisted of two immense flywheels, driven by a waterfall and "capable of storing an enormous amount of energy." Of course this would not have worked in reality, but in the story the brick moon is launched inadvertently, carrying a number of people with it into orbit. Since it had not yet been whitewashed it was difficult to find in the sky, but when the people on the ground finally located it they learned that the people in the brick moon were not only alive but well and happy.

Chronologically, the next mention of artificial satellites is that by Lasswitz in 1897 (see pages 42–44), but unlike the Reverend Mr. Hale, who

[1] The February 1870 issue of the *Atlantic* contained a sequel, "Life in the Brick Moon"; both stories were reprinted in Hale's book *His Level Best and Other Stories* published in 1872 and have recently been reprinted in *Masterpieces of Science Fiction* edited by Sam Moskowitz (Cleveland and New York: World Publishing Company, 1967).

had a true earth-centered orbit in mind, Lasswitz kept his two manned space stations immovable over the geographic poles of the earth.

Next in line was Ziolkovsky, who talked seriously about an earth-orbiting spaceship in his second publication (first printed in 1911 in the aviation journal *Vestnik vosdukhoplavaniya*), a publication which he always regarded as "Part II" of his first publication in 1903. After describing the initial ascent and a positioning maneuver, Ziolkovsky wrote:

> the rocket is now prevented from falling; it assumes the velocity normal to the radius vector, i. e., a velocity along a circle like the moon and, like the latter, it will rotate around earth indefinitely.
>
> Now we can become completely calm since the rocket has obtained a stable position; it has become earth's satellite.
>
> From the rocket we can see the huge sphere of the planet in one or another phase like the moon. We can see how the sphere rotates, and how within a few hours it shows all its sides successively. The closer it [the earth] is to the rocket, the larger it appears. . . . This distance can be so small that travel around it can be carried out in two hours, and we shall observe various points on the surface of the earth for several minutes from different sides and very closely. This picture is so majestic, attractive, and infinitely varied that I wish with all my soul that you and I could see it. Every two hours the rocket is eclipsed and enters the earth's shadow and night. This latter lasts for less than an hour; then for over an hour the sun shines before giving way again to darkness.
>
> If we wanted to make use of a larger quantity of light, i. e., of a longer day, we must either move further away from earth or rotate not in the direction of the equator but in the direction of the meridian, so that our path would intersect the poles of earth. . . .[2] (NASA TT F-243)

Except for some of the peculiarities of Ziolkovsky's style this passage could have been written very recently, but it was written in 1910.

About a dozen years later came Oberth's first publication. Oberth had never heard of Hale and at the time knew of Ziolkovsky only by hearsay. But the earth-orbiting station was simple scientific logic, so Oberth developed his own ideas. In the last chapter of his short book he estimated that a rocket carrying two men would weigh at least 400 metric tons. Then he continued, "If we let such rockets of the largest size [mentioned] move around the earth in a circle they will behave like a small moon. Such rockets no longer need to be designed for landing. Contact between them and the earth can be maintained by means of smaller rockets so that the large ones (let's call them observing stations) can be rebuilt in the orbit the better to suit their purpose. If the continuing state of apparent weightlessness should have undesirable consequences, which, however, I doubt,

[2] The quotation is from NASA's Technical Translation (1965) of all Ziolkovsky's works, based on the Russian edition published in Moscow in 1947 by the Publishing House of the Defense Industry. My rendering of Ziolkovsky's name is the form that he himself employed in his letters to me. The first letter of his name as written in Cyrillic characters cannot be expressed in English by just one letter, and the last letter poses a similar difficulty. NASA has settled for "Tsiolkovskiy."

one could connect two such rockets by wire ropes a few kilometers in length and make them rotate around each other."

This really is much more than the mere germ of an idea; it is the whole concept without engineering detail: the ship which stays in space, not designed for landing; the supply by transport rockets; the gradual improvement with materials brought up by these transport rockets; the trick of having centrifugal effects substitute for gravity, if needed. Oberth then proceeded to outline a few possible uses:

> With their powerful instruments they would be able to see fine detail on earth and could communicate by means of mirrors reflecting sunlight. [This, it must be remembered, was written in about 1920 or 1921 when radio was very much in its infancy.] This might be useful for communication with places on the ground which have no cable connections and cannot be reached by electric waves. Since they, provided the sky is clear, could see a candle flame at night and the reflection from a hand mirror by day, if they only know where and when to look, they could maintain communications between expeditions and their homeland, far distant colonies and their mother land, ships at sea, etc. . . . The strategic value is obvious especially in the case of war in areas of low population density; they might either belong to one of the two countries at war or else sell their services at high rates to one of the combatants. . . . The station [at this point the term "station" is used for the first time] would notice every iceberg and warn ships. . . .

After making the inadvertently false statement that "the disaster of the *Titanic* in 1912 would have been avoided by such means" (the *Titanic* had iceberg warnings enough but just did not pay any attention since she was "unsinkable"), Oberth added another completely new idea which had not been voiced before anywhere:

> All this constitutes a certain amount of practical advantages. But an even greater one could be gained in the following manner: one could spread a large circular wire net simply by rotating it around its center. Small plane metal mirrors could be fitted into the spaces between the wires and their position relative to the wire net could be controlled electrically from the station. The mirror as a whole should rotate around the earth in a plane which forms a right angle with the plane of the earth's orbit. The wire net would be inclined to the direction of the sun's rays by 45 degrees [Fig. 61]. By proper adjustment of the positions of the single facets, one could either concentrate the reflected sunlight on specific points of the ground or could diffuse it over large areas, or, if not needed, make the whole beam miss the earth. If, for example, the mirror is 1000 kilometers [600 miles] distant, the image of the sun from each facet would have a diameter of 10 kilometers; if they are made to coincide, the energy would be concentrated in an area of 78 square kilometers. Since the mirror can have any size desired, it could have colossal effects. It would be possible, for example, to keep the shipping lane to Spitsbergen and the North Siberian ports ice-free by such concentrated sunlight. With a mirror 100 kilometers in diameter a diffused reflection could make large areas in the north habitable; in the middle latitudes it could prevent sudden drops of temperature in the Spring and in the Fall and could, especially in Spring, prevent freezing temperatures at night, thereby

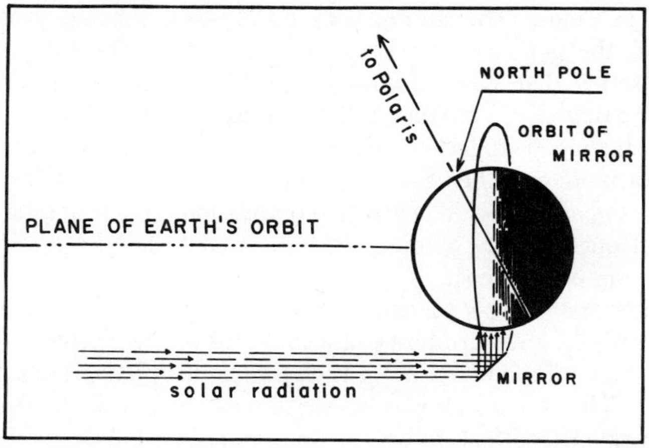

Fig. 61. Oberth's space mirror. The earth is assumed to be near aphelion (July) with the mirror passing over the Antarctic regions. This diagram is not taken from Oberth's book but it illustrates his principles.

saving the fruit and vegetable crops of whole countries. It is especially important that the mirror is not stationary over any one point of the earth and is therefore capable of rendering all these services.

After a short discussion of the most suitable material for the mirror (deciding in favor of sodium because of its light weight) and the estimated costs (far too low) Oberth continued:

I wish to mention one more thing: the observing station could also be a refueling station. If the hydrogen and oxygen [the fuels Oberth had in mind] are shielded against solar radiation they'll keep for any length of time in the solid state. A rocket which is refueled at the station is no longer hampered by air resistance and not much by the earth's gravitation. . . . Furthermore it no longer needs a high initial velocity [this should read acceleration, obviously a slip of the pen that escaped detection in proofreading] to leave the earth's gravitational field. In the first place the potential of the earth is lower at the distance of the station; in the second place the rocket needs only to make up the difference between the required final velocity and the velocity, of about 6 kilometers per second, of the station. If we now connect a large sphere of sodium metal which was assembled and filled with fuel in the station's orbit with a small solidly constructed rocket which pushes the fuel sphere ahead and draws its fuel from it, we have a highly efficient apparatus which should easily be capable of flying to another planet. There the disconnected rocket would land while the fuel sphere is left in an orbit around the planet. After re-ascent of the rocket the connection is re-established so that the apparatus may return [to the earth].

This is as far as Oberth went in 1923.

Six years later there appeared a book which was largely devoted to the space station; one might even say that it was a book about the space station except that its author still felt it necessary to explain rockets and rocket propulsion in detail. The author was an Austrian named Potočnic who wrote under the pen name of Hermann Noordung. The title page of his book also states that he was a captain of the reserve and a graduate engineer. To this day I haven't been able to find out whether these two statements belonged together—meaning that he was a captain in the engineer corps—or whether the "engineer" part represented his peacetime occupation and the other a wartime commission. In fact, I haven't been able to find out anything about Potočnic, except that he died soon after the publication of his book.

Noordung's station was to consist of three separate units, connected to each other only by pressure hoses (for air) and electric cables. These three units were the "living wheel" (*Wohnrad*), the powerhouse, and the observatory. The Wohnrad was to be a wheel-shaped structure with a diameter of about 100 feet. It was to spin around its hub so as to substitute centrifugal force for gravity in the rim. The counter-rotating hub was correctly drawn as an airlock. He also stated correctly that there would be a slight difference in pseudo gravity between the head and the feet of a standing man and that this would have to be compensated for while walking around. He wanted to obtain power from the sun by means of collecting mirrors and boiler tubes, with the condenser pipes located in the back of the mirror (Fig. 62). Along with these essentially correct thoughts there ran a number of strange mistakes. For example, he was afraid of "cold" space and every window was a somewhat convex lens to concentrate sunlight in the room. Moreover, outside every window, or at any rate most of them, there was a plane mirror to reflect even more sunlight into the lens. He wanted to spin his Wohnrad at the rate of once every 8 seconds so that there would be a full *g* of pseudo gravity in the rim. Actually ⅓ *g* would be enough, for the main purpose of pseudo gravity would be to make things stay put, so that, say, a plate of food placed on a table would stay there and cigarette smoke would rise to the ceiling and the air-conditioning outlet, instead of forming a dense near-spherical cloud around the smoker's head. One very important reason for advocating only ⅓ *g* of pseudo gravity is that the whole station can be lighter and therefore cheaper to carry into an orbit piecemeal. Noordung drew all his units massive enough for exhibit on the ground; they would even have survived being tossed around on an angry ocean without much damage.

The second unit, the observatory, was not described in detail. Noordung merely said that it would be like a cylindrical boiler in shape and contain all the necessary instruments. It was to be heated by piping air of the right temperature into it through a flexible hose. The electric cable from the Wohnrad was to provide power for the instruments. As Noordung went along the description grew more and more vague; the powerhouse was just a large paraboloid mirror with a set of boiler pipes coiled up in the focus.

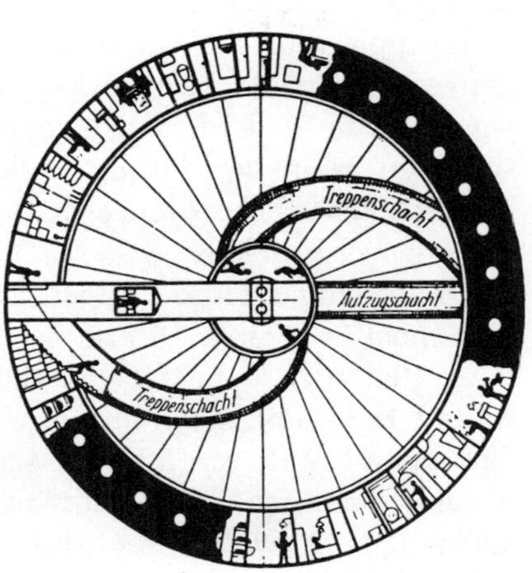

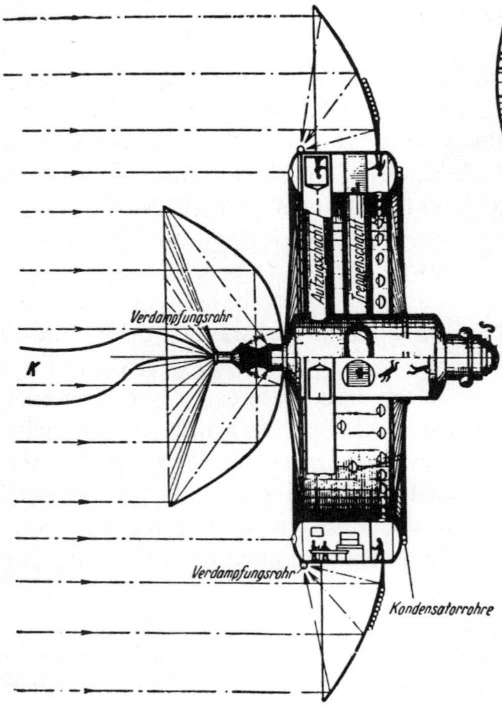

Fig. 62. Noordung's *Wohnrad.* In the diagram at left *K* designates the electric cable to the "observatory," and S the airlock. *Kondensatorrohre* means condenser pipes, and *Verdampfungsrohr* boiler pipe. Above, *Treppenschacht* means stairwell, and *Aufzugschaft* elevator shaft.

The current produced was to be supplied to the Wohnrad for either immediate use or storage.

Shortly before Noordung's book appeared in the bookstores, the VfR's monthly journal *Die Rakete* (vol. II, 1928) had run several articles by von Pirquet on the functions of a manned space station. These had not included design suggestions as did Noordung's book, but they had arrived at a number of interesting results. Baron von Pirquet had been playing around with figures for a voyage to Mars and had come to the conclusion that a spaceship for a direct trip from the earth to another planet posed insurmountable problems as far as size and fuel consumption were concerned. But it seemed likely that such a ship could be put into an orbit around the earth. If it were left in such an orbit and other ships brought up fuels and supplies—at the price of 68 tons of fuel for every ton of payload delivered—it could be refueled for a trip to Mars. In fact, much less fuel would be needed for a trip to Mars from the orbit than for putting the ship into orbit; von Pirquet called this the "cosmonautical paradox" and declared that the existence of this paradox shed new light on the whole space-travel problem.

Establishing a large manned artificial satellite was not only a necessary preliminary operation, it also was the most difficult step, after which everything that was to follow would be simple by comparison. These articles shifted the emphasis from "observing station" (Oberth's main point) to "spaceport," but von Pirquet knew perfectly well how important (and possibly lucrative) the observing function was. To combine the various advantages he evolved the idea of a triple station. It was to consist of an Inner Station (mostly observation) in a 100-minute circular orbit, an Outer Station (mainly spaceport) in a circular 200-minute orbit, and a Transit Station in an elliptical orbit with an orbital period of 150 minutes. This Transit Station was to approach the two stations in their circular orbits at regular intervals to a distance of about half a mile; the velocity differences between the Transit Station and either of the others at the points of closest approach would have been about 0.75 miles per second.

No details of design or even appearance were given, and the postscript to the triple station is that the author himself abandoned the idea at a later date.[3]

Since the concepts expressed by Oberth, Noordung, and von Pirquet had been derived from ideas about manned space travel it had been assumed without a second thought that the artificial satellites would be space stations, i.e., manned satellites. That telemetering was in its very early stages, without any clearly expressed thoughts of where it would lead, was a contributing factor. During the twenties you could visualize that a man might take pictures while in orbit, but you could not imagine these pictures reaching the ground without an actual landing.

Just when and where the idea of unmanned earth satellites was first proposed is not clear. It may have been in Peenemünde, for Dornberger mentions in his book *V-2* that in after-hours talk about the future a suggestion was made that the space pioneers be honored by entombing their embalmed bodies in glass spheres sent into orbit around the earth.

Thoughts about artificial satellites hardened into a more concrete form by considering the very long-range missiles that Dr. Vannevar Bush had said should not be thought about at all because they could not be built. In June 1946 there was a (classified) lecture on artificial satellites at the Infantry School at Fort Benning, Georgia. A short summary of that lecture

[3] Baron von Pirquet died in 1966 at Castle Hirschstetten, near Vienna, at the age of 90.

was published some time later (for example, in *The New York Times,* August 21, 1946). There was much surprise and much comment when the year-end report for 1948 written by the late James Forrestal in his capacity as Secretary of Defense contained a paragraph stating that official studies concerning a satellite rocket had been made. Since this study cannot possibly contain anything that is now "secret," it would be desirable to have it released as a historical document.

Soon afterward, members of the various rocket societies in different countries began to work on the problem. This became clear at the Second International Congress on Astronautics (London, September 1951) where most of the technical papers presented dealt with satellite vehicles, but with one exception they all considered various problems of the manned space station. The exception was a paper by three members of the British Interplanetary Society, K. W. Gatland, A. M. Kunesch, and A. E. Dixon, entitled "Minimum Satellite Vehicles." Three were investigated on paper, all of them three-step rockets, called Scheme "A," Scheme "B," and Scheme "C." In "B" and "C" the third stage was to carry a payload of 100 kilograms (220 pounds); in Scheme "A" there was no payload. The take-off weights came out as 16,800 kilograms (37,000 pounds) for Scheme "A"; 62,400 kilograms (137,570 pounds) for Scheme "B"; and 90,900 kilograms (200,400 pounds) for Scheme "C." A fourth, Scheme "D," tentatively advanced as being "of a more speculative nature," had the same take-off weight as Scheme "C" but a final payload of 220 kilograms (485 pounds).[4] During 1952 and 1953 published articles on unmanned satellites slowly grew more numerous. Among them was one by Wernher von Braun in *Collier's* (June 27, 1953).

At the Fourth International Astronautical Congress in Zürich in 1953 a new word could be heard. It was "MOUSE," standing for Minimum Orbital Unmanned Satellite of Earth. Professor S. Fred Singer of the University of Maryland declared that we could have a MOUSE now![5] While most earlier papers and articles had dealt with the propulsion part of the problem and had concentrated on getting a third stage into space as an unmanned satellite, Dr. Singer's paper concentrated on the payload. He examined the questions of just what such an orbital rocket could explore, what instruments would be needed for it, what they would weigh, and how the information could be transmitted to the ground. Dr. Singer's MOUSE was the first artificial satellite in the modern sense, a separate instrument package, spherical in shape, which would be separated from the third stage. The MOUSE, weighing about 100 pounds, was to be stabilized by spinning so that it would not tumble when circling the earth. The axis of the spin-

[4] Published in *JBIS,* no. XLIII, November 1951, and in *The Artificial Satellite,* edited by L. J. Carter (see Bibliography).

[5] Published in *JBIS,* no. LVII, March 1954. The term "MOUSE" was coined jointly by Dr. Singer and by Arthur C. Clarke and A. V. Cleaver of the B.I.S.

stabilized instrument package was to point in the direction of the sun and that "pole" of the sphere was to contain equipment for gathering solar energy to power the transmitter. The orbit was to go over both geographical poles, which had the side effect that if MOUSE was vertically over a given point on the ground the local time for that point would be either 6 A.M. or 6 P.M. It was to be a 190-mile orbit with an orbital period of 90 minutes.

The main reason for the circumpolar orbit was that there were two definite points where the transmission could be received, namely the poles. The instrument data would be recorded on a slow-moving tape which progressed at the rate of 2 inches per minute so that 45 minutes worth of information could be stored on 7½ feet of tape. At the pole an airplane serving as telemetering receiving station would be in the air when MOUSE came overhead. A radio signal from the plane would trigger the transmitter in the satellite and the tape recording would be "squirted" out in half a minute, to be recorded in the plane. As part of the transmission technique the tape would be erased so that it was ready to record the next 45 minutes of data. Transmitting for only 30 seconds every 45 minutes was a device for saving power.

In the spring of 1954 the Space-Flight Committee of the American Rocket Society had worked out a satellite proposal which had been submitted through various channels. The time was ripe; the government, in particular the Office of Naval Research, had already realized that scientific instruments were needed in the outer fringes of the earth's atmosphere. Therefore the response was quick. When Frederick C. Durant, president of the International Astronautical Federation, called the Office of Naval Research to tell them that Wernher von Braun would be in Washington late in June, a meeting date was set for June 25. The meeting place was room 1803, T-3 Building of the Office of Naval Research. Present were, in addition to von Braun, Fred Durant, Dr. Fred Singer, Professor Fred L. Whipple of Harvard, Commander George W. Hoover (U. S. Navy and also A.R.S.), David Young of Aerojet-General and, of course, officers of the Office of Naval Research.

At that time there were quite a number of high-altitude research programs, but no satellite program was on the books—yet. The question was whether a satellite, just large enough to be detected, could really be shot into an orbit, say, 200 miles up, within a reasonable length of time. What was a reasonable length of time? About two or three years. "Sooner than that," said von Braun, and developed an idea which he had probably worked out in his spare time. (I must mention that from about 1950 on the favorite pastime of everybody interested was combining existing rockets on paper to see whether they might reach satellite-carrier velocities.) If you took a Redstone rocket as the first stage, von Braun explained, and then

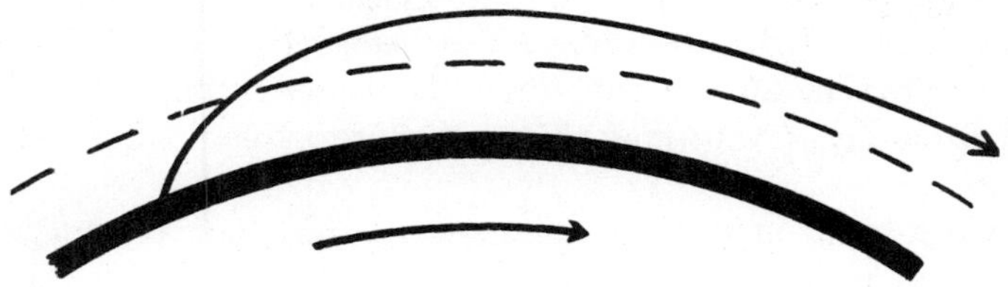

Fig. 63. Oberth's "synergy curve," showing the most economical take-off maneuver possible.

placed successive clusters of Loki[6] rockets on top, you should add up enough velocity to get the top Loki into an orbit around the earth.

This scheme had the major advantage of very nearly predictable cost, since it operated with existing rockets. Every one at the meeting then contributed to the discussion along the lines of his own specialty, and a preliminary decision of the following nature was reached: Dr. Singer's MOUSE, though no doubt a useful device, had to be considered as the second phase of the program. The first phase should be a much simpler and also a lighter satellite.

A visit of Navy representatives to Redstone Arsenal followed, and after everybody up to and including the Chief of Army Ordnance, General Leslie E. Simon, and the Chief of Naval Research had been notified and consulted, Commander Hoover was assigned as project officer of Project Orbiter.

The Redstone rocket, developed by Wernher von Braun and his team at the Redstone Arsenal near Huntsville, Alabama, was the direct successor of the V-2. It was taller—69 feet 6 inches—but its diameter was only 1 inch more than that of the V-2, probably just a coincidence. The take-off weight was 60,000 pounds, the horizontal range around 200 miles, and the payload capacity several tons. While the Redstone was a powerful rocket, it just barely qualified as a satellite-carrying vehicle. To be certain, every advantage that could be found anywhere had to be utilized, including the fact that the earth rotates on its axis.

Oberth had written the theoretical prescription many years earlier and had called it the "synergy curve" (see Fig. 63 and Appendix 3). To take off with the least fuel expenditure, Oberth had reasoned, a rocket should take off horizontally and move eastward. By moving eastward it would utilize the velocity which the launching pad has as part of the turning earth. But taking off horizontally would subject the rocket to too much air

[6] The Loki was a development of the unfinished Peenemünde Taifun, but for solid fuels. It has since been dropped from the list of potential or operational weapons and transformed into the HASP (High Altitude Sounding Projectile).

resistance. The compromise was to take off vertically for 5 or 6 miles until the densest layers of the atmosphere had been left below. Then the rocket was to be tilted in an easterly direction, leaving the atmosphere more or less horizontally. (Note how this procedure agrees with Newton's conditions.)

Of course a point at the equator moves faster than a point at latitude 45°. To get all possible assistance from the earth's rotation, the Redstone rocket should take off from the equator. Since Orbiter was a joint Army-Navy project everything worked out well; the Army would provide the rocket and the Navy would provide a suitably equipped ship to carry the rocket to a point at the equator in the Atlantic for a shipboard launch. One V-2 and one Viking had proved that a shipboard launch, in a reasonably calm sea, could be carried out. If Project Orbiter had gone ahead as planned, the United States would have placed a satellite in orbit during the summer of 1956.

But in the meantime other mills had been grinding elsewhere. The A.R.S. proposal had also gone to the National Science Foundation which was involved in the planning for the International Geophysical Year. Scheduled to start on July 1, 1957, the IGY was to run for 18 months, to the end of 1958. Its purpose, in which all the nations of earth, except a few very small ones, participated, was the exploration of our own planet.[7] A satellite shot that would gather information obtainable in no other way could, and even should, be a part of the IGY activities.

Plans for the IGY satellite shot began with an examination of Project Orbiter. The Redstone, even by using all the tricks of the trade, could not put into orbit a satellite that could hold all the instruments that were desirable. A heavier satellite—which naturally implied a more powerful rocket—was wanted. The Department of Defense, then headed by Charles E. Wilson, had rejected many well-considered proposals for a more powerful rocket, always insisting that the proponent show "an immediate military need" for such a rocket. The phrase is meaningless; if there is "an immediate need" it is evidently already too late. But it was a fine phrase for clubbing planners that advanced anything more radical than a change in Defense Department bookkeeping procedures.

The U.S. IGY project, being financed by the National Science Foundation, could neatly by-pass Secretary Wilson. And if a new rocket were built, it would not even be necessary to ask the armed services for missiles which were still in short supply.

Incipit tragoedia!—as used to be said in old books, "Here begins the tragedy." It was a tragedy compounded of fallacious reasoning in some

[7] The date of the IGY was chosen because, 25 years earlier, there had been a South Polar Year, also with international cooperation. And the South Polar Year had been preceded, 50 years before it, by a Polar Year in which those countries near the Arctic Ocean cooperated in the exploration of the Arctic regions.

quarters, overconfidence in others, callous indifference in still others, and an over-all lack of understanding.

The IGY satellite project was to be known as Project Vanguard, with the Navy in charge. White House Press Secretary James C. Hagerty made the official announcement on July 29, 1955.

Project Orbiter had been pushed into the background and was later shelved.

Let us begin with a description of the Vanguard rocket. It should be said at the outset that the *design* of the Vanguard was superior to anything else in existence at the time. The Vanguard would use less fuel per pound of satellite weight than any other rocket, it was to have a superior guidance system, and it would incorporate a number of new concepts. In short, it was to be a *new* rocket—and for an idea of what a new rocket may do, just reread the Viking section in Chapter 9.

The first decision that had to be made was what instruments should be housed in the satellite, a rather difficult decision that had no precedent to go by, since this was the first artificial satellite to be actually built. The only satellite design available, Dr. Singer's MOUSE, was much too heavy for the Vanguard. The satellite was to be spherical and at first its diameter was announced as an even yard. Later this was reduced to 24 inches, with a weight of 21.5 pounds. Of course since there were to be several shots, the instrumentation in the different satellites might not always be the same, but there should always be a total payload of 22 pounds (what the extra half pound was for is described shortly). Three stages would be needed to accelerate that payload to satellite velocity.

The first stage was based on the design of the Viking rocket and used the same liquid fuels. The main innovation for the first stage, aside from its greater length, was the fact that it was without fins, relying on the gimbaled motor for steering. One other innovation not visible from the outside was the lack of guidance equipment in the first stage—all the guidance equipment for the whole rocket was located in the second stage. The second stage would guide the first stage while the two were still connected, and it would also position the third stage for the final push. The second stage also was a liquid-fuel rocket, but used different fuels. Its oxidizer was nitric acid and the fuel was a substance known to chemists as unsymmetrical dimethyl hydrazine ("Dimazine" became the trade name); these two liquids constituted one of the hypergolic combinations, bursting into flame when touching each other without the need for a special ignition device. The third stage, finally, was to be a solid-fuel rocket.

There was a practical reason for building a three-stage rocket with three different fuel systems. As a rocket using liquid oxygen must be fired soon after the fueling, the fueling procedure might have been quite complicated if all three stages had used liquid oxygen. Since the solid-fuel

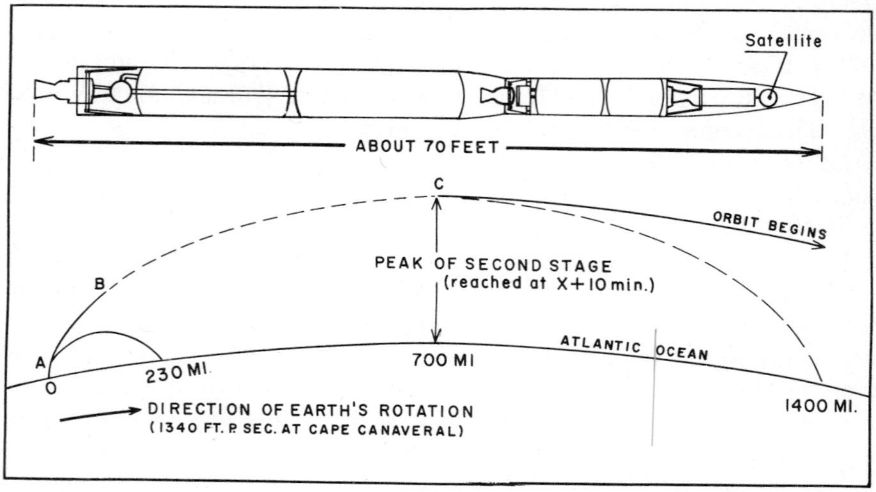

Fig. 64. Take-off performance of the Vanguard satellite carrier. Brennschluss of the first stage is at *A*, that of the second stage at *B*; the third stage begins firing at *C*. The dotted line is the trajectory of the second stage after its Brennschluss.

rocket (third stage) presented no fueling problem and since the fuels of the second stage could stay in the rocket for a considerable length of time without harm, the difficulties, if any, would be confined to the first stage.

The take-off procedure to be followed was also carefully planned (see Fig. 64). The firing site was to be Cape Canaveral which, being located at latitude 28° 28′ N., would contribute 1340 feet per second of rotational velocity. The job assigned to the first stage was vertical lift-off and tilting the whole rocket through an angle of 45 degrees. At the moment of separation the rocket would be 36 miles above sea level; the distance from the firing site, measured horizontally, would be a little less—about 32 miles. After separation the burned-out first stage would continue to climb to about 110 miles and would finally splash into the ocean 230 miles from the firing site.

The second stage was to start burning immediately after separation and to continue the tilt. Simultaneously with the beginning of burning the flight-programming equipment in the second stage would jettison the nose cone that protected the satellite on top of the third stage. At an altitude of 36 miles the satellite would no longer need protection and there was no reason for using up fuel to carry the nose cone any farther. By the time the second stage had used up its fuel supply, the altitude reached would be 140 miles and the direction of motion nearly, but not quite, horizontal. Still connected, the second and the third stage would coast to a peak altitude of between 200 and 300 miles above sea level, the peak altitude being different for different shots. Peak should be reached 10 minutes after lift-off vertically above a point in the ocean 700 miles from the firing site. At the peak, where the direction of motion would be horizontal, the second separation would take place. During the unpowered climb from 140 to 250 miles there would be some loss of velocity which would have to be made up by the third stage. In fact the third stage had to contribute a little more than 2 miles per second. But long before the second stage splashed into the ocean 1400 miles from the firing site, the third stage and the satellite would be in orbit.

Then a final maneuver would take place, involving that extra half pound of the payload, which was a device for separating the satellite from the third stage. It was thought then that if it was connected with the third stage the satellite might not be able to function properly. The separation mechanism was essentially a tensed spring that was "locked" by a small weight. Though the relatively mild accelerations of the two liquid-fuel stages would not dislodge that weight, the high acceleration of the third stage would. Then, after a suitable delay, the expansion of a pressure bellows would release the catch of the spring, and the spring would push the satellite away from the third stage at a rate of only about 3 feet per second. Of course, the third stage, having orbital velocity, would stay in orbit too.

For a while there was a plan for a subsidiary satellite, a plastic balloon covered with aluminum foil which would be carried into orbit folded and inflated after separation. This "subsatellite" was to have the same diameter as the satellite, but while the satellite would weigh 21½ pounds the subsatellite would weigh less than 1 ounce. Both, however, would present the same cross section to any residual air resistance in the orbit. If even a minute amount of resistance were encountered, the subsatellite would be retarded far more because of its lesser mass. The difference in retardation would establish the amount of resistance left. This would have been a very instructive experiment; unfortunately it was never carried out.

This was the design.

Then the muddle started.

Or, more accurately, muddles, for there were several even before the crowning touch of Russian competition materialized.

The first move was a pious announcement to the effect that the American satellite accomplishment would be all the greater because it would be done by means of an "entirely peaceful" rocket that could not be used as a missile. It is said that President Eisenhower was especially pleased with this, but it was as meaningless as the phrase "immediate military need." "*Bon,* not only are we going to get an American satellite, but a *virtuous* American satellite, probably as virtuous as prohibition," a French magazine jeered.

Seriously, nobody throughout all history had ever cared whether a discovery had been made by civilians or by men in uniform. The fact that the eastern tip of Asia and the western tip of the North American continent are separated by water was discovered in 1728, by a civilian, the Danish explorer Vitus Bering. Later during the same century the *Endeavour,* commanded by Captain James Cook of the Royal British Navy, made the first circumnavigation of Antarctica, but would anybody have been annoyed if the Bering Sea had been charted by the *Endeavour,* a military vessel? Or, to cite an actual example: during the last few decades of the nineteenth century many islands and island groups in the Pacific were charted by American destroyers, British cruisers, and German corvettes. Was there ever a sea captain who rejected these charts on the grounds that they had been made with the aid of warships? On the contrary, navigators were happy that the work had been done and willingly paid eighty-five cents, two shillings sixpence, or one mark and forty pfennig for the charts. Or, even closer to home, who had objected to the attempt to explore the upper atmosphere with V-2 rockets?

Of course, Project Vanguard enjoyed tremendous publicity all through the second half of 1955 and through all of 1956. There was so much publicity that the chief of the project, Dr. John P. Hagen, warned publicly that there might be only one success for six attempts. The statement was dismissed by many as "protective official hedging"; actually Dr. Hagen well knew that with so many things which could go wrong the probability of success for a specific shot was small indeed. The final score turned out to be three successes for eleven attempts.

This overwhelming amount of publicity was not caused merely by the fact that this was a new project; there was another reason for it. Both the Army and the Air Force began testing new rockets at Cape Canaveral, and a large crowd of enthusiastic "birdwatchers" with expensive optical equipment began to assemble at a suitable spot along the beach. The White House countered with an angry order that no information at all was to be given to the public, not even the names of new missiles. However, Vanguard had been specifically declared *not* to be a missile and since it was not a missile it could be talked about. The Navy had stressed the nonmissile nature of Project Vanguard by referring to the Vanguard rocket as a "vehicle," which resulted in a large volume of mail inquiring who was going to ride in this vehicle, plus a few dozen offers to ride it.

Of course Project Vanguard also was discussed in Russian journals, one of which carried an article by Professor Kyrill Stanyukovitch that contained a significant phrase. After describing Vanguard, Professor Stanyukovitch said that Russian engineers "believe it is possible to build larger satellites than those now being discussed in the Western press." (Excerpts from this article were printed in *Aviation Week*, October 31, 1955.)

But during an international conference dealing with the organization of the IGY in Paris in December 1955 the Russian delegates stated that the Soviet Union did not have a satellite project for the IGY.[8]

Working on Project Vanguard must have been an unbelievably frustrating experience. The demands were so high that the chance of success was marginal indeed. The timetable was too tight. The original appropriation had been too small and additional appropriations were parsimonious and slow in coming. Even the jurisdiction was not as clear as one had a right to expect. And then there was another source of delay about which nobody had the courage to scream aloud in public, because to scream, or even to moan, would have been "un-American." This was the policy, set by somebody who probably could not be trusted to change a light bulb or to drill a hole through a piece of pinewood, that in "the spirit of democracy" the project must "have the full participation of industry."

This is a fine rule for production, but it leads to disaster when experimental work is involved. In practice, this is the way it worked. A dozen or two specimens of a scientific device consisting of a small set of instruments and a number of switches were needed. No single component was completely new; on the other hand, no single component could have been bought in precisely the size, or of precisely the function, that was needed. There is a government-owned research laboratory, well equipped and employing skilled mechanics and competent scientists; its staff could have built the necessary devices in three or four weeks and the Vanguard people would have been absolutely sure of their own project. But no, industry had to participate. Instead of having the job done in three weeks, they had to write *forty* contracts, in quintuplicate, engage in the correspondence that goes with forty contracts, and then check and test the pieces as they came in.

Small wonder that Project Vanguard moved slowly during 1956; the fact that it moved at all was due to the determination of a handful of men who stubbornly did the job they had said they would do. Around the middle of 1956 rumors of Russian long-range missile firings began to be heard. The existence of Russian missiles was officially denied until a courageous senator—I don't recall just who it was—stated the fact in public and was probably reprimanded behind the scenes for having violated security.

On September 20, 1956, Redstone Arsenal, which had meanwhile become ABMA (Army Ballistic Missile Agency), achieved a great success. A new rocket, known around ABMA as "Missile 27," had made a record flight over a distance of 3100 miles (Fig. 65).

It was a Jupiter-C rocket and it came to its name in a roundabout way.

[8] This probably was a truthful statement; the Russians' intercontinental missile program and their satellite program were two sides of the same coin. By December 1955 they might not have known how far their missile program would progress during 1956 so that they could not tell whether a satellite program was feasible.

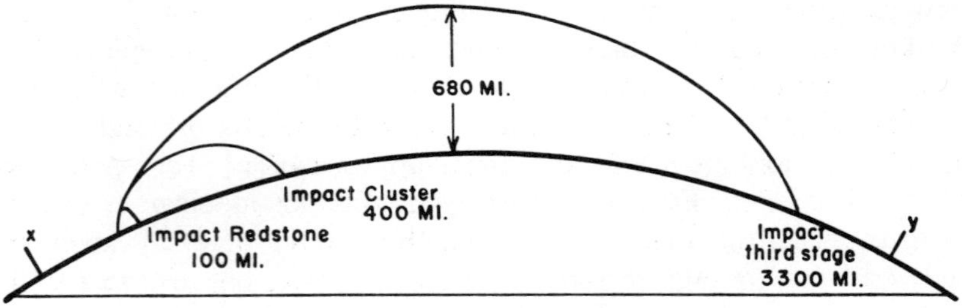

Fig. 65. The Jupiter-C shot of September 20, 1956. The two short lines marked x and y are two radii of the earth, forming a geocentric angle of 60 degrees.

Early during that year ABMA had been charged with the development of a missile that was to have a range of about 1500 miles. Its name was to be Jupiter. Of course new components had to be test-flown and the re-entry nose cone for the Jupiter missile created a special problem. Since ABMA had Redstone rockets available a number of them were assigned for testing Jupiter components. But a very minor amount of cheating was needed to have the testing program move fast. As General John B. Medaris, then commanding officer of ABMA, put it, in his book *Countdown for Decision:*

> I must confess here that the nomenclature of these different missiles was—and is—confusing, but there was a reason for it. In testing components for Jupiter, we were constantly using the Redstone missile as a vehicle. But Jupiter had a much higher priority than Redstone on firing dates at the Cape, and in other ways. We therefore decided to label these Redstones used for testing Jupiter components "Jupiter-A." They were not Jupiters, of course, but the label identified them properly with the program, and gave them the necessary priority.
>
> Similarly, when we put together a composite missile for testing nose cones, we called it "Jupiter-C." The first true Jupiter missile was not fired until May 1957.
>
> Actually, for this first long-range shot we were readying two identical missiles, no. 27 and no. 29. We wanted to have a spare in case the first shot failed. These two missiles looked—and in fact were—exactly like the satellite carrier that we were to use over a year later,[9] except that for this test shot we were using a dummy fourth stage. If we had put a solid propellant into the fourth stage instead of the inert material we were using to get it to the right weight, we could have fired that particular missile into orbit as a satellite. We didn't do this for the simple reason that we were forbidden to do so. We had no mission for putting up a satellite—Vanguard had that assignment. And nobody had any intention of giving us that mission. So we put sand aboard the fourth stage instead of powder.
>
> (Medaris)

[9] The Jupiter-C missile had an elongated Redstone as its first stage; the second stage consisted of 5 solid-fuel Recruit rockets arranged in a circle around a large center hole, the third stage was a cluster of 3 Recruit rockets fitting into that hole, and the fourth stage was a single such rocket on top of the third stage. The whole was enclosed in what was called the "launch tub," or "the basket," a cylindrical metal container on top of the Redstone. Prior to take-off the "basket" was rotated rapidly by an electric motor.

Actually, as many readers may have guessed, the Jupiter-C was the same concept as the original Orbiter, except that the Redstone had been made longer and that the new Recruit rockets were more powerful than the old Lokis.

While Missile 27 was an outstanding success, the spirits of the officers and scientists of ABMA were dampened by a strict order from Washington that this shot must not be mentioned. Fortunately for American morale several magazines (among them *Life*) did learn about it and published the facts. What they did not know at the moment was that ABMA had pointed out that long-range testing of nose cones might result in an artificial satellite and that a strict order saying "no accidental satellites" had been issued in reply.

Just before the news about the flight of Missile 27 became known, the rumors of Russian preparations became more definite. The magazine *Aviation Week* (October 29, 1956) published excerpts from an article in the *Moscow News* by Professor Georgi Pokrowsky, which stated that the Russian satellite would have a diameter of 24 inches (actual: 23.8), that it would weigh over a hundred pounds (actual: 184), and that the perigee of the orbit would be at 185 miles and the apogee at 810 miles (actual: 156 and 560.) Of course anybody could assume that Professor Pokrowsky had just published the results of some private calculations, but if one did not make that assumption the article proved that the Russian design had progressed to the point where figures could be mentioned.

In December 1956 the newspapers proclaimed triumphantly that "Vanguard" had made its first flight on December 8 and that this had been a successful flight. Well, yes and no. It had been the first flight in the Vanguard program and it had been successful, but the rocket had been a Viking XIII, used to test instrumentation. Even if the rocket that took off had been a Vanguard rocket, the program would have been behind schedule.

In the meantime the whole picture had been beautifully confused by an order of the Secretary of Defense stating that the Army was responsible for ballistic missiles up to a range of 200 miles, while longer ranges were the responsibility of the Air Force. No matter what the reason for this order, it should have had the result that the Army's development of the 1500-mile Jupiter missile would be stopped and that the Air Force's development of the 1500-mile Thor missile would go ahead. But in reality, the Jupiter program was not stopped, it was just slowed down. Nobody has ever been able to understand the true meaning of that order; as far as subsequent developments went it might just as well not have been issued. All it really did was to lower morale by creating a situation of uncertainty. A committee created to decide between Thor and Jupiter found itself in about the same situation as the German committee that had to decide a dozen years earlier between V-1 and V-2. In fact the situation was strikingly similar: the Air Force missile clearly carried the favor of the policy makers, but the Army missile was further advanced. Since the situation was the same, the outcome was the same: development of both missiles would continue.

With the coming of spring in 1957 the pace of events seemed to quicken. On May 1 the second test shot in the Vanguard program was fired; it was Viking XIV, carrying the third stage of Vanguard for testing. The second stage was not yet ready.

On June 2, 1957, *The New York Times* reported on an article in *Pravda* by Professor Alexander Nesmenyanov, which contained the statement: "We have created the rockets and all the instruments and equipment necessary to solve the problem of the artificial earth satellite." On June 10 the Russians reported to IGY headquarters in Uccle (Belgium) that the satellite project was ready and that the satellite would be fired into an orbit as close to the meridian as feasible. This report was distributed in sixty-four countries and carried by most major newspapers in most of these countries, but after Sputnik-I had gone into orbit many prominent people, including even a few Congressmen, asserted that the Russian satellite shot had "come as a complete surprise." In June the Air Force distributed a memo (not classified) stating that there was every reason to believe that the Russian satellite shot would be made on the hundredth anniversary of the birth of Konstantin E. Ziolkovsky.

Ziolkovsky's birth was registered as having taken place on September 5, 1857, as stated on page 93 of this book. But that was an "old style" date (Julian calendar) and the Russians, soon after their revolution, had adopted the Gregorian calendar. Consequently they now refer to earlier events in their history as if the Gregorian calendar had been in use in Russia at that time, with the slightly disconcerting result that an event known as the "October Revolution" is commemorated on November 7. As applied to Ziolkovsky's birthday the change meant that it fell on September 17, so that September 17, 1957, was clearly the intended firing date. But nothing happened.

On August 3, 1957, Reuter's reported from Russia that Professor Evgenii Fyedorev had been appointed head of the satellite program. He had announced that the firing would take place in the early morning hours (it did) and that the orbital period of the satellite would be 90 minutes (actual: 93 minutes).

There was no news from Project Vanguard but Washington took the official position that the United States was *not* involved in a satellite race with the Soviet Union. One can only hope that this pronouncement was a considered political lie, for if it was not it was plain stupidity. But at that time Secretary Wilson said in so many words that he did not care if the Russians succeeded with a satellite program ahead of the United States.

ABMA, of course, followed the Vanguard program carefully and General Medaris took a preliminary step:

> On August 21, therefore, I issued a directive to stop the Jupiter-C test program at once, and to put all Jupiter-C hardware into protected storage so that it would suffer no deterioration. The original program had called for twelve shots. We had used three[10] and had nine precious missiles, in various stages of fabrication, to hold for other and more spectacular purposes. After looking the missiles over carefully, and evaluating the state of readiness of each, I advised General [James M.] Gavin, Chief of Research and Development of the Army, that we could hold the two most advanced missiles in such condition that one satellite shot could be attempted on four months' notice, and a second one a month later.
>
> (Medaris, *l. c.*)

In September 1957 the Russians broadcast the wavelengths to be used by their satellite to the countries in the Far East and on September 30 the Soviet Embassy distributed the same information to the American press. (But some people, later, still called the satellite shot a surprise.)

The first Russian satellite went into orbit when it was evening in the Western Hemisphere on October 4, 1957. Washington had a deservedly rude awakening. It still tried to stick to the slogan, "we are *not* in a satellite race," but public opinion shouted down its own government. American scientists were fairly unanimous in denouncing a policy that had caused such a setback for their country. General Toftoy, who happened to be in Madrid at the moment, simply said what he thought; General Medaris decided that his ABMA would fare best if he kept quiet. He had his special reason: it had been announced that Secretary Wilson was resigning and that Neil McElroy would be his successor. And by a weird coincidence McElroy had been at ABMA when Sputnik-I began orbiting the earth.

Before dinner Wernher von Braun spoke to McElroy, undiplomatically but truthfully, all the pent-up resentments tumbling out: "We knew the Russians were going to do it! Vanguard will never make it. We have the hardware on the shelf. For God's sake turn us loose and let us do something. We can put up a satellite in sixty days, Mr. McElroy! Just give us a green light and sixty days."

Medaris interposed: "Ninety days."

It turned out to be eighty.

In addition to creating general turmoil and unhappiness in the West, the Soviet satellite added a new word to the English language: sputnik. Actually the Russian term was *sputnik zemlí,* pronounced *SPOOT-nik zem-LEE;* the Rusian plural is *sputniki. Sputnik* is a standard Russian noun, composed of *s* (meaning "with"), *put* (pronounced *poot,* meaning "road"), and the

[10] President Eisenhower, in a TV appearance, referred to one of these shots as "an artificial comet" but never explained what he had meant by that term.

suffix *nik,* which can refer to either an object or a person. The meaning, then, is "travel companion." Russian astronomers have been using *sputnik* since the turn of the century as a term for natural satellites, especially if the satellite is small in size. As for the second word *zemlí,* it is derived from the Russian word for "earth" (*zemlyá,* as in the name of the island Novaya Zemlyá, "new land") so that the whole means "companion of earth"; the use of *sputnik* to mean an artificial satellite originated with Ziolkovsky.

Some time after the launching, the astronomer Fred L. Whipple suggested an international system for designating artificial satellites, modeled after the astronomical system of designating comets. Whipple's system made the first sputnik 1957 alpha; the next satellite would be 1957 beta if fired in 1957. Since 1957 alpha consisted of several bodies, a finer distinction became necessary. Naturally the orbiting rocket casing, being much larger than the satellite, had a greater visual brightness. The orbiting bodies were labeled according to their brightness as alpha-1 (the rocket), alpha-2 (the satellite) and alpha-3 (the nose cone, which was also in orbit). Orbital decay proceeded as expected, but a little bit faster, indicating that the residual air resistance at altitudes above 100 miles was greater than had been expected from calculations. Since the rocket body was larger than the satellite, it had a more pronounced orbital decay and assumed shorter orbits sooner. When alpha-2 (alias Sputnik-I) made its 345th revolution, alpha-1 was on its 346th. Alpha-3 (the nose cone) seems to have re-entered and burned up unnoticed; alpha-1 experienced burn-up during the first week of December 1957, over the northern Pacific.

Alpha-2 transmitted on two wavelengths—15 meters and 7.5 meters, corresponding respectively to 20.005 and 40.002 megacycles; the choice of wavelengths was probably determined by the availability of receiving equipment on the ground. After having completed 326 revolutions (22 days after launch) the batteries ran down and Sputnik-I fell silent. It re-entered and burned up during the first week of January 1958; tracking indicated that it broke into several pieces during its last few revolutions.

The latter part of 1957 was devoid of good news for Americans. One of the very few American space experiments made miscarried during September and October. This was Project Farside, which apparently had started as a moon-shot project but had been cut back to a shot to 4000 miles, or one earth radius. The Farside rockets, of which six were built, were four-stage solid-fuel rockets, launched from a Skyhook balloon, or more precisely, through a Skyhook balloon. The first stage of the assembly was a cluster of four Recruit rockets, the second stage one such rocket, the third a cluster of four Asp motors, and the fourth stage a single Asp with instrumentation. The rockets were fired by radio command when the balloon had reached the necessary altitude. Because considerable drift could be expected, the firings took place over the Pacific Ocean in the vicinity of Eniwetok atoll.

Of the six shots four were complete failures; all the stages seem to have ignited more or less simultaneously. One seems to have worked but could not be tracked because its radio transmitter failed. Only one was a limited success; at an altitude of 2700 miles, while the rocket was still climbing, the transmitter ceased to work. The rocket probably reached a peak altitude of 3000 miles.

Meanwhile, what had happened to Project Vanguard, *the* official American satellite project, while all this was going on? The first shot of the actual Vanguard rocket, first stage only with dummy second and third stages, took place on October 23, 1957. The official designation was TV-2 (for test vehicle), and during the preliminary tests it had given trouble at every stage of the procedure—so much so that the engineers working on it called it "a lump of garbage" and worse names. But it did make a beautiful flight, proving that the design was workable and really needed nothing but time, the one commodity that was not available.

The successful shot was a very mixed blessing for the Vanguard engineers. Public reaction was: "See, it works—even if a little late. Now for the real satellite shot." Official reaction, though phrased differently, was basically the same. In the meantime somebody had had a new thought. The whole Vanguard program consisted of "test vehicles," or TVs, and Satellite Launch Vehicles, or SLVs. The two Viking rockets had been labeled TV-0 and TV-1, and after TV-2 there were to be at least three more TVs. Of course the actual launching of the American IGY satellite had to wait for the SLVs, but the TVs could launch an "experimental satellite" weighing 3¼ pounds—the one that Khrushchev later called "the grapefruit"—that emitted only an electronic whistle. Since this satellite would *not* be the American IGY satellite, the fact that it would carry no instrumentation was not important.

The chief of Project Vanguard, Dr. John P. Hagen, was summoned to the White House for a conference with President Eisenhower to explain the Vanguard situation. Hagen emphasized that TV-3 was strictly experimental and that the orbiting of the test satellite should not be announced beforehand. Then he left for the West Coast to confer on the second stage—and the White House handed out a press release that TV-3 would launch the American satellite! The country was now clearly nervous from the top on down.

Before TV-3 was even ready for a full-scale test, the Russians launched Sputnik-II, which received the astronomical designation 1957-beta. It was much larger than Sputnik-I and the satellite was not separated from the top stage of the launch vehicle. The whole orbited as one body, with a weight of about 7000 pounds. The satellite section consisted of two parts, one a duplicate of Sputnik-I, and the second an animal capsule holding an 11-pound dog named Laika ("barker"). The perigee of 1957-beta was 145 miles, the apogee 1056 miles, the orbital period 103.7 minutes. Among other things the heartbeat of Laika was telemetered; it had gone up considerably during the ascent but settled back to normal within a number of minutes after the satellite had attained orbit. As had been hoped by some, and doubted by others, the weightlessness of an orbiting body did not interfere with the bodily functions of the animal.

After about 100 hours the batteries in Sputnik-II ran down and the dog was painlessly put to death. Because of its great mass the orbital decay of 1957-beta was slow, even though its perigee was lower than that of 1957-alpha. It stayed in orbit for 2366 revolutions and was seen by many millions of people. It finally re-entered over the Caribbean Sea late in the evening of Sunday, April 13, 1958, which happened to be Russian Easter, old style.

During the first days of December 1957, the time for the launch of "America's answer to the sputniks," the little test satellite in the nose of TV-3, drew near. Two American launch crews were busy; the Huntsville crew readied the fourth Jupiter-C for firing, while the Vanguard crew worked on TV-3. The Army had not set a date yet for its Jupiter-C shot, but it was clear that Vanguard would be first. Press, radio, TV, and plain spectators assembled around the Cape en masse.

It so happened that I had a lecture engagement in Baton Rouge, Louisiana, for the evening of December 4. Press and radio reported during that day that the countdown was proceeding—it had been started during the very early morning hours. The chairman of the lecture committee stationed his teen-age son backstage with a radio set and told him to interrupt me as soon as Vanguard had reached orbit. I was not interrupted; the launch attempt had been "scrubbed" (called off) after a countdown that lasted so long that some of the launch crew members simply fell asleep from utter exhaustion.

The next morning it was learned that the launch had been set for December 6. I waited around in Baton Rouge during the 5th, spending part of the day finding out how to get from Baton Rouge to the Cape without hiring an air taxi—the weather was such that I had no confidence in single-engined airplanes, though it was good enough for normal operations of large airliners. From Baton Rouge I could get a plane to New Orleans and another from there to Miami, but I would risk a so-called "lay-over" of more than 4 hours in Miami. It was a better bet to fly via New Orleans to Atlanta and south from there. I landed in Atlanta at about 12:30 P.M. on the 6th. While looking for the information counter I ran into the local manager of Eastern Air Lines, whom I happened to know. "Where are you bound?" he greeted me. "For Cape Canaveral if I can get a connection." "Why? Haven't you heard? Vanguard blew up an hour ago." "In that case . . . please confirm my return flight to New York."

Again the countdown had started during the very early morning hours. The earlier long countdown had been plagued by many small leaks in the system, but this one seemed reasonably troublefree. The firing button was pushed at 11:44:35 A.M. on December 6, 1957. The rocket engine of the main stage of TV-3 caught fire, burned a little half-heartedly for about half a second, and then steadied, and the rocket began to lift. After about 2 seconds the launch crew saw a stiletto-like flame issuing from the side of the rocket, in the area of the upper part of the rocket engine. Simultaneously the thrust dropped below the weight of the rocket, which settled down, then toppled over. The fuel tanks split open, and the launch pad became a mass of flames. After it was all over the launch crew found to their amazement that the solid-fuel third stage, though surrounded by flames, had not ignited and that the little test satellite was whistling away as if it were in orbit.

"America's answer to the sputniks" had turned into the most publicized failure in history. The country was shocked; there were the customary cries for an investigation (as if such an event could come to pass without being investigated); and the foreign press was scornful, even the British press. If it had been the first attempt at a satellite launching anywhere, there might have been some sympathy, but two Russian satellites were orbiting overhead. A West German magazine compared the triumphant advance announcement to the "discovery" that Easter Island had disappeared beneath the waves.[11]

The actual reason for the failure has never been fully established. Of course the cause had been a fuel leak somewhere between fuel tank and rocket engine and the most probable cause of a fuel leak is a misalignment. The engineers from General Electric (which had built the engine) blamed the Martin Company (which had built the rocket); the Martin engineers took the opposite view. While the argument was never resolved it is worth noting that, if the rocket had worked properly, it would have been the *very first flight of the second stage,* which had never been test-flown.

A few days after my return to New York I faced the television cameras of the ABC network and its commentator Quincy Howe. To what did I ascribe the failure of Vanguard? To the fact that it is a new and untried rocket. Did I think the Jupiter-C rocket that had been announced as being readied for a satellite shot would make it? I said I did not doubt that it would—having in mind that a Jupiter-C rocket had already almost made it. *When,* in my opinion, would the Jupiter-C make it? This was a difficult

[11] In 1922 or 1923 an American destroyer made port with the announcement that it had failed to find Easter Island and that the island therefore was assumed to have "sunk." The fact was that, in very bad weather, the destroyer had simply failed to find it.

question and I thought fast. Christmas was only a few weeks away. I doubted that the rocket could be ready that soon, and for some reason I felt sure that nobody would shoot between Christmas and New Year's. I said: "In January, Mr. Howe, in January."

There was one interesting factor of which I was not aware at that time. If "Missile 29" had been permitted to go into action in 1956, it would have produced the first artificial satellite but this would have had nothing to do with the IGY satellite which was to be launched as part of the Vanguard program. Now, when nobody believed any longer that Vanguard would ever make it, the Jupiter-C shot became the IGY shot and the satellite had to carry instruments. The Vanguard satellite, spherical in shape, was not too heavy for the Jupiter-C, but it was too large. The instruments had to be rearranged to fit into a cylindrical tube.

Everything was ready late in January 1958; except for one factor the shot could have been made on January 28. That factor was the "jet stream," a very fast constant current of cold clear air that moves at high altitudes above the United States. It travels always from west to east, but over different latitudes, sometimes farther north than at other times. Late in January 1958 it happened to be exceptionally far south. If the rocket had to go through the jet stream the shot might be ruined. On the 30th it was decided to make the rocket ready to the point where the liquid oxygen had to be put into the tank, and then, on the basis of the jet-stream weather forecast, decide whether to go ahead. Balloons were flown into the jet stream and the computer at ABMA ran an analysis.[12] At 9:20 P.M. the message came: *Highly marginal—we do not recommend that you try it.* The firing was canceled. The next morning the jet stream started shifting northward; in the evening its velocity over the Cape was expected to be down to about 110 miles per hour. The Jupiter-C, it was known, could cope with a 110-mile-per-hour jet stream, but the night before it had been about twice that.

Jupiter-C no. 4 lifted off at 10:48:16 P.M. on January 31, 1958. Two minutes and 36 seconds later the first stage had Brennschluss, 60 miles up. The upper stages coasted to the peak of the trajectory to be fired by radio command. Two minutes after this command the top stage was in orbit. The men at the Cape as well as those in Washington could know at the moment only that the top stage had the necessary velocity. The instrumentation that could be carried did not reveal whether it was also pointed in the right direction. The ground tracking network had not yet been developed to the point where one can tell within 10 minutes that orbit has been attained; a wait of about 1½ hours was required until one could be certain. But orbit had been attained. The word came: "Goldstone has the

[12] General Medaris, the launch chief Dr. Kurt Debus, and the launch crew were at the Cape, while Wernher von Braun and Dr. Pickering, who was responsible for the instrumentation in the satellite, were in Washington. In view of the political importance of the shot this was necessary, but it resulted in an incredible number of long-distance telephone calls.

bird," meaning that the big Goldstone radar in California had picked up the satellite as it crossed the West Coast, having nearly completed its first orbit.

At 1 A.M. on February 1, America was jubilant. The Huntsville radio urged all citizens to come to the town square for a celebration. They came, with signs reading "Our Missiles Never Miss"; they also hanged ex-Secretary Wilson in effigy, apparently not wishing to have their feelings misinterpreted, and changed a sign at the city limits from *The Missile Center of the USA* to *Space Capital of the Universe.*

At 1:30 A.M. a long-distance call from Washington woke me up; the office of the Associated Press told me that the satellite was in orbit, adding: "We thought you would like to know." They had also ruined my sleep for the next four or five hours. I sent a wire to Wernher von Braun, saying: "Congratulations! And thank you for having kept me an honest man by the margin of one hour and twelve minutes. I had told everyone that Jupiter would jump into space in January."

The first American satellite received the name Explorer-I and the astronomical designation 1958-alpha. As with Sputnik-II, there was no separation of the instrument package from the top stage of the carrier rocket; they orbited as a unit. The first apogee was at 1573 miles, the perigee at 224 miles, and the orbital period 114.8 minutes. The lifetime in orbit, predicted from these figures, was 3 years, but Explorer-I had a number of surprises in store. One was that it would still be in orbit in 1967, though by that time the orbit had decayed to some extent. The figures for Explorer-I for January 1967 read: apogee down from 1573 miles to 890 miles, perigee down from 224 miles to 210 miles, orbital period 102.6 minutes.

The Vanguard rocket labeled "TV-3 back-up" was on its pad when Jupiter-C jumped into space. It had been scheduled for February 3, because the IGY committee had informed the National Science Foundation that they could handle the tracking of satellites only if they were fired at least three days apart. But valve trouble forced a postponement to February 5. For the second stage of Vanguard this was still the first flight, and again the second stage had no chance to prove itself. "TV-3 back-up" lifted off on February 5 and looked beautiful, but 57 seconds after lift-off it broke apart in midair and exploded.

The next Jupiter-C shot, carrying Explorer-II, was scheduled for March 5. There was an unexpected "hold" for 10 minutes in the countdown, because the first Explorer was passing overhead and existing instrumentation would have had difficulty in disentangling the signals from two satellites relatively close together in space. But Explorer-II did not reach space, in spite of a perfect take-off. After the necessary wait it became clear that there was no orbiting satellite. For unknown reasons the top stage had not ignited and the shot had turned into an unwanted repetition of the shot of "Missile 27."

On March 17 Vanguard TV-4 was ready. This time the first stage behaved, the second stage made a flawless maiden flight, the third stage performed as expected, and the little test satellite became Vanguard-I and 1958-beta. It was put into such a high orbit (perigee 409 miles, apogee 2453 miles) that it would never re-enter on its own accord. After Explorer-I finally decays, Vanguard-I will be the oldest artificial satellite in orbit and will remain the oldest until it is removed by a manned spacecraft to be put on exhibit in the National Air and Space Museum in Washington.

On March 26 another Jupiter-C put Explorer-III into orbit as 1958-gamma.

Meanwhile, where were the Russians?

The Russians were getting ready to show the world, and especially the United States, what they could do. The orbiting test satellite of the Vanguard program weighed 3¼ pounds; the Explorers weighed about 31 pounds. Sputnik-III, fired on May 15, 1958, weighed 2925 pounds. The total weight in orbit was well over 8000 pounds.

The American public was flabbergasted once more. Everybody actively engaged in the space program received a large volume of mail which could not even be dismissed as "crank mail" because in most cases the letter writers were honest—and worried—citizens who sincerely tried to help. Of course they talked nonsense in 90 per cent of the cases, but this was due merely to lack of scientific or engineering knowledge. I was not a member of any government organization or even of a semiofficial committee—in fact I had no connection with the space shots that were going on or being planned except that most of the people who were shooting Jupiters and Vanguards were good personal friends—but I got my share of such mail too.

Of course I don't remember all the letters, but those I do remember had one thing in common: their authors had sat back and "thought." They had "reasoned out" what the Russians were using and what we, therefore, should also use. A fairly common conclusion was that the Russians must be using an enormous catapult to throw their rockets into the air. My reply to this kind of letter was almost standardized: I pointed out that a catapult capable of imparting a useful velocity to a rocket probably weighing 100,000 pounds would have to be very large and enormously expensive, provided that it could be built at all. Another common conclusion was that the Russians were using atomic energy, and why did our atomic energy "boys" sit back, resting on their laurels? If it was not atomic energy, then it was a new fuel. One letter writer who "knew" that the Russians were using a new fuel also knew that our own chemists were deliberately ignoring the problem, being clandestine members of the Communist Party. He had reasons for saying so; he had written to the research department of a big chemical company and the company had answered that there was no reason to believe that the Russians were using an unknown fuel. One man wrote that all the people who were guessing at new fuels or catapults were barking up several wrong trees. The secret of the Russian successes was that they had discovered a metal that did not weigh anything!

Not many of the letter writers were women, but the few who were introduced a number of fascinating thoughts. One wrote to say that her nephew, who was in college at the time, had heard the true story from a fellow student who was a German refugee. The Germans, late in 1944, had developed a fuel that was five times as powerful as any fuel we used because it was a solid, not a liquid. And the man who had this fuel had been captured by the Russians and collaborated with them because he was a Communist. My suggestion that this German refugee had probably amused himself by inventing a tall tale was brushed aside with the hint that I was probably a Communist too.

I thought, then, that this letter represented a kind of apogee, but some years later I had to change my mind. Another woman wrote, enclosing a reprint of an article by the Russian physicist Pyotr Leonidovitch Kapitsa. Professor Kapitsa's article dealt with ball lightning, a rare electrical phenomenon. He tried to evolve a theory of ball lightning, to find what conditions have to prevail during the appearance of a lightning ball. Kapitsa did not succeed and said so in the introductory paragraph of his article. But never mind what he said; according to my correspondent, the secret of the Russian space shots was that they don't use ordinary fuels at all. They were propelling their rockets by ball lightning!

I don't recall how often I wrote (or said after lectures during the question and answer period) that the "secret" was not a new metal or a new fuel, and certainly not a catapult or a vertical tunnel in a mountain. The secret, I kept repeating, was that the Russians had bigger rockets. I hope that at least a few people believed me.

12 The Shot into Infinity

Yon rising Moon that looks for us again—
How oft hereafter will she wax and wane;
How oft hereafter rising look for us. . . .
—*Omar Khayyám (who also was an astronomer)*

By October 1958 seven—or possibly eight—artificial satellites had been orbited. The seven of which one could be sure were Sputniks-I, -II, and -III, Explorers-I, -III, and -IV, and Vanguard-I; the eighth was a doubtful case. The U. S. Navy had experimented with satellite launches from high-flying jets. The jet carried a multistage, solid-fuel rocket which was ignited while the jet was traveling at its peak altitude and highest speed, going due east. Six such experiments were made and the Navy felt sure that one of its test satellites had attained orbit. But it was only a small uninstrumented sphere that could not be tracked with the means then available. Later, when space tracking had become a fine and accomplished art, it could not be found; evidently it had re-entered in the meantime.

The public, which had just grown used to the idea of artificial satellites, wondered what would come next. A shot to the moon? Or was a moon shot still something for science fiction magazines only? Actually the first shot to the moon was to take place soon and it would be a Russian shot, but before that happened the United States scored a great success with a rocket that was actually called SCORE.[1]

SCORE was one of the then relatively new Atlas rockets. It carried a tape-recorded Christmas message and a transmitter for broadcasting it, but it was not the first "communications satellite" because it could not relay a message from one point on the ground to another point. However, it was a great success insofar as it represented the first heavyweight satellite (8750 pounds) put into orbit by the United States.

The story of the Atlas rocket is interwoven with the stories of three

[1] The letters were *interpreted* to stand for Satellite Communications Orbital Relay Experiment.

other rockets—namely Thor, Jupiter, and Polaris—with a fourth one, Titan, lurking in the background. But since the Atlas design began with yet another rocket that is generally considered to be its ancestor, we have to begin with the classified Project MX-774.

In October 1945, at the time that Army Ordnance men were busy sorting through the piles of V-2 components (see page 233), the U. S. Air Force sent a letter to various companies in the aviation industry, inviting proposals for a ten-year program of research and development on missiles of various types, with a maximum range of 5000 miles. One of the companies asked was Consolidated-Vultee Aircraft Corporation (later called Convair), and one of their scientists, Belgian-born Karel J. Bossart, began to think about missiles. The ultimate range of 5000 miles which the Air Technical Service Command had mentioned could be accomplished in two ways: by what amounted to an unmanned jet plane or by a rocket. But if a rocket, there had to be a smaller test rocket; the company asked the Air Force whether it would be permitted to build a number of test missiles. The Air Force agreed, specifying that the number of test missiles should not exceed ten. Project MX-774 was then broken down into three designs, labeled A, B, and C. Design A was the subsonic self-navigating jet plane, design C was the 5000-mile rocket, and design B was the test missile that was to try out the concepts that were to go into design C. One of the engineers, William Lester, then provided names for the three designs: Design A was dubbed Teetotaler because it was the only design of the three not using alcohol; design B became the Old Fashioned because it was to be an Americanized version of the old V-2; and design C became the Manhattan, with reference to the Manhattan Project since it would probably carry an atomic bomb.

The Old Fashioned, based on the design of the V-2 and slated to use the same fuels, was to show (and did show) the following changes. To begin with, the warhead would be separated from the rocket body immediately after Brennschluss. Having a higher density it would travel a better trajectory, less influenced by air resistance, than the empty rocket body. The V-2's graphite vanes were discarded and a gimbaled rocket motor was to be used. There would be no separate fuel tanks, because the rocket's skin was to hold the fuel.[2] Finally, the skin was to be so thin that it would be stiffened by internal pressure only, like an automobile tire. This pressure stiffening had been incorporated by Oberth in his patent application in 1928, but Bossart thought of it independently. All these features were evolved, discussed, and calculated during the year 1946, and some preliminary drawings were made. Reaction Motors, Inc., was requested to develop a rocket engine; they made a four-chambered engine like the one for the X-1 research airplane (see Appendix 1).

[2] These two features, as we have seen, were later adopted by the designers of the Viking rocket.

On July 1, 1947, the Air Force canceled the whole project, having had its own missile budget cut sharply. But the company had some money left over from the original study contract. Could they use this money to build the planned test missiles?

The Air Force agreed and three test missiles were built. The finished rocket stood 31 feet 7 inches tall, with a diameter of 30 inches and a dry weight of 1200 pounds. In November 1947 the first of the test missiles was carted to Point Loma at San Diego Bay for a static firing. Everything was ready on November 20, but when the firing button was pressed, nothing happened. The cause could be found quickly; the igniter squibs had failed to work because the connections had corroded.

The squibs were cleaned up and the connections replaced in the evening, and the rocket was refueled the following day. Immediately after pressing the firing button the engineers saw that something was wrong. Instead of the four fiery jets that should have come to life there were only two, and seconds later the engine blew up and the rocket was hidden by flames. A system of water sprays, rigged up for just such an emergency, doused the flames. Even without an investigation it was quite clear what had happened. Again two squibs had failed, but the fuel and oxygen was fed to all four chambers and after the chambers had filled up with alcohol and oxygen, the two that had not ignited properly were ignited by the exhaust flames of the other two. Fortunately the rocket itself could be saved. Since the weak spot was known, the testing group was especially careful with the igniters and succeeded in obtaining six flawless static tests, one in January 1948, two in February, and three in March.

Then all three MX-774 test missiles were brought to White Sands, but it cannot be said that they were great successes. The first one, fired on July 13, 1948, rose to 6200 feet, turned over, and crashed only 600 feet from the blockhouse. The second, on September 27, 1948, climbed to 10 miles before the engine cut off for no known reason. The rocket then coasted to 29 miles and the testing crew hoped to recover their missile, as it was equipped with a parachute for recovery. But the parachute was not ejected, presumably because of battery failure. When the engine had cut off, the oxygen tank had not yet been empty, so on the way down oxygen pressure built up and ruptured the rocket. It broke apart in midair.

The third test missile made its flight on December 2, 1948. This time everything seemed to be going well. At X + 45 seconds the exhaust flame was still bright and sharp. At X + 51 seconds it cut off. Parachute ejection was ordered and the observers on the ground could see the parachute come out, but it did what the parachute over the Raketenflugplatz had done fifteen years earlier, it tore off. The rocket crashed.

The over-all result of these tests, the static tests as well as the three not fully successful flights, seemed to be that the thin stainless steel tanks performed well and that the swiveling of the motors for steering could be made to work. But a good deal of work and an unguessable number of static tests and flights would be required, and the available funds were exhausted.

For some years thereafter the only experimental programs underway were those described in Chapter 9—the V-2 and Viking programs with the Aerobee to follow. The Army quietly developed the 200-mile Redstone missile as one of the "tactical missiles." During those years—from 1948 to 1953—ideas about missiles slowly crystallized and three classifications were born. The first was "tactical missiles," with a range anywhere from 5 to 200 miles. The second was missiles of "intermediate range"—between 1000 and 2000 miles. The third class was "intercontinental missiles" with a range of 5000 miles. At first it was not definite that all these missiles would be rockets and a few cruising missiles with jet engines were actually developed: the Matador with a 500-mile range and the Snark with a 4000-mile range. But it was clear that such cruising missiles were vulnerable; they could be intercepted, as the defense of London against the V-1's had shown. Naturally the chances for a successful interception increased in proportion to the range. So the long-range missiles became rockets, or ballistic missiles (strictly on paper, that is), and the designations became IRBM for intermediate range and ICBM for intercontinental range.

The delay in actual design, construction, and testing was the result of two dissimilar reasons. The first was that preliminary estimates of such specifications as price, performance, and destructive power demanded that the IRBMs and the ICBMs carry atomic warheads, though high-explosive warheads might do for the tactical missiles. However, the weight of atomic warheads was then too great for rockets. The specialists stated that they hoped to reduce these weights considerably and fairly soon. But since a rocket designer has to know the weight of the payload before he can even begin, the project simply had to wait until the atomic-energy experts could provide definite figures. Of course, another course would have been possible: the designer could obtain the best estimates of the weight of future nuclear warheads, and then design, build, and test missiles for these weights, in the hopes that the actual weights would come close to the estimates.

This is what the Russians did during that same period. They did not know the weight of future warheads either, but they made guesses, and it seems as if their guesses were too high. However, they built rockets in accordance with these guesses and were therefore able to blossom out with big rockets in 1957 and 1958. In the United States this procedure was impossible for the second of the two reasons that caused the general delay. The United States, its President said, had to balance the budget. Balancing the budget was the paramount goal of the administration and for this reason future projects remained future projects. Of course planners continued to live in the realm of future projects since that is their natural habitat, but several events had to occur before they could do anything about these projects.

As far as the Air Force planners were concerned, they kept thinking about their ICBM which already had the name Atlas. Its proposed range was 5000 miles; its warhead was to be a hydrogen bomb. In 1951 the Air Force decided on a new range, namely 5500 *nautical* miles, or 6325 statute miles. A design sketch of the Atlas, made with a still undisclosed assumption of warhead weight, led to a rocket 90 feet tall, with a diameter of 12 feet, five main rocket motors, two vernier rocket motors for fine velocity adjustment, a take-off thrust of 650,000 pounds, and a take-off weight of around 480,000 pounds.

On March 1, 1954, a new type of hydrogen bomb was exploded. It proved that a fusion explosion could be accomplished with much smaller weights than that of the first hydrogen "bomb," which weighed between 50 and 60 tons, and the dimensions of the Atlas could therefore be pared down to a height of 75 feet (60 feet of this was tank section), a diameter of 10 feet, and a take-off thrust of 360,000 pounds that could be furnished by three main rocket motors. At about the time that the actual design of the new Atlas was tackled, the Atlas project was joined by Krafft Ehricke who had been with General Dornberger in Buffalo and who was greeted in San Diego as the third "Peenemünde German."[3]

By that time the main design decisions had been made. The tank of the Atlas consisted of stainless steel which was no thicker than a dime at any point. Generals—and later congressmen—worrying about what would happen if somebody dropped a wrench on such a tank were conducted to a test version of the "stainless steel balloon" that was stiffened by pressure and offered a choice of assorted mallets to see whether they could dent it. In each case the general or congressman grew tired before he had even succeeded in producing a mark that could be seen.

The light weight of this tank, not quite 2 per cent of the fuel weight that it could hold, led to another innovation. The Atlas was not a two-stage missile, but it was not truly a single-stage rocket either, so that it was jokingly referred to as a one-and-a-half-stage rocket. This term meant: all the fuel of the rocket was in the single tank, divided by a bulkhead to keep the oxygen from the highly refined kerosene fuel. The rocket motor, located below the center line of the fuel tank, was dubbed the "sustainer engine." But the sustainer was flanked on both sides by "booster engines" attached as a special tail-end structure that supported the rocket. The booster engines received fuel from the same tank as did the motor, but after they had done their job of lifting and initial acceleration, the tail

[3] The other two from Peenemünde who had joined earlier were Dr. Walter Schwidetzky and Dr. Hans R. Friedrich, the latter a specialist in guidance systems which then were still called autopilots.

structure holding the two booster engines dropped off, while the sustainer kept burning. In some of the early tests of the Atlas only the two boosters were ignited, but the normal procedure is a take-off with all three engines roaring.

What had happened to the IRBMs while the Atlas took shape?

The main problem in the background was which service should operate IRBMs. That the ICBM would complement SAC (the Strategic Bomber Command) was clear. Whether the IRBM should do the same was less clear, even if the Air Force thought so. The decision had to come from the Secretary of Defense, of course, and on November 8, 1955, it came: all three services should have IRBMs. The Air Force was to develop its Thor missile and the Army its Jupiter, while the Navy was to adapt the Jupiter for shipboard use. General Bernard A. Schriever was in charge of the Thor (among his other responsibilities); General John B. Medaris was in charge of the Jupiter; Admiral William F. Raborn was in charge of the seagoing Jupiter. Admiral Raborn felt that the Navy had been ill treated by this directive. Not only had he, as he phrased it, "been shoved into bed with the Army," but he had also been told that the Navy would not get any extra funds for its work in developing or adapting the Jupiter for Navy use; this had to come out of its normal allocation of funds. The Budget, with a capital B, still had to be balanced.

Raborn sent a representative, William Hasler, to Huntsville where the Jupiter was to be developed and the collaboration actually began in February 1956. Remember that the V-2 program had included Operation Pushover; a fully fueled V-2 was made to topple to see what would happen. The rocket was toppled on the simulated deck of a ship, and the result was such that most Navy men came to the conclusion that if a ship suffered such an accident it might have to be given up. So the Navy was not happy with the idea of a seagoing liquid-fuel rocket, especially since the Army's Jupiter was to stand 65 feet tall, with a diameter of 95 inches. The thought of fueling such a missile at sea in any but the calmest of weather was frightful to Navy men. First they tried paring down the size, but if the seagoing Jupiter stood only 55 feet tall its diameter would have to be increased to 105 inches. The difficulty was not the size of the missile—a ship large enough could certainly be built, provided one could get the appropriation.[4] At that point somebody thought of the new solid-fuel charges that had been developed. The problem was the fueling operation. If one could use solid fuels, the missiles would be taken aboard fully fueled. Quick calculations were made for a Jupiter-S (S standing for "sea") with solid fuels. But the solid fuels available then were not yet

[4] The Navy found out later that the appropriation was not the last hurdle, for Congress can appropriate money but cannot force the administration to spend it. That happened with a number of Polaris submarines; the money was appropriated by Congress but not released by the administration.

powerful enough. The liquid-fuel Jupiter would weigh 110,000 pounds; Jupiter-S would weigh 160,000 pounds.

Admiral Raborn and his scientific advisers decided that the Navy needed a seagoing missile developed as a seagoing missile. They started out with the name Polaris, since the star Polaris has always guided mariners, and with the conviction that solid fuel had to be used. New developments in the field of solid fuels made a missile with a take-off weight of 28,000 pounds, and a range of 1200 miles, seem possible. And that missile would be only 28 feet tall; this could be carried below deck and even in a submarine. Late in 1956 the plans for the Polaris had progressed to the point where one could say how it would look and what it would be able to do. Now the Secretary of the Navy had to approve and, after him, the Secretary of Defense. The Secretary of the Navy did approve, but what would Secretary Wilson say? Admiral Raborn explained the reasoning to Wilson, being well aware of the fact that Wilson never paid any attention to technical difficulties. Wilson said nothing throughout the presentation. Raborn ended his speech with the statement that the Polaris program, with Polaris submarines, would be cheaper than Jupiter-S, with the ships to carry it, by half a billion dollars. This was the tide-turning argument. Wilson signed a directive for a solid-fuel Polaris missile on December 8, 1956.

So there were four different missiles under development at the same time. The Air Force developed two, the ICBM Atlas (with the Titan as a back-up) and the IRBM Thor. The Army developed the Jupiter, and the Navy the Polaris. Atlas, Thor, and Jupiter were all ready for their first test firings in 1957. The Polaris, having been authorized last, did not appear at the Cape until 1958.

As has been said in the preceding chapter, 1957 was a year that offered no comfort to Americans. The first Thor missile, designated as "101," had been ready for firing on December 20, 1956. But when the countdown reached zero the missile just stood on its pad; a faulty relay had prevented ignition and the missile quietly vented the oxygen that had reassumed its gaseous form. On January 24, 1957, Thor 101 was on the pad again. This time the flight was "scrubbed" one minute before ignition because one beacon did not work.

On January 25 everything seemed to go well, but on ignition the missile did not lift; instead it seemed to melt itself down in its own fire. The launching pad was ruined, too. The malfunction could be traced to a valve failure, and although for a different cause the same sequence of events would befall Vanguard at the end of the year. Thor 101 had risen by 6 inches when the valve misbehaved; then, as the thrust dropped below the missile's weight, it settled back and was consumed by its own fuel.

The fate of the next Thor, dubbed "102," was even worse. Every missile test-flown at the Cape carries a number of high-explosive charges that can be set off by a radio command. The responsibility for sending, or not sending, this command rests with the Range Safety Officer, usually referred to as RSO. He watches the take-off on several TV screens activated by cameras in different locations. He also watches the impact predictor, which shows from moment to moment where that missile would land. The main concern of the RSO is the protection of people and property on the ground; the test flight as such interests him only privately.

On April 19, 1957, after the launch pad had been repaired, "102" was ready to go. The countdown did not offer any special problems and the test crew was pleased with the smooth take-off. But 35 seconds later "102" became a ball of flame; the RSO had pulled the "destruct" switch. The reason: somebody had made a mistake in the wiring of his console; while the Thor was actually heading out to sea, his instruments showed the missile as going in the opposite direction, heading inland over Florida.

Still, 35 seconds of perfect flight indicated that the rocket could perform well and "103" was put on the pad with good cheer in May 1957. On the 21st it was ready and the countdown, which was troublesome, got under way. Then the missile exploded on the pad at T — 4 minutes! That a rocket might explode some time after ignition was a sad but known fact. That it might explode on ignition was also known. But how could a missile explode four minutes *before* ignition?

It took days to find the answer.

Thor "103" had had a very long countdown. Three times something had gone wrong—something different each time—and the countdown had to be stopped until repairs had been made. Then the countdown had been resumed at some point prior to the failure that had caused the "hold," a process known as re-cycling. All along something had been wrong with the main fuel valve but the signal that it did not work properly had been either misinterpreted or overlooked because attention was focused on the other problems. In any event the faulty valve caused an overpressurization of the tanks and by the time the final countdown approached its end the tanks could no longer stand the overpressure.

The next rocket to be test flown was an Atlas. The designers of the Atlas had built a number of rockets which they referred to as the A series. If tests of the rocket showed the need for modification, the next series would be called the B series, and so forth. The first Atlas to be flown was no. 4 in the A series, hence called 4-A. The countdown had been practiced endlessly—there were 25 rehearsals—and in the process the test crew had invented an abbreviated language. No time for long explanations—something was either "go" or "no go." The true countdown lasted only 3 hours and 20 minutes, beginning at 5 A.M. on June 11, 1957. Atlas 4-A was exploded after 1 minute of flight; it had become unstable and the RSO had to do his job.

The public was dismayed, to put it mildly. The Atlas incident had been visible to everybody for miles around, unlike the Thor failures which could be kept secret (though the fact that something had gone wrong became known anyway). Having not yet been informed about the range safety arrangements, the public only knew that the rocket had exploded. The public also did not realize—and doesn't to this day—that such a test flight is made for the sake of the information it will yield and therefore the outcome of the flight itself is not as important as it seems. Lacking such information newspaper reporters, editorial writers, radio commentators, and others inevitably drew the wrong conclusions. A few commentators took the position that such a big rocket simply could not be made to work; they did not mention the fact that about half a century earlier some airplane crashes had also been taken to be proof that "flying machines" could not work. Some wondered whether the explosion could have been engineered by Russian spies; one writer speculated that the Russians might be able to make our rockets explode by "jamming" from a submarine. And one simply blamed "the unions."

Since the true purpose of the flight was to obtain information about the functioning of the rocket, the flight of the 4-A was by no means a "failure." General Donald N. Yates, then commander of the Atlantic Missile Range, had an evaluation of the 4-A prepared and saw, probably to his own surprise, that the rocket had accomplished 95 per cent of its mission. If the goal had been, let's say, 200 points of information, 190 points had been obtained. But even to those who knew all the events, the picture in August 1957 did not look rosy. No Thor missile had made a successful flight, a Jupiter missile had malfunctioned, and the only Atlas flown had to be exploded.

The Russians chose this month to announce that they had an ICBM. The announcement probably contained a fair amount of exaggeration, but it deepened the gloom. And no relief was in sight. Thor "104," on August 30, 1957, broke apart after 92 seconds of flight. Atlas 6-A, on September 25, was a repetition of Atlas 4-A. At least Thor "105," on September 20, had made an 1100-mile flight, but Thor "107" ("106" was never flown), on October 3, 1957, lost thrust only seconds after ignition, fell back, and burned up. That was the day before Sputnik-I. Every one of the engineers concerned with the Thor needed only a few minutes to calculate the size of the Russian rocket that had put the satellite into orbit. It had to be much bigger than their Thor, but indirectly—and in contrast to the general public—the engineers could derive some solace from this fact. If the Russians had built a much bigger rocket that did work, this proved that very large rockets were possible. And what one engineering team could do, another engineering team could duplicate, and, with time and some luck, surpass.

Thor "108," on October 11, 1957, traveled very nearly its full range.

Thor "109," on October 24, 1957, was stripped of most of the instrumentation it normally carried for test flights, and traveled much farther than the final product was supposed to go. It splashed into the sea 2700 miles from the Cape. But since the policy then was to keep successes as secret as failures, nobody learned about the accomplishment of "109."

On December 7, one day after the Vanguard catastrophe, Thor "112" was ready to go. It had to be exploded 107 seconds after take-off because it became unstable. At that time two Atlas missiles, 10-A and 12-A, were on their pads. Rocket 10-A was to be the third Atlas test flight, but 10-A had troubles and Convair's manager of the Atlas program at the Cape, B. G. MacNabb, decided to bypass 10-A temporarily and to fire 12-A instead. It was supposed to fly only a few hundred miles and the mission would be considered 100 per cent successful if the Atlas hit the selected target area. It was fired December 17 and was a complete success. Two days later Thor "113" made a perfect flight. There were still more failures to come for both missiles, but by the end of 1957 the worst was over; the percentage of successful shots increased all the time.

Nobody that I know of made a Fourth of July speech in 1958 in which the emphasis was on the new big rockets of the United States. But there were four of them: Redstone, Jupiter, Thor, and Atlas. Of these the Redstone could be considered "fully reliable." The other three were successful most of the time but they were not yet perfectly reliable. It was reasonable to think of space again. One of those who did was Berlin-born Krafft Ehricke at Convair. He had been "converted to space travel" as a twelve-year-old by the movie *Frau im Mond.* "That picture knocked me right into orbit. I saw it twelve times," he told San Diego newspapermen later. As a fourteen-year-old he had watched some of the experiments at the Raketenflugplatz, unnoticed by anybody. In San Diego, in 1958, when he was forty-one, Krafft Ehricke engaged in his favorite activity: calculations, especially calculations of mass-ratios. Now this Atlas rocket, after it had dropped the two booster engines and the tail structure, weighed so and so much. It should be possible to save several hundred pounds at the front end, especially if no unnecessary testing devices were to be carried. With such a lightened Atlas the fuel available should produce such and such a velocity, enough velocity to orbit. It should be possible to put an Atlas into orbit. Krafft Ehricke sent a memo about his calculations up to management and thereby thinking was started at higher levels.

The decision was to try. The missile selected was Atlas 10-B.

The launch crew could see that 10-B was somehow different from 12-B which was being readied at the same time and was actually fired (successfully) ahead of 10-B. Atlas 10-B had a different nose cone; a man from the Army Signal Corps installed equipment no other Atlas had ever carried. Some things were missing and some were "wrong." All this was duly reported to manager MacNabb who waved it aside or promised that this would be taken care of during the next practice countdown. Only a very small group of people was in on the secret and these saw to it that the number did not grow. On December 6 Dr. Hans R. Friedrich, who had contributed so much to the Atlas missile, died of a heart attack in San Diego, and missile 10-B was unofficially dedicated to his memory.

The countdown began at 4 P.M., local time, on December 18, 1958. At 6:02 P.M. Atlas 10-B lifted off. For the first 120 seconds, the burning time of the booster engines, nobody suspected anything unusual, except for one guidance-station operator, who called in to say that his impact predictor *did not predict any impact at all.* It was one of the cases in which a computer reasons better than a man. The operator *expected* an impact somewhere; the computer, going only by heading and velocity and without preconceived opinions, stated that there would be no impact. Of course not, for Project SCORE was going into orbit. It stayed in orbit until January 21, 1959. Its perigee was only 110 miles above sea level and the orbital decay therefore proceeded relatively fast. But while it lasted, SCORE made up for Vanguard's failure one year earlier.

Krafft Ehricke and the Convair group had not been alone in thinking about space in 1958. General Schriever, the Air Force, and the Douglas engineers who had developed the Thor had been doing the same. If the Thor did not have to carry a warhead and instead was equipped with an extra stage, it should be able to reach the moon. Project Vanguard had supplied the rocket for the upper stage—the second stage of the Vanguard rocket, which had been modified and given a new name: Able. The whole rocket, logically, became the Thor-Able. If everything worked well the Able rocket, boosted by the Thor, should reach escape velocity.

Escape velocity—7 miles per second—embodies an interesting theoretical concept. As was explained in Chapter 10, a vertically rising rocket, when the active propulsion period is over, constantly loses velocity because of the earth's gravitational field. In the case of escape velocity the momentum of the rocket at any moment is always a little greater than the force exerted by gravity at the distance where the rocket happens to be. Mathematically speaking, the earth's gravitational field would bring the rocket to a standstill relative to the earth when, after an infinite time, the rocket has reached infinity. A shot with escape velocity, therefore, is a shot into infinity. It would actually work that way if the earth were the only body in space; the existence of other bodies has very interesting results, as we will soon see. But a shot to the moon can be described as a shot into infinity which is fired in such a direction and with such timing that the moon will get in the way of the rocket.

A Thor-Able rocket intended to do just that was ready at the Cape on August 17, 1958. The Thor was no. "127," standing 88 feet tall with its extra stage. Lift-off was at 7:18 A.M. For 76 seconds everything went well. During the next second the Thor exploded, having reached an altitude of about 50,000 feet. The failure was due to a malfunction in the turbopump gearbox, a failure that had ruined other Thor shots too.

On October 11, 1958, another Thor-Able (Thor no. "130") was ready for a try at the moon. It carried a lunar probe designated as Pioneer-I. This time the first stage performed perfectly, but one of the upper stages did not do well. Pioneer-I climbed to a higher altitude than any man-made device. At first the peak altitude was given as 67,850 miles, but a recalculation of the telemetered signals showed that it had actually been 70,700 miles. At that altitude the momentum proved to be less than the strength of the earth's gravitational field and the probe fell back. It re-entered the atmosphere 43 hours and 17½ minutes after take-off over New Guinea (Fig. 66).

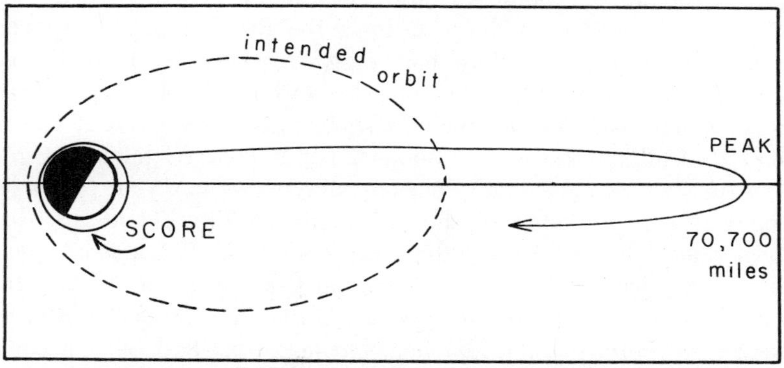

Fig. 66. Trajectory of Pioneer-I. The dotted line shows the orbit Pioneer-I might have assumed if its retro-rocket had fired. The orbit of SCORE is added for comparison.

Pioneer-I carried a retro-rocket which had been planned to be fired when the space probe reached the vicinity of the moon. This would have reduced the velocity to permit the comparatively weak gravitational field of the moon to exert some influence on Pioneer-I and, with luck, make it orbit the moon. When Pioneer-I fell back the retro-rocket was still in place and presumably in working order. The trajectory of the space probe was elliptical, of course, but the ellipse was so narrow that the earth did not fit inside it. Firing the retro-rocket might "widen" the ellipse enough that Pioneer-I, even though it had failed to reach the moon, would orbit the earth as an artificial satellite. But the retro-rocket did not respond, apparently because of battery failure. The battery failure, in turn, was ascribed to an unexpectedly high heat loss of the space probe. This explanation, however, can hardly be correct; take-off had been at 3:42 A.M., not long before dawn. Therefore the rocket had been in the earth's shadow for only a short time and had then been exposed to the sun's rays for well over 24 hours. Obviously the battery had failed, but "chilling" cannot have been the reason.

The life of Pioneer-II was much shorter than that of Pioneer-I. The Thor-Able took off on November 8, 1958. The Thor performed, but the Able rocket did not ignite. It climbed on the momentum imparted by the Thor to a peak altitude of 963 miles, then fell back, and re-entered the atmosphere 42.4 minutes after take-off.

Lunar probe Pioneer-III, on December 6, 1958, was carried by a different rocket, called Juno. This was a Jupiter missile with extra upper stages, conceived and built along the lines that had worked so well with the Jupiter-C rockets. But this time success eluded even the Huntsville team. Pioneer-III reached an altitude of 66,654 miles, fell back, and re-entered over equatorial Africa 38 hours and 6 minutes after lift-off. A successful American shot to the moon was still three months in the future.

In the meantime, on January 2, 1959, the Russians shot a space probe past the moon. Because they had bigger rockets at their disposal the Russians, then and for a number of years to come, were able to do everything first. The rocket used for this shot was designated CH-10 and was probably an adaptation of the Russian T-3 ICBM. Its take-off thrust was later estimated to have been 660,000 pounds, which is surprisingly close to the first design sketch of the Atlas; maybe the CH-10 also had five main engines. The space probe bore the name *Metchtá,* a word best translated as "daydream." Even now we don't know whether Metchtá was supposed to strike the moon or miss it. It did miss by 4600 miles and the Russian news services naturally behaved as if this had been according to plan. The element of doubt is created by the fact that radio signals from the space probe stopped quite soon after the moon's orbit had been crossed, which looks as if battery power had been supplied for the time needed for an impact shot.

Up to the first Russian moon shot the story of the beginning of the Space Age could be told more or less in chronological order. But beginning in 1959 activities diversified to such an extent that a strictly chronological accounting would be like trying to listen to three military bands simultaneously. Therefore we now have to proceed with one theme at a time. The first theme is the story of Polaris.

The test flights of the actual missile were preceded by Flight Test Vehicles (FTVs). Lockheed had built these, using the bottom stage of their X-17 re-entry test rocket. The first of them, dubbed FTV 1-7, was launched October 15, 1957. In addition to the steering device to be tested it had large stabilizing fins. Since the X-17 had already outgrown its own development problems, FTV 1-7 made a beautiful flight. So did FTV 1-9 on December 10. It flew without stabilizing fins, to the amazement of engineers from other projects who did not expect an object shaped like a wine bottle to fly true. All in all, there were sixteen test vehicles.

Then came the first Polaris, called AX-1.

Eventually the Polaris was to be launched from submerged submarines by compressed air, to ignite itself automatically as soon as it had broken the surface of the ocean. But shipboard launches and underwater launches came later. The first round of Polaris missiles was to be fired from dry land. So on September 24, 1958, the AX-1 stood on a launch pad at Cape Canaveral in an area chosen by General Yates to keep the Polaris as far away as possible from all other projects; large solid-fuel missiles were new and probably untrustworthy. When the AX-1 was ignited it seemed for a short while as if General Yates had been unduly pessimistic. But the missile blew up at an altitude of 40,000 feet.

AX-2 was ready on October 15 and performed in a manner that had never been seen before and probably not since. At "zero" there was a loud roar and a lot of smoke—"something" had gone up, but the roar on the launch pad continued. When the roar ceased, it was comparatively easy to figure out what had happened. The fuel charge of the first stage was of the type shown in Fig. 33 on p. 178, with a center hole going all the way through the charge. Within a split second of the ignition of the first stage, the second stage had been ignited and had taken off by itself. The first stage, still on the pad, had no thrust at all, because the exhaust gases escaped not only through the four exhaust nozzles at the bottom but also from the now open top. Preventing a repetition of this malfunction was easy, but the next set of troubles was more difficult to remedy.

AX-3 took off on December 30 and climbed nicely. But just at about the time the second stage was to separate itself from the first stage, the first stage went into strange gyrations. Nevertheless, the second stage did separate, but it seemed to lurch sideways. At T + 45 seconds the RSO threw the "destruct" switch.

AX-4, on January 19, 1959, lasted for 30 seconds.

AX-5, on April 20, 1959, was expected to fail, but some of its components had to be tested and the tests could be made during the first 10 seconds of flight. The experimenters had 27 seconds; then the missile exploded. By that time, the cause could be recognized—engineers' slang called it a "hot bottom," meaning that the bottom section of the rocket could not withstand the heat generated by the fuel charge for longer than about half a minute. A new construction material had to be, and was, found,[5] and in April 1959 AX-6 made a 300-mile flight. Before the Polaris could be called operational another set of tests was made with another test missile, the so-called "dolphin." They were of precisely the weight and shape of a Polaris missile, but were filled with water. Their purpose was to test the compressed-air-ejection mechanism of the missile tubes of the Polaris submarines; their nickname was derived from their appearance when they broke surface.

[5] Later the Polaris cases were made of spun Fiberglas.

Though the Polaris is plainly and simply a military missile, it has had an influence on space activities. An early version of the Polaris became the first stage of the Scout rocket, a four-stage, all solid-fuel, satellite-launch vehicle. The first stage, after modification, was given the name Algol, the second stage of the Scout is a Castor rocket, the third stage an Antares, and the fourth stage an Altair. (See Appendix 3 for a tabulation of these rockets, which are known collectively as "off-the-shelf hardware.") The Air Force's version of the Scout is very similar to the NASA version and is called the Blue Scout. The Navy's version, the Sea Scout, consists of a two-stage Polaris rocket without the warhead, with an Antares as the third stage and an Altair as the fourth. All three versions can place a 200-pound payload into an orbit 300 miles above sea level.

During the years 1958 through 1962 the number of satellites increased in proportion to the number of rockets available. Since these rockets were missiles and therefore produced in quantity, Professor Whipple's scheme of astronomical designations soon turned out to be inadequate. During 1958 the satellite count only went up to theta, the eighth letter of the Greek alphabet. In 1959 it went to xi, the twenty-second letter; in 1960 it went to tau, the nineteenth letter. But in 1961 the Greek alphabet was not long enough; after they had run through all twenty-four letters from alpha to omega they had to start over again with A-alpha and went up to A-lambda, the eleventh letter. In 1962 American-Russian competition ran through the Greek alphabet from alpha to omega precisely three times.

Beginning with January 1, 1963, a new system was started. The first satellite in the new year became 1963–01, the next one 1963–02 and so forth. If the carrier rocket of 1963–02 was brighter than the satellite, it was called 1963–02 A and the satellite 1963–02 B. The old astronomical designations of the earlier years were not changed for those still in orbit, but going back to the beginning, a running catalogue number was assigned to each piece that was orbiting or had orbited. This made Explorer-I no. 004. To give an example of a higher catalogue number, weather satellite Tiros-VII (put into orbit on June 19, 1963) and its carrier rocket became 1963–24 A and 1963–24 B, respectively; the catalogue numbers are 604 and 605. The first shot of the year 1967 was a communications satellite dubbed Intelsat-II, F-2; its designation is, obviously, 1967–01A; its catalogue number is 2639.

As the number of satellites increased, the technique of putting them into orbit developed, too. In 1959 everyone was completely happy if an orbit had been attained at all. Later, specific orbits were attempted and most of the time achieved. It is quite understandable that a satellite with one specific purpose requires an orbit quite different from that of a satellite with another purpose.

Let us look at an example.

Explorer-I discovered the Inner Van Allen Belt, a zone around the earth

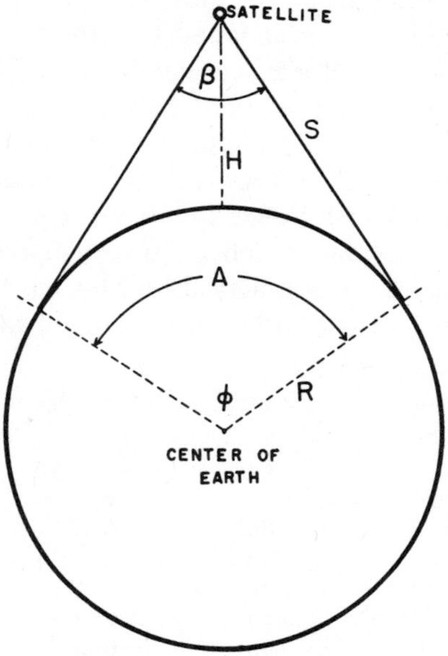

Fig. 67. Area of the earth visible from an orbiting satellite. The table shows that a distance of 4000 miles represents the optimum. (*After Robert P. Haviland.*)

where subatomic particles from the sun—electrons and protons—have been "trapped" by the earth's magnetic field. This belt extends from about 40 degrees northern to 40 degrees southern latitude. Its lower edge is about 600 miles above sea level, and the thickness of the belt a little less than its distance from the ground. Soon after the existence of this zone of subatomic particles had been established, Pioneer-III during its unsuccessful moon flight discovered an Outer Belt beginning about 4000 miles from the surface. It is now known that the Van Allen Belts are not always of the same intensity, and the Outer Belt at least also varies in extent. It seems larger at some times than at others, and there was once a week during which it could not be found at all. If a satellite is wanted for monitoring the intensity and extent of both belts all the time, it has to have a perigee inside the Inner Belt—that is, at 400 miles or less—and an apogee above 30,000 miles. This can be accomplished by giving the satellite a velocity of 6.6 miles per second at perigee; it would have an orbital period of 15 hours.

Even for nearly circular orbits the demands can be different, depending on the contemplated use of the satellite. For example, a satellite that is to report on the weather should be close enough to earth so that its pictures show reasonably fine detail, but it also should be far enough out so that it can "see" a sizable area. The relationship between these two factors is shown in Fig. 67 and the accompanying table. As the last column in the table shows, the area of earth that can be seen and photographed increases rapidly with distance but reaches an optimum at 4000 miles; from then on increasing distance adds very little. Actually a distance of between 400 and 600 miles was found best for weather satellites. Only 20 to 25 per cent of the earth's surface can be seen at a glance, but more useful detail is contained in pictures taken at such comparatively low levels. But if the purpose of the satellite is to transmit communications (long-distance telephone calls, pictures, or TV coverage of important events) the most important factor is the area covered at a given moment. Given the necessary power supply, which is now available, communication satellites should be as far out as is practicable. The synchronous orbit at 22,300 miles, thought to be best from the outset, proved to be so.

VISION ANGLE β (DEGREES)	GEOCENTRIC ANGLE ϕ (DEGREES)	HEIGHT H (MILES)	SLANT RANGE S (MILES)	VISION ARC A (MILES)	PER CENT OF EARTH VISIBLE
160	20	57.3	705	700	8
140	40	257	1,445	1,400	17
120	60	620	2,310	2,100	25
100	80	1,210	3,350	2,800	32
80	100	2,222	4,750	3,500	38
60	120	4,000	6,920	4,200	43
40	140	7,650	11,000	4,900	47
20	160	19,000	22,700	5,600	49
17	163	22,300	26,800	5,800	49.5

This brings us to the question of power supply for satellites. The first ones, both American and Russian, had to rely on batteries, which would run down after a time. The Vanguard test satellite pointed the way to the future: it carried the then new silicon solar cells that can convert sunlight into electric current. In the test satellite the current was used directly, but in later and heavier satellites the silicon cells, located on the outside of the satellite or on special panels, charge a silver-cadmium battery and the battery supplies the power to the instruments and to the transmitter. Of course there have been attempts to use atomic energy for this purpose and a great deal of research has gone into these units, which are collectively called SNAP (System of Nuclear Auxiliary Power). The existing SNAP units which produce a reasonable power output are all rather heavy, but atomic power in some form will undoubtedly take over one day.

All artificial satellites can be divided into two categories. The first consists of all those satellites that produce new findings (such as the discovery of the Van Allen Belts) about the environment of space or astronomical phenomena. These are the "research satellites." The other consists of satellites that serve a specific purpose, such as communications, weather watch, or navigational aid. Unfortunately this category lacks a good collective name; the satellites in it are variously called "working satellites," "utility satellites," and "technology-application satellites," none of which terms is entirely satisfactory.

The fact that there are so many types of satellites accounts for the large number of launches. Each type needs experimental models, prototypes, replacements for failures, and so forth. Fortunately many of the early satellites did not go into permanent orbits and experimental satellites that had to work for a few weeks only were purposely put into orbits with a low perigee and therefore a short lifetime. Quite a number were deliberately removed from orbit—"de-orbited" is the established if unbeautiful term.

But, each launch also puts some unwanted matter into orbit. In addition to the top stages of the carrier rockets that have orbital velocity, there are metal rings that held satellite and rocket together, discarded protective covers, and even lens caps and similar items. All this unwanted hardware goes under the drastic name of "space junk"; fortunately most of it reenters and burns up after a while.

The following tabulation shows the number of successful satellite shots and the space junk they carried with them, from Sputnik-I to the last shot in 1966.[6]

YEAR OF LAUNCH	TOTAL NUMBER	STILL ORBITING (Dec. 31, 1966)	SPACE JUNK
1957	2	—	—
1958	6	2	2
1959	11	3	2
1960	19	9	18
1961	37	11	14
1962	54	13	26
1963	49	10	66
1964	90	22	65
1965	116	62	103 (+ ca. 80)
1966	106	35	108
Totals	490	167	404 (+ ca. 80)

Of these satellites, 311 were sent into orbit by the United States, 171 by the Soviet Union, and 8 by other countries (2 British, 2 Canadian, 3 French, and 1 Italian satellite). A few officially unexplained explosions of satellites in orbit (both Russian and American) produced over 200 small pieces in 1961, all of which have re-entered, and 370 pieces in 1965, of which about 80 are still orbiting.

To give a brief rundown of the "firsts" in satellite development, let us start with military satellites, which are politely referred to as "surveillance satellites." Their story is a fine example of why even a carefully thought-out plan must often be abandoned. After the Atlantic Missile Range had been established, the old White Sands Proving Ground was to serve for troop training with smaller missiles and, if necessary, for experimental shots of small missiles with ranges of less than 100 miles. Everything big that needed testing was to go to the Atlantic Missile Range, the later Kennedy Space Flight Center, in Florida. The space agency was to use Wallops

[6] Based mainly on *Satellite Situation Report,* vol. 6 no. 24, by the Goddard Space Flight Center in Greenbelt, Md.

Island, off the coast of Virginia, for minor scientific shots. The Air Force converted the Army's old Camp Cooke in California into Vandenberg Air Force Base, which was to serve for troop training with large missiles, mainly the Thor.

But then directive "WS-117L," also called Project Sentry, came along, and geographical realities ruined the neat scheme. The project involved, among other things, the use of the Thor with the new Agena rocket as a top stage for putting satellites into orbits passing over both poles. This was experimental, of course, and should have gone to Florida. But one could not fire a satellite into polar orbit from the Cape. When a satellite is to go into a polar orbit, it can be fired due north or due south—the direction of the take-off does not matter. What did matter, in this case, is that every shot begins with a first stage that returns to the ground. If one tried for a polar orbit from the Cape, and the missile were directed northward, that first stage would be likely to land somewhere in Georgia. If it were directed southward, the first stage might fall on Cuba, and even if Cuba were a friendly nation that would strain relationships.

The situation was different at Vandenberg. Even from there one could not shoot northward; the first stage would fall somewhere in Oregon. But for a southward shot there was only open ocean between Vandenberg and Antarctica. Hence the experimental shots of Project Sentry had to be made from California. Project Sentry itself very quickly split into three different projects. The first involved the Discoverer satellites, which were to have many different functions. The second was "Midas" (Missile Defense Alarm System) and the third "Samos" (Satellite and Missile Observation System) —a far cry from the normal connotations of the name of this classical Greek isle floating in Homer's "wine-dark sea."

Discoverer was ready first, and the first one took to space on February 28, 1959. No signal was received, but radar proved that it was in orbit, a rather elongated orbit with perigee at 99 miles and apogee at 605 miles. Naturally, with such a low perigee the orbit decayed fast; Discoverer-I re-entered March 5, 1959. The second Discoverer followed on April 13, 1959, and went into a better orbit, with perigee at 142 miles and apogee at 220 miles. Now the aim of the Discoverer shots was to put into orbit a satellite that could be retrieved. A capsule was ejected "backward" so that it would re-enter immediately. However, the Air Force researchers had to be patient. Shot after shot failed to provide the hoped-for retrieval, until Discoverer-XIII was fired on August 10, 1960. Its capsule was de-orbited the following day and fished from the Pacific. Discoverer-XIV, on August 18, 1960, gave an even better result: an airplane equipped with two booms and a rope stretched between them succeeded in catching the capsule in midair.

The two other sections of the original Project Sentry were also successful: Midas-II went into orbit on May 24, 1960, and Samos-II on January

31, 1961. Another military satellite system, bearing the strange name Vela Hotel, was first orbited October 17, 1963. It went into an enormous orbit, with a perigee at 63,442 miles and apogee at 69,057 miles, and with an orbital period of 4 days, 10 hours, and 10 minutes. Later Vela Hotel satellites—they have been made known to be radiation detectors—were put into nearly the same orbit.

Another military satellite caused the printing of about as many words as did Sputnik-I—and it was not even a true satellite. It was an Air Force project named West Ford. When Midas-IV went into orbit on October 21, 1961, the rocket also carried a canister that was filled with a naphthalene-like substance into which 300,000 copper wires, each about ½ inch in length, were imbedded. The reasoning behind this unusual payload was this: in orbit the naphthalene would be exposed to space and it would slowly evaporate. In evaporating it would carry the copper wires (dipoles) along so that they would be spread throughout the whole orbit. This ring of copper wires around the earth was to be used for military communications: it was to be a reflector for radio messages that could not be damaged by enemy action. Astronomers, radioastronomers, and scientists in general protested vehemently; such a ring would interfere with observations, especially with research by means of radio telescopes, and astronomical work would be ruined for years to come.[7] But the Air Force considered reliable military communications more important than astronomy and went ahead with its project.

Then nothing happened. The orbiting canister could be found by radar, but the dipoles had not spread. Whether the canister had not opened or whether the matrix had failed to evaporate is unknown; in any event there was no ring of dipoles and everybody cheered. Against a background of more protests, the Air Force announced that the experiment would be repeated, and it was, with the orbiting of Midas-VI, on October 26, 1962. This time the matrix did evaporate and the "needles" did spread, and then it turned out that everybody had been wrong. The dipoles did not interfere with astronomical work (astronomers could just find them if they tried hard) and the Air Force found that after only a few weeks the transmission via orbiting dipoles became very poor. Project West Ford died, because it had not worked.

If it had worked, it would have been a so-called "passive communications satellite," one of the two possible types of communications satellites. A "passive" satellite is something in orbit that will produce an echo with radio waves. It has the advantages of being simple, cheap, and almost invulnerable, but since the echo is weak, very sensitive ground equipment is needed. An "active" satellite receives the signal, amplifies it, and rebroad-

[7] There was a time limit to the complaint because the Air Force had estimated that the dipoles would gradually be forced-into the upper atmosphere by the pressure of solar radiation and that within six or seven years they would all have re-entered.

casts it immediately. Such a satellite is more complicated, needs a power supply, is more expensive, and could conceivably be damaged, but the equipment on the ground can be less sensitive.

The first passive communications satellite was the 100-foot Echo balloon, orbited on August 12, 1960. Its very thin skin consisted of aluminized plastic (Mylar) and its orbit was very nearly circular, 1018 miles at perigee and 1160 miles at apogee. As expected, Echo, having a large surface and very little mass, was strongly influenced by solar radiation pressure but, again as expected, the effects tended to cancel out. Echo was also expected to be struck by micrometeorites, but if these made holes in the balloon that would not matter, since all a hole could accomplish would be to cause the internal pressure in the balloon to drop to zero. Since the external pressure 1000 miles up is also zero, the balloon would keep its shape. Several years later, Echo was wrinkled and no longer a perfect sphere, but it still fulfilled its function.

The first active communications test satellite had the designation Courier. Courier-I went into orbit on October 4, 1960, with perigee at 500 miles and apogee at 745 miles. It proved that active communication satellites could do what was expected of them, and on July 10, 1962, the first commercial communications satellite went into orbit. This was Telstar-I which caused a public sensation by transmitting films, live TV shows, photographs, and newspaper frontpages.[8] It also silenced a fear that had been voiced in several quarters, namely that the Van Allen Belt might interfere with transmission through it. Since the apogee of Telstar-I was at 1962 miles and its perigee at 593 miles it was in the Inner Belt practically all the time.

Since one could transmit through the Van Allen Belt, there was no longer anything in the way to prevent trying communication satellites in the synchronous orbit. The first satellite to be gradually nudged into this orbit was Syncom-I (from *syn*chronous orbit and *com*munication). Fired on February 14, 1963, it did reach orbit but was a disappointment otherwise. The first "syncom" to work well was Early Bird, orbited April 6, 1965.

The first weather-watch satellite was Tiros-I, orbited on April 1, 1960, with perigee at 429 miles and apogee at 468 miles. Its active life in orbit was only 78 days, but during that time it transmitted 22,852 cloud-cover pictures to the ground, giving meteorologists their first comprehensive view of weather over the whole planet.

The year 1960 also first saw an equivalent of Hale's "Brick Moon" in the sky—the first navigational satellite, named Transit-I B. It was orbited on April 1, 1960, with a perigee of 239 miles and an apogee of 472 miles. Of course it was far superior to the Brick Moon because it did *not* rely on

[8] The Russian communication satellites are named *Molniya* ("lightning"); the first was orbited April 23, 1965. The long series of Russian *Kosmos* satellites comprises many different types of satellites. Some are known to be weather satellites, some are believed to be surveillance satellites, and some were prototypes of manned space capsules.

being visible. As every sailor, professional or amateur, knows, navigation by the stars is a fair-weather proposition. No such navigating can be done if there is a fog bank over the surface or a cloud cover higher up. Fog and clouds block visible light but not radio waves, and because Transit uses radio it is "visible" to the proper instruments at any time. The satellite navigational system was declared to be operational in 1965.

Of course there were also satellites for very special purposes; only one of these shall be mentioned, namely ANNA. The name stood for Army, Navy, NASA, Air Force. ANNA was put into a precisely circular orbit (perigee and apogee both at 660.5 miles) on October 31, 1962. Its purpose was to enable the four services incorporated in its name to check and calibrate their ground equipment.

While writing all this, I kept hearing a small boy's voice in my mind. During the winter of 1959–1960 any lecturer on space activities still had to explain carefully how an orbit comes into being. After one of my lectures a boy, aged probably about eight, came up to me and said with tears in his voice: "I thought you wanted to shoot to the moon!"

The time has now come to heed this voice. We can now proceed to the moon shots and the shots beyond the moon. Let us recapitulate the performance of the first moon shot, Russia's Metchtá of January 2, 1959. It missed the moon by 4600 miles, possibly intentionally, possibly not. If Metchtá had hit the moon its trajectory would have looked somewhat like Fig. 68.

The next moon shot was made by the United States, using one of Wernher von Braun's Juno rockets, which took off on March 3, 1959, carrying Pioneer-IV. It was intended to be a near miss, but Pioneer-IV missed by 37,300 miles, more than seventeen times the moon's diameter. Of course the instruments aboard Pioneer-IV could not register anything at that distance.

The United States was ready for another moon shot in September 1959. The Able rocket that was to carry an instrument package to or past the moon was backed up by an Atlas rocket as its first stage, a far more powerful rocket than the Thor that had been used in October and November 1958. Hopes were soaring high, much higher than the rocket as it turned out, for when the firing switch was thrown on September 10, 1959, the Atlas-Able exploded on the pad. Two days later the Russians sent their Cosmic Rocket no. 2[9] to the moon. This time they did hit; the probe struck 35 hours after take-off, halfway between the crater Autolycus and the mountain chain known as the Lunar Apennines. The Russians stated that they had come within 160 miles of the aiming point, with an impact velocity estimated at 2.1 miles per second. The weight of the space probe

[9] Cosmic Rocket no. 1 was Metchtá. The term "Lunik" is a nonsense word concocted by American newspapermen by analogy with sputnik; there is no such word in Russian.

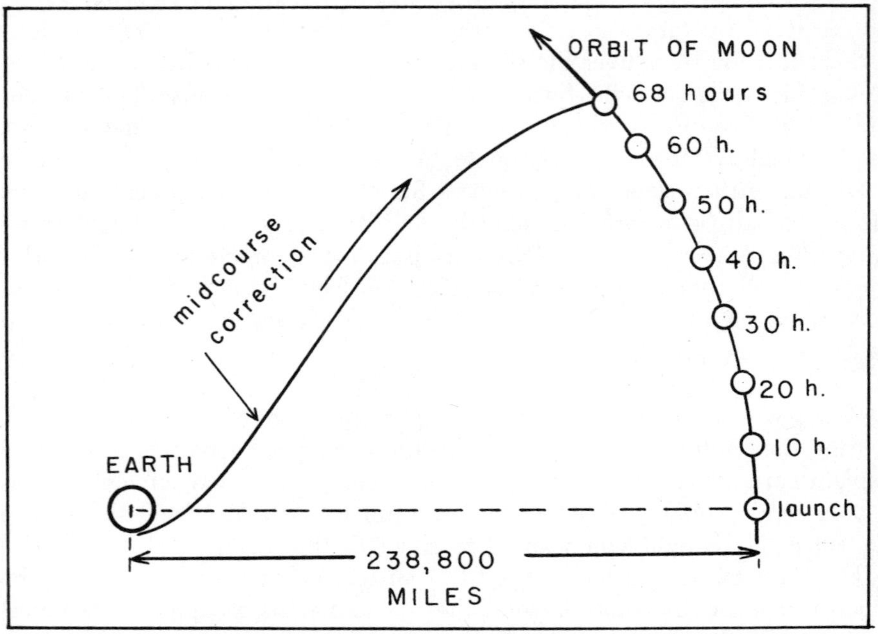

Fig. 68. Trajectory of a shot to the Moon. (The trajectory shown is the one actually flown by Ranger-VII on July 28, 1964.)

was 860 pounds, and a Czech astronomer believes that he saw the dust cloud produced by the impact. The big British radiotelescope at Jodrell Bank tracked the probe, but later on no impact crater could be found. Apparently the impact of something weighing 860 pounds and striking at a velocity of 2.1 miles per second is not enough to produce a crater visible from earth. At a later date three American lunar probes, each weighing around 800 pounds, also failed to produce visible impact craters, even though the places where impact occurred are known with enough precision that astronomers could tell which areas to check.

On October 4, 1959, on the second anniversary of Sputnik-I, Cosmic Rocket no. 3 (Fig. 69) went into space. It was an interesting cross between an artificial satellite and a moon probe. In essence Cosmic Rocket no. 3 described an elliptical orbit around the earth, but the ellipse was so elongated that its apogee was at 291,870 miles—or beyond the orbit of the moon. While the moon probe passed the moon a camera took pictures, the shot having been timed in such a manner that a portion of the far side of the moon that can never be seen from earth could be photographed. Of course the moon's gravitational field strongly distorted the orbit of the probe, but the general statement that it was in an elliptical orbit around the earth still held true. The perigee of the new orbit was 24,840 miles

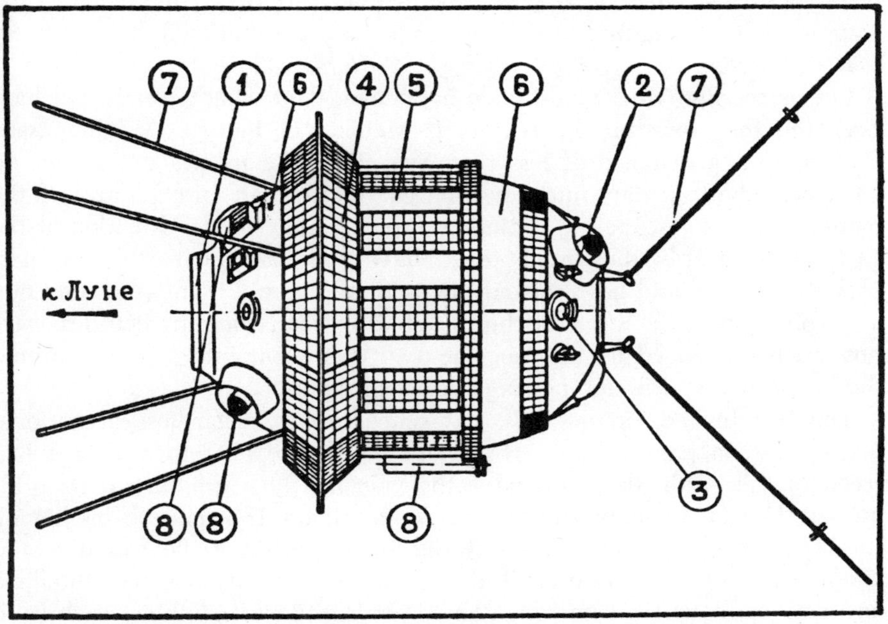

Fig. 69. Russia's "Cosmic Rocket no. 3." Drawing, supplied by Russian news agencies, of satellite 1959 theta, which took the first pictures of the moon's far side. The Russian letters above the arrow mean "to the moon." The numbered features are: (1) camera windows, (2) attitude-control nozzles, (3) solar sensor, (4) silicon cells for electric power, (5) temperature-control panels, (6) heat-insulating panels, (7) antennas, (8) external sensors for instruments inside the spacecraft.

away, but the space probe could be expected to stay in its new orbit for some time to come because from then on it would cross the moon's orbit a long way from the actual position of the moon.

The plan had been to take the pictures during the first passage near the moon and to transmit them to the ground every time the probe passed through perigee. That way a minimum of a dozen transmissions could be expected and if a few pictures were marred by static at one passage they would not be at another passage. But luck was not with the experimenters all the way. The camera stuck after only about a dozen pictures had been obtained, and after the first transmission the transmitter failed. The pictures were of such poor quality, that some people (wrongly) doubted that they were genuine. On May 19, 1960, on its fifteenth orbit, the space probe ran into the earth.

But what happened to Metchtá and to Pioneer-IV after they missed the moon? They did not fall back, as some commentators expected them to at the time, because they had escape velocity and the earth's gravitational field could not force them back. But neither did they continue on their way to infinity, because they were still in a gravitational field, namely that of the sun.

Let us see first what could have happened, then what actually did happen, and then look for the reason. If Metchtá or Pioneer-IV had passed the moon at a distance of less than 100 miles, the moon's gravity would have bent the trajectory into a hairpin curve after half an orbit around the moon, with the "escape leg" pointing approximately in the direction of the earth. Conceivably they might even have run into the earth, but most probably they would have missed the earth and would then have gone into an orbit around the sun. As things actually were, their trajectories were not much influenced by the moon and so they went into an orbit around the sun directly, without passing the earth again.

The fact to keep in mind is that space probes, regardless of national origin, were a part of the earth prior to take-off. Therefore, even when freed of the earth, they still have the velocity the earth has in its orbit around the sun, amounting to 18.5 miles per second. If a space probe having this velocity—so to say, as its birthright—were orbiting the sun it would simply stay in the earth's orbit. But its birthright velocity has been modified and now we have the same situation as with artificial satellites: each orbit has its own particular velocity. If the velocity does not match the orbit, there has to be a change in the orbit. The only difference is that we are now talking about orbits around the sun, not orbits around the earth.

To understand what must—and does—happen, let us consider a purely theoretical case not involving the moon. We have a spacecraft capable of reaching escape velocity and the guidance system has been programmed to keep the rocket on a strictly vertical path. What happens to this spacecraft will then depend on only one factor, the time of the take-off. The "front side" of the earth, the one that points in the direction of its orbital movement, is the "dawn side." For a take-off at dawn the rocket would add its velocity, for a take-off at dusk it would subtract it. We may refer to these two cases as the "dawn rocket" and the "dusk rocket."

What would happen is shown by Fig. 70 which is based on calculations made by Professor Werner Schaub of the German Gesellschaft für Weltraumforschung e. V., the postwar successor of the VfR. The dawn rocket, being "too fast" for the earth's orbit, will drift outward in the solar system, in the direction of the orbit of Mars. But in drifting outward, the rocket climbs "up" in the sun's gravitational field which gradually slows it up. Some 6½ months after take-off the sun would win out. The rocket would be at its aphelion point, between the orbits of earth and Mars, and would then "slide down" again, gaining velocity in the process. Another 6½ months later, or 13 months after take-off, the dawn rocket would reach its perihelion point, which is located where it took off originally, namely in the orbit of earth at the point where the earth was 13 months previously.

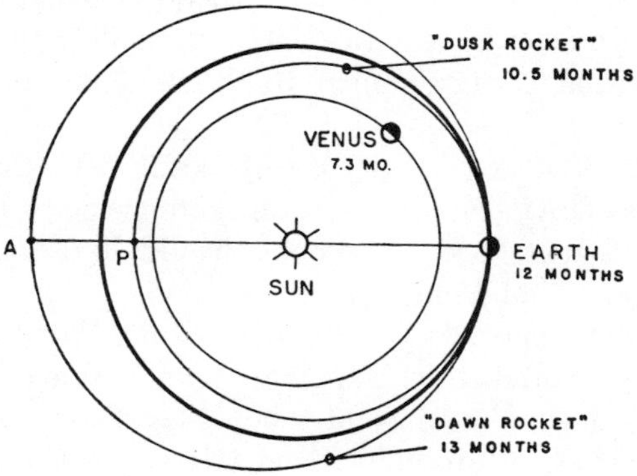

Fig. 70. "Dusk rocket" and "dawn rocket." If two rockets left earth with escape velocity, one at dusk, the other at dawn, they would describe the two orbits shown, and after a few revolutions around the sun crash on earth again. *P* marks the perihelion point of the "dusk rocket," which has its aphelion at the position of the earth. Conversely, the position of the earth marks the perihelion of the "dawn rocket" and *A* marks its aphelion.

The earth, of course, would be the distance of 1 month's orbital movement away from that point.

The dawn rocket would then continue, swing out again along its orbit, reach aphelion again another 6½ months later, turn back, and behave, in general, just like an independent celestial body with a somewhat elongated orbit. That independence, however, would last for only eleven complete revolutions, because after eleven full revolutions the rocket would meet the earth—which had meanwhile completed twelve revolutions—at its perihelion point and would crash.

The dusk rocket, having subtracted its own velocity from that of the earth, would slide inward in the solar system. For the dusk rocket take-off point would be aphelion, with perihelion somewhere between the orbits of earth and Venus. The dusk rocket, too, would be an independent planet for a while, but for an even shorter while than the dawn rocket, for after eight complete revolutions of the rocket the earth would have completed almost seven—to be precise, 6.96—revolutions. Both earth and rocket would have almost the same velocity, would travel in the same direction, and be quite close. The dusk rocket might simply run into the earth, or, if it did not do that, would be thrown into still another orbit by the earth's gravitational field.

Both Metchtá and Pioneer-IV, after passing the moon in relatively undisturbed orbits, were in the position of the dawn rocket, close to the earth's orbit but with a velocity that was too high. Their actual orbits now are quite close to that of Professor Schaub's theoretical dawn rocket. And on March 11, 1960, Pioneer-V was put into an orbit between the orbits of earth and Venus that is very similar to the orbit of the theoretical dusk rocket.

Well, the dawn rocket would *almost* go to Mars' orbit and the dusk rocket *almost* to Venus's. Obviously just a little more velocity added to, or subtracted from, the earth's orbital velocity would have brought the rocket to the orbit of Mars or of Venus.

This seems obvious now after it has been done, but it took a lot of talk to convince people that the best time to send a space probe to another planet, say Mars, is not the moment when Mars happens to be closest to the earth. The theoretical groundwork was laid half a century ago by Dr. Walter Hohmann, who decided during the First World War that he would be unable to find a publisher for his book. He has now been officially honored by having a certain orbit from planet to planet called "a Hohmann." Let's follow his reasoning and see how he arrived at the orbit named after him. The fundamental fact with which Hohmann started is that all the planets of the solar system move around the sun in the same direction, counterclockwise as seen from the celestial north pole. An orbit from planet to planet would have to move in the same direction. If one tried to travel opposite to the general direction of rotation in the solar system one would first have to overcome the earth's gravity, then reduce the earth's orbital velocity to zero, acquire velocity in the opposite direction, and then meet the other planet on a collision course, requiring once more the reduction of the ship's velocity to zero, and then acquire the target planet's orbital velocity, all this leading up to a landing against the planet's gravity. This would require such a fantastic fuel expenditure that Dr. Hohmann ruled out such orbits as "impossible."

Since orbits moving against the general direction of rotation of the solar system were "impossible," orbits following the general direction of rotation were labeled "possible" (Fig. 71.) Now the question was which of the possible orbits should be considered best, and the best orbit obviously was the one with the least fuel expenditure. To gain an idea, Dr. Hohmann drew five orbits which he called *A, B, C, D,* and *E,* using a hypothetical trip to Venus as an example. Orbit *A* *touches* the orbits of both Venus and earth, orbit *B* touches the orbit of Venus but crosses the orbit of earth, while orbit *C* touches the orbit of earth but crosses that of Venus (Fig. 71, right). Orbits *D* and *E* (not shown) were between *C* and *A* and between *A* and *B.* Calculating the fuel requirements for these five orbits, Dr. Hohmann obtained the following figures for a one-way trip (without even landing on Venus):

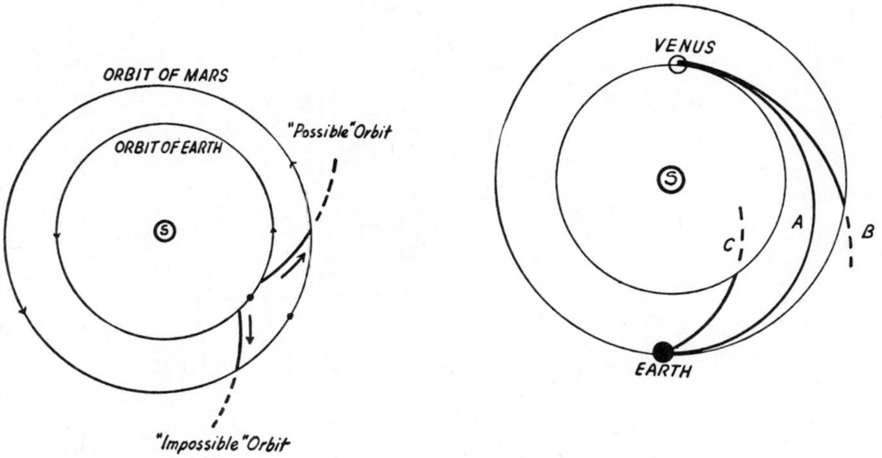

Fig. 71. Hohmann orbits.
Left: A "possible" orbit, which follows the general rotation of the solar system, and an opposing orbit, labeled "impossible" because of the fantastic fuel expenditure it would require.
Right: Three "possible" orbits, of which the one labeled *A* takes the longest time but also involves the smallest fuel expenditure.

ORBIT USED	DURATION OF ONE-WAY TRIP (DAYS)	ORIGINAL MASS OF SHIP (METRIC TONS) FOR THREE DIFFERENT EXHAUST VELOCITIES		
		9800 FT/SEC	13,100 FT/SEC	16,400 FT/SEC
A	146	49	34	27
B	75	530	200	104
C	69	5900	1060	417
D	109	141	70	48
E	102	172	83	55

This table made it perfectly clear that only *A* orbits could be considered, even though they took longest, and it is the *A* orbit that is now called by Hohmann's name. Using Dr. Hohmann's method of calculation[10] the Orbit A to Mars would take 258 days.

But this did not mean that an expedition to Venus would take twice 146 days plus the time spent near or on the planet. The planets of the solar system all move in the same direction, but not with the same velocity (Fig. 72). If a ship were put into an *A* orbit for Mars it would, assuming that everything functions properly, reach the orbit of Mars after 258 days, but whether the planet is at that point of its orbit when the ship gets there is another matter. Obviously the take-off must be timed in such a manner that the ship and the planet arrive at the same point at the same time. The

[10] He assumed the orbits of the planets to be circular, at their mean distances from the sun, and also assumed all orbits to be in the ecliptic.

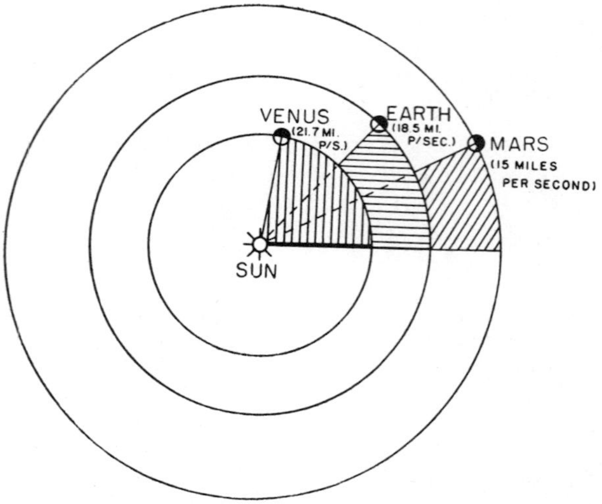

Fig. 72. The movement of Venus, earth, and Mars during a fifty-day period.

period during which a successful flight to Mars can be started is now called the "launch window." From the strictly mathematical point of view a launch window would last only as long as the planet takes to move the distance of its own diameter along its orbit. But minor course corrections are possible, and for that reason a launch window for Mars, with the rockets and upper stages now available, lasts just about one month. Once a launch window has been missed, one would have to wait until the relative positions of earth and Mars are repeated, which is, on the average, after two years and two months.

For the purpose of visualizing the requirements, Dr. Hohmann reasoned as follows: the assumed ship travels through an angle of 180 degrees of arc in 258 days—that is, 0.7 degree of arc per day, as seen from the sun. Earth needs 365 days to move through 360 degrees of arc (a full circle); hence it moves through 0.983 degree of arc per day. Mars, being slower, needs 687 earth days to move through 360 degrees of arc, which means a daily motion of 0.523 degree of arc per day. Now, if we count backward from the moment the ship reaches the planet, the earth, at take-off, must be 180 degrees of arc from the rendezvous point, while Mars must be 258 times 0.523, or 135 degrees of arc from the rendezvous point. This means that at take-off Mars must be 45 degrees of arc "ahead" of the earth. But during the time which the ship needs to go to Mars, the earth will travel 258 times 0.983, or 253.6 degrees of arc, which is 73.6 degrees of arc beyond the rendezvous point that was 180 degrees of arc ahead when the flight began. Hence, at the moment of rendezvous the earth will be 73.6 degrees of arc "ahead" of Mars.

Thus, if the ship returned at once, it would only reach the orbit of earth; earth would be elsewhere. In order to reach the earth, the ship, which is on Mars or orbiting Mars, would have to wait until the earth, having completed more than one full revolution around the sun, is 73.6 degrees of arc "behind" Mars. This waiting period amounts to 455 days. Then the return trip to earth can begin, and it also lasts again 258 days, so that the whole expedition to Mars would require 258 + 455 + 258 = 971 days, or about two years and eight months.

Applied to the planet Venus, the calculation resulted in 146 + 470 + 146 = 762 days, or two years and about a month. Dr. Hohmann concluded with some satisfaction that expeditions to the neighboring planets would involve absences of two years and a few months, "about the same time as a prolonged expedition to the South Pole."

This is how things stood in 1916 when he put his manuscript away. Nobody then had anticipated television, course corrections by radio command, tracking by radar, and all the other electronic activities we now take for granted, and therefore manned expeditions had to be assumed. With unmanned planetary probes we are far better off, since it does not matter if the probe is lost in space after it has performed its duty. Hence, we only have to pay attention to the launch window, and we don't have to worry in case the probe, when it reaches the target planet, does not have precisely the same velocity as the planet, or the probe's orbit crosses the planetary orbit at a slight angle. For this reason the actual transit times of our planetary probes turned out to be somewhat shorter than those calculated by Dr. Hohmann.

Compared to what can, and often does, happen in reality a discussion of theory can be quite relaxing, but the relaxation period is now over and we have to return to firing schedules, firings, and—unfortunately—misfirings.

On November 26, 1959, another lunar probe sat on top of an Atlas-Able assembly, waiting for the countdown to reach its end. The assembly took off beautifully, but during the climb through the dense atmosphere the protecting nose-cone shroud tore off and 45 seconds after take-off, Atlas-Able II became a ball of flames. By then space activities were under the auspices of the National Aeronautics and Space Administration (NASA), which was then only a little over a year old, having been officially established on October 1, 1958. (For some details about the establishment of NASA see Chapter 13.) NASA decided to delay the next attempt to shoot to the moon, and in the meantime quite a number of artificial satellites were launched successfully. On March 11, 1960, Pioneer-V, already mentioned, was put into orbit around the sun; the experiment yielded the first record for long-distance transmission through space, a distance of more than 20 million miles.

But on September 26, 1960, Atlas-Able III failed: the upper stages did not produce enough velocity to go to the moon and the space probe fell back and burned up on re-entry. On December 15, 1960, Atlas-Able IV failed, exploding at an altitude of 40,000 feet.

In February 1961, while Sputnik-VIII was in orbit around the earth, the Russians tried something new. It had been previously suggested in theoretical discussions that it might be wise to place a moon shot or a planetary probe into an orbit around the earth first. When the earth-centered orbit was fully established by careful tracking, the deep-space device could be fired from orbit when it pointed in the right direction. An orbit used for this purpose came to be called a "parking orbit." On February 12, the Russians fired a Venus probe from Sputnik-VIII in parking orbit. From what is known the spacecraft must have passed the planet on May 20, 1961, but the experiment was a failure, for radio contact was lost soon after the probe had left the parking orbit. Though the Russians, at that time, still had the bigger rockets, they were to suffer a streak of transmission failures.

At the Cape the unsuccessful moon shots continued without respite during 1961 and 1962, as can be seen from the listing in Appendix 3. But while the moon shots continued to go wrong in all possible ways, preparations for a shot to Venus were made; there was a launch window for Venus lasting from the latter part of July to the end of August 1962. An Atlas-D/Agena-B rocket, carrying planetary probe Mariner-I stood ready on the first day of the launch window, July 20. Various difficulties delayed the take-off until the morning hours of July 22, and lift-off took place at 4:21:23 A.M., Eastern Standard Time. But 3½ minutes later the Range Safety Officer detected a veering to the northeast, indicating that the Atlas would fall into the North Atlantic shipping lane, if he permitted it to go on. Besides, with such an unforeseen factor in the take-off path it was doubtful that the proper orbit could be attained in space. The RSO pulled the destruct switch.

But another Mariner spacecraft was ready and the duration of the launch window was long enough for all the necessary preparations to be made. The same Atlas-Agena combination with Mariner-II lifted off at 1:53:13.9 A.M. on August 27, and this time everything worked well. The spacecraft first went into a parking orbit that led it nearly halfway around the earth at a distance of 116.2 miles. Then the Agena was restarted and Mariner-II began its 129-day trip to Venus.

Many people have likened the Mariner-II spacecraft to a dragonfly in appearance, and though one might object that a dragonfly has four wings and Mariner-II only had two solar-cell panels, the comparison is the best one available. The weight of Mariner-II was 447 pounds; its structure was a hexagonal frame made of magnesium with a 77-pound girder "tower," carrying an antenna on top. The two panels of solar cells, weighing 48 pounds, were 60 inches long and 30 inches wide, their total area was 27 square feet, and they held 9800 solar cells, which fed their current into a sealed 33-pound silver-zinc battery. Scientific experiments (this term refers to automatic equipment), electronic circuitry, and a propulsion system for course correction accounted for the remainder of the weight.

While Mariner-II was drifting toward the orbit of Venus there were a number of tense moments. At one point one of the solar-cell panels stopped working, but then it recovered, both stoppage and recovery remaining unexplained. There was concern lest Mariner-II be overheated by the sun, which would have rendered the instruments inoperative. But this did not happen. On December 14, 1962, Mariner-II, with all instruments working, passed the planet Venus at a closest distance of 21,598 miles, reporting that the surface temperature of the planet was an incredible 800 degrees Fahrenheit. At the same time it reported that Venus had no magnetic field that could be detected at this distance and that no evidence of rotation of the planet (now known to be very slow) could be detected. One other interesting result was that the total amount of radiation received by Mariner-II during its voyage was so low that an astronaut, if there had been one aboard, could have survived many times the radiation dosage actually encountered. Mariner-II also broke Pioneer-V's record for long-distance transmission of data, bringing this up to 53.9 million miles.

While Mariner-II was still on its way to Venus, the Russians used the opportunity of a November launch window for Mars to launch their probe Mars-I. It left its parking orbit on November 1, 1962, and was due to pass near the planet in June 1963. It probably did, but there is no information. On March 31, 1963, radio contact was suddenly lost; Russian scientists believe that their probe was struck by a meteorite larger than grain size and started tumbling so that the radio beam no longer was directed.

In April 1964 there was again a launch window for Venus and the Soviet Union had ready a new type of space probe called Zond ("sonde"). Zond-I left on April 2, 1964, but no data were received at all; something must have gone wrong with the electronic system immediately. The Russians asked Great Britain to use the radio telescope at Jodrell Bank to track their probe, but even Jodrell Bank could not find it.

November 1964 brought another launch window for Mars. The United States had Mariner-III and Mariner-IV ready for the event, and the Soviet Union had Zond-II poised. Heavier than Mariner-II, Mariner-III weighed 575 pounds, having four solar panels instead of two. Because Mars is at a greater distance from the sun than Venus the sunlight would be weaker and the extra panels would be needed to generate enough electric power. Mariner-III was put into the proper orbit on November 5, 1964, after a fine take-off from the ground and well-executed maneuvers in parking orbit. And then something happened that nobody had foreseen: the protective shroud for the spacecraft could not be jettisoned. Mariner-III remained inside its covering and the solar panels could not unfold—the package remained packaged and completely useless.

Fortunately there was Mariner-IV as a "back-up." It went into space on November 28, 1964, during the same launch window.

Lift-off was at 9:22:01 A.M., Eastern Standard Time, and all the critical operations took place during the first hour. At 9:27:20 the Agena, carrying Mariner-IV, separated from the spent Atlas. At 9:27:23 ejection of the shroud was ordered and it was ejected. From 9:28:00 to 9:30:00 the Agena burned to add velocity and was then cut off. At 10:02:00 the Agena was ignited again and cut off at 10:04:27, Mariner-IV was then 122.8 miles from earth's sea level, 127.7 million miles from Mars, and going in the right direction, but it was going 7 miles per hour too fast, something that could be corrected later. At 10:05:51 the Agena with Mariner-IV entered the earth's shadow; at 10:07:10 they were separated. At 10:15:05 the solar panels were unfolded and the "sun sensor" was ordered to be ready. At 10:17:35 Mariner-IV left the earth's shadow, and by 10:31:00 the sun sensor had "locked" on the sun and the four panels were wide open to the sun. Mariner-IV was on its way.

On November 30. 1964, Russia's Zond-II followed Mariner-IV. Both American and Russian scientists in their ground stations had problems with their crafts that were drifting through space. Mariner-IV, in order to work properly, had to have a certain "attitude" along its orbit. The sun sensor saw to it that the four solar panels were wide open to the sun for the sake of power. But there was another sensor which was supposed to turn the spacecraft in such a way that it pointed at the fixed star Canopus. Only if it did would the camera point at Mars during the fly-by. The Canopus sensor behaved unexpectedly; it locked itself on stars far less bright than Canopus and showed a preference for the star Regulus, which is a nice bright star but not the one wanted. After repeated attempts the "lock" on Canopus was finally accomplished. The Russian scientists were satisfied with the performance and attitude of Zond-II, but they soon noticed that the power supply of their spacecraft was only half of what it should have been. Since they had allowed for some weakening this was not yet necessarily fatal, but several weeks later the power supply of Zond-II dropped to zero; the spacecraft was useless.

Mariner-IV passed Mars on July 14, 1965. Just before disappearing behind the planet, as seen from earth, the camera took twenty-two pictures; while Mariner-IV was "behind" Mars it sent out a radio signal that would go through the Martian atmosphere and furnish clues to its density. After this "occultation" by the planet, Mariner-IV started transmitting the pictures obtained. Because of the long distance (close to 135 million miles) the

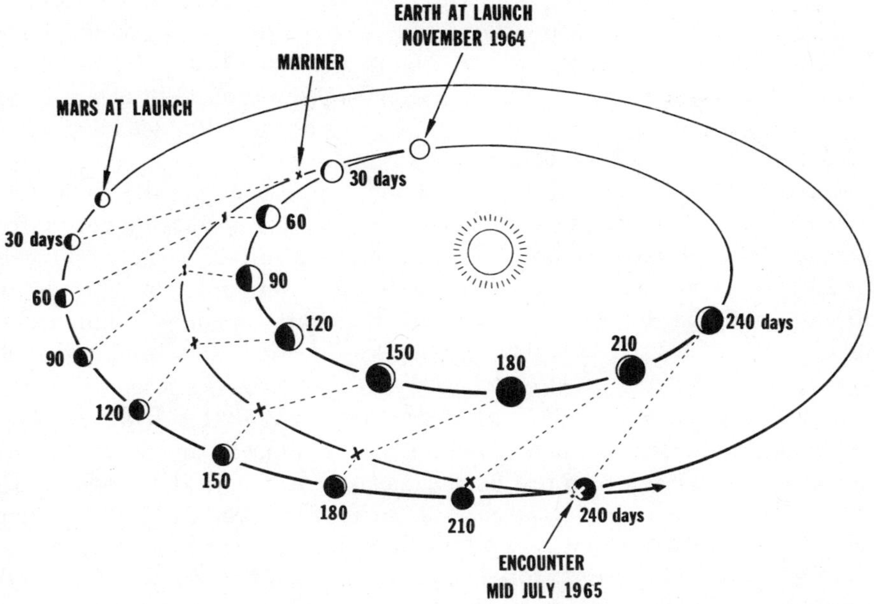

Fig. 73. Comparison of the Hohmann A orbit to Mars with the orbit actually flown by Mariner-IV.

pictures had to be transmitted point by point, so that the transmission of each picture took about 6 hours (Fig. 73).

The main results were: Mars also does not have a discernible magnetic field and therefore, like Venus, lacks Van Allen Belts. The density of the atmosphere at ground level is only 1 per cent of the density of our atmosphere at ground level. Of the 22 pictures taken the last 3 are useless, since the camera tried to photograph the night side of Mars. About ten of them, taken over a slant range from 8900 miles to 7400 miles (the closest, nos. 17 and 18) and each covering an area measuring about 170 by 150 miles, showed that there are many impact craters on Mars—sections of it look like pictures of sections of our moon. The pictures from no. 13 to the end of the sequence show white patches on crater rims, hoarfrost or snow.[11]

Meanwhile the United States had finally achieved success in the area of moon shots, after one especially disappointing failure. The failure was that of Ranger-VI, on January 30, 1964. It had made a perfect flight and made impact on the moon where it was supposed to, but its television cameras had unaccountably failed to work.

[11] Of course these pictures do not settle the question of life on Mars; over 20,000 Tiros pictures of earth, taken from a distance of only 450 miles, "prove" that the earth is a lifeless planet.

Ranger-VII, fired on July 28, 1964, reached the moon after 68½ hours of flight and took and transmitted 4316 pictures during the last 15 minutes of its flight. The last picture before impact was taken from a height of only 1000 feet above the lunar surface. Ranger-VIII, fired on February 17, 1965, was another success, transmitting 7137 pictures, and Ranger-IX, fired on March 21, 1965, sent back 5814 pictures.

On July 18, 1965, the Russians also achieved a success with their Zond-III. The spacecraft was fired to miss the moon and to take pictures while near it. Three million square miles of the lunar surface, mostly of the far side, were photographed by Zond-III. It then went into an orbit around the sun and the Russians waited until the craft was more than 100,000 miles from the moon before they ordered transmission of the pictures, which were good.

After three American Rangers and Russia's Zond-III had all photographed the moon's surface from satellite distances, the next step was obviously to obtain pictures of the lunar surface at the surface. The Russians had been working on such a "soft-landed" camera for some time, but they too had a series of failures. Finally, on February 3, 1966, their Luna-IX (fired January 31, 1966) landed gently on the moon. Four months later, America's Surveyor-I, fired on May 30, did the same. The results will be discussed in more detail in the next chapter; at this point we need mention only that the American and Russian engineering teams used two different approaches.

The problem was that, in order to land a spacecraft gently on the airless moon, its fall had to be braked by retro-rockets, but the exhaust blast of retro-rockets would disturb the lunar surface at the point of landing, and the pictures should, of course, show an undisturbed lunar surface. The Russians solved this problem by a last-minute separation. Their Luna-IX was a 200-pound spherical capsule that opened up like the petals of a flower after striking the ground. The capsule was carried in a structure that held the retro-rockets and the guidance equipment: at about 100 feet above the surface the capsule was ejected sideways from the landing structure so that it fell into an undisturbed area. Surveyor-I, which looked somewhat like a Mariner-II but with four landing legs, stayed in one piece during landing, but the retro-rockets were cut before their exhaust flame touched the ground. Surveyor-I then fell the remaining distance, which under the light lunar gravity was the equivalent of a drop of 15 feet on earth. The impact was absorbed in the shock absorbers of the landing legs. Surveyor-I had separated from its carrier rocket which weighed 2194 pounds, and after its retro-rockets had used up their fuel it weighed only 620 pounds. Over 10,000 pictures were taken and transmitted by Surveyor-I before night fell where it had landed. Against expectations, it "survived" the two-week lunar night and transmitted a few score more pictures afterward.

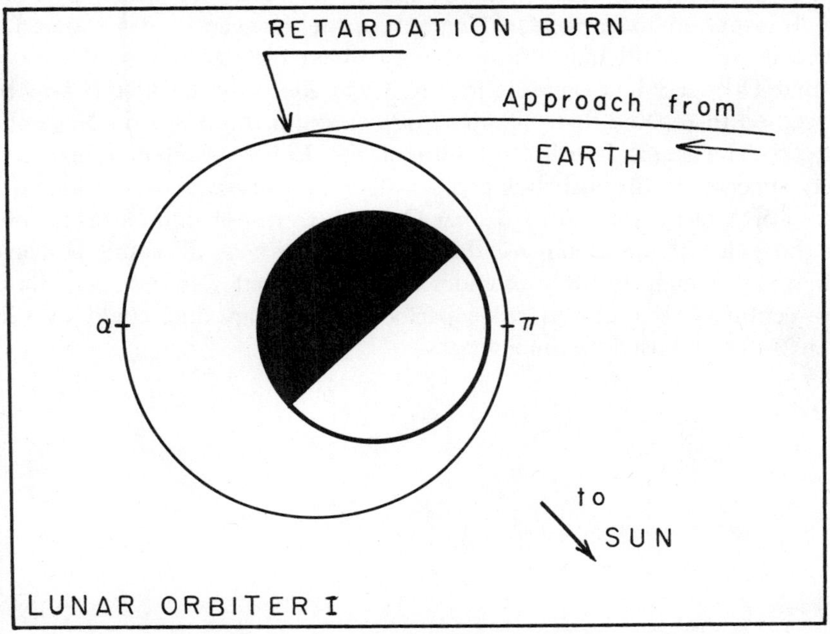

Fig. 74. Diagram of the performance of a Lunar Orbiter.

Both the United States and the Soviet Union made preparations for over a year to put a spacecraft into orbit around the moon. The maneuver needed to accomplish this is to shoot for a near miss and then, when the spacecraft is near the moon, to reduce its velocity by means of retro-rockets so that the moon's gravity can force the spacecraft into a closed orbit. The velocity must be reduced to a little less than 1 mile per second, and if the retro-rocket did not fire, the spacecraft would pass the moon with roughly twice that velocity. The plan for the American Lunar Orbiter I (which was carried out) was first to put the spacecraft into a temporary orbit around the moon and then to fire retro-rockets once more to produce an orbit that would approach the surface to within 28 miles for picture-taking purposes (Fig. 74).

Once more the Russians did it first; their Luna-X took up an orbit around the moon on April 2, 1966, but was not equipped to take pictures. Its successor, Luna-XI, achieved an orbit around the moon on August 27, 1966.

Meanwhile, America's Lunar Orbiter I, fired on August 10, 1966, had achieved orbit around the moon on August 14. Even though one of its two cameras failed to work properly, the results were good. Every time the spacecraft was close to the lunar surface, pictures were taken that looked very much as if they had been taken by a high-flying aircraft. While Lunar Orbiter I was behind the moon, as seen from earth, the pictures were developed automatically and when the spacecraft became visible again they were transmitted to earth. On August 29, the spacecraft was ordered to reduce its speed still more, until it crashed on the far side of the moon.

Lunar Orbiter I had done its job. At Cape Kennedy, Orbiter II was being readied to replace it. It achieved orbit around the moon on November 10, 1966. The third, fourth, and fifth of the Lunar Orbiters followed in steady succession, the fifth being the last of the series. (See the listing at the end of Appendix 3.) Some Lunar Orbiters were put into an orbit going over the poles of the moon. All did well, and the over-all result is a map of the moon which not only includes its far side that was unknown for so many centuries, but also is far superior to any map that could ever be drawn by earth-based astronomers.

13 Man in Space

Illi robur et œs triplex
Circa pectus erat, qui fragilem truci
Commissit pelago ratem primus.

Oak and brass of triple fold
Encompassed sure the heart, which first made bold
To the raging sea to trust a fragile bark.
—Quintus Horatius Flaccus, *Odes*, I

Johannes Kepler, in an enthusiastic moment after reading Galileo Galilei's *Sidereus nuncius,* wrote: "When ships to sail the void between the stars have been invented there will also be men who come forward to sail those ships."

Three hundred and fifty years later, in 1960, about thirty artificial satellites had gone into orbits around the earth, two American devices and one Russian device were in orbit around the sun, and one Russian rocket had struck the moon. This was enough evidence for most people that the time for "the ships that sail the void" had come close. It was a logical conclusion that the men who would sail such ships were walking among us, though still unidentified.

Strange to say, there were still scoffers. In 1959 the newly appointed Astronomer Royal of England, Australian-born Richard van de Riet Wooley, told the British press that space travel was "utter bilge." Ever since, the British Interplanetary Society has had a fine time giving him reports such as "An American named Carpenter has penetrated utter bilge for the fourth time." In the United States Dr. Vannevar Bush, then seventy years old, told *The New York Times* (April 7, 1960) that the idea of putting a man into orbit was only "a stunt" which he opposed. Unperturbed by this opposition a Russian, Yuri Alekseyevitch Gagarin, orbited the earth one year and five days later, on April 12, 1961. Since the landing place of the spacecraft Vostok[1] was about 200 miles to the west of the take-off site, Gagarin's

[1] The name looks like "ВОСТОК" when printed in Russian letters. It means "east" but the reason for the name was that the first Russian sailing ship that circumnavigated the earth bore that name. It is pronounced *vo-STOCK,* with the accent on the second syllable.

flight was not technically a full orbit, but he had been higher than any other man, the apogee of his orbit having been 203 miles above sea level.

By the time Gagarin made his flight, the United States already had a manned space-flight project, named Mercury after the swift winged messenger of the gods.

No specific date can be given as the starting point for Project Mercury, for it was the result of a gradual development. During the early part of 1958 it had become clear that *somebody* should carry out a United States space program. The Air Force had conducted a study that some people called *MISS* (from Man in Space Soonest), but that never was its official name. The Air Force volunteered to carry out this program. But the Army had something to say about it too; after all the first American satellite had been orbited by an Army rocket. Numerous observers, deeply interested but not directly engaged in these discussion, felt that something new should be created for the space program, either a new service, the U. S. Space Force, or else a civilian agency along the lines of the Atomic Energy Commission. President Eisenhower, having to make the final decision, ruled against the Air Force on August 18, 1958 and decided in favor of a civilian agency. After all, we did have a civilian agency for aeronautical research, NACA. Eisenhower decided to have NACA reorganized with a very considerable broadening of its scope. The results was NASA, the National Air and Space Administration.

I fail to understand why this decision was so heavily criticized at the time. The interesting point is that some of the old NACA men themselves were not happy—they would have preferred to keep their organization as it had always been. As NACA they had done the scientific groundwork for aviation developments, and the Air Force, or the Navy, had then done the development work based upon NACA's findings. As NASA they were supposed to do the research *and* the development. But NASA was established and on September 30, 1958, NACA, after 43 years of faithful and successful service, ceased to exist. T. Keith Glennan became NASA's administrator, while Dr. Hugh L. Dryden, the long-time Director of NACA, became the deputy administrator.

As regards Project Mercury, Administrator Glennan cannot recall just when it started officially. He recalls that there was a week-long series of meetings late in September and it was this set of meetings that formulated the project in outline. Later on—presumably while trying to answer questions from the press—nobody seemed to be able to recall when, or even whether, the President had charged NASA with Project Mercury. Apparently NASA simply assumed that Project Mercury was included in its duties. In any event, by the end of September the meetings had resulted in a short report of only 2½ pages of typescript, called "Objectives and

Basic Plan for the Manned Satellite Project." The first two paragraphs[2] read as follows:

I. OBJECTIVES
The objectives of the project are to achieve at the earliest practicable date orbital flight and successful recovery of a manned satellite, and to investigate the capabilities of man in this environment.
II. MISSION
To accomplish these objectives, the most reliable available boost system will be used. A nearly circular orbit will be established at an altitude sufficiently high to permit a 24-hour satellite lifetime; however, the number of orbital cycles is arbitrary. Descent from orbit will be initiated by the application of retro-thrust. Parachutes will be deployed after the vehicle has been slowed down by aerodynamic drag, and recovery on land or water will be possible.

The third paragraph was entitled "Configuration" and began with the words: "The vehicle will be a ballistic capsule with high aerodynamic drag. . . ."

As can be seen, the statements were general, but they were not vague, which indicated that a lot of thinking had been done even before the conference members sat down to spell out the outlines of Project Mercury. Of course there had been a great deal of preliminary thinking. There had been studies applying the results of other experiments to manned flight; there had also been attempts of visualizing a manned space flight in detail, for the purpose of finding what problems existed and should be investigated.

The earliest concern, from the prehistory of space travel, had been about meteorites. Since the early part of the nineteenth century curators of natural history museums had tried to accumulate collections of meteorites, because the public found meteorites even more fascinating than dinosaur bones. Fossil bones reflected past ages of the earth and were indeed impressive. But meteorites had not originated on earth; they had come from somewhere in the universe and (with the few exceptions of those that had been seen to fall) at an unknown time. By, say, 1925, a visitor to such a collection could see a tray of pebbles, stated to be "stones from the Pultusk fall." Or he could see a jagged piece of iron labeled, "Fragment of a prehistoric (?) meteorite fall, found in Sicily, weight 48 pounds." He would also see large irregularly shaped iron masses, weighing 3½ tons. And he might remember having read that we are protected from these cosmic missiles by our atmosphere. A ship in space would lack this protection. Every meteorite seemed to weigh at least 5 pounds. What will happen to a ship that is struck by a 5-pound meteorite with a collision velocity of 20 miles per second or more? Can anything be done about this?

[2] Quoted from *This New Ocean—A History of Project Mercury* by Lloyd S. Swenson Jr., James M. Grimwood, and Charles C. Alexander (NASA SP-4201, Washington, D.C., 1966). A fully documented official history of 681 printed pages.

Of course nothing can be done about preventing such a collision. The result cannot even be predicted, because everything would depend on what section of the ship is struck and on the construction of the ship at the point of impact. Thinking about a collision with a meteorite weighing several pounds is an exercise in futility.

It is futile because the question is poorly stated; the question is not what would happen but whether it is likely to happen. And here the answer tends to be negative. The first fact that must be remembered is that the collections in our museums show the most easily recognizable—roughly the same as "biggest"—meteorites found in over a century of collecting and that many of them must have fallen centuries before it was known that meteorites existed. This in itself shows that large meteorites are comparatively rare. Then it must be realized that our atmosphere does *not* protect us from large meteorites. Anything that weighs several pounds when entering our atmosphere will reach the ground, though with greatly reduced velocity. And from the last few centuries we do have reports about buildings as well as ships at sea that have been damaged by meteorites, even people have been hit and injured (and sometimes killed) in that manner. However, the number of such reports is small; if large meteorites were common there would be many more reports.

Every once in a while meteorites do produce a notable display in the night sky. The most famous example is the fall of the so-called Leonids in 1866. In that year our earth passed through an especially dense sector of that swarm of tiny bodies. But from the available descriptions it could be calculated that even in that dense portion of the Leonid swarm the distance between two bodies exceeded 110 kilometers (approximately 70 miles).

In order to get a more recent estimate, one of the members of the VfR questioned Professor Graff of the Observatory near Hamburg in 1928. Graff replied, "Even in very dense swarms there is hardly one particle weighing less than a gram [1/28 of an ounce] in a space of 100 cubic kilometers, a space immense beyond imagination. As far as large meteorites are concerned, the danger of collision becomes zero."

Considering that these were meteorite *swarms,* that did not sound very frightening. But it was not possible at that time to arrive at even reasonably reliable estimates; there was a considerable difference in expert opinion even about the number of meteorites that strike the earth daily. "The number may be as high as 100,000 per day," said some, while others claimed that one or two million would be closer to the truth. Both factions agreed that the vast majority of all these meteorites burned up in the atmosphere and that only about three a day reached the ground.

The first good figures for the whole planet earth were given in 1941 by Professor Fletcher G. Watson in his book *Between the Planets* (a book on meteorites and comets, not about space travel, as one might think from the title). And, based on Watson's figures, G. Grimminger did an excellent piece of theoretical research in 1946 for the Douglas Aircraft Company.[3]

As for the whole earth, it is pelted daily by 28,000 meteorites with a diameter of ½ inch, and a few larger ones. If you go down to a meteorite size corresponding to a diameter of 1.1 millimeter (which is about the thickness of a dime) the total number has grown to 75 million. A meteorite of such a size has what astronomers call a visual magnitude of 5; it can just be seen in the dark of the night with the unaided eye. The smaller the meteorites, the more numerous they are, of course, and for a visual magnitude of 30 (far below anything visible even with a telescope) the average size turns out to be 0.0005 millimeter in diameter. The total number of all meteorites down to that size hitting the earth daily must be $75 \cdot 10^{16}$. (See Appendix 3.)

The grains of very fine sand correspond in size to a visual magnitude of 15, and the finest particles of the best clay only to visual magnitude 25. The reason for choosing visual magnitude 30 as the limit is simple: anything smaller than that would be pushed out of the solar system by the pressure of the sun's rays. Of course there is a fantastic difference in target size involved. The earth (with atmosphere) is a ball over 8000 miles in diameter with a very considerable gravitational field. Even a 100-ton satellite is invisible by comparison, and its theoretically existing gravitational field would not make the faintest difference.

In order to have a figure on which to base his calculations Grimminger assumed an exposed area of 1000 square feet which would be hit by a meteorite of visual magnitude 20 every 3½ hours. Considering all meteorites of that magnitude and larger, a hit could be expected every 2 hours. But visual magnitude 20 means a diameter of 0.01 millimeter which is the particle size of fine silt on the bottom of a lake. Visual magnitude 20, it goes without saying, is nothing to worry about. Visual magnitude 10 means a diameter of ¼ millimeter. One of those would hit the rocket every 33,800 hours; if you consider everything of that magnitude and larger you could count on a hit every 20,400 hours, or every 2 years and 4 months. For visual magnitude 8 (roughly ½ millimeter in diameter) the time between hits grows to 15 years. And for visual magnitude zero (diameter ⅕ inch) it is 23,800 years.

Grimminger's figures were doubted as being too low, but now they look somewhat too high. Artificial satellites, beginning with Explorer-I, were equipped with meteorite-detecting devices. As expected, the satellites were struck by cosmic dust grains, and these hits were more frequent in the

[3] Published later under the title "Probability That a Meteorite Will Hit or Penetrate a Body Situated in the Vicinity of the Earth," in *Journal of Applied Physics*, vol. 19, no. 10, Oct. 1948. See also *JBIS*, no. XXIX, July 1949 article by A. C. Clarke, and *Science*, no. 2696, Aug. 30, 1946, article by F. G. Watson.

immediate vicinity of the earth; in open space beyond the orbit of the moon they were rare indeed. The tentative conclusion drawn from the impact statistics is that the earth, in the course of time, has acquired a very faint dust halo. There is, however, an alternate suggestion—namely that the instruments are influenced by the earth's magnetic field and react as if a hit had taken place when, in fact, none had.[4]

In any event no satellite has ever sustained a hit that caused it to malfunction, so that the meteorite worry is a thing of the past. This does not mean that the possibility of a hit does not exist, but merely that this possibility is remote.

The next set of problems all dealt with the reaction of the body of the space pilot—the term "astronaut" was still in the future. It was easy to see that each space flight had three distinctive phases. During the take-off the pilot would be subjected to the acceleration of the rocket, which would cause him to feel as if he weighed several times his normal weight. Normal weight, the weight we feel just sitting still, is the result of the earth's gravity and is used as a unit and called *g*. Logically, all the accelerations that occur are measured in terms of *g* and are referred to as *g*-forces.

The period of acceleration with high *g*-forces would be followed by a second period of apparent weightlessness, better called zero-*g*. Then, during re-entry into the earth's atmosphere, there would be another period of high *g*-forces, this time caused by the deceleration of the spacecraft. While the duration of the deceleration might be different from the duration of acceleration during take-off, and while the *g*-forces might differ in intensity, the first and the third periods could both be considered as periods of increased *g*-forces. The increased *g*-forces occur because the rocket accelerates against the pull of the earth's gravity.

If a rocket takes off in such a manner that its acceleration going up is equal to the acceleration of a falling body, the pilot would be subjected to 2 *g*. One *g* would be due to gravity and the second to the rocket's acceleration. In such a rocket the final acceleration might be as high as 8 *g* if the rocket is still climbing vertically near the end of its burning period. But the peak load of the *g*-forces could be adjusted by adjusting the rate of fuel flow; if 5 *g* was the highest a human body could stand, the rocket designer could take this requirement into account. The first problem, obviously, was to find out the limits of human endurance. This was done in precisely the manner suggested by Oberth in 1923—namely, by building a

[4] Whether such a dust halo exists will be decided by the three Pegasus satellites (described in Chapter 14), which carry panels that can later be detached by astronauts and brought back to the ground for microscopic examination. A comparison between the hits visible on those panels and the record of telemetered hits will show whether there is such an instrumental error in the count.

XXXII. Scout rocket carrying Italian San Marco satellite into orbit from Wallops Island. Firing date December 15, 1964, 3:20 P.M. EST. (*NASA Photo*)

XXXIII. The first San Marco Satellite, Italian-designed and Italian-built, launched by an all-Italian crew with an American Scout rocket. (*NASA Photo*)

XXXIV. Balloon satellite of the Echo type, 135 feet in diameter, undergoing tests in hangar at Weeksville, N. C. *(NASA Photo)*

XXXV. Weather satellite Tiros-VI before it was placed in its protective shroud. Tiros-VI was orbited September 18, 1962. *(NASA Photo)*

XXXVI. Atlas-Agena rocket at take-off from Cape Kennedy. This is the launch that sent the successful Mariner-IV probe to Mars. (*NASA Photo*)

XXXVII. Planetary probe Mariner-IV undergoing checkout at JPL in Pasadena, California. (*NASA Photo*)

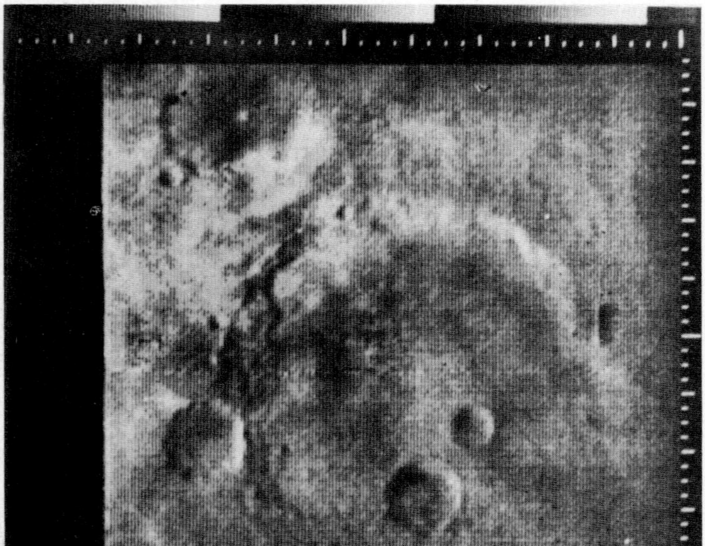

XXXVIII. The best picture (no. 11) returned by Mariner-IV, showing overlapping impact craters on Mars. The slant range from the probe to the craters was 7800 miles. (*NASA Photo*)

XXXIX. The X-15 rocket-propelled research aircraft on the ground at Edwards Air Force Base in California. (*NASA Photo*)

XL. Cross section of the X-15 research aircraft. (*U.S. Air Force Drawing*)

XLI. The first seven American astronauts. Front row, left to right: Walter H. Schirra, Jr., Donald K. Slayton, John H. Glenn, Jr., M. Scott Carpenter; back row, left to right: Alan B. Shepard, Jr., Virgil I. Grissom, L. Gordon Cooper. (*NASA Photo*)

XLII. Redstone rocket carrying Mercury capsule and escape tower, on display at Marshall Space Flight Center, Huntsville, Alabama. (*NASA Photo*)

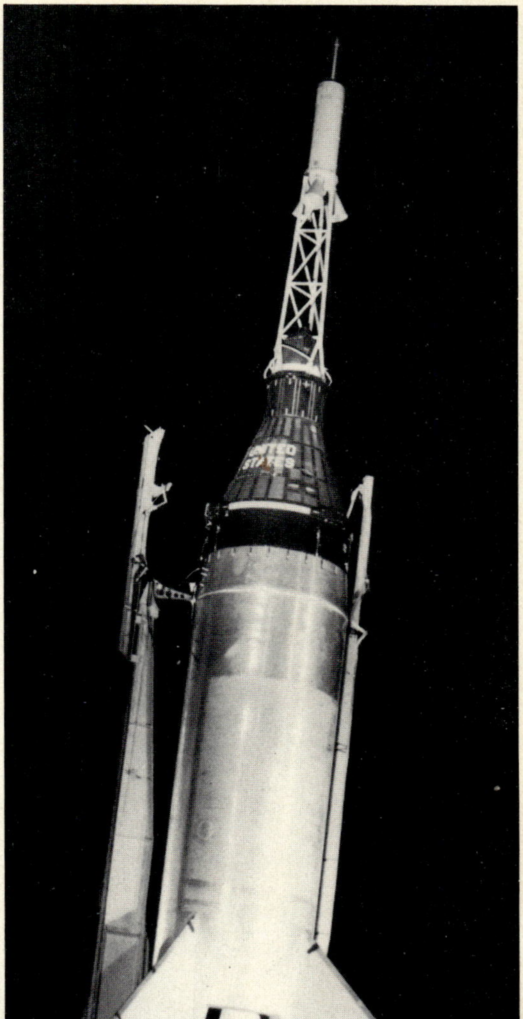

XLIII. Little Joe VI rocket, carrying Mercury capsule and escape tower, being readied for flight. (*NASA Photo*)

XLIV. Take-off of GT-IX on June 3, 1966, from Cape Kennedy. GT-IX was to make rendezvous with the Augmented Target Docking Adapter (ATDA) which had meanwhile turned into an "angry alligator" (see Plate XLVI). (*NASA Photo*)

XLV. Spacecraft Gemini-VII in orbit, photographed from Gemini-VI. (*NASA Photo*)

XLVI. The "angry alligator" as it looked to astronauts Stafford and Cernan in Gemini-IX. (*NASA Photo*)

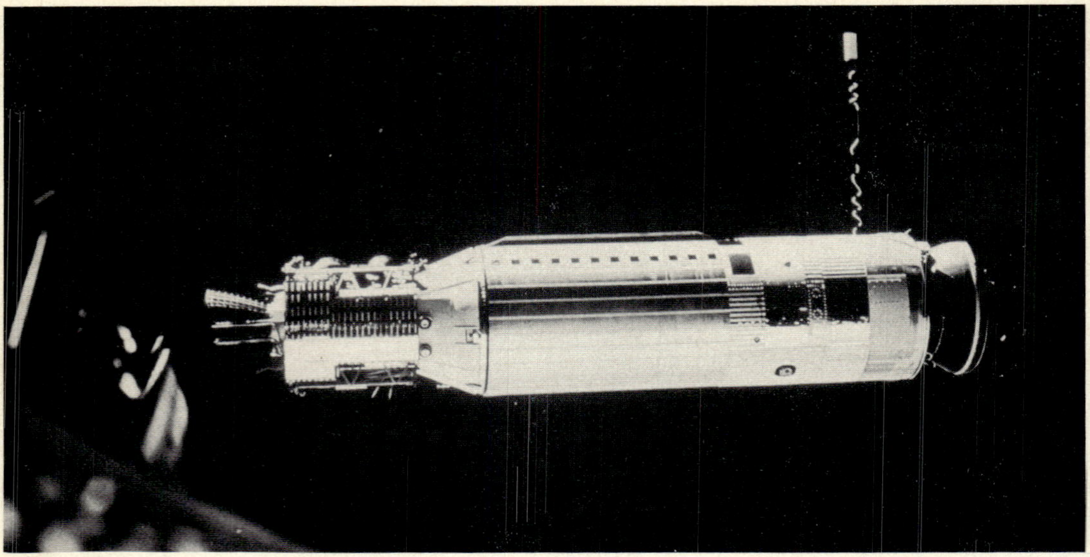

XLVII. Agena target vehicle in orbit, as photographed
from Gemini-XI. (NASA Photo)

XLVIII. Pegasus meteoroid detection satellite: (1) the Service
Module after separation from the Saturn rocket;
(2) Pegasus satellite emerging from Service Module;
(3) detection panels beginning to unfold; (4) detection panels
fully extended. (*NASA Drawing*)

XLIX. Splashdown of Gemini-XII, the last flight in the Gemini Program. Splashdown took place in the Atlantic Ocean November 15, 1966, at 2:21 P.M. EST, only 3 miles from the rescue vessel, the USS *Wasp*. (*NASA Photo*)

L. Booster stage of Saturn undergoing final checkout in its hangar prior to shipment to Cape Kennedy. (*NASA Photo*)

LI. S-IV stage of a Saturn rocket departing from Sacramento, California, for its flight to Cape Kennedy in a special cargo plane. (*NASA Photo*)

LII. The first Saturn take-off. This take-off, which inaugurated a long series of successful flights, took place on October 27, 1961. (*NASA Photo*)

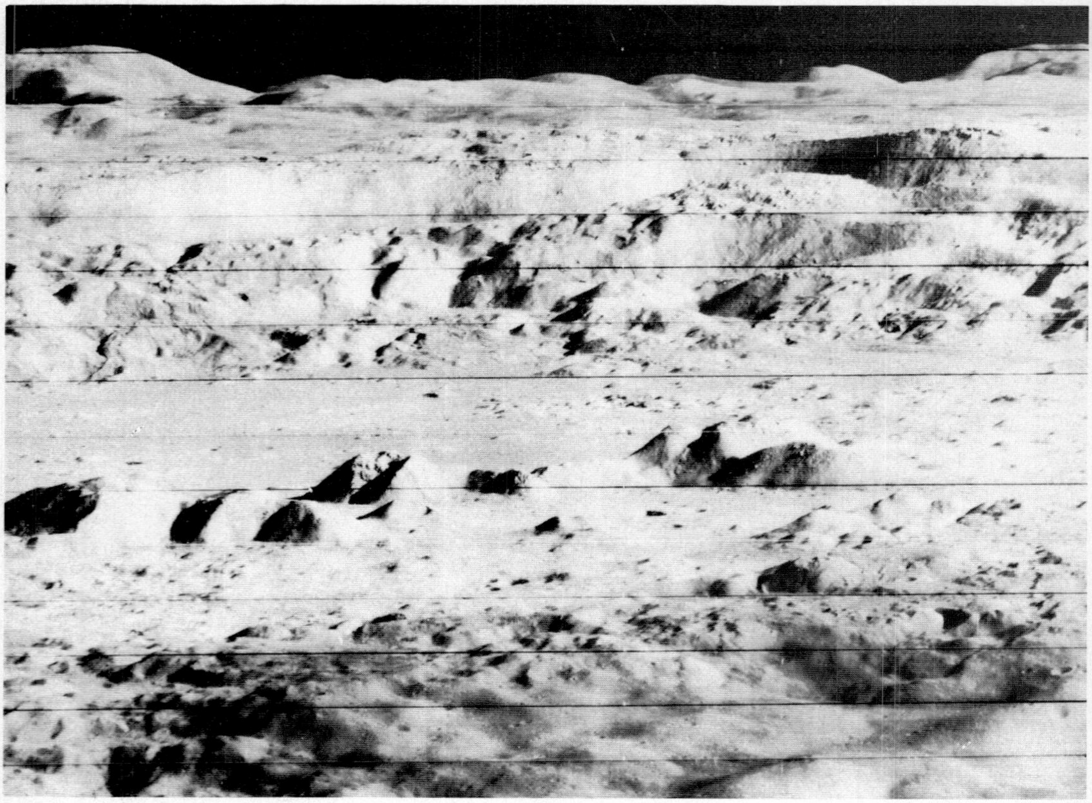

LIII The lunar crater Copernicus, photographed November 23, 1966, from Lunar Orbiter II. (*NASA Photo*)

LIV. An unnamed crater in the Oceanus procellarum, photographed by Lunar Orbiter III. The crater in the center of the picture is approximately 500 feet across and gives the impression that it was formed very recently by a meteorite impact. Debris thrown out by the impact is clearly visible. (*NASA Photo*)

LV. A 260-inch solid-fuel unit in its test pit in Florida prior to firing. (*NASA Photo*)

LVI. Night firing of a 260-inch solid-fuel "short length" rocket. The rocket developed 3.6 million pounds of thrust. (*NASA Photo*)

LVII. Take-off of a Titan III-C rocket, June 16, 1966. The rocket carried seven military communications satellites and one research satellite. The engines of the liquid-fuel rocket in the center were not ignited until after the jettisoning of the two solid-fuel boosters that accomplished the take-off. The rocketlike device that can be seen strapped to one of the boosters (a second one on the other booster is hidden by the booster itself) is a container for compressed gas. *(U.S. Air Force Photo)*

large centrifuge. Oberth, being the son of a doctor, had thought about such problems; Ziolkovsky had taken human endurance for granted.

The reasoning behind the actual centrifuge experiments was the following: The highest speed that Man will need for quite a number of years to come will be escape velocity. If you accelerated slowly in order to avoid strain, so and so many minutes of acceleration would be needed to build up to escape velocity. If one could accelerate a little more enthusiastically the time would be shorter but the strain would be greater. The times required to attain escape velocity for different accelerations are given in the following table:

ACCELERATION (g)	TIME REQUIRED TO ATTAIN ESCAPE VELOCITY
3	9 min, 31 sec
4	6 min, 21 sec
5	4 min, 45 sec
6	3 min, 48 sec
7	3 min, 10 sec
8	2 min, 40 sec
9	2 min, 20 sec
10	2 min, 6 sec

Small mammals, it was already known, could survive these accelerations; as for men it looked doubtful. The highest accelerations experienced by men had been those in fighter aircraft in sharp turns and fast pull-outs. There 4 *g* had been established as the safety limit; at 6 *g* a pilot "blacked out." But no pilot had had to stand 4 *g* for minutes. In the early days of rocket discussions 3 *g* had always been assumed to be the most a man could be expected to undergo.

The men who underwent the tests were volunteers. To avoid the side effect caused by the rapid whirling the test was conducted in the dark, with a dim light in the center (fulcrum) for the man to focus his eyes on. As an additional precaution the test subject had a switch which turned the power off—providing him the means (and the psychological assurance) of terminating the experiment at any moment. The first test, run at 3 *g*, seemed to substantiate the original expectations. Nobody was hurt, but the men complained bitterly. In some manner their time sense had been distorted; they seriously thought that they had been in for 20 minutes and more. Then the 4-*g* test was made. To everybody's surprise it was easier. The men went through the 5-*g* test and the 6-*g* test and on, all the way to the 10-*g* test. It was no fun, but it could be endured without "blacking out." One man even subjected himself to an incredible 17 *g* for a full minute, without harm, without black-out.

The reason the experience of the wartime fighter pilots had been misleading was found quickly. In the fighter plane the pilot sat upright; if he made a pull-out the acceleration he experienced was in the same direction as his spine, draining blood from the brain into the large blood vessels of the body and causing a black-out. In the centrifuge (and later in the actual spacecraft) the acceleration operated at nearly a right angle to the spine. There was strain, but the distribution of blood inside the body remained virtually the same as always.

At first the deceleration during re-entry had been expected to lead to smaller *g*-forces than the acceleration at take-off, but for a longer time. For this reason a 20-minute test with 1½ *g* was run and then a test of similar duration with 2 *g*. Again the test subject did not have any severe problems, but the tests proved unnecessary because re-entry turned out to cause the high *g*-forces for a shorter duration. In fact, the centrifuge tests were more severe than the reality proved to be. In the first place, they built up to a theoretical velocity of 7 miles per second, whereas only 5 miles per second were needed for orbital flight. Furthermore, in the centrifuge tests, the acceleration was constant, while in reality the acceleration stays at a fairly low level, reaching a high peak for only a number of seconds just before cut-off.

The second phase of an orbital flight—at the time when all this was just theory with no practical experience—was the most puzzling, to the expert as well as to the layman. But expert and layman were puzzled for different reasons. The layman did not understand why there should be a period of weightlessness when the spacecraft was only a hundred miles or so above sea level, evidently still in the grip of the earth's gravity. The answer to that was comparatively simple: the apparent weightlessness is the result of not offering any resistance to gravitational pull. But, as I know from countless questions, this answer is meaningless to most laymen so we have to go back a little further.

The misunderstandings surrounding the concept of weightlessness were caused by a wrong assumption. The layman assumed that the human body can "sense" speed and also "feel" gravity. The body can do neither. It cannot sense speed at all, as millions of passengers of jet aircraft now know. Unless there is turbulence the passenger of a jet that races across the sky at a rate of 550 miles per hour cannot feel that he is moving, and if there is a monotonous cloud layer far below and a clear sky above he cannot tell that he is in motion even if he tries.

Similarly we cannot "feel" gravity. The force we feel when we stand still is not gravity, but our resistance to gravity, because the floor, or pavement, prevents us from following the pull of gravity. If we could follow the pull of gravity freely, without feeling air resistance and without finally hitting something, we would not feel any "pull." There is no experience in daily life that produces zero-g and consequently there is no word to describe the sensation. The nearest an average person can come to experiencing a sensation resembling zero-g is to float in water of body temperature.

The layman was puzzled about the concept of zero-g, but the expert was puzzled about what zero-g could be expected to do to the body. It would not interfere with respiration because that is based on the difference between the pressure in the expanded lungs and the pressure in the surrounding atmosphere. It should not interfere with swallowing food or drink because that is done by muscular contractions; remember that all grazing animals swallow "upward" and that every schoolboy has at least once indulged in the trick of eating a sandwich against gravity while hanging from a bar by his knees. It would not influence heart action to any great extent, for most of the work done by the heart is not the "lifting" of the blood from below the heart against gravity but consists of moving the blood around in the body against the friction in the blood vessels. As far as one could find out from just thinking about it, zero-g should not interfere with the gross bodily functions.

Another puzzle for the expert, though a less severe one, was how one could experiment with zero-g. The first experiment to be devised was an animal experiment.

Two monkeys and two white mice took a trip in a specially constructed animal capsule carried by an Aerobee rocket. The two white mice were enclosed in a transparent plastic drum which rotated slowly and was filmed for the whole duration of the trip; the two monkeys were strapped to foam-rubber couches built to their measurements. Various instruments watched their heartbeat, their breathing, and so forth (and the monkeys had to be slightly anesthetized because they considered it part of the experiment to pull the instruments off their bodies), and after the Aerobee had passed its peak, the capsule was disengaged from the rocket, with a parachute that opened later. The films proved that the white mice were confused when zero-g set in; electrocardiograms of the monkeys which were taken during the flight hardly differed from those taken the day before and the day after. Moreover, as soon as they were released from the capsule, one of the monkeys indicated that he would be receptive to an offer of some food and ate a banana with relish and dispatch.

The monkeys chosen for the experiment were young—if they had been people they would have been called teen-agers—for reasons other than that younger animals should prove more resistant. If the researchers had taken old monkeys and one of them had died a few months later there would have been no end to the discussions of whether the rocket flight was to blame. Young animals, if not harmed by the flight itself, could be expected to live out their natural life span. The monkeys did, and for years they were on exhibit at the Smithsonian Institution's zoological garden in Washington, D.C.[5]

The monkeys proved to everybody's satisfaction that a few minutes of zero-*g* left no physiological impression, but they could not describe how it felt. But by the time the monkey experiment was reported at a meeting of the Aero-Medical Association the brothers Fritz and Heinz Haber, then both members of Professor Hubertus Strughold's Department of Space Medicine, had found a way of subjecting a man to zero-*g* for a short time. You take an airplane, the faster the better, and put it into a reasonable power dive. Then you pull out and point the nose of the airplane up. Simultaneously you kill the engine, or preferably reduce its power to such a degree that all it will do is overcome air resistance. The airplane will then move along a vertical "hump," closely resembling a parabola. As long as it is on this "hump," the pilot will be under zero-*g* (Fig. 75). The reason I said "the faster the better" with respect to the aircraft is simply that a faster plane will produce a bigger hump and a longer period of zero-*g*.

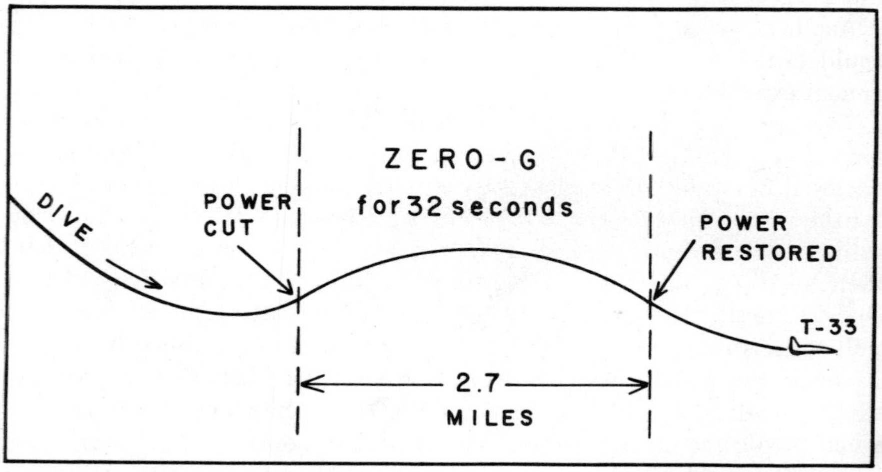

Fig. 75. "Parabolic flight."

[5] During an international meeting in Paris at the École Nationale Supérieure de l'Aéronautique (December 3–8, 1956), Alexei Pokrovsky, head of the Aero-Medical Research Center near Moscow, reported on experiments with dogs carried by Russian sounding rockets to altitudes of about 60 miles. They flew a total of nine dogs, three of them twice. Each dog at first had his own sealed capsule with a film camera built in. From the time the capsule was sealed until the time it was recovered and the animal released, a fairly long time—as much as three hours—elapsed. In another flight the dogs were enclosed in the canine equivalent of a space suit, apparently reinforced by a frame. One of the two dogs was expelled at a height of 56 miles; the other fell with the rocket to 25 miles. But the parachute for this second dog was set not to open until it was 2 miles above the ground. No change in the life processes or behavior of the dogs could be observed. (See *JARS*, vol. 27, numero 2, part 1, February 1957.)

The experiment was performed for the first time by Captain Charles E. Yeager, and the pilot's first report was not promising. Yeager felt as if he were sitting on a large globe which seemed to be spinning in several directions at once. He did notice that a pencil which happened to be lying on the instrument panel rose up and hovered in midair while he performed the maneuver. But he broke up the test after some 13 seconds by feeding more fuel to his plane's engine. Captain Yeager's report only brought more problems. Was zero-*g* really unbearable, or was this an individual reaction, possibly even an individual reaction caused by circumstances which would not hold true another day? The answers could be found only by having others repeat the test. The reports of other pilots were more of the nature of "Oh, I didn't mind"—which of course was better but only by comparison. The Department of Space Medicine kept on and at the Rome Congress in 1956 another one of Professor Strughold's researchers, Dr. Siegfried Gerathewohl, could report on quite a number of such tests:

> Practical weightlessness was produced by flying a Lockheed T-33 type aircraft powered by a J-33A-35 jet engine developing 4600 pounds thrust, along a parabolic course. Two flight maneuvers were used: the first yielded radical accelerations of 3 g and periods of 25 to 30 seconds practical weightlessness in consecutive sequence; the second yielded 10 to 15 seconds of weightlessness without any noticeable increase in acceleration prior to and after the weightless state. . . . By and large the experiences reported by the subjects can be classified in the following way:
> 1. *Sensations of Comfort and Pleasantness.* The majority of the subjects felt fine during the states of weightlessness, did not report any untoward effects but enjoyed the situation including changes of acceleration. . . .
> 2. *Sensations of Motion.* The subjects reported rather neutral sensations of motion (neither pleasant nor adverse) such as falling, floating, being rotated or suspended in midair.
> 3. *Sensations of Unpleasantness and Psychosomatic Symptoms.* A small group of subjects reported experiences of discomfort and showed symptoms of vertigo and nausea, which were rather typical for the motion sickness syndrome.
> (Gerathewohl)[6]

The obvious conclusion, also drawn by Gerathewohl, is that there is a very wide range of susceptibility and adaptability from individual to individual. Zero-*g* apparently will make one individual scream for the power lever, while another individual will find the same situation most enjoyable; Gerathewohl found one man who said so in so many words. Later on, another man who had gone through a number of such flights wrote to a friend: ". . . contrary to this being a problem, I think I have finally found the element in which I belong." The writer of this letter was Colonel John H. Glenn, who was to become the first American astronaut to go into orbit.

When speaking about these early "parabolic flights" the pilot of the air-

[6] See also *Astronautica Acta,* Vol. 11, Fasc. 4, 1956, and *The Journal of Aviation Medicine,* Vol. 28, No. 1 (February 1957).

plane with which volunteers were tested for their reactions must not be forgotten; after all, he became weightless, too. He was Major Herbert D. Stallings, M.D., who by April 1958 had, by his own estimate, made 4000 such flights, adding up to 37 hours of weightlessness.

Of course everybody realized that a spacecraft would have to have a hermetically sealed cabin, but the only hermetically sealed cabins that had ever been built had been submarines. In 1952 Professor Strughold persuaded the Air Force to have a ground-based sealed cabin built for medical experimentation. The cabin was delivered to the Department of Space Medicine (then still at Randolph Air Force Base near San Antonio, Texas) in 1954. After a short time Strughold realized that the term "sealed cabin" was so noncommittal that nobody paid any attention to the innovation. He then named it for its purpose; the "sealed cabin" became the "space cabin simulator" and then it did attract attention.

Some of the research done with the sealed cabin dealt with the question of what kind of atmosphere there should be in a spaceship cabin. Our air, at sea level, has a pressure of 14.7 pounds per square inch. Oxygen in the free atmosphere amounts to slightly more than 20 per cent, so that the "partial pressure" of oxygen is, rounded off—3 pounds per square inch. The not quite 80 per cent of nitrogen acts mainly as a dilutant, but nitrogen has a side effect. Normally some nitrogen is dissolved in the body fluids, particularly the blood. As long as the external pressure stays at an established level, the saturation of the body fluids with nitrogen is normal and of no consequence. But, if for some reason the pressure drops fast, the nitrogen comes out of solution, forming bubbles. And these bubbles cause the extremely painful and often fatal condition known as "caisson sickness" or "the bends."

The Navy had inevitably had much unpleasant experience with the bends; around the year 1950 it began some experiments on ways to avoid this condition. One solution was to "dilute" the oxygen with helium instead of with nitrogen. Helium also dissolves in the body fluids, but to a much lesser extent than nitrogen. Diving experiments with a helium-oxygen atmosphere had very good over-all results, but the strange side effect that the diver's voice became squeaky and sometimes incomprehensible. Of course the divers had had in their diving suits the high gas pressures that are needed to overcome the water pressure. If the pressure was lower, the results might be different. Actually there was no reason that the spaceship cabin should not have a lower pressure than that at sea level—perhaps a total pressure of only 6 pounds per square inch, 3 pounds of oxygen pressure and 3 pounds of helium pressure, would work.

All this could be investigated with the space-cabin simulator.

Some researchers wondered whether oxygen by itself might not be simplest, with a pressure of somewhat over 3 pounds per square inch, say

4½ pounds or 5 pounds. This seemed to work out best, medically speaking. In 1956, the Air Force decided to test it during very high balloon flights with sealed cabins that were called Operation ManHigh. The first ManHigh flight was made on June 2, 1957, with Captain Joseph W. Kittinger aboard. The balloon had a pure oxygen atmosphere and hovered at altitudes above 90,000 feet for 7 hours. ManHigh II, a number of weeks later with Dr. David Simons aboard, stayed aloft for 32 hours, 5 of these hours above 101,000 feet. A pure oxygen atmosphere could be used.

When NASA was established in September 1958 all the information from such preliminary studies was available, along with some plans. In April 1958 the Air Force had drawn up a "Manned Military System Development Plan" in four phases. These are likely to sound very familiar by now, because most of this study was incorporated in the later space program. Phase I of the Air Force plan was Man in Space Soonest and had three subphases: (1) orbiting instruments; (2) orbiting primates—first monkeys, then chimpanzees; (3) orbiting men. Phase II was Man in Space Sophisticated, involving a two-week orbital flight. Phase III was Lunar Reconnaissance, ending with the soft landing of instrument packages and TV cameras. Phase IV was Manned Lunar Landing and Return, again with monkeys preceding men.

The Army did not have such a comprehensive program but offered the idea of using a Jupiter rocket to carry a man to an altitude of 150 miles; the name of this proposal was Project Adam. It was later turned down by Dr. Hugh L. Dryden, who compared it with the circus performance of having a man shot from a cannon. But he did not think that it could not be done, only that the information from such a shot would not be worth the expense. Nevertheless, such "suborbital flights" were later incorporated into the Mercury Program.

The Mercury Program planners had to make up their minds about a number of questions. One was what rockets to use, and here the choice was restricted, since only the Atlas (of the D series) was powerful enough to orbit a manned capsule. Preliminary suborbital flights could use the Redstone rocket. But the capsule itself had to be tested, which would need about half a dozen test shots. The Redstone could have done the job but was a bit expensive; the price of a Redstone flight was then stated at a round million dollars. Could a cheaper rocket for these flights be designed in a hurry? It could, by putting four Castor rockets into one airframe. This design received the name Little Joe, first as a nickname. Then four Recruit rockets were built in with the Castor rockets. This would produce a maximum thrust of 230,000 pounds for a sufficient length of time to lift a payload of 4000 pounds to an altitude of 100 miles.

The next question confronting the Mercury planners was the selection of the men who would go into space. Even in that field there had been some preliminary thinking. An astronaut would not be a very young man; in order to learn everything he had to learn, both from books and instructors and by doing, he could hardly be less than 28 years old. He had to have a high I.Q., say 125 or better. He had to be in perfect physical condition, which implied that he must not be too tall. He had to have experience with machinery, and preferably should also have had experience in extricating himself from difficult situations, a requirement that spelled "test pilot."

In order to make the search for such men less complicated, Eisenhower decided around Christmas 1958 that the astronauts should be selected from among military pilots. A military pilot, in order to qualify for examinations for the status of astronaut, had to fulfill seven conditions: (1) age less than 40; (2) height less than 5 feet 11 inches; (3) excellent physical condition; (4) Bachelor's degree or equivalent; (5) graduate of test pilot school; (6) 1500 hours total flying time; and (7) qualified jet pilot.

A total of 508 service records were screened and about every fifth man met the requirements—110 in all. There were 58 Air Force pilots, 47 Navy pilots, and 5 Marines; no Army pilot met condition 5. Only 12 were to be selected. The list of 110 names was arbitrarily split into three groups. The men had to volunteer and personal or family reasons were expected to limit the number of volunteers. But of the first group of 35 pilots not less than 24 volunteered. As a result only the first two groups were called in at all; the third group was disregarded. Of the 69 asked to report, 36 passed the preliminary tests, written examinations, interviews and medical checkups. Of these 36 pilots 32 volunteered once more and became candidates. More testing resulted in 18 finalists who expected that only one out of three would be chosen, since in the meantime the number of Mercury astronauts had been reduced to 6. But 7 were chosen: 3 Air Force Captains—Donald Kent Slayton, Leroy Gordon Cooper, Jr., and Virgil I. Grissom; 3 Navy men, Lieutenant Commanders Walter Marty Schirra, Jr., Alan Bartlett Shepard, Jr., and Lieutenant Malcolm Scott Carpenter; and a Marine Lieutenant Colonel—John Herschel Glenn, Jr., who found himself the senior astronaut both in age and in date of rank.

While the astronauts went into their special training, the capsule that was to carry them was being developed and tested.

The main problem in the design was to make the capsule survive the re-entry into the atmosphere. Air behaves in a peculiar manner at high velocities, as is explained in some detail in Appendix 1; at this point we need only say that a body returning from orbit would compress the air in its path. Compressing heats the air and the heated air, in turn, heats the body that does the compressing. The problem was to find a shape that would deflect most of the hot air so that as little heat as possible would be transferred to the re-entering body. A conical capsule, entering with its blunt end forward, was one solution. The blunt end would have to be protected by a heat shield, and existing experience with re-entry nose cones for ballistic-missile warheads helped in the choice of the material.

In actual operation the capsule would leave its orbit by being retarded by the firing of three solid-fuel retro-rockets; this would also reduce the speed enough for the heat shield to be able to withstand the heat produced by the re-entry. The designers warned that the heat shield would not be able to survive a re-entry resulting from a natural orbital decay. Nor would it survive re-entry if only one of the retro-rockets fired. If all three fired there would be no problem; the firing of two presented a doubtful case.

But the capsule still needed a safety device in case the booster rocket malfunctioned. This device was the "escape tower"—a girder structure above the narrow end of the capsule, with another set of solid-fuel rockets on top. If a malfunction of the booster rocket occurred during take-off, the escape tower would pull the whole capsule away from the rocket and lift it high enough to enable the astronaut to use the capsule's parachute for an emergency splashdown in the ocean. If everything went well, the escape tower would be jettisoned while the booster was still on its way up.

The two missiles that were to carry the capsules—the Redstone and the Atlas—also had to be modified to some extent. The Atlas needed only an adapter for carrying a Mercury capsule instead of a warhead. The Redstone's tank section had to be elongated by 96 inches. The chief effect of the changes was to make both missiles taller, as shown here:

	TACTICAL REDSTONE	MERCURY REDSTONE
Over-all height	69 ft	83 ft
Motor and tank section	40 ft 4 in.	59 ft
Upper section and warhead	28 ft 8 in.	——
Capsule with escape tower	——	24 ft
Diameter	70 in.	70 in.

	ICBM ATLAS	MERCURY ATLAS
Over-all height	ca. 80 ft	95 ft 4 in.
Motor and tank section	67 ft 4 in.	67 ft 4 in.
Warhead	ca. 12 ft	——
Adapter	——	4 ft 3 in.
Capsule	——	6 ft 10 in.
Escape tower	——	16 ft 11 in.
Diameter	10 ft	10 ft

In telling the story one inevitably makes it sound as if all these activities —capsule design, astronaut selection, booster changes, and so forth—took place in sequence. In reality, of course, they all overlapped; while one section of NASA worried about capsule detail, another section drew up plans on just what should be taught to the astronauts. Then the astronauts made suggestions about capsule detail, while the George C. Marshall Space Flight Center in Huntsville made ready a Redstone rocket so that astronauts could "ride" a Redstone on the test stand for a full duration firing, to see how they would react to the noise and the vibration. Elsewhere test flights of so-called "boiler plate" capsules by means of Little Joe rockets took place.

Needless to say, decisions on the number of unmanned orbital flights were made and unmade; the desire to go into orbit as early as possible and the desire to do it as safely as possible conflicted on more than one occasion. Shifts in "projected schedules" usually hinged on the punctuality of deliveries, but the "luck of the launch pad" played a role too.

The very first shot in the Mercury program was Little Joe-1, ready for firing August 21, 1959, on Wallops Island. Half an hour before the planned take-off time there was suddenly the noise of a rocket roaring and smoke enveloped the launch pad. When the smoke cleared, literally speaking, Little Joe was still standing there; the escape tower and the dummy Mercury capsule had taken off without it. Of course the event had to go down in the books as a failure, but nobody grieved too much, the main interest lay in preventing a repetition.

In contrast, "Big Joe" of September 9, 1959, caused nothing but jubilation, especially, strange as it may seem, because a few things had gone wrong. "Big Joe" was an Atlas rocket (Atlas 10-D) that was to put a Mercury capsule through a test that Little Joe could not accomplish. The Atlas was to climb to 100 miles, being in a horizontal position when that altitude was reached. Then it was to pitch slightly downward. The separated capsule was then to turn itself around by means of its own thrusters (compressed nitrogen was used for the thrusters in that experiment) and re-enter the denser layers of the atmosphere, heat shield first. While the flight was underway, telemetry gave strange indications, and the impact predictor stated that the capsule would fall short by 500 miles. Six destroyers of Flotilla Four raced full power to the new impact point. A Navy plane spotted the capsule in the Atlantic while the nearest destroyer was still 100 miles away. The destroyer *Strong* reached it first and took it aboard; then the details of the shot that had gone vaguely wrong could be analyzed from the tapes aboard.

These were the findings. When the Atlas pitched slightly downward the capsule had failed to separate from the then boosterless rocket. The thrusters used up their whole nitrogen supply in an effort to turn the capsule *and* the Atlas around. Of course they were too weak to do so, nor would it have improved the situation if they had succeeded. At an altitude of 345,000 feet the capsule broke free, falling toward denser air. But the shape of the capsule was such that it turned itself into the proper position just because of the aerodynamic forces and entered the denser layers of the atmosphere heat shield first. And the heat shield did the job for which it had been designed. This indicated that even in case of thruster failure a safe re-entry might be made. Incidentally, on the same day another Atlas rocket, no. 12-D, made the first full-range ballistic-missile flight from the Pacific Missile Range in California.

The Little Joe rocket that had been left behind on the launch pad on August 21 was ready again on October 4, 1959; meanwhile it had acquired the designation Little Joe-6 or LJ-6. It stood 55 feet tall and had a take-off weight of 20 tons. This time the main purpose of the flight was for Little Joe to prove itself. After all, there had been no Little Joe flight so far. It went up to 37 miles and was destroyed 2½ minutes after lift-off in order to test the destruct system. The next Little Joe, dubbed LJ-1A, took off precisely one month later, on November 4. The flight looked fine, but the engineers were not completely happy because the escape system had fired a full 10 seconds later than intended.

Another month later, on December 4, LJ-2 took off. The press was especially happy about that flight, for LJ-2 had a passenger in the capsule, a rhesus monkey named Sam.[7] The rockets of the Little Joe booster were used up at an altitude of 100,000 feet; then the escape rocket lifted the capsule off the booster and added some more speed in the process, so that the peak altitude became 53 miles and the range 194 miles. The trajectory was such that Sam had been under zero-*g* for about 3 minutes, longer than any man had been at the time. Sam experienced one minute less of zero-*g* than expected, had a milder re-entry but a harder splashdown than expected, and had to bob in the Atlantic for 6 hours until rescued by a destroyer. He survived all this in good health.

The next flight, on January 21, 1960, was LJ-1B with "Miss Sam" in the capsule. For this flight there was to be an "accident" at 9 miles altitude when the speed was slightly over 2000 miles per hour. The booster was to "explode" and the escape system was to "rescue" the capsule. Everything went as planned, but the simulated accident resulted in a short flight—the altitude was only 9 miles and the range 12 miles. But through it all, "Miss Sam" did her job, which was to pull a lever when a certain light blinked on. The capsule was picked up by a helicopter almost immediately after splashdown.

Little Joe had performed well on the whole, and "Big Joe" had been an interesting success, but the time for disappointments was at hand. The first of these was MA-1 (Mercury Atlas no. 1), using Atlas 50-D. The plan was to carry a production-line Mercury capsule, heavily instrumented, to an altitude of 112.7 miles and a distance of 1500 miles. Everything was ready on July 29, 1960, and only the weather was not favorable—it was raining heavily at the Cape. But as the cloud ceiling was quite high, there was no need to postpone the test because of the weather. At 9:13 A.M. the rocket came to life, climbed upward through the rain, and disappeared in the clouds, while the people on the ground continued to hear its roar. About a

[7] Because he was a "graduate" of the School of Aviation Medicine.

minute after take-off, when the rocket was about 32,000 feet up, all telemetry stopped without warning. Only the signals from the capsule kept coming, indicating violent motion.

The Atlas must have exploded one minute after take-off. The capsule separation systems were not to be made ready until three minutes after take-off; therefore the capsule remained attached to what was left of the rocket and the debris fell into the ocean about 7 miles offshore.

November 8, 1960, was Election Day and also the day that Little Joe-5 misbehaved. At T + 16 (seconds), while the booster was still burning, both the escape rocket and the tower jettison rocket were ignited by a stray current in the system. Since the escape rocket could not lift anything off the booster while the latter was still accelerating furiously, everything stayed together and struck the ocean 13 miles off shore two minutes later.

Mercury Redstone no. 1 (MR-1) was scheduled for flight on November 12, but *pas un sans trois,* as the French say to express the conviction that things happen in threes. MR-1 was a complete failure too. Later it was dubbed "the 4-inch flight" and "the day they launched the escape tower," but at the time events were very unfunny.

At the end of the countdown the Redstone's rocket engine started up with the characteristic roar. But almost immediately the sound stopped and observers in the blockhouse saw the rocket settle back on its tail fins, after having lifted itself to a maximum of 4 inches. But at the same moment the escape tower, disengaged from the capsule, shot into the air, and a few seconds later the Mercury capsule ejected its parachute, which fluttered to the ground, followed by the reserve chute that dangled from the opening alongside the rocket. The failure was complete, and also complicated, with an explanation about as complicated as the failure itself. The rising rocket disconnected a plug as it was supposed to do. But this happened faster than planned, with the result that an automatic relay sent a cut-off signal to the fuel pumps. This explained the engine shutdown. But the short moment of lift apparently had been enough to activate all the other systems. A repetition of this performance could be avoided by making sure that the rocket remained electrically grounded for a short period after it disconnected the plug.

The repeat flight, called MR-1A, took place on December 19, 1960. It was as complete a success as MR-1 had been a failure. The rocket carried the capsule to an altitude of 130.7 miles and splashed down at a distance of 235 miles. Duration of the flight had been 15 minutes and 45 seconds, 35 minutes after launch a helicopter from the carrier *Valley Forge* retrieved the capsule, and 48 minutes after launch it was secure on the flight deck of the carrier.

After all these preliminaries the flight program could get underway. This was to be a succession of suborbital shots—the old Project Adam—to be followed by orbital flights of longer and longer duration. And in each case

THE MERCURY PROGRAM

NUMBER	NAME	ASTRONAUT	DATE	ORBITS	FLIGHT DURATION (HR:MIN:SEC)	ZERO-g (HR:MIN:SEC)	MAX. g DURING FLIGHT	PERIGEE/APOGEE (MI)
MR-2	——	chimpanzee Ham	Jan. 31, 1961	suborbital (range 422 mi)	0:16:39	0:06:40	14.7	157 (peak)
MR-3	Freedom 7	Shepard	May 5, 1961	suborbital (range 303 mi)	0:15:28	0:05:16	11.0	116.5 (peak)
MR-4	Liberty Bell 7	Grissom	July 21, 1961	suborbital (range 302 mi)	0:15:37	0:05:18	11.1	118.3 (peak)
MA-4	——	robot	Sept. 13, 1961	1	1:49:20	1:32:28	7.7	98.9/142
MA-5	——	chimpanzee Enos	Nov. 29, 1961	2	3:20:59	3:04:38	7.7	99.5/147.4
MA-6	Friendship 7	Glenn	Feb. 20, 1962	3	4:55:23	4:48:27	7.7	100/162.2
MA-7	Aurora 7	Carpenter	May 24, 1962	3	4:56:05	4:39:32	7.8	99.9/166.8
MA-8	Sigma 7	Schirra	Oct. 3, 1962	6	9:13:11	8:56:22	8.1	100/175.8
MA-9	Faith 7	Cooper	May 15, 1963	22½	34:19:49	34:19:49	7.6	100.3/165.9

The orbital period for all manned Mercury flights was about 88 min 30 sec; longest orbital period was that of Sigma 7 with 88 min 55 sec.

chimpanzees were to be first; since a chimpanzee could not take an active part in the flight, the chimpanzee flights would show whether the capsule with its automatic programming could bring back an astronaut even if, for some unforeseeable reason, he was unconscious.

MR-2, on January 31, 1961, carried a male chimpanzee named Ham, then 3 years and 8 months old. Throughout the flight he had to pull two levers, a right-hand lever when a white light went on and a left-handed lever when a blue light showed. Interestingly enough the reaction time of a chimpanzee is somewhat slower than that of a man—0.7 second, while a man averages 0.5 second. Ham did not know it, of course, but his rocket was nearly 1100 feet per second faster than planned, his peak altitude was 42 miles higher than planned, his zero-*g* lasted for 6.6 minutes instead of 4.9 minutes and the *g*-forces also climbed 3 *g* higher than planned. He worked through all this, having only two timing errors with his right hand and none with his left. The splashdown was somewhat too hard, and his capsule began taking on water—when a helicopter lifted it from the ocean an estimated 800 pounds of seawater was inside. When he was released from the capsule about 3 hours after lift-off, Ham placidly accepted an apple and half an orange, unperturbed by the arrival of the space age.

A few more test shots followed. MA-2 carried a capsule 1432 miles downrange on February 21, 1961. At the urging of Wernher von Braun, another Redstone flight was made on March 24, 1961, with a peak of 113.5 miles and splashdown 307 miles from the take-off site. It did not receive a running number like the other shots but was labeled MR-BD, Mercury-Redstone Booster Development. Von Braun wanted to be very sure of the performance of his elongated Redstone before a human astronaut risked his life.

After this flight it was certain that the Redstone could be trusted with a suborbital manned flight. Not quite three weeks later, on April 12, 1961, Yuri A. Gagarin orbited the earth. The leading men at NASA had said all along that the Russians were likely to succeed ahead of the Americans. But Congress was dismayed, the public was dismayed, and some bright publisher of a political "newsletter" told the American people that Gagarin had not actually orbited the earth but had just made a parachute jump from a high-flying jet. It is a sad fact that this nonsense persisted for years among a segment of the American public.

It was probably Gagarin's flight that changed the mission of MA-3. MA-3 had been planned to make a long suborbital flight, with a dummy or robot in the capsule, a robot designed to use up as much oxygen as a man and to release the proper quantities of carbon dioxide and body heat—in short, a device for testing the life-support systems of the capsule. A few days before MA-3 was on the launch pad (April 25, 1961), the mission was changed to an orbital flight. But 40 seconds after take-off the Range Safety Officer was obliged to throw the destruct switch. The inertial guidance system in the Atlas had failed to tilt the rocket into the proper path.

On May 5, 1961, there followed MR-3, the first manned suborbital flight of the Mercury program. It was a perfect shot. The Redstone rocket cut off after 142 seconds and immediately afterward the tower-jettison rocket fired. The temperature inside the capsule was 91 degrees Fahrenheit, but the temperature inside astronaut Shepard's spacesuit was only 75 degrees Fahrenheit. Retro-rockets and parachute ejection worked fine. Just 11 minutes after splashdown a helicopter deposited the astronaut and the capsule on the flight deck of the carrier *Lake Champlain* and Shepard had to reassure physiologists that the 5 minutes of zero-*g* had not ruined his body (in fact he had not noticed when weightlessness started) and psychologists that zero-*g* had not affected his mind. In 1961 misconceptions about the effects of zero-*g* were still numerous, fanciful, and powerful.

The second suborbital flight was set for July 19, 1961. As far as external ballistics went, MR-4 was to be a repetition of MR-3. But as a result of MR-3 the activities of the astronaut, Virgil I. Grissom, were somewhat different. Shepard said that he had been overloaded with duties during the 5 minutes of zero-*g*. Grissom therefore had fewer such duties and was advised to spend the free time gained in looking out through the window to test the visual abilities of an astronaut in flight above the atmosphere. Bad weather postponed the launch for two days.

The actual flight of MR-4 went as smoothly as that of MR-3. Helicopters were hovering over the recovery area and when the capsule, dangling from its parachute, came into sight, they flew in the direction of the descending spacecraft. Inside the capsule astronaut Grissom had taken off his helmet and unbuckled his harness and when a helicopter pilot asked him whether he was ready for pickup, Grissom asked for 5 more minutes, he was busy recording cockpit panel data. All of a sudden the hatch cover blew off into the sea and water started pouring into the capsule. Grissom scrambled out, his spacesuit provided good buoyancy at first. The first of the recovery helicopters hooked on to the capsule, while another one came for Grissom.

The first helicopter's pilot reluctantly had to decide to abandon the capsule; full of water it weighed 5000 pounds, 1000 pounds more than the helicopter could carry. The capsule sank in seconds. The helicopter pilot dropped a marker in that spot, thinking that the capsule might be recovered later. But one glance at a sea chart on board the aircraft carrier took care of that idea—at that point the water is 3 miles deep.

In Washington decisions were being made. Originally two more suborbital flights had been planned but the last of them had already been canceled by the time Grissom took off. Now the third suborbital flight was canceled too, since it looked as if nothing more could be learned from such flights. In fact, the planners were already busy with "Mercury Mark II," later renamed Project Gemini.

On August 6, 1961, Gherman Stepanovitch Titov orbited the earth for the Soviet Union, landing safely after 17½ orbits and a flight duration of 25 hours and 18 minutes. Titov's flight caused various reactions. Anything the Mercury astronauts might now do would only be a duplication of the Russian feat. On the other hand, Titov's flight proved once more that one could send a man into orbit and recover him safely. But then Titov said that for the last eleven orbits he had felt "vaguely sick"; maybe zero-g did prove to be dangerous after a few hours.

Project Mercury went ahead with a few odds and ends before sending an astronaut into orbit. The Mercury-Atlas flight that was to orbit a robot still had to be made. Atlas 88-D was groomed for this flight, which then received the designation MA-4. The capsule that had carried the "man simulator" on the ill-fated MA-3 was overhauled, repaired, and used again. The flight was performed on September 13, 1961, and the capsule was recovered after one orbit. But after the two successful suborbital flights there was a general feeling that MA-4 had not really been necessary. The public as well as the astronauts themselves felt that an American should go into orbit now, before the Russians did it for the third time. The real value of the MA-4 flight was not evident to the public: it had served to test the world-wide tracking network of the Mercury program and also provided the Navy with some practice in recovery techniques.

Before MA-4 went into orbit, a flight of a capsule on top of a Scout rocket had been planned. The Air Force had provided a Blue Scout and was eager to launch it in order to gain launch experience. NASA reluctantly supplied a capsule and the Scout was made ready on Wallops Island. MS-1, as it was called, was launched on November 1, 1961, but because of a mistake in wiring the rocket behaved erratically and at T + 43 seconds the destruct switch had to be thrown.

By that time Atlas 93-D had been checked out at the Cape for something more interesting; it was to become MA-5 and carry the chimpanzee Enos through three orbits. Take-off was on November 29, 1961, and during the first orbit both Enos and the capsule performed well. During the second orbit the capsule temperature began to rise, and the attitude of the capsule, which a man could have controlled, became erratic. Opinion was divided. From the medical point of view, Enos could have safely completed the third orbit. From the engineering point of view, the capsule was not so trustworthy. The decision was to break off the flight after two orbits. Enos landed safely.

Atlas 109-D was at the Cape, ready to produce MA-6, the first American manned orbital flight, with Lieutenant Colonel John H. Glenn as the astronaut. Hopes that the flight still could be made in December 1961 remained only hopes. The flight date was set for January 23, 1962, but had to be postponed because Florida had very bad weather. The new date was January 27. In spite of a heavy overcast the countdown was carried through until T — 20 minutes, when the flight had to be canceled. There was no chance that the weather would improve on that day and the last day of January became the new target date.

Early in the countdown a technician discovered that fuel had accumulated between the bulkheads of the rocket, a condition that can only result in an explosion. This was so serious that the flight was postponed for two weeks. Then February 20 became the new date and newspapermen tried to guess what the cause for the next postponement might be. But there was none. MA-6, after delays that added up to several hours, took off at 9:47 A.M. After insertion into orbit Glenn sent cheerful reports on how happy he felt in zero-g and commented repeatedly on the wonderful view. He complained that the window was too small and remarked later that what he really wanted was a capsule that was all glass.

During the second orbit the attitude of the capsule was not quite what it should be and Glenn switched from system to system to find out how to maintain the proper attitude with the smallest fuel consumption. Then he became suspicious as tracking station after tracking station began to ask him to make sure that the landing-bag deploy switch was in the OFF position.

On the ground a telemetry signal had been received which, if correct, meant that both the landing bag that absorbs the impact on the water and the all-important heat shield were loose. Of course the straps of the retro-rocket package still held both in place and a re-entry might be accomplished by not jettisoning the retro rockets after burning them. But—and that was important—this could be done only if all rockets fired. If one did not fire, the retro-rocket package would have to be jettisoned, as an unfired retro-rocket would be ignited by the re-entry heat and would most likely explode. The retro package and its straps would burn away in that case, but by the time they did, the aerodynamic pressure would be great enough to hold the heat shield in place, even if it was loose. When, halfway through his third orbit, Glenn was told not to jettison the retro package, he realized what the ground crew was worried about.

All the retro-rockets fired—Glenn said that it felt as if they wanted to push him back to Hawaii—and during the re-entry glide across the United States he kept his capsule as steady as he could. He saw glowing pieces of the retro package fly past his window but everything that happened confirmed his belief that his heat shield was in place and that something had been wrong with the indicator.

Splashdown took place in the Atlantic, but 40 miles short of the predicted point, because the capsule was lighter on re-entry than when it went into orbit. In the meantime thruster fuel had been used up, a small quantity of oxygen had leaked from the capsule into space, and, of course, the solid fuel of the retro-rockets also was gone before re-entry. A destroyer, the *Noa*, happened to be only 6 miles from the splashdown point and the crew had seen the capsule. Seventeen minutes later the *Noa* was alongside and the capsule, with Glenn still inside, was hoisted aboard. The first words of the hero of the first American manned orbital flight were not the kind that are later put into the mouth of a hero in a play. When crew members assisted Glenn to leave the capsule he simply said, "It was hot in there."

Glenn was hot, perspiring copiously, momentarily exhausted, thirsty but not hungry, and not inclined to talk until he had showered, but he was alive and healthy and the only physical harm he had suffered was a bruised knuckle when he opened the hatch cover.

MA-7, to be orbited by Atlas 107-D, had been scheduled to go into orbit during the second week of April. But as a result of Glenn's flight some changes were found useful, some new instrumentation had to be installed, and the recovery ships had to take part in a tactical exercise of the Atlantic Fleet. This combination of factors postponed the flight by about five weeks.

MA-7 also was to be a three-orbit flight, but while during the first orbital flight the emphasis had been on engineering, the second flight was to be devoted mainly to scientific observations. The astronaut was Scott Carpenter and when he was awakened at 2:15 A.M. on May 24, 1962, the Cape was covered by dense fog. Aircraft radioed the information that there were broken clouds overhead, which could not be seen from the ground. The weather forecasters promised that the sun would dispel the fog and that above the fog layer the weather really was quite good. By 5 A.M. Carpenter was in the capsule, the hatch had been bolted shut, and the fog was still hiding everything from view.

The ground fog was still persisting at 6:30 A.M. and because of it there were three consecutive 15-minute holds. Then the sun won out—quite suddenly—and MA-7 took off at 7:45 A.M. Except for the fog it had been "the smoothest countdown ever." Everything that followed, booster separation, jettisoning of the escape tower, separation from the Atlas, and turn-around in orbit went just as smoothly. Carpenter thoroughly enjoyed himself and was surprised when, some time later, the Canton Island tracking station inquired solicitously whether he was very uncomfortable. Their telemetry reported the astronaut's body temperature as 102 degrees Fahrenheit. Carpenter told them that that could not be true. "I can't imagine that I'm that hot." Since the instrument kept reading high and Carpenter kept insisting that he felt comfortable, the medical staff on the ground agreed that the instrument must be at fault.

Turning the capsule sideways for taking pictures and turning it in such a manner that the earth seemed to be overhead—to assess the psychological reaction, if any—kept Carpenter busy but also made great inroads in his supply of maneuvering fuel. By the end of the second orbit he had less than half of it left. For most of the third orbit the capsule just drifted, which was pleasant for Carpenter and an interesting experiment for the flight directors on the ground: this information might be useful in planning sleeping periods during longer flights.

When the time came for the retro firing, things were not so perfect. The capsule was not well aligned with the flight path and there was no time for adjustment. When the first retro-rocket fired, the capsule was canted, in fact, more canted (25 degrees) than Carpenter estimated. Because the retro-rockets worked at an angle, instead of opposing the motion directly, the capsule was not slowed down as much as it should have been. This did not constitute any danger but contributed to the fact that Carpenter overshot his intended splashdown point. Later the factors causing the overshoot could be added up: the faulty position during retro fire contributed 175 miles to the overshoot; the fact that the retro-rockets had been fired 3 seconds late contributed another 15 miles; and analysis of the retrofire results showed that the retro-rockets had been weaker by 3 per cent than the design figures called for, adding another 60 miles to the overshoot.

For some time during re-entry the capsule is surrounded by a so-called "plasma sheath"—partly ionized air through which communication is impossible. The ground stations, therefore, were out of touch with Carpenter. But the beacon signal from the capsule had been received by two airplanes after this period so that it was certain that the capsule had survived the re-entry. The actual splashdown point could be calculated with fair certainty, but the nearest ship, a freighter, was 31 miles away and the nearest naval vessel, the destroyer *Farragut,* was 75 miles away, long distances for seagoing vessels.

Meanwhile Carpenter, after assuring himself that his capsule was watertight, had opened the top hatch and squeezed himself through, pushing an uninflated life raft ahead of him. He climbed down the outside of the capsule and inflated the life raft. It was upside down, but he flipped it over and settled down on it to wait. A little more than half an hour later he was spotted by two airplanes and another 20 minutes later he was joined by two Air Force frogmen, bringing their own rafts. Another half hour later a seaplane arrived. The pilot was willing to try a landing on the choppy sea but was ordered not to, as the sea was a bit too rough for a safe landing. Carpenter and the Air Force pararescue men just continued to wait. After three hours a helicopter picked up Carpenter and deposited him on the carrier *Intrepid,* 4 hours and 15 minutes after splashdown. The capsule was secured by the destroyer *Pierce.*

The next Mercury flight, MA-8, using Atlas 113-D with Walter M. Schirra as the astronaut, was planned to be a six-orbit flight and was tentatively scheduled for September 1962.

On August 11 Russia's Vostok-3 went into orbit with Major Andrian G. Nikolayev. On the following day Vostok-4 with Lieutenant Colonel Pavel

R. Popovitch took to space. The two spacecraft approached each other within a few miles; the men certainly could see each other. But, though the opposite was reported at the time, they apparently did not attempt a rendezvous. The chiefs of NASA and the astronauts, preparing for a six-orbit flight, could read in the papers that Nikolayev completed sixty-four orbits and Popovitch forty-eight.[3]

Take-off of MA-8 was at 7:15 A.M. on October 3, 1962.

Ten seconds after lift-off the Atlas rocket developed a roll (about its longitudinal axis) and for a short while it seemed that the escape tower might have to be tested in earnest. But the rocket stabilized and there was a flawless insertion into orbit and flawless turn-around, accomplished with the smallest fuel expenditure ever, only 3/10 of a pound of hydrogen peroxide was used. But soon Schirra felt hot in his suit, and this time it was not a false signal. The condition was bad enough that Ground Control considered cutting short the flight after one orbit. Slowly, carefully, Schirra adjusted the cooling unit of his suit, with success; the flight was not terminated. Later the malfunction was found to be due to dried-out lubricant in a needle valve.

After this difficulty had been overcome, the flight became, as Schirra put it, a textbook flight during which nothing went wrong. Like Carpenter, Schirra had nothing good to say about the capsule periscope. It was excess baggage in daytime and useless at night. Schirra had been told that he might see the Echo balloon satellite during one of his orbits, but he did not. For a long time Schirra flew, as he expressed it, "in chimp configuration"—meaning that he exerted no more control over his spacecraft than a chimpanzee would. One of the things that surprised him was the amount of detail on the ground that he could see in those areas not obscured by clouds. He had known that, when in orbit, he would be ten times as far from the surface as he had ever been before and had expected to see much less—"but it was the same old deal; I might as well have been in a jet 40,000 or 50,000 feet up."

Helicopters from the carrier *Kearsarge* were in the air when the capsule parachuted down into the Pacific. Pararescue men attached a flotation collar to the spacecraft but when they offered to help Schirra out, he radioed that he preferred to stay inside and requested that a boat be sent to tow him to the carrier. The *Kearsarge* dispatched a whaleboat to the capsule, which was gently bobbing about half a mile away. Precisely 9 hours and 54 minutes after launch the capsule was hoisted aboard the carrier.

The Mercury program had been started in the hope of ending with an eighteen-orbit flight, but after Schirra's flight there were suggestions to terminate the Mercury program with MA-8 and to leave the longer flights

[3] For a tabulation of the flights of Russian cosmonauts see Appendix 2, section 10.

to the two-man capsules of the Gemini program, which had been publicly announced and named on January 3, 1962. Extensive ground tests had shown, however, that the Mercury capsule could make a 30-hour flight and the decision went in favor of MA-9, to be orbited by Atlas 130-D. The astronaut for the long flight was L. Gordon Cooper. Since such a long flight might be terminated earlier and a splashdown might occur in any of several possible recovery areas, the rescue service was enormous in size. It consisted of 28 ships, 171 aircraft, and 18,000 servicemen. But, significantly, the medical specialists only numbered 84, less than half of the number for Glenn's flight when nobody knew what to expect.

Scheduled take-off date for MA-9 was May 14, 1963. But early in the morning the Bermuda tracking station complained about malfunction in its radar. By about 8 o'clock repairs had been made, the countdown was resumed, and finally the Gantry was ordered to roll back to free the rocket. Then the Diesel engine failed to work! Two hours were lost because something as commonplace as a Diesel engine refused to run. By that time the Bermuda radar was misbehaving again and the launch was postponed until the next morning.

It took place on May 15, 1963 at 8:04:13 A.M.

On the third orbit Cooper launched a 6-inch sphere with xenon strobe lights from his spacecraft; he was to test his ability to spot the flashing light. He could not see it and was somewhat doubtful whether he had succeeded in launching it, but then he spotted it on the fourth orbit. He saw it again during the fifth and sixth orbits. About 10 hours after take-off he was officially informed that he had a "go" for seventeen orbits, later extended to twenty-two orbits. The sleeping period was scheduled for orbits 9 through 13, but Cooper did not succeed in sleeping for a normal time as later astronauts did. He slept an hour, then woke up because his suit temperature was too high; for the next 6 hours he dozed intermittently, always being waked up by the misbehaving temperature regulator of his suit, which made it either too hot or too cold. In spite of much picture taking and the many medical checks he had to perform—this long flight was mainly for space-medical purposes—time seems to have dragged a bit for Cooper.

On the twentieth orbit the attitude-control system began to give false indications. Soon there were no attitude readings at all. During the next orbit, a short circuit occurred in a busbar and the automatic stabilization and control system was without electric power. Cooper had to make a manual re-entry, Glenn, from the ground, gave him a 10-second countdown for the first retro-rocket. The recovery area was again in the Pacific and the *Kearsarge* was again the prime recovery ship. Cooper splashed down 4 miles from the *Kearsarge*, 34 hours and 20 minutes after lift-off. Forty minutes later the spacecraft was hoisted aboard and the hatchcover was blown off. Cooper stayed inside during an 8-minute medical examination.

When he emerged he had a short spell of dizziness but regained his equilibrium and went unassisted to the carrier's sick bay.

Project Mercury was over.

Project Gemini was to come.

But there was to be an interval of a little less than two years during which time the Russians had space, and the headlines that go with it, to themselves.

Vostok-5 went into orbit on June 16, 1963, for an 81-orbit flight.

While Vostok-5 was still orbiting, Vostok-6 took off on June 16 for a 48-orbit flight, causing extra excitement because the cosmonaut was Valentina Tereshkova, the first woman in a spacecraft. Only a few months later, on October 12, 1963, the Russians demonstrated their new spacecraft, named Voskhod (pronounced *vos-KHOD* and meaning "sunrise.") It carried three men and only the commander wore a spacesuit; the other two were dressed as if they were taking a scheduled airliner for a routine trip.

During 1964 there were no manned space flights, but in preparation for a coming American event a Gemini capsule was carried into orbit by a Titan-II rocket, staying in orbit for four days. The Titan-II was not only a more powerful rocket than the Atlas. It also had the advantage over Redstone, Atlas, and Titan-I of not using liquid oxygen, which meant that a hold in a countdown would not automatically lead to another hold for the purpose of replacing the liquid oxygen that had evaporated during the first hold. For a while it had been doubted whether the Titan-II could be used to carry manned spacecraft because it had what became known as the "pogo stick effect"—an up-and-down vibration. But this vibration could be eliminated.

Most of the preliminaries that had been necessary for Mercury were not needed for Gemini. There was no need for animal flights, or for manned suborbital flights. The only suborbital flight made (called GT-II for Gemini-Titan no. 2) was a 2127.1-mile down-range flight for testing various systems. GT-III, the first manned flight, was scheduled for March 1965, but Russia's Voskhod-2 beat it by five days. Voskhod-2 performed only seventeen orbits, but one of the two men aboard, Alexei A. Leonov, left his spacecraft for 10 minutes of EVA—extra-vehicular activity, dubbed a "space walk" by the newspapers.

In appearance the Gemini spacecraft (the term "capsule" had been dropped late in the Mercury program) suggested a larger Mercury vehicle, which, of course, it was. The seats for the two astronauts were arranged side by side so that both faced in the same direction. There were now four rockets in the retro package, the parachute was larger and the interior somewhat roomier. Since a "cabin atmosphere" of pure oxygen with a pressure of slightly over 5 pounds per square inch had worked well all through the Mercury program, the same system was used throughout the Gemini program. In the Mercury program each astronaut had been on his

THE GEMINI PROGRAM

NUMBER	CREW (FIRST NAME IS THAT OF COMMAND PILOT)	LAUNCH DATE	REVOLUTIONS*	DURATION
GT-I	unmanned	Apr. 8, 1964	64	4 d
GT-II	unmanned	Jan. 19, 1965	suborbital	(range 2127.1 mi)
GT-III	Virgil I. Grissom; John W. Young	Mar. 23, 1965	3	4 hr 53 min
GT-IV	James A. McDivitt; Edward H. White II	June 3, 1965	66	4 d 1 hr 59 min
GT-V	L. Gordon Cooper; Charles P. Conrad	Aug. 21, 1965	120	7 d 22 hr 59 min
GT-VII	Frank Borman; James A. Lovell	Dec. 4, 1965	206	13 d 18 hr 35 min
GT-VI	Walter M. Schirra; Thomas P. Stafford	Dec. 15, 1965	16	1 d 1 hr 52 min
GT-VIII	Neil A. Armstrong; David R. Scott	Mar. 16, 1966	7	10 hr 42 min
GT-IX	Thomas P. Stafford; Eugene A. Cernan	June 3, 1966	46	3 d 0 hr 21 min
GT-X	John W. Young; Michael Collins	July 18, 1966	43	2 d 22 hr 47 min
GT-XI	Charles P. Conrad; Richard F. Gordon	Sept. 12, 1966	44	2 d 23 hr 17 min
GT-XII	James A. Lovell; Edwin E. Aldrin	Nov. 11, 1966	59	3 d 22 hr 36 min

* For explanation see below.

own, but for the Gemini flights, with two men aboard, somebody had to be in command; hence on each mission one of the two astronauts was called the command pilot. As far as possible the command pilot was an astronaut who had been in space before.

Another innovation in the Gemini program—beginning with GT-V—was the switchover from orbits to revolutions. The distinction, which does not hold true for all orbits[9] can best be explained by visualizing a model consisting of a globe and a wire loop with a bead around it. The wire loop is the orbit, the bead is the spacecraft, and the beginning of an orbit by the spacecraft is marked on the wire with a spot of paint. When the spacecraft returns to the starting point on the orbit, it has completed "one orbit." But in the meantime the earth has turned so that the spacecraft is over a different meridian, some distance to the west of that over which it was inserted into orbit. For the Gemini flights the 80th meridian west of Greenwich, which touches the Florida coast near Palm Beach, was chosen as the reference meridian. When the spacecraft crossed this meridian again it had completed a "revolution," which occurred about six minutes after the completion of an "orbit."

[9] In an orbit going over both poles "orbit" and "revolution" are the same.

Except for GT-III the Gemini spacecraft were not given names as the Mercury capsules had been. GT-III did have a name; it was called the Molly Brown. The name was chosen by the command pilot of the flight, Virgil Ivan Grissom. Grissom (generally known as "Gus") had had to swim to safety when his Mercury capsule *Liberty Bell 7* sank after the second suborbital flight. At the time GT-III was being readied, there was a successful musical show on Broadway called *The Unsinkable Molly Brown,* and Grissom decided that if a name could help to keep his second capsule afloat that would be the one. NASA officials were a bit reluctant to accept "Molly Brown" but when Grissom suggested "Titanic" as his alternate choice they gave in.

The take-off of GT-III took place at 9:24 A.M. on March 23, 1965. Among the witnesses at the Cape was Wernher von Braun, who celebrated his fifty-third birthday by watching the performance of a large rocket he had not designed himself. Eight minutes after lift-off GT-III was in orbit. Since this was mainly a test of the spacecraft only three orbits had been scheduled. During these three orbits Grissom and his co-pilot Lieutenant Commander John Watts Young made various experiments to prove the maneuverability of the spacecraft. They changed their orbit, and even the inclination of the orbit, and everything worked fine. Grissom also did something unscheduled: he ate a corned-beef sandwich (network commentators changed it to a bologna sandwich for reasons best known to themselves) which Young had brought aboard. Young had been supposed to consume some specially prepared food, but Grissom, acting as a control, was not supposed to eat at all.

At 2:18 P.M. the capsule splashed down to the east of Bermuda, nearly 60 miles short of the aiming point where the carrier *Intrepid* was waiting. This was the result of an incorrect estimate of the lift the spacecraft would produce during re-entry.

Because the Gemini spacecraft was heavier than the Mercury capsule a different rescue procedure had been worked out. Frogmen attached a "flotation collar" to the spacecraft; then the hatches were opened and the astronauts lifted out by helicopter. The spacecraft itself was salvaged later. The two astronauts of the GT-III flight were put aboard the *Intrepid* at 4:01 P.M., less than 2 hours after splashdown.

Before GT-III was inserted into orbit, editorial writers had been telling their readers that the Gemini program was "12 to 18 months behind schedule," and expressing the conviction that for that reason the program would go poorly. While GT-III did fly later than was originally estimated, what followed was one success after another. Not that everything went right on every flight—there were a number of failures, some minor and some severe—but every time the astronauts aboard managed to convert a possible failure into a success. The Gemini program proved that astronauts are not just "passengers" aboard their spaceships but serve an important purpose: that of taking over when the machines go wrong.

GT-IV, with James Alton McDivitt as command pilot and Edward Higgins White II as co-pilot, had the first long-duration mission in the Gemini program. The flight began with a smooth take-off on June 3, 1965. Among the experiments planned was a close approach to the orbiting second stage of the Titan-II rocket that had put the Gemini spacecraft into orbit and a so-called spacewalk for the co-pilot. Since this was a long-duration flight there had to be sleeping periods for the astronauts; these were so scheduled that one would sleep while the other was on duty. The astronauts found out that this scheme did not work well; no matter how careful the waking member of the crew was, he always disturbed the other's sleep, so that both astronauts slept very little during their four days in space. Consequently, on later long missions the sleeping periods were scheduled to have both astronauts sleep at the same time.

During the flight McDivitt saw a satellite, probably Pegasus-I though the picture he took of it did not show anything. The experiment to make rendezvous with the orbiting second stage did not work out; the rocket was still 2 or 3 miles away when half of all the maneuvering fuel had been used up. But White's spacewalk was very successful. He stayed outside the spacecraft for a little over 20 minutes, the time it took to cross the United States, and some banter was exchanged near the end of the period, when White told his command pilot that he would not come in: "This is fun." Of course he then obeyed the order. Splashdown came on schedule at 12:13 P.M. 4 days after take-off. Again it was short of the target area—46 miles from the waiting carrier *Wasp;* the aerodynamic characteristics of the Gemini spacecraft were still not completely known. But the astronauts were deposited on the deck of the carrier 56 minutes after splashdown. They were somewhat dehydrated; McDivitt had lost 4 pounds (preflight weight 156 pounds) and White 8 (preflight weight 173 pounds); and they had not had enough sleep. But the medical team, to its pleasure and also to its surprise, did not find anything wrong with them. They had lived through 4 full days of zero-g without any change at all.

GT-V was scheduled to go into space late in August 1965 and was to stay in space about twice as long as GT-IV had done. In interested circles this was dubbed the "fuel cell flight," since the newly developed fuel cells were to be tested in space for the first time.

A fuel cell is a device that reverses the schoolroom experiment in which water is decomposed into its constituent elements oxygen and hydrogen by means of an electric current. In a fuel cell hydrogen and oxygen are carried along as supercold liquids and are combined to form water, producing electric current in the process. Fuel cells had been developed because they offered a possibility of saving on battery weight. Unfortunately batteries of any type are quite heavy and do not hold, or produce, enough electric current to justify their weight except on the ground. The fuel cells built into the aft section of GT-V were known, from many ground tests, to produce about twice as much current as the best batteries, weight for weight, and they would yield water as a by-product though this water would not be utilized during that particular flight.

The command pilot on GT-V was L. Gordon Cooper, who had made the longest of the Mercury flights, and Charles P. Conrad was co-pilot. The take-off date was scheduled to be August 19, 1965. During the countdown a number of minor problems developed, causing holds totaling a little over 3 hours. When the countdown came close to its end there had to be another hold, this time because a thunderstorm moved across the Cape. All of a sudden there was a surge of electrical power through the launching complex and the spacecraft, probably caused by a bolt of lightning. In the spacecraft the onboard computer was knocked out and its electrical memory was scrambled. The flight had to be postponed. The actual take-off, two days later, was flawless. Three minutes after lift-off the two stages of the Titan-II rocket separated and 3 minutes later the spacecraft separated from the upper stage, the spacecraft being in an orbit with a 100-mile perigee and a 215-mile apogee. The first stage, which had fallen into the Atlantic Ocean, produced a welcome addition to knowledge. It did not sink, as had been expected, and was salvaged by the destroyer *Dupont* for thorough study.

GT-V was to make rendezvous with an orbiting "vessel," and since nothing suitable was in an orbit the astronauts could reach, the spacecraft carried a "radar evaluation pod" (REP) to be ejected from the spacecraft and used later as the rendezvous target. This was done during the second revolution, 2½ hours after take-off. Ejection of the REP was the last item in the original program that took place on schedule.

Even before the ejection Conrad reported to the ground that the fuel cell was misbehaving. The pressure in the oxygen tank of the fuel cell, which should have been between 800 and 900 pounds per square inch, was more than 150 pounds per square inch below normal. By the end of the second revolution the pressure was down to 180 pounds per square inch. Thinking of the reason for this low pressure was simple. The liquid oxygen was heated to produce the necessary pressure, the heating element being an ordinary glow wire worth about 15 cents, the same kind that is used in electric household heaters. Evidently the wire did not glow, some contact must have been broken. If anybody aboard the spacecraft or on the ground had decided that safety was paramount, the failure of this 15-cent item would have led to a re-entry after two or at most three orbits. Cooper told the flight director on the ground, Chris Kraft, that he "had to re-enter early or else power down." Kraft ordered them to forget about REP and to shut down as many current-consuming systems as possible. Cooper shut down what could be shut down and then turned the spacecraft around, thinking that full sunlight on the oxygen tank of the fuel cell would help to increase the pressure so that more current would be produced. It did not; the pressure continued to drop. When the flight command next asked for a pressure reading it was down to 125 pounds per square inch and while the flight command was talking to the spacecraft the pressure dropped by another 30 pounds.

The astronauts were still safe, because their back-up batteries had enough power to handle one and a half orbits plus re-entry. But re-entry, to reach the prime recovery area near the Bermuda islands had to be made at the end of the sixth revolution or wait until after the eighteenth. Re-entry after any of the intervening revolutions would land the spacecraft in inconvenient locations. By the time the decision had to be made, the pressure was down to 71 pounds per square inch *but it had stabilized at that point.* Chris Kraft issued a "go" for a total of eighteen revolutions. During these additional revolutions the situation improved because the spacecraft itself was giving off heat some of which was absorbed by the fuel-cell tank.

The "go" was repeated on a day-by-day basis and GT-V had the quietest spaceflight in history. But some experiments could still be performed. Rendezvous with the REP could not even be contemplated any more, because the batteries in the pod had run down in the meantime. But GT-V made rendezvous with an imaginary vessel assumed to be passively orbiting. GT-V approached it within 17 miles, as established by a computer on the ground. Another set of experiments dealt with establishing what an astronaut could see on the ground. The reports of earlier astronauts had caused frank disbelief among eye experts. For GT-V "eye charts" were laid out on the ground in New Mexico and in Australia. Cooper and Conrad had no trouble recognizing the signs, even those they were not supposed to recognize.[10] Conrad told ground control that he thought he saw an aircraft carrier and a destroyer at the entrance of the harbor of Jacksonville. A check revealed that the "carrier" was just a tug and the "destroyer" a lighter.

In spite of trouble with the fuel cells and a spastic air-conditioning system, Cooper and Conrad flew their full mission. Or almost—they re-entered after 120 instead of the planned 121 revolutions, because hurricane Betsy churned up the waters in the recovery area for the last of the planned revolutions. Astronauts and spacecraft were recovered by personnel from the aircraft carrier *Lake Champlain.*

GT-VI was planned to be the first flight that made an actual rendezvous with another vehicle, an Agena rocket fired into orbit by an Atlas for just this purpose. But things went wrong, and as a result GT-VI and GT-VII have to be treated as a unit, which became the high point of the Gemini

[10] Some eye experts have since evolved the theory that zero-g improves vision. This theory may be unnecessary, since we are dealing with men with perfect vision who are used to flying at high altitudes.

program. On October 25, 1965, two rockets were counted down simultaneously at the Cape. One was a Titan-II rocket with GT-VI on top and command pilot Walter M. Schirra and co-pilot Thomas P. Stafford waiting for lift-off. The other was an Atlas, carrying an Agena. It took off 101 minutes before the countdown for GT-VI reached zero and it was to take up a circular orbit 185 miles from the ground. The Atlas roared off, but there was no telemetry from the Agena. Hopes dropped, but not yet to zero, since it was possible that the Agena was performing well but had suffered a transmitter failure. One had to wait until it would pass over Australia where the tracking station has "skin capability." (This refers to a radar that would find the Agena by radar echo from its skin.) The Australian station should locate the rocket at 6 or 7 minutes before noon (Cape time) but it did not. The Australian at the microphone summed everything up in two words: "No joy," he chanted, "no joy." Since there was no target, the countdown of GT-VI was stopped immediately. The little that is known leads to the conclusion that the Agena exploded on ignition.

As no other Agena was ready, GT-VI had to be postponed for an indefinite period. But in the meantime GT-VII passed all the check-outs and took off on December 4, 1965. Command pilot was Frank Borman; co-pilot was James A. Lovell. The flight plan called for a long-duration flight of about 2 weeks. With GT-VII nicely in orbit there suddenly was a target for GT-VI, if GT-VI could be readied in time. The take-off was scheduled for December 12, but the rocket did not leave the ground. An electric plug which the rocket should have pulled out on lift-off was jarred loose 2½ seconds too early and the safety systems went to work to prevent ignition. Then it was found that the rocket would not have taken off anyway, because a plastic dust cover had been left on a fuel line. GT-VII, in orbit, sent reassuring messages to the crew of GT-VI, saying "We're here, waiting for you." The wait did not take too long, GT-VII lifted off at 8:27 A.M. on December 15 and went into a preliminary orbit with perigee at 100 miles and apogee at 161 miles. GT-VII was orbiting at 185 miles. With a series of short burns GT-VI enlarged its orbit to 185 miles and at 2:30 P.M. (Cape time) the two spacecraft were 1½ miles apart. By 2:35 Stafford told ground control, "We are 120 feet apart and sitting." But apparently they "sat" only to take a deep breath; by 2:45 the distance had shrunk to 20 feet and 4 minutes later the vessels were an estimated 10 feet apart. Measurements of the photographs that were made later show that Schirra or Stafford had overestimated the distance—the two ships were not 10 feet apart but only 2 or 3 feet.

All this had used up less than half of the maneuvering fuel aboard GT-VI. Schirra flew completely around GT-VII and the two spacecraft stayed together in orbit for 5½ hours. Then Schirra reduced his speed, went into a lower orbit, re-entered, and splashed down in the morning hours of December 16 after 16 revolutions. GT-VII stayed in orbit until the morning hours of December 18, re-entering after 206 revolutions. After two weeks of zero-g both astronauts of GT-VII were in perfect condition —in their opinion because they had flown for a long time without their spacesuits on.

GT-VIII took off on March 16, 1966, with Neil A. Armstrong as command pilot and David R. Scott as co-pilot. The flight plan contained the first actual "docking" in orbit, physical contact between two orbiting vehicles. It did not include a supreme test of the resourcefulness of astronauts in orbit, but one occurred. Again there was the simultaneous countdown of two rockets, and this time the Agena rocket took on its assigned circular orbit 168 miles above sea level. It had been supposed to do this on March 15, but during the fueling of the Atlas kerosene had sloshed over into the rocket's interior, requiring a thorough clean-up and the replacement of a few valves. On the 16th both rockets blasted off on schedule, the Titan-II following the Atlas-Agena at 11:41 A.M. The first orbit ran between 100 and 168 miles, and then the perigee altitude of GT-VIII was increased to 123 miles, after which the orbit was made circular at 167 miles from sea level.

During the fifth revolution, while the two vehicles were over the Brazilian coast, the Gemini spacecraft docked with a relative speed of 1½ miles per hour, a very slow walk. There was no electric discharge from one vessel to the other, as had been feared. The jolt of docking was so minor that it could hardly be felt. Careful turning of the Gemini spacecraft proved that both vehicles were locked together. Not quite half an orbit later—40 minutes after docking—the joined vessels began to roll and pitch uncontrollably. The astronauts did not then know the reason for the trouble, but they feared that the joined ships might break apart and that GT-VIII might be disabled in such a break-up. There was also the possibility that the fuel tanks of the Agena, still more than half full, might split open and the fuel explode. They separated from the Agena as fast as they could, and Flight Control on the ground suggested an emergency landing. GT-VIII re-entered during its seventh revolution and splashed down 550 miles from Okinawa, only 1 mile from the calculated landing point. Frogmen, parachuted from a Douglas C-54, ascertained that the capsule was floating in the right position, that it was watertight, and that both astronauts were well. Astronauts and spacecraft were later rescued by the destroyer *Mason*.

Telemetered signals later on told what had gone wrong. A short circuit had developed in the control system of thruster no. 8. The thruster fired for several seconds, then stopped for two or three seconds, again fired for a few seconds, shut off once more, came on again and stayed on until the astronauts disconnected the whole system. If the source of the trouble had been known at the time, they could have gone on after separation from the Agena for the full planned 71 hours; since nobody could tell at the moment what it was, the flight lasted not quite 11 hours. During the next day, when it was known that nothing had gone wrong with the Agena, its rocket motors were ignited by radio command and it was slowly maneuvered into a 253-mile circular orbit to serve as a target for a later flight.

The mission of GT-IX, with Thomas P. Stafford as command pilot and Eugene A. Cernan as co-pilot was more or less repetition of what GT-VIII had accomplished, plus those items the crew of GT-VIII had not had time for. The date set for takeoff was May 17, 1966, and an Atlas was to carry an Agena into orbit ahead of time. A Northeast Airlines passenger jet flying from Miami to New York was scheduled to pass the area at the time of the Atlas take-off. There was no danger of collision, but the Federal Aviation Agency informed the captain of the airliner, Richard Hubble, that he might see the Atlas climbing into the sky. Captain Hubble, in turn, informed his passengers and probably everybody on board became a witness. "We first spotted the rocket's vapor trail at about 15,000 feet as it broke through a cloud layer," Captain Hubble said later. "We watched the rocket rise vertically directly in front of us and suddenly there was a great yellow eruption and we knew something unusual had happened . . . an object fell from the base of the yellow cloud and dropped toward the ocean."

Of course the countdown of GT-IX was canceled immediately and no new plans were made until the cause of the accident was ascertained. It was as unpleasant as it was simple. A short circuit in the Atlas had caused one of the booster engines to swivel into the "hard-over" position, making a sharp loop which caused the rocket to break up. The Agena, apparently unignited, fell into the ocean.

No other Agena was ready or would be ready for some time, so NASA officials decided to use something called ATDA instead. This device, the "augmented target docking adapter" had been built by the McDonnell Aircraft Corporation which also built the Mercury and Gemini spacecraft. ATDA, which existed in only one specimen, had no propulsion system of its own. All it needed was a conical shroud to serve as a nose cone for the trip through the atmosphere and an Atlas rocket to carry it. ATDA was sent into orbit on June 1, 1966. There was no double countdown, because space officials wanted to be sure that it was in orbit before they made GT-IX ready. Tracking reported that the target was in orbit, so GT-IX took off the two days later. But when it came close to the target the astronauts saw a strange sight. Stafford told ground stations, "We have an angry alligator on our hands." What had happened was that, because of a mistake in wiring (the ground crew had never handled such a device before), the shroud had only split open lengthwise and otherwise remained attached to the target, making docking impossible. There was a short consultation as to whether GT-IX should try to knock the shroud loose, but because that might have damaged the spacecraft the docking was struck from the schedule. Instead they made three different rendezvous maneuvers.

Then Cernan went outside and maneuvered his way around to the back section of the spacecraft where a backpack called AMU (astronaut maneuvering unit) was waiting for him. But Cernan found work in weightlessness exceedingly difficult. He had to pause for rest repeatedly and his face plate kept fogging so badly that he could hardly see. Finally, after he had been outside the spacecraft about 2 hours, command pilot Stafford ordered him back in. The AMU unit was not actually used and could not be examined afterward because it had to be jettisoned along with the back section prior to re-entry. GT-IX splashed down in the Atlantic only half a mile from the target point and about 3 miles from the waiting carrier *Wasp*. The final descent by parachute could be clearly seen from shipboard, and television cameras sent it via the Early Bird communications satellite into countless millions of homes.

Since "simple" rendezvous had been accomplished repeatedly, the tasks grew more complicated as the program progressed. For GT-X they were quite breathtaking. John W. Young, who had been Grissom's co-pilot for GT-III, was command pilot of GT-X; Michael Collins was the co-pilot. The Atlas-Agena rocket took off at 3:39 P.M. on July 18, 1966, and the Agena put itself into an orbit with an apogee of 190 miles and a perigee of 183 miles. GT-X took off precisely 101 minutes later—the leeway for lift-off was only 40 seconds and missing this very short launch window would have caused a 2-day delay. After only a few revolutions, GT-X made rendezvous with the orbiting Agena during the late evening hours, and paced it for about 40 minutes before the actual docking. However, this was costly in terms of maneuvering fuel; 770 pounds should have been left after docking but the supply was down to 385 pounds.

But docking had been accomplished and the Agena had fuel left in its tanks. (Nobody had ever assigned running numbers to Agena rockets, except in the factory, so the target Agena for GT-X was referred to as Agena-10.) After checking the orbiting rocket, astronaut Young fired the Agena's rocket motor from his control panel and the Agena's fuel served to carry the joined vehicles into an orbit with a perigee of 184 miles and an apogee of 474 miles, the highest manned flight that had been made at that time. Then more orbit adjustments had to be made to get into the orbit of "Agena-8"—that is, the Agena left in orbit by the GT-VIII flight. Agena-8 was then 1400 miles ahead, but the astronauts saw a very bright object which was probably that rocket. However, before they could overtake it, Collins had to do some scientific photographing. Standing on his seat, with head and shoulders outside the spacecraft, and secured to it by only a nylon cord, he took photographs of certain stars, with some difficulty, because he could see only the very brightest of them. After about 45 minutes fumes entered the suits of both astronauts. Collins climbed down and the hatch was closed so that the astronauts could open their face plates and put drops in their eyes.

During the afternoon of July 20, after having been docked for 39 hours and 40 minutes, Young separated his GT-X from Agena-10 and went after Agena-8, 160 miles ahead and 7 miles farther out. Within 3 hours GT-X had bridged the gap and the two vehicles were orbiting side by side. Because the fuel supply was low, there was no attempt to dock with Agena-8. But Collins went outside, succeeded in strapping himself into the backpack unit, and after two unsuccessful attempts reached Agena-8. He returned after 27 minutes, half the time scheduled for his "EVA," because stabilizing GT-X had used more fuel than anticipated. On the third day of the mission the hatch of GT-X was opened once more, this time to throw out unneeded equipment that cluttered up the interior. This action put twenty-two pieces of space junk into orbit, but most of it must have re-entered since. GT-X splashed down 2½ miles from the target point and 5 miles from the small carrier *Guadalcanal* that sent helicopters to effect the recovery.

By then the Gemini program had only two flights to go and the things still to be done were refinements of earlier accomplishments rather than things that had never been done before. For example, GT-XI was to make rendezvous and dock with an Agena on its first orbit. This required precision firing of both the Atlas-Agena and the GT-XI Titan rocket. Each of the rockets caused a day's delay. An electronic "sniffer" reported that the Titan leaked nitrogen tetroxide and careful inspection showed that there was a pinhole leak in the oxidizer tank. It was sealed with a drop or two of "waterglass" (sodium silicate) strengthened by an aluminum patch after drying. As for the Atlas, a stray signal indicated a possible malfunction that did not really exist; in any event the Atlas performed flawlessly on September 12, 1966. Command pilot of GT-XI was Charles P. Conrad who had been co-pilot on GT-V; the co-pilot was Richard F. Gordon. They docked with the orbiting Agena 94 minutes after lift-off and stayed docked through two sleeping periods.

Using the Agena's fuel, Conrad put the joined vessels into a new orbit with a perigee at 175 miles and an apogee at 850 miles, the highest manned flight to date. From this apogee the astronauts could see an area five times the size of the United States. Then, still using the Agena's fuel, they returned to an orbit 180/190 miles from sea level. The next experiment to be performed was to disengage the GT-XI from the Agena but to tie the two vessels together with a tether so that they could be rotated around their common center of gravity for the purpose of creating "artificial gravity" in the Gemini spacecraft via the centrifugal effect. The task of doing this fell, as usual, to the co-pilot, who had an EVA of about 2 hours on his schedule. Gordon found movement outside the spacecraft enormously tiring; he perspired so much that he could not see out of his right eye. The command pilot ordered him back into the spacecraft after about 10 minutes, but the 100-foot tether had been fastened to the Agena and the rotation experiment could begin. The thrusters of the Gemini set the ships spinning, at the rate of one full rotation in 9 minutes. The artificial gravity produced was too weak to be felt by the astronauts; it was estimated that it would cause an object to fall 3 inches in 10 seconds. Four separate docking maneuvers were carried out and Gordon photographed several stars in Scorpio and Orion, using a filter that was transparent to ultraviolet light only. These pictures have meanwhile yielded interesting astronomical results.

GT-XI was the first spacecraft to re-enter by computer only, without any overriding control by the astronauts. Again the final phases of the splashdown were visible to TV cameras on the carrier—in this case the *Guam*—and were sent out via the Early Bird satellite. The actual splashdown occurred at 8:59 A.M. on September 15, 1966. This time the recovery force outdid itself, accomplishing the entire recovery in only 25 minutes.

The last flight of the Gemini program, GT-XII, took off on November 11, 1966, preceded by an Atlas-Agena. The command pilot was James A. Lovell, who had made the endurance flight in GT-VII while the co-pilot was Edwin Eugene Aldrin, a major in the Air Force with a Ph.D. in science from the Massachusetts Institute of Technology; his thesis had been on orbital mechanics and rendezvous maneuvers. But during the flight it was his skill as a pilot and not his scientific knowledge that counted. While chasing the Agena he noticed that the computer did not show any changes in the distance to the Agena. Command pilot and co-pilot looked at each other for a minute, then Lovell said, "Ah, it couldn't happen to us," and the two made rendezvous with the Agena visually and under manual control.

Aldrin, knowing that an extended EVA was going to be on his schedule, had discussed the problem of exhaustion during EVA with the astronauts who had experienced it. When his turn came he thought he knew how to avoid this exhaustion—namely, by making only very slow movements. It worked. His longest EVA period was 2 hours and 9 minutes; counting the shorter periods too, he exposed himself to space for a total of 5½ hours. He did everything he was supposed to do, unscrewing bolts and tightening them, making and breaking electrical connections, and so forth. Before re-entering the spacecraft he carefully wiped the windshield, causing Lovell to remark from the inside, "Change the oil, too." The program included taking pictures of stars in Orion (especially "hot" stars of great interest to astronomers) and photographing an eclipse from orbit. French scientists fired two high-altitude Centaure rockets for the benefit of GT-XII, both releasing clouds of sodium vapor. Aldrin dutifully photographed the area at the right time but failed to see the sodium vapor clouds because the Sahara desert from which the rockets were fired provided too bright a background.

Though the Agena rocket held a good deal of fuel, the GT-XII astronauts were advised not to use it, as there were indications that the bearings in the Agena's fuel pump were untrustworthy. Hence there were no major orbital changes during the GT-XII flight and after 59 revolutions the spacecraft re-entered the atmosphere and splashed down; the recovery vessel was again the carrier *Wasp*.

On November 15, 1966, the Gemini program was over. There had been ten manned flights and ten successes and even the emergency landing of GT-VIII had not been necessary.

One of the strangest aspects of the two-year Gemini program was the absence of Russian cosmonauts. The Russians did not make a single manned flight during this period. Of course it was known that they were not idle. Among the large number of Russian Kosmos satellites orbited during these two years there were several so large that they were obviously tests of a new space capsule, the successor to Vostok and Voskhod.

And, of course, NASA was conducting not only the Gemini program. Surveyor-I had been landed on the moon; Lunar Orbiters were in orbit around the moon and sent back fantastic pictures. On the ground the components of the Apollo spacecraft were built and tested, especially the so-called "command module" for three astronauts and the LEM (lunar excursion module) which will land two astronauts on the moon.

Postscript: "If we die. . . "

At one time during the Gemini program astronaut Grissom told the press: "If we die, we want people to accept it. We are in a risky business, and we hope that if anything happens to us it will not delay the program. The conquest of space is worth the risk of life."

The tragic fact is that Grissom did die, along with Edward H. White II, who had been the first American to leave a spacecraft in flight, and Roger B. Chaffee, who was still looking forward to his first trip above the atmosphere. The three were to be the first crew of the first orbital flight of the Apollo command module, planned to be a two-week flight with the tentative take-off date of February 21, 1967. They died in their command module, but not while they were halfway to the moon or 800 miles above the Indian Ocean. The command module was merely perched on top of a Saturn 1-B rocket, about 220 feet above the ground.

The test of January 27, 1967, was to be "just an exercise," entered in the books as "Apollo 204." The astronauts worked in a capsule pressurized to 16 pounds per square inch with pure oxygen, with a lesser pure oxygen pressure in their suits. At 6:31 P.M. a voice (probably Grissom's) said, "Fire in the spacecraft." Then another voice shouted, "Get us out of here," and there was a cry of pain. Two technicians, standing on the Gantry structure at the level of the spacecraft, saw a blinding flame inside the cabin. Other technicians, some wearing face masks and asbestos gloves, tried to open the hatch but were driven back by the intense heat and the smoke that began seeping from the capsule. Finally, 6 minutes after the first alarm, the hatch yielded and hot air and greasy smoke escaped.

The three astronauts must have died almost instantly. The capsule was a total loss, the rocket was only slightly damaged. A board of inquiry was appointed immediately but the hope of finding the actual reason for the disaster was slight. The board found that the workmanship of detail in the capsule had been often below standard, and a careful check of the second Apollo capsule led to a wholesale condemnation of the inspection practices of the manufacturer, North American Aviation. This, in turn, led to almost wholesale management changes in the company. But what the public wanted to know, with some Congressmen asking countless searching questions, was, "the reason" for the catastrophe, and this could not be fully established, except that the fire probably started with a short circuit.

The question about "the reason" was wrong, because there were several reasons. If the spacecraft had had a hatch that could be opened instantly, the astronauts, temporarily protected by their suits, might have survived the fire. If everything in the capsule had been tested, not just for combustibility, but for combustibility in pure oxygen, the fire might not have spread.

Of course the fact that there was a pure oxygen atmosphere received the main blame. In principle this was correct, but the accusers disregarded the whole record of the space program.

Yes, there is a very pronounced fire hazard in a pure oxygen atmosphere. In pure oxygen a cigarette will burn, rapidly, with a bright flame. In pure oxygen a plain iron wire will burn like a waxed thread. But that admitted fire hazard had been successfully avoided so often. There had been the ManHigh flights of the Air Force with a pure oxygen atmosphere. There had been two Mercury flights with chimpanzees, six Mercury flights with human astronauts, and the ten manned Gemini flights, all in pure oxygen atmospheres. In one of the Mercury flights there had even been a major short circuit, with no other results than more work for the astronaut.

The greatest irony is that the short circuit that initiated the disaster would probably have been harmless if it had taken place in orbit. On the ground a burning candle, say, keeps burning because hot air is lighter than cold air. The heated air rises, taking the carbon dioxide produced by the combustion with it; new cold air with oxygen flows in and keeps the combustion going. In orbit hot air has the same weight as cold air, namely none; hence a candle lighted in an orbiting spacecraft would suffocate in the carbon dioxide which it produces. And if a fire did spread, the opening of the hatch to the vacuum of space would end the menace within seconds.

The result of "Apollo 204," as the death of the three astronauts is usually referred to, is going to be normal atmosphere in the capsule for ground tests and oxygen under reduced pressure after orbit has been reached.

"Apollo 204" brought the total of astronaut casualties to six. One, Theodore Freeman, had crashed with his jet on a solo flight, while two others, Elliott M. See and Charles A. Bassett (who were to have been the crew of GT-IX), crashed in a two-seater jet while trying to land at St. Louis in rainy weather with poor visibility.

The Russians had lost two of their cosmonauts—not in orbit as an irresponsible European news service claimed at one point, but from accidents. One died in an automobile crash and the other was killed in a parachute jump.

In April 1967 the Soviet Union lost a third cosmonaut, Colonel Vladimir M. Komarov, who had been command pilot of Voskhod-1 in October 1963. On Sunday, April 23, at 3:35 A.M. Moscow time (Saturday, 7:35 P.M. New York time) a new Russian rocket had lifted the third type of Russian space vehicle, named Soyuz-I,[11] into orbit. Though the Soyuz-type spaceships are believed to have room for at least four men, Komarov was the sole occupant. The orbit had perigee at 120 miles and apogee at 135 miles. Rumors that had been circulating weeks before the take-off said that Soyuz-I would be joined in orbit by at least one other ship and that they would exchange crews in space. Since the Soviet spaceport at Baikonur in Kazakhstan is forbidden territory to foreigners (and to the overwhelming majority of Russians, too), nobody can say whether other ships were ready at the time.

Komarov prepared to re-enter after about 24 hours in space, or after about 17 orbits. A Russian-language broadcast monitored in California said that Komarov had had trouble controlling his spacecraft during the last three orbits; his problem may have been similar to that of GT-VIII. Presumably Ground Control advised him to land. Re-entry maneuver completed, Komarov released his main parachute at an altitude of about 23,000 feet. But the lines of the parachute somehow snarled, so that the spacecraft fell with great speed from that altitude and Komarov was killed by the impact.

[11] To be pronounced *so-YOOZ,* meaning "union."

14

Open-End Story: The Big Rockets

". . . a technological victory, in our civilization, has become just as important as a victory on the battlefield. Conversely a technological defeat becomes as costly as a lost battle."

—Wernher von Braun (1959)

The Redstone was bigger than the V-2, the Jupiter bigger than the Redstone, the Atlas bigger than the Jupiter, the Titan bigger than the Atlas, the Saturn bigger than the Titan—and then there will be the Shape of Ships to Come, bigger than the ships that came before them.

The point where we have to enter this succession is Saturn, not the ringed planet but the eight-engined rocket that in ten flights scored ten technological victories but existed only for the purpose of paving the way for still bigger rockets.

Talk about a very large rocket to come goes back to about 1957. The name Nova was often used for this future rocket, but it must be emphasized that different people used the name Nova for different concepts that had nothing much in common except large size and a general lack of crystallization. "One million pounds of take-off thrust" was talked about with confident awe as something that would one day be accomplished. Even for that 1 million-pound figure opinions differed. Some rocket engineers insisted, or merely took for granted, that a 1 million-pound engine would have to be developed. But a preliminary study made by the advance planners of ABMA, in 1957, envisioned a rocket with a number of rocket engines of identical construction.

The one-engine people derided this idea and called it a makeshift (but traveled to the meetings where they delivered their papers in makeshift four-engine airliners) and thought it proper only that a rocket with a given take-off weight should be propelled by an engine with a thrust equal to about 1½ times that take-off weight. If the defenders of the single-engine concept were in favor of liquid fuels they often conceded that one might do well to look around for new and more powerful fuels. The replacement of liquid oxygen by liquid fluorine or by a mixture of liquid oxygen and liquid fluorine was a favorite topic of discussion for a while. Those single-engine men who were in the field of solid fuels advanced a slightly different argument. We are close to several promising, powerful, and reliable types of solid fuels, they said. Let us say that we get one that is nearly ideal. The thrust developed by such a fuel would depend only on the length and diameter of the charge and it will be easy to make casings of the proper dimensions so that we'll have a solid fuel propulsion unit for almost any conceivable purpose.

Wernher von Braun, who made himself the spokesman for the multi-engine concept, argued as follows: If we develop an engine for liquid fuels of a practical size and performance, we can build this engine in as many specimens as desired. The more of them we build and test, the closer we will come to an absolutely reliable engine. Once we have such a reliable engine we can use three or five or seven of them for a specific rocket. And if rockets grow so heavy that too many of these engines would be needed, we can start all over again with the development of a single engine of the thrust of five or six of the old ones. And then we can, if necessary, use several of the bigger engines in one rocket.

That is precisely what happened.[1]

The first sketch of what was to become the Saturn rocket was based on a rocket engine that had not been developed at the time. It was the engine called E-1, which was to have a thrust of 360,000 pounds. The new rocket was to have four of these engines. But then it looked as if there might be a long wait before the E-1 was operational and reliable. The smaller and weaker (188,000 pounds thrust) H-1 engine could be had more quickly. The H-1 was itself a development of the very reliable S-3 engine that powered the Thor and Jupiter missiles. The main difference between the two engines was that the turbopump of the H-1 was mounted on the outside of the engine instead of above the combustion chamber. The H-1 was therefore more compact because engine and pump assembly formed a unit. This shortening also offered some advantages in construction of the engine supports. The thrust of eight H-1 engines would add up to just about 1½ million pounds. The engines were to be arranged in two "squares" of four engines each. The inner set could be mounted rigidly, since it only had to provide push; the four outer engines were mounted on gimbals for control and steering.

[1] This is also precisely what happened in Russia. During the Paris Airshow in the Spring of 1967 the Russians exhibited the Vostok rocket. The main body of the rocket had a cluster of 4 rocket motors plus two vernier rockets. The body was surrounded by four take-off boosters, with identical propulsion units. The Russians had said some time earlier that the Vostok carrier had five "engines," which we would call assemblies. Actually the rocket, at take-off, was propelled by 20 identical rocket motors with 10 vernier rockets for velocity adjustments.

The tanks of the new rocket were designed with an eye to existing manufacturing facilities. In tooling up for the production of tanks for large liquid-fuel rockets the diameter of the tank is the most important consideration. Going from a diameter of, say, 48 inches to one of 60 inches requires new tooling, but making a 48-inch-diameter tank 40 feet long instead of 35 feet requires only a modification. Hence the Saturn's eight tanks, arranged in a circle, were planned to have a diameter of 70 inches each, the same as those of the Redstone. Four of them were to hold kerosene and four liquid oxygen. But since to burn kerosene completely more than an equal volume of oxygen is needed, these eight tanks surrounded another oxygen tank with a diameter of 105 inches, that of the tank of the Jupiter missile. The over-all diameter of the tank cluster was 21½ feet, its over-all length 82 feet.

This assembly, however, was not the Saturn rocket but only its first stage, usually called the Saturn booster. From the beginning it had the designation S-I, but the whole rocket was first called Saturn C-1. Later some of the designations were changed and the rocket as a whole became Saturn I. (In the following description only the new designations, introduced in February 1963, will be used.)

The complete Saturn I rocket consists of three stages, called S-I, S-IV, and S-V, and adding up to a total height of 162 feet. In addition to the features already described, the S-I has, on top of the tanks, forty-eight spherical pressure vessels made of Fiberglas, which contain compressed gaseous nitrogen for pressurizing the kerosene tanks. The oxygen tanks are pressurized by gaseous oxygen which is obtained by passing some of the liquid oxygen through heat exchangers that are part of the engine package. At their bases, all fuel tanks and all oxygen tanks are interconnected so that the level during burning stays the same in all tanks. This arrangement serves a particular purpose, which was a main talking point in defense of the cluster concept. If one engine malfunctions after lift-off, the interconnection of the tanks serves to feed the fuel and oxygen originally intended for the dead engine to the other seven. Of course the thrust will fall off by one-eighth, but the seven remaining engines will have a longer burning time, so that the final velocity of the rocket will be the same, within very narrow limits. During one of the test flights one engine was deliberately cut off to see how the arrangement would work out; it worked out well.

The second stage of the Saturn I rocket, the S-IV, is 18 feet in diameter and about 40 feet in length. The S-IV uses the most powerful chemical fuels known—liquid oxygen and liquid hydrogen. It has six engines (type RL-10) with a thrust of 15,000 pounds each. All six of them are gimbal-mounted.

The third stage, the S-V, is also hydrogen-burning and is propelled by two RL-10 engines. Actually the S-V is the upper stage of another rocket, the Centaur, modified to fit on top of the S-IV. In the test-flight program

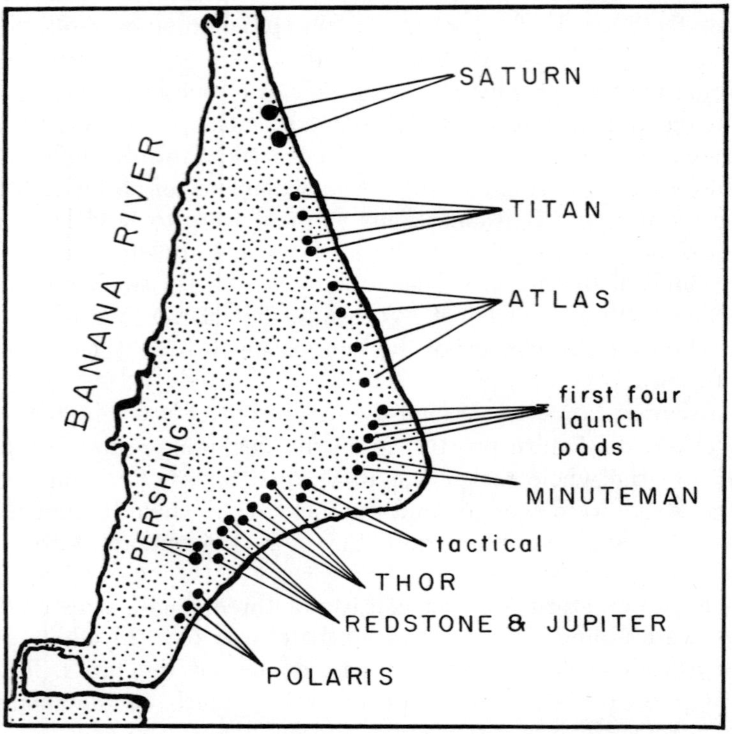

Fig. 76. Map of Cape Kennedy.

the S-V stage was present only theoretically. The first flights were made with only the booster stage in operating condition; the two upper were dummies of the right size and shape, ballasted to have the right weight. Later, when the second stage (the S-IV) was operational, experimental payloads were flown and in these flights the payload took the place of the third stage, with a special instrument section between it and the second stage.

One complication of the whole program resulted from the enormous size of the stages, especially the booster stage. That was much too large for railroad transport and while it might be possible to get a stage from Huntsville to Cape Kennedy with special road vehicles, all other traffic on the roads would have had to stop. Moreover, road transport is never quite gentle enough for large and sensitive equipment. The answer was water transport. A small fleet of enormous barges—some of them converted Navy vessels—had to be assembled. Rockets capable of flying thousands of miles per hour when under their own power had to travel thousands of miles on waterways at a pace somewhat slower than a brisk walk. The stages built at Huntsville traveled a waterway detour until they reached the Mississippi. The Mississippi carried them to the Gulf, then they traveled around the southern tip of Florida and up the coast to the Cape. The stages built in California had an even longer voyage, because they had to travel via the Panama Canal and then north to the Cape.

The day of the first flight test—after repeated static firings at Huntsville—was October 27, 1961. Of course everybody was worried, partly because everything had gone so well at Huntsville. If experience has taught anything at all, it has taught that a new rocket has to have "bugs." None had shown up at Huntsville; therefore, experience said again and loudly, they would show up in the flight tests. The countdown went smoothly and when it reached zero the engines ignited properly and the enormous rocket rose slowly and majestically into the air. The take-off weight was 925,000 pounds, of which the fuels accounted for 600,000 pounds. The black and white rocket, clearly visible even from a distance, did not falter for a moment; at the proper time it tilted eastward. The flight lasted 8 minutes; the range could be estimated quickly at 215 miles with a peak altitude of about 90 miles.

A few days later NASA could provide more specific data about this first flight, which was referred to as flight SA-1. On October 31, 1961, the George C. Marshall Space Flight Center in Huntsville sent out a release that read (*in toto*) as follows:

> Analysis of the extensive data which were radioed from the first Saturn space vehicle (SA-1) launched from Cape Canaveral October 27 revealed that the rocket attained a range of 214.727 statute miles and a peak altitude of 84.813 statute miles.
>
> Maximum velocity was 3,607 miles per hour.
>
> The vehicle's four inboard engines burned 109.37 seconds, whereas the four outboard engines burned 115.15 seconds.
>
> The period of "hold-down" on the launch pedestal—from the moment of initial ignition to lift-off—was 3.97 seconds. The engines were ignited in pairs, requiring a total of 3/10 of a second.
>
> Final weight of the rocket, at lift-off, was 927,000 pounds.
>
> The length of the flight was 8 minutes and 3.6 seconds.
>
> Thrust achieved at lift-off was 1.296 million pounds. The maximum thrust at altitude, which occurred just before cutoff, was 1.515 million pounds.

Experience then said: Well, all right, everybody was not only bright and efficient but also lucky; just wait until next time. Next time was April 26, 1962, and experience was wrong again. There was a half hour "hold" near the end of the countdown, but that was because a Navy destroyer and a commercial tanker happened to be in the danger zone where the rocket would make impact and had to be given time to get away.

SA-2 therefore took off at 9 A.M. (EST) instead of 8:30. At an altitude of 35 miles the engines stopped burning and at 2 minutes and 40 seconds after lift-off the rocket was at an altitude of 65 miles. At that moment the range safety officer pressed his "destruct" switch. Nothing was wrong with the rocket; this was another experiment. As was mentioned earlier, the dummy upper stages had to carry ballast, which the Florida beaches could provide in abundance. But in SA-2 the ballast consisted of 95 tons of water. Scientists were interested to see how water would behave at an altitude that water vapor never reaches under normal conditions, and the ballast for SA-2 therefore turned into Project High Water. The water, suddenly released into the virtually zero pressure of 65 miles altitude, formed a large disk-shaped vapor cloud in which lightning flashed. Some of the water just froze to form chunks of ice on the dummy top stage.

SA-2 had also been a complete success.

The scientists responsible for Project High Water were so surprised at the lightning that they asked for a repeat. SA-3 obliged. This was the first test flight in which the fuel tanks were charged to capacity, so that the rocket had 750,000 pounds of fuel at lift-off. The extra fuel increased the burning time by about 30 seconds.

SA-3 performed on November 16, 1962.

SA-4 performed on March 28, 1963—this was the flight in which one of the inboard engines was shut down some time after lift-off.

SA-5 was the first launch in which the second stage was the actual S-IV rocket and the upper stage was planned to go into orbit. For this reason the third stage, still a dummy, was ballasted with sand, since water, after several days in orbit, could be expected to freeze into a solid lump, which might be big enough to survive re-entry and land somewhere in one piece. Launch date was set for January 29, 1964, but had to be postponed for one day because a cover had been left on a fuel line. On January 30 SA-5 took off, with an 85-minute delay because a radio broadcast that was interfering with the tracking equipment had to be located by Air Force planes.

The weather was chilly and windy, but the take-off, at 11:25 A.M., was smooth. About 2 hours later a sign listing the precise performance was displayed in the lobby of the main building of the Huntsville Space Flight Center. It read:

T −	3.1	Ignition
T −	0	Lift-off
T +	15	Tilt starts
T +	70	Range 3 mi, altitude 8 mi.
T +	107	H_2 pre-start flow in S–IV
T +	134	Tilt stops
T +	139	S–IV O_2 pre-start flow
T +	141	S–I inboard engines cutoff
T +	147	S–I outboard engines cutoff
T +	147	S–IV ullage rockets fire
T +	148	Separation command, bolts explode
T +	149	S–I retro-rockets fire, S–IV engine ignition

T + 150	S–IV ullage rockets cutoff
T + 153	S–I retro-rocket cutoff
T + 169	Ullage rockets jettison, tilt resumes
T + 630	Tilt ends, S–IV engine cutoff
T + 630	Rocket goes into orbit 1300 miles downrange
T + 57 min.	First apogee reached.

Note: The retro-rockets mentioned in this list are small rockets that "hold back" the first stage so that the upper stage can lift iself off. The ullage rockets of the upper stage are small rockets that "shake down" the fuel in the tanks for proper performance.

The burned-out S-IV stage plus the ballast made an artificial satellite weighing 37,700 pounds. The first determination of its orbit read: perigee 160 miles, apogee 400 miles, orbital period 1 hour and 43 minutes, orbital velocity 16,650 miles per hour. Actually the perigee was at 167 miles and the apogee at 470 miles. Lifetime in orbit for the satellite was estimated at 500 days, but the actual date of its re-entry into the atmosphere was April 30, 1966.

SA-6, on May 28, 1964, carried the first actual payload, a "boiler plate" version of the Apollo spacecraft. This time the weight set orbiting was 37,300 pounds and the orbit had a perigee of 114 miles and an apogee of 149 miles. The spacecraft re-entered after 3 days and 8 hours.

SA-7 repeated the experiment on September 18, 1964. Weight in orbit was 37,000 pounds, perigee was at 115 miles, apogee at 140.5 miles, and re-entry took place on September 21, 1964. (Two barnacle-encrusted camera capsules from this flight were recovered from the Atlantic seven weeks later; both were in good condition and the pictures they had taken could be developed.)

SA-9 took off at 8:37 A.M. on February 16, 1965. It preceded SA-8 because SA-8, which was built by Chrysler (the others had been built at Huntsville), was not yet ready when SA-9 had passed all the ground tests. It carried the first of three Pegasus satellites. The Pegasus consists essentially of two enormous "wings" that unfold when orbit has been attained. They measure 96 by 14 feet and stay attached to the S-IV stage. They are not of course wings but panels for the detection of meteoroids in space.

The weight placed into orbit by SA-9 is 23,000 pounds; the orbit of Pegasus-I has a perigee of 308 miles, an apogee of 462 miles, and an orbital period of 97 minutes. Each meteoroid impact on the panels is registered by the satellite and telemetered to the ground. Most of the aluminum panels have a thickness of 1/60 inch, others have only half that thickness, while a few are as thin as 1/600 inch. This enables the satellite to register three different degrees of penetrating power. After 10 months of operation the Pegasus-I satellite indicated that 1 square foot of aluminum 1/600 inch thick would be penetrated seven times per year, while it would take about 9 years to have one penetration of a 1/60-inch aluminum sheet. These are, of course, preliminary figures, but they agree with earlier theoretical studies about the meteoroid hazard. With metal panels thick enough to have the structural strength they need, penetration would be very unlikely.

Pegasus-II was sent into orbit by SA-8, the ninth launch in the series, on May 25, 1965. The orbit runs from 316 miles at perigee to 459 miles at apogee, with an orbital period of 99.6 minutes. Pegasus-III followed on July 30, 1965, with SA-10. Its orbit is nearly circular (from 320 to 332.4 miles) with an orbital period of 95.2 minutes.

The first Saturn test-flight program of ten vehicles had come to an end. Experience notwithstanding, all ten had been successful.

Of course the people of the Marshall Space Flight Center were elated. But did they talk about their ten Saturn flights? No, their minds were now on Saturn IB and Saturn V, the next two rockets to be tested. Boosters of both these rockets were undergoing static firings in October of 1965, the year that had seen the conclusion of the first test-flight program.

Both Saturn IB and Saturn V are parts of the Apollo moon-flight program. The difference between these two rockets is that IB can only put the Apollo spacecraft into orbit around the earth, while Saturn V will make the actual trip to the moon in 1969. Since the payload of both rockets will be the Apollo spacecraft, that must be described first.

The Apollo spacecraft consists of three sections. The one that houses the three astronauts is the conical "command module," 12 feet high and 13 feet in diameter at the base. The second section is the "service module," which is essentially a rocket. The third section is the LEM (lunar excursion module), which will land on the moon with two astronauts; the third astronaut will meanwhile orbit the moon in the command module. The LEM itself has two sections. When the time comes to return to the orbiting command module, the lower section of the LEM will serve as the launching pad for the upper section. The upper section will then make rendezvous with the orbiting command module, the two astronauts will transfer to the command module, the upper section of the LEM will be left in orbit around the moon, and the command module, propelled by the service module, will return to earth. The service module will be left in space while the command module makes re-entry and lands.

The Apollo spacecraft is almost as tall as an Atlas rocket and is slightly heavier than an empty Atlas rocket. Portions of the Apollo mooncraft (minus the LEM) had already been flown in the first Saturn test-flight program; in fact, each of the three Pegasus satellites had been folded into the empty service module of the Apollo spacecraft, though they were attached to the S-IV stage.

Saturn IB, as far as the booster stage is concerned, is a modified Saturn I. Redesigning a number of parts trimmed its structural weight by more than 10 per cent, and a more powerful version of the H-1 engine is used. The result is a lighter and more powerful rocket with a take-off thrust of around

1,600,000 pounds. The second stage is the same as the third stage of the Saturn V. (The designation for that third stage is S-IVB.) Its diameter is 21½ feet and its length 58 feet. It also is a hydrogen-burning rocket but with only one engine of type J-2 which develops 200,000 pounds of thrust.

The Saturn IB with the S-IVB as the upper stage and a payload totaling 45,900 pounds, made its maiden flight on February 26, 1966, after two postponements caused by bad weather. The payload was an Apollo spacecraft, without the LEM, but with a charged service module. At an altitude of 310 miles the upper stage put the Apollo spacecraft into the right position and the rocket engine of the service module ignited to drive the command module into the atmosphere with a speed resembling that of a return from the moon. The command module, hanging from three parachutes, splashed into the ocean about 5000 miles downrange from the Cape and 38 miles from the recovery carrier, the USS *Boxer.*

The next launch, on July 5, 1966, was designated AS-203 (Apollo/Saturn 203). It carried a 58,500-pound payload and placed a partly filled hydrogen tank into orbit for the purpose of ascertaining how liquid hydrogen will behave under orbital conditions.

The thirteenth launch of a Saturn rocket took place on August 25, 1966. This was planned to be a very long suborbital flight for the Apollo spacecraft. The Apollo's service module was also charged and it lifted the command module to a maximum height of 700 miles. The suborbital flight carried the spacecraft across southern Africa and northern Australia to a splashdown in the Pacific, nearly 18,000 miles from the Cape. The spacecraft was recovered. The designation of this flight was AS-202.

This was to be the last experiment before "Apollo 204."

Saturn V is more or less the Nova rocket of earlier discussions. Its booster stage (S-IC) has a diameter of 33 feet and a length of 138 feet. It has five engines, one mounted rigidly in the center and the four others spaced around the rim and gimballed. But this is a new and bigger engine, the F-1, with a thrust of 1.5 million pounds, so that the take-off thrust of Saturn V will be 7½ million pounds, a feat that has been accomplished repeatedly in static tests. The second stage of the Saturn V is called S-II. It has the same diameter as the first stage and is 82 feet long. Propulsion is by five J-2 hydrogen-burning engines, delivering a total thrust of 1 million pounds. Again one engine is rigidly mounted in the center while the other four are gimballed. The third stage is the S-IVB rocket which has already proved itself in combination with the Saturn IB.

With the Apollo spacecraft as payload, Saturn V stands 363 feet tall. The first flight of the Saturn V was designated "Apollo 4" and scheduled for 7 A.M. (EST) on November 9, 1967. The gigantic rocket rose from the launch pad within seconds of the scheduled take-off time, shaking the ground for a few thousand feet around it. The unmanned Apollo command module splashed into the Pacific Ocean 8 hours and 37 minutes after lift-off. Splashdown was near Hawaii; the capsule was recovered in good shape.

Readers who are in the habit of remembering dates will have noticed that the Saturn test flights spanned just about the same time period as the Gemini flights—1965–1966. But Gemini and Saturn, plus Ranger shots to the moon, do not exhaust the story of the space activities during this period. The Air Force was testing, and using, a fabulous large rocket of its own, called the Titan III-C.

The Titan-III rocket is a Titan-II with a third stage, called the "transtage." This rocket, which is sometimes called Titan III-A for greater clarity, is the "core rocket" for the Titan III-C. In the III-C version, the liquid-fuel core rocket is flanked by two solid-fuel boosters of great power. A complete Titan III-C stands 127 feet tall. The three stages of the "core" all burn the same liquid fuels, namely a 50/50 mixture of hydrazine and unsymmetrical dimethyl hydrazine with nitrogen tetroxide as the oxidizer. The thrust of the bottom stage is 470,000 pounds, that of the second stage almost precisely 100,000 pounds, and that of the transtage 17,000 pounds.

The two solid-fuel propulsion units have diameters of 120 inches and lengths of 75 feet. A nose cone to reduce air resistance adds another 11 feet. The fuel is a synthetic rubber (polybutadien) plus acrylic acid and an addition of pulverized metallic aluminum; the oxidizer in the fuel is ammonium perchlorate. Each of the solid-fuel boosters produces a take-off thrust of 1.3 million pounds.

The first static test firing of a Titan III-A took place at the Cape on August 21, 1964. The problems were minor and the first test flight could be scheduled for early in 1965. The rocket did take off on February 11, 1965, and first assumed a parking orbit with perigee at 112 miles and apogee at 125¼ miles. Then the transtage fired and climbed to a near-circular orbit with perigee at 1730½ miles and apogee at 1733 miles. The orbital period is 145.4 minutes and the lifetime is indefinite. Among the ballast there was also a payload, a 69-pound satellite (called LES-1) for communication experiments.

Four months later, on June 19, 1965, the first flight of the III-C version took place. There was a certain amount of apprehension around, because one of the solid-fuel boosters had exploded in the test stand on May 2, 1964, but the engineers were satisfied that this had happened only because the welding seam of that particular casing had been defective. They were right. Titan III-C, burning only its two solid-fuel boosters, lifted off very well. After burnout the boosters were jettisoned on schedule and the core rocket ignited itself at an altitude of 24 miles, the largest rocket ever to have mid-air ignition. Two minutes later the first stage of the core was exhausted and the second stage took over, it burned for 3½ minutes. Then, after a short planned delay, the new transtage started burning and kept going for four minutes.[2] Finally it put itself and 21,000 pounds of lead ballast into a 115-mile orbit, a total weight of 29,300 pounds. The lead ballast consisted of deeply scored blocks that would break up and burn up on re-entry. Re-entry took place later the same month, the precise date is either not known or has not been released.

The final purpose of the Titan III-C will be to put the Air Force's Manned Orbiting Laboratory (MOL) into orbit. This will be a cylindrical structure 10 feet in diameter and 25–40 feet in length, providing ample room for experiments in a zero-*g* environment that is protected from space hazards. The front end of the MOL is a Gemini spacecraft with a hatch in its heatshield. After two or four weeks in orbit the astronauts will enter their Gemini capsule and use it for re-entry, leaving the MOL in orbit for future experiments.

In the meantime the Titan III-C has been kept busy putting Department of Defense satellites into various orbits. On June 16, 1966, a III-C put eight satellites into an orbit 20,941 miles above sea level, where the orbital period is so close to being synchronous with the earth's rotation that the satellites need 12 days to complete one "revolution"—that is, to appear over the same meridian again. Of these eight satellites, one was a "gravity gradient" research satellite; the others were IDCSP[3] satellites.

Another Titan III-C carrying eight satellites exploded in midair 80 seconds after take-off on August 26, 1966. The cause was traced to a failure of the shroud enveloping the satellites.

A Titan III-C flight that took off on November 3, 1966, was a test in the MOL program. The rocket carried a simulated MOL (an old fuel tank) with a used Gemini capsule on top, the whole payload measuring 49 feet. The tall payload did not disturb the balance of the rocket. The unmanned Gemini capsule, which had a 24-inch circular hatch in its heatshield, was separated from the simulated MOL at an altitude of 127 miles to begin its return trip to the ground, or rather the ocean. After traveling a distance of about 5500 miles the capsule made splashdown and was recovered by the USS *LaSalle;* it had survived the journey and was theoretically re-useable. Meanwhile the transtage had been ignited twice and had placed the empty fuel tank into a 185-mile orbit. There three satellites were ejected in succession, the first being a classified scientific experiment, the other two being a pair of communication satellites, one a receiver and the other a trans-

[2] The name transtage was chosen to indicate that this stage is maneuverable and therefore capable of transferring itself and its payload from one orbit to another. It demonstrated this capability after a Titan III-A take-off on May 6, 1965. Being placed into a nearly circular orbit at about 115 mi., it broke out of this orbit to go into an elliptical transfer orbit, then changed into a near-circular orbit (1729 to 1743 mi.), left an artificial satellite behind in that orbit, and went into a fourth orbit with perigee at 1725 mi. and apogee at 2317 mi.

[3] The initials stand for Initial Defense Communication Satellite Project.

mitter, to test communications from satellite to satellite through the ionosphere.

On January 18, 1967, the experiment that had gone wrong in August 1966 was repeated. This time the flight was flawless, depositing eight IDCSP satellites into a near-circular orbit 21,000 miles above the equator. Finally, on April 28, 1967, another Titan III-C orbited three research satellites and two Vela satellites; the latter are nuclear-blast detectors orbiting the earth at a distance of 69,000 miles.

The solid-fuel boosters of the Titan III-C are the largest solid-fuel rockets actually in use, but like their liquid-fuel relatives each type is only a prototype of a larger one to come. The temporary goal for solid-fuel boosters was one of 260 inches, or a little over 21½ feet. In 1950 a solid-fuel rocket 21½ feet long would have been acclaimed as a masterpiece, but in 1965 when NASA said 260 inches this referred to the diameter.

The program for the development of unusually large solid-fuel units had been initiated by the Air Force in 1963. Later the Air Force retained only the 120-inch units (for Titan III-C) and the 156-inch units, for possible "up-rating" of the Titan III-C. Still larger units, because of problems of mobility, were given to NASA.

The contract for the 260-inch rocket was won by Aerojet-General. The units to be built were referred to as SL (for "short length"), because a unit actually to be used for a space mission would have double the length, and also double the thrust of the SL motors tested. Even so their sizes were impressive. The exhaust nozzle was 20 feet tall and had a throat diameter of 6 feet. The casing weighed 60 tons before charging and presented a number of problems. Transportation to Florida from the Sun Shipbuilding Company in Chester, Pennsylvania, was the least of them; it could be shipped by barge down the East Coast and via the Inland Waterway. But a steel casing normally had to be annealed and there was no furnace of the necessary size anywhere in the country. Nor would the budget permit construction of an annealing furnace for this purpose alone. The problem was solved by using a new steel alloy, called maraging steel, which contains 18 per cent nickel.

The gigantic charges were placed into a concrete-lined pit with only the nozzle showing; logically they had to be fired upside down. In order to ignite the charge (which would burn at the rate of 6 tons per second) another rocket was needed, having a thrust of 250,000 pounds and producing an 80-foot flame. This rocket, after jetting its flame for about half a second into the big charge, was permitted to fly out through the nozzle but was tethered with steel cables which forced it to end its flight in a nearby "recovery pond." Of course the "igniter" also needed an igniter—a solid-fuel rocket with a weight of 100 pounds and a thrust of 4500 pounds.

One firing of the 260-inch short-length motor was carried out in 1965 and one in 1966. The results, and a comparison with a 156-inch firing, are shown in the following table:

	156-IN. (THIOKOL)	SL-1 (AEROJET)	SL-2 (AEROJET)
Place of firing	near Brunswick, Ga.	Dade County, Fla.	Dade County, Fla.
Date of firing	Feb. 27, 1965	Sept. 25, 1965	Feb. 23, 1966
Weight, total	910,000 lb	1.86 million lb	1.86 million lb
propellant	800,000 lb	1.68 million lb	1.68 million lb
Length	100 ft	60 ft 3 in.	60 ft 3 in.
Chamber diameter	13 ft	21 ft 8 in.	21 ft 8 in.
Burning time	54.6 sec	112 sec	114 sec
Thrust, peak	3.2 million lb	3.57 million lb	3.6 million lb
average	3.0 million lb	2.95 million lb	near peak

The third (and last) of the 260-inch SL units was fired in Florida on June 17, 1967, and attained a maximum thrust of 5.7 million pounds. The test was marred by the fact that during the last few seconds of burning, the nozzle gave way and parts of it were carried up by the exhaust blast. Fortunately no one was injured, but the accident proved that full reliability is still to be achieved.

Having followed the growth of both solid-fuel and liquid-fuel rockets we now come to those types of rockets that do not yet exist—in the form of flyable vehicles, that is; they do exist as working models and as prototypes. There are three different kinds of such rockets: the hybrids, the "atomic" rocket, and the "electric" (or ion) drive.

The hybrid rocket is precisely what its name proclaims, a hybrid of liquid and solid fuels. Research on hybrids is being actively pursued by a number of companies in the United States and by French and Swedish research teams.[4] An explanation of how a hybrid works is best furnished by a description of a small working model that has been shown to engineering meetings for years by representatives of the United Technology Center in California.

This model, mounted on a table, had a container holding gaseous oxygen under pressure at one end. At the other end there was a thick-walled tube of Plexiglas (the fuel) and the exhaust nozzle. Between these main components there was the ignition mechanism which was operated by pulling a trigger. When the trigger was pulled the oxygen valve opened and a small amount of propane gas was introduced into the oxygen flow and electrically ignited. The burning propane which carried a large excess of oxygen ignited the Plexiglas tube along its center hole and the Plexiglas kept burning until release of the trigger closed the oxygen valve.

[4] For a report on Swedish work written in English, see the article on hybrids by Björn Ankerswärd of Svenska Flygmotor AB in *Interavia,* December 1964, pp. 1838 ff.

For a hybrid rocket that is to do more than just demonstrate the principle, a solid fuel and an oxidizing agent that produce hypergolic ignition would be chosen so that no separate ignition device would be needed. The hybrid rocket, like the liquid-fuel rocket with a hypergolic combination, permits instantaneous shut-off and nearly instantaneous re-ignition. This capability makes hybrids look attractive for thrusters in manned spacecraft and braking rockets. Whether a hybrid could compete in the role of a take-off booster or as a second stage seems to be mostly a question of development and operating cost.

The so-called "atomic" rocket,[5] under development since about 1960, does use the rocket principle for propulsion, but the energy is not derived from the combustion of a fuel. A nuclear rocket consists essentially of a tank holding the "reaction mass" to be expelled and an atomic reactor which is run at a temperature as high as the reactor will stand without melting down. The reaction mass in the experimental nuclear rocket is hydrogen in liquid form. In operation the liquid hydrogen is pumped into a jacket surrounding the reactor, just like the cooling jacket of a liquid-fuel engine. But in this case the purpose is not to cool the reactor but to heat the hydrogen so that it becomes a gas. The hydrogen gas then flows through a multitude of holes in the reactor core where it is further heated. It emerges as a very hot gas which is then expelled through an exhaust nozzle.

If such a rocket were used in the lower atmosphere the hot hydrogen would then burn with atmospheric oxygen, but this would be just a side effect which would not contribute to the propulsion of the rocket. However, because of radiation problems and some other factors, nuclear rockets are under consideration only for upper stages that will remain inert until orbit has been reached. Nuclear rockets are attractive because they are expected to be about twice as powerful as rockets with chemical fuels, but they do have their drawbacks. Because of the great temperature difference between the hydrogen and the reactor they must be started slowly. Similarly the reactor, when shut down, will stay hot for some time, and even after it has cooled off it is still "hot" in the radioactive sense of the word. This is relatively unimportant in the case of an unmanned mission but quite important in manned missions.

On March 3, 1966, an experimental NERVA engine was run under full power for the first time; before that there had been eleven test runs with reduced power, adding up to a total of 61 minutes of operation.

And finally: the "ion drive."

Even in the ion drive the principle of propulsion is still that of the rocket: the vehicle accelerates in space because mass is expelled from it. The difference lies in the nature of the exhaust and in the manner in which it is

[5] Project NERVA; Nuclear Engine for Rocket Vehicle Application.

accelerated. Ions, to begin at the beginning, are atoms or molecules carrying an electric charge. Because they do carry a charge they can be accelerated by electric fields and it was quite obvious from the outset that the velocity of expulsion of ions would be several times the exhaust velocity of a chemical rocket. But while the exhaust velocity would be high, the mass of the exhaust would be quite small; the latter factor could be ameliorated to some extent by using ions with a high mass—that is, atoms of an element with a high atomic weight. But here the choice was quite limited, since the heaviest naturally existing atoms are also naturally radioactive and this would interfere with the operation of the engine. Therefore the element that was to provide the ions had to meet three demands. Its atoms must be heavy; they must not be radioactive; and the element itself must be in reasonable supply at a reasonable price.

Two elements met these conditions—cesium and mercury.

To transform a quantity of cesium or mercury into propulsion it is first necessary to change the metal into a vapor, with (under ideal conditions) each particle ionized. The vapor then is accelerated by electrostatic fields and acquires a high velocity. But both the vaporizing of the metal and the maintaining of the fields requires energy that has to be obtained somehow. An atomic reactor would provide the necessary energy, but in space energy from the sun is available all the time and in virtually any desired amount. Hence an ion-propelled spacecraft can be designed in several different ways.[6] Experimental research on ion engines is complicated by the fact that the molecules of the air also interfere with the operation, so that real testing will have to be done in orbit. Small ion engines, one using cesium and one using mercury, have been lifted into space by a Scout rocket and have proved to work under space conditions.

Because an ion engine could never lift its own weight off the ground—even if the atmosphere did not interfere with its operation—these can serve only as the final stage of a space mission. The amount of thrust obtained from an ion engine is small—one preliminary calculation based on solar radiation as the energy source indicated 1 pound of thrust per ton of equipment weight—but the engine can operate for hours and days, and possibly even weeks. Under space conditions even a weak thrust, if applied for many hours, can build up considerable velocities.

Since an ion engine will need so much time to achieve high velocities it cannot be used for nearby targets such as our moon. Even for missions to Venus and Mars simple coasting flight, as used in the Mariner space probes, will get the device to the target planet as fast as an ion drive

[6] For an outline of theory and design possibilities see the article "Electric Propulsion" by W. E. Moeckel (of NASA's Lewis Research Center in Cleveland, Ohio) in *Science*, October 11, 1963. Dr. Ernst Stuhlinger of the George C. Marshall Space Flight Center in Huntsville, who has devoted years of research to the ion drive, has not yet published a major work on his researches.

would. But for long missions, say a flight to the major moons of Jupiter, the ion drive looks superior to the other means of propulsion we now have.

Of course there is no proper ending to the story of rockets and spacecraft to come, any more than there was a proper ending to a story of exploration of our own earth when the North Pole had not yet been reached, Antarctica was merely known to exist, and the interior of Asia was forbidden territory. By now, of course, the exploration of our own planet may be said to have been completed.

In the exploration of space no such end can be predicted. There is, first, our moon to be explored. Then the inner planets of the solar system. Then the outer planets. And beyond our solar system there are the stars, millions and millions of suns, most of them with planetary families of their own.

The exploration of space will go on forever and ever. . . .

APPENDIX 1

Rocket Airplanes and Rocket-Assisted Take-Off

Pour aller plus haut, et l'homme voudra aller plus haut, il faudra adopter un principe différent. Le principe de la fusée est tout indiqué. Le moteur à réaction s'en déduit. L'homme sera enfermé dans une enceinte ou l'air respirable eu sera fabriqué. A vrai dire, il ne montera plus une machine volante, mais plutôt un projectile dirigeable.

—Capitaine F. Ferber de Rue (1911)

The story of the rocket airplane is the story of a sideline of rocket theory, and the rocket airplane itself is in a manner of speaking a by-product of rocket research. It may sound like an obvious thought to take a rocket motor and to build an airplane around it. Actually this idea, which occurred again and again to literally scores of people, happens to be unworkable.

Such a flat statement may seem all the harder to defend if it is admitted at the same time that the Germans flew a rocket airplane operationally during the Second World War and that the American rocket-propelled research plane X-1 passed the speed of sound a few years later. Still, in its original form the problem of the rocket airplane is unsolvable, if one requires a solution to be an efficient solution.

What I just called the "original form" is what was advocated more than a score of years ago by Max Valier who liked to think that a good, sound, fast rocket airplane could be produced by simple substitution of rocket motors for internal-combustion engines in existing airplanes (the airplanes of 1928!) and that it was merely a question of adding rocket motors and of reducing wing surface in order to arrive at a man-carrying space rocket in the end. Even though Valier's presentation of what he took to be a technological problem was universally rejected, the first experiment in all history of a man-carrying rocket-propelled airplane can be traced back, in the end, to Valier's publicity campaign.

That experiment took place in the summer of 1928, and it was an offshoot of the Opel experiments initiated by Valier. The airplane was a glider of the then novel canard type—which to the uninitiated always looks as if it were flying backward—and it made its first and only flight on June 11, 1928, from the Wasserkuppe, one of the Rhön Mountains in western Germany.

The experiment was a joint venture: Sander manufactured the power rockets used, von Opel paid for them or at least contributed part of the expense, and the Rhön-Rossitten Gesellschaft, a German glider society which derived its name from the sites of its two most important camps, furnished the glider. It also provided a number of glider models which were tested before the full-sized glider was risked. The experiments were directed by A. Lippisch who reported on them in a German aeronautical journal[1] while the pilot of that first rocket glider, Friedrich Stamer, was present and assisted during the preliminary tests. The assortment of rockets furnished by Sander for the experiments consisted of five types, three for the model gliders and two for the full-sized glider.

TYPE	BURNING TIME (SEC)	THRUST (KG)	(LB)
For the models			
Bored rocket	3	75	165
Bored rocket	3	175	385
Solid brander	30	5	11
For the glider			
Bored rocket	3	360	790
Solid brander	30	20	44

Naturally the aviators made the model tests first. The model chosen for the experiment was a tailless type called *Storch* (stork) with a wing span of a little over 7 ft and a weight of about 30 lb. (It is not stated whether the rocket was included in that weight.)

For the first experiment one of the powerful 165-lb-thrust rockets was used. The result was what should have been expected: the powerful rocket regarded the wings and ailerons of the model as a mere hindrance to proper performance; it rushed the model vertically up into the air until it was burned out and then let it drop. And it did drop, since it was in an unstable position. When the second experiment, with one of the small solid branders, was almost as bad a failure, Lippisch realized that the models were poorly adjusted for that type of propulsion.

In the third experiment, also with a small solid brander, the model was launched from a small wooden rack by a rubber rope which was released automatically. The result looked encouraging: the model made a long and perfect flight. The fourth model experiment, as well as the fifth, was made with the most powerful rockets that were available for the models; they had to stand an acceleration of 12 *g* and there was some doubt whether they would.

Test 4 was very much a repetition of test 1. The model left the rack "like a projectile," as Lippisch put it, and climbed to about 300 ft. Alone, the rocket would have attained about ten times that height; the wings did nothing but provide a lot of unnecessary air resistance which cut down efficiency. For some reason they stayed on and held together. At 300 ft, when the rocket was burned out, the model turned on its back, flew for a few seconds in that position, and then slipped over one wing, assuming normal position which resulted in a long glide. In test 5 the wings gave up; they were not designed for an acceleration which made the model attain 350 mph in less than 3 sec. They simply broke and the model fell vertically to the ground like a stone when the rocket was exhausted.

These tests gave a few hints of what was to be expected with the glider, and the experimenters followed the most obvious of them: they discarded the 790-lb-

[1] See *ZfM* (Munich), vol. 19, no. 12, June 28, 1928.

thrust rockets, using only two solid branders of 26-lb and 33-lb thrust, respectively. Since it was expected that the pilot might be excited and make mistakes, the electric ignition of the rockets was rigged in such a manner that he could not ignite both of them simultaneously, no matter what he did wrong. This was a nice precaution, and the planned procedure was that the glider should be launched by the customary rubber rope. The pilot was not to ignite the rockets until he was free of the rope and safely in the air.

In spite of all these arrangements the first two attempts failed. In the first something went wrong with the rubber rope and Stamer did ignite one of the two solid branders before he was in the air. It just burned out and nothing happened. That it burned without any mishap may have quieted the pilot somewhat, because during the second test he did get off the ground. But he found trouble balancing the glider and landed after a flight of some 700 ft without igniting the second brander. The glider was then brought back to the starting point and the second brander was taken out. After inspection of the ignition system the glider was fitted with two solid branders of 44-lb (20-kg) thrust each.

The quotations which follow are from Stamer's report:

> The plane took to the air by means of the rubber rope, aided by one rocket [brander]. After flying for about 200 yards in a straight line—I noticed a slight climb—I made a curve of about 45 degrees to the right and flew another 300 yards. Then I turned again to the right, again about 45 degrees. Immediately after the [second] turn the first rocket ceased burning and I ignited the second. . . . This time I flew about 500 yards in a straight line, made a turn to the right of about 30 degrees, and landed the machine after another 200 yards, just a few seconds before the second rocket was exhausted.

The distance covered was just about one mile and the time needed was a little over one minute. Asked about his impressions, Stamer declared that the flight had been "extremely pleasant" and added, "I had the impression of merely soaring, only the loud hissing sound reminded me of the rockets."

The second flight was intended to go over a small mountain.

> The launching went all right and while the plane took to the air I ignited the first rocket. After one or two seconds it exploded with a loud noise. The nine pounds of powder were thrown out and ignited the plane instantly. I let it drop for some sixty feet to tear the flames off. After succeeding in that I landed the plane without mishap.

Just after landing the second brander caught fire but fortunately did not explode. The fire had damaged the glider to such an extent, however, that extensive repairs were necessary. The Rhön-Rossitten Gesellschaft did not resume the experiments; it seems that the leading members had come to the conclusion that rockets did not serve their purpose.

The members of Rhön-Rossitten Gesellschaft did have a definite purpose in mind when making these tests; they had not been made just to please von Opel or anybody else. Following the trend of an oft-made suggestion, they had hoped to use rockets in lieu of the rubber rope or the car tow to get their gliders into the air. Both these methods, the rubber rope and the car tow (the latter was rarely used by that society because of the mountainous territory in which it worked), were good and inexpensive methods, but they had a disadvantage: a glider pilot could not go out for a flight alone. Without help he could not get into the air; he needed either at least eight men with the rubber rope or else a car and a driver. If rockets had proved satisfactory that problem would have been solved. And since the rockets did not weigh very much it would have been possible, theoretically, to carry two or three of them along for other take-offs after intermediate landings.

The main reason for discontinuing the experiments was not so much the fact that the first experiments had not come off too well as the ever-present danger of explosion which increased with the age of the rockets. Since then, as explained in Chapter 7, the problem of the explosion-proof powder rocket has been solved satisfactorily for military purposes, so it might be possible to resume glider experiments with these improved rockets.

After the Rhön-Rossitten Gesellschaft had abandoned the project it was taken up by the Raab-Katzenstein Airplane Company at Kassel. This firm set out to build a tailless plane similar in design to the small Lippisch stork model, but large enough to carry a pilot and possibly even a passenger. For some unknown reason the first tests were failures and Raab-Katzenstein gave up; in an interview Raab said the Army intervened. But Fritz von Opel, who had somehow been connected with that attempt too, was not so quickly subdued; his early doubts had been replaced by the conviction that difficulties could be overcome by sufficient perseverance. He had a new rocket glider built for himself.

It was ready for flight testing on September 30, 1929. A wooden take-off rack some 70 ft long was used, but there was no rubber rope or other launching device; the take-off was to be accomplished by the rockets too. The first two tests, made in the early forenoon of the day, were not successful. The rocket batteries did not develop enough power to get the plane into the air; it just made short jumps from the end of the cradle. After lunch, the rocket battery having been replaced in the meantime, von Opel tried once more, this time successfully. He made a flight lasting about 10 min during which a top speed of about 100 mph was attained. But something went wrong in landing. The plane had a high landing speed (70 mi per hr) for its weight, which was about 600 lb including the pilot, and the construction was apparently not strong enough for the abuse it had to take. Probably it would have been ruined by the landing anyway, but apparently also the wings caught fire in the air; in any event the plane was very badly damaged during landing and was useless afterward. An eyewitness stated that it seemed like a miracle that Opel could have even escaped with his life; actually he was not even injured.

That was Opel's last attempt in rocket research.

A few similar experiments followed in the wake of the Rhön-Rossitten and Opel flights in Italy three years later. In 1931 reports came that an Italian engineer named Ettore Cattaneo had built and tested a 620-lb glider-like airplane at the airport of Milan. Since these reports spoke of high-power rockets for take-off and of less powerful rockets for maintaining the flight, Cattaneo probably followed Opel's example closely. In one of the tests the craft remained in the air for 34 sec, traveling a distance of 1 km or about ⅔ mi.

It goes without saying that these rocket gliders of Lippisch, von Opel, and Cattaneo were not the first projects of their kind; they just were the first to be actually built and flown. There had been many earlier projects, not even counting those mentioned in Chapter 4 which were conceived before the internal-combustion engine existed and where the choice of a power plant was, consequently, not a matter of free choice.

Among the earlier projects was one by a Russian engineer, Fyodor Geschwend of Kiev, who dreamed of a winged railroad car propelled by a steam jet. There was another one by a German inventor named Wilhelm Gaedicke who wrote under the *nom de plume* of "Engineer Crassus" and who advocated a helicopter with a multi-bladed rotor, driven by the reaction of an exhaust jet of air and illuminating gas. A similar jet, attached to the suspended cabin, was projected for forward movement. There was also the flying torpedo of Alexander Gorokhov, with three exhaust jets on either side of the body and with wings so small that they were more like large stabilizing fins.

More important were the suggestions of René Lorin, published in a series of loosely connected articles in the French aviation magazine *l'Aérophile,* beginning in the year 1908. As far as the aerodynamic features of his aircraft were concerned Lorin had been groping around, as was logical for that time. But he did have a sound idea of what he wanted to use for a rocket motor. It was a normal six-cylinder in-line engine which Lorin wanted to build so flat that it would fit into the wing of the airplane—what is now called a "pancake engine." An exhaust nozzle was to be attached to each cylinder and there was to be one engine in each wing, so that the plane would be propelled by a constant series of exhaust blasts.

Lorin's critics had to admit that the scheme looked as if it would have a number of minor advantages. There were no drive shafts and no gears. There were no propellers, and since the wooden propellers of those days were hardly very trustworthy their elimination looked like another advantage. It all apparently added up to a weight-saving power plant. There was only one thing wrong with it: the Lorin scheme, with its several minor advantages, lacked a major one. That major advantage was on the side of the engine with drive shaft and propeller.

Lorin had been caught on that dangerous ground where the two sciences of rocket propulsion and of aerodynamics are supposed to merge but show, instead, a remarkable tendency to clash. The facts that got in the way of the early proponents of reaction propulsion for aircraft are now well understood and, for this reason, have become tractable to a large extent. The problem is the same one of over-all efficiency that has been mentioned repeatedly in earlier discussions in which it was stated that one must strive to match the velocity of the rocket to the velocity of the exhaust. In plain rockets of the type of V-2 or Viking this comes close to happening at the end of the burning period and for such rockets it is actually mostly a question of over-all size. In a small rocket it can't be done no matter what. In a large rocket of long burning time it will occur more or less automatically; the designer will do his best to make it occur as soon after take-off as feasible. But when it comes to airplanes the problem is complicated by several factors, the most important being the existence of the wings which produce both lift and drag, and therefore are fine devices for holding the speed down.

In Lorin's scheme there was a kind of early jet engine, producing a very fast exhaust jet of low mass. Lorin could not know why this had to be inferior to propeller aircraft, where the propeller creates a "jet" of large mass and low velocity. But only a few years later engineers began to realize the real reasons, the gaping abyss between exhaust velocity on the one side and airplane speed on the other.

There were two ways to bridge that abyss. One was to increase the speed of the plane. The other was to reduce the velocity of the exhaust. Complete bridging would probably require both.

In 1917 O. Morize of Châteaudun, France, proposed the first power plant for airplanes which was supposed to help to bring the two, airplane and reaction propulsion, together. By means of an engine-driven compressor, fuel injection pumps, and a combustion chamber with exhaust nozzle, Morize first produced a jet. But then he used a thrust augmentor on it. A thrust augmentor is a device which resembles a set of curved lamp shades, growing successively larger and held together by flat braces so that air can pass between them. When a high-speed blast goes through the center of this device the fast jet sucks air in through the spaces between these nozzle rings, and the air adds itself to the fast jet, increasing its mass and slowing it down at the same time.

Morize, it appears, never built such a power plant, but it sounded like a possible solution of the problem of utilizing an exhaust jet, and another Frenchman, Henri F. Mélot, made the Morize scheme practical. He dispensed with most of Morize's equipment and took two cylinders which he fitted end to end. Fuel inlet ports and spark plugs were placed at either end of the two-cylinder assembly. Inside there was a piston, but it had no connecting rod, it merely shuttled back and forth to provide compression. The exhaust from either end of the double cylinder escaped through a tube into one common "buffer chamber" to which the exhaust nozzle was attached. The result was an intermittent rocket blast achieved by simple means —means simpler than those of Morize, at any event—which then was sent through a thrust augmentor.

Mélot did more than draw a nice design and write about it in *La Science et la vie*. He actually built his motor. He actually made it work too. It was difficult to get it started, but once it was running it kept running. After some practical results had been accumulated, Mélot calculated that two somewhat enlarged engines of his kind would be capable of propelling an airplane of then customary dimensions. He experimented during the period from 1918 to 1920—and resumed experimentation again a few years later—but it seems that he never progressed much beyond his original success. There were numerous speculations at that time as to the reason for no further and impressive news from Mélot. These speculations ranged from a suspicion of plain failure to an assertion of marvelous success with resulting purchase by the French War Ministry and inevitable military secrecy. The actual explanation of the great mystery is quite simple: there was no reason to replace the well-known and steadily improving internal-combustion engine with a new and insufficiently tested device. Airplanes cruising at speeds from 100 to about 130 mph at their very best did not need a new type of engine which at some future date might *possibly* prove superior to the then-known engines.

Up to the time of Mélot the story of the attempts at rocket-propelled airplanes had progressed in a straight line, from inventor to inventor, from project to project, from theoretical improvement to theoretical improvement. But then developments began to branch out, mainly because the efficiency problem had been recognized.

One set of projects held on to the rocket as propulsive mechanism and saw the solution of the efficiency question in a sufficiently high speed. The other set of projects accepted whatever speeds seemed reasonable and, like Mélot, tried to adjust the rocket to the airplane rather than the airplane to the rocket. The latter resulted in what is now commonly known as jet propulsion.

While the story and the problems of jet propulsion are outside the scope of this book, a few words about the differences and similarities of rocket and jet propulsion are needed.

In both, the movement of the airplane takes place in accordance with Sir Isaac Newton's Third Law of Motion. The difference is that the jet engine is an air-breathing engine while the rocket motor is not. One might say that the jet engine is a rocket motor which gets its oxygen out of the air and which for this reason is not only fairly complicated mechanically but also limited as to altitude. The statement can also be turned around to read that a rocket motor is a jet engine which is greatly simplified mechanically by carrying its own oxygen along and is therefore not limited by altitude.

The long gap between von Opel's rocket-propelled glider and the modern work on rocket airplanes is filled by just one name: that of the Austrian engineer Dr. Eugen Sänger. He was indubitably the first among rocket-aircraft designers who did not grope and hope but attacked the problem systematically. He began his career as a rocket expert with an extensive series of rocket-motor tests conducted in the laboratories of the *Technische Hochschule* (College of Engineering) in Vienna.

During his lifetime Dr. Sänger published only short résumés of his experimental work, and his style, if I may be forgiven for saying so, was not always above reproach: one could not always tell what he had done at a given moment and what he hoped to do next. But he did keep a laboratory notebook during the Viennese period, and this notebook, comprising nearly four hundred handwritten pages and sketches, has been preserved. His widow made a photographic copy available to the National Air and Space Museum in Washington, and the assistant director of the museum, Fred C. Durant III, permitted me to use this copy.

The notebooks run from October 1933 (when Sänger submitted a proposal to the Austrian Ministry of Defense) to the end of 1934 when the Dean of the College forbade further experimentation. In the proposal Sänger advocated a research program aimed at the development of a rocket motor for airplane propulsion so that airplanes with speeds of 600 mi per hr and flight ceilings of more than 10 miles could be realized. He stated that a rocket motor should be used not like a piston engine but to provide "propulsion during the ascending leg of the flight path which is then followed by a long glide." The rocket motor to be developed should furnish a thrust of about 22,000 pounds for a duration of 20 minutes with absolute reliability of operation.

On February 3, 1934, the Austrian Ministry of Defense replied in the form of a sharp rebuke. The letter, only about ten lines long, states that the whole proposal does not deserve consideration because the main item of the proposal, namely the liquid-fuel rocket motor, cannot be realized because of the "detonation-like character" of the combustion. I had always wondered why Sänger, in his few publications, always stressed that "the continuous detonation inside a rocket motor," or the "explosion-like burning of a rocket," can be controlled. Now I know why he used these terms.

After receiving this rebuff, he began to work in earnest. At first he thought that the rocket motor had to be lined; one of his earliest sketches shows, from the inside out, a layer of high-density graphite (which would gradually be worn away), then a thick layer of soft graphite, then a layer of sheet steel, then a space through which the fuel would run in order to cool the sheet steel, and finally a layer of thin sheet metal to confine the fuel. He must have discovered soon that the graphite layers were unnecessary. The fuel was to be a thin fuel oil; though the oxygen would be liquid in the later phases of his research, he began with gaseous oxygen. Similarly, though his very first sketch shows fuel-cooling, he first used water-cooling on the test stand.

Experiment no. 25, conducted on March 15, 1934, is called "first successful experiment; internal pressure 15 atm., duration 30 sec." Several weeks later Sänger, while still using water-cooling, drew a sketch of a rocket motor in which the cooling liquid is inside a thin pipe wrapped around combustion chamber and exhaust nozzle.

In June 1934 Sänger made some calculations about metals as rocket fuel "either in pure form or as additives." According to his table, lithium gave the best over-all result; beryllium released about twice as much energy as lithium but required more than twice as much oxygen to do so. Early in August 1934 he obtained a thrust of 30 kilograms (about 66 pounds) and made a sketch showing the "shock diamonds" in a rocket exhaust, the first mention of this phenomenon that I know of.

After the continuation of the experiments had been forbidden, Sänger wrote an article for the December 1934 issue of the Austrian aviation journal *Flug (Flight)* giving a picture of a fuel-cooled rocket motor (Fig. 77). It may be noted that this picture differs in a number of respects from the sketches in his notebook; I assume that this is what he had intended to build next.

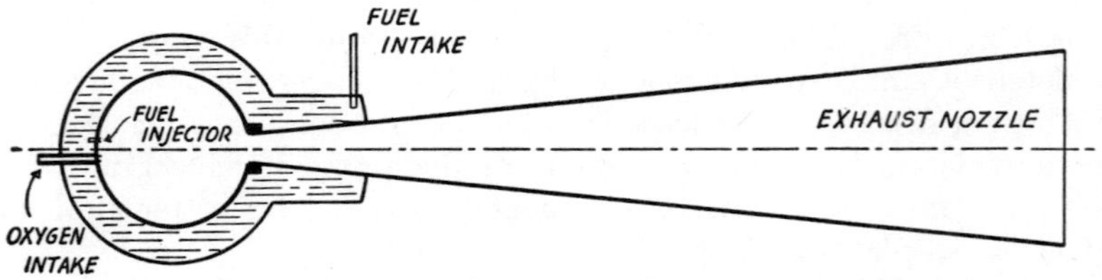

Fig. 77. Dr. Sänger's experimental rocket motor.

The next step was to work out the requirements for the rocket airplane itself. Oberth had worked a little on that problem, largely in defense of rocket theory against Valier's loud claims, and had showed that a rocket-propelled plane could get range only by a very steep take-off, by leveling off at a high altitude, getting up to maximum speed by using up all its fuel as quickly as possible, and then operating as a high-speed glider. Sänger came to essentially the same conclusions, but approached the problem more from the point of view of an airplane designer. He decided in favor of a rather shallow take-off (for a rocket, that is) of 30 degrees but aside from that his method was the same as that outlined by Oberth. For an

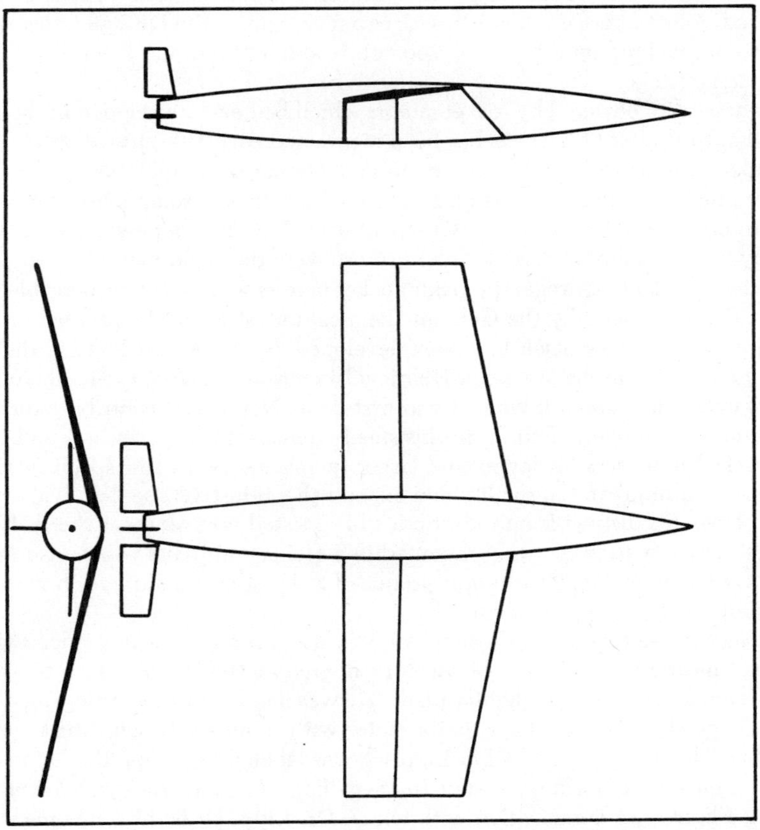

Fig. 78. Dr. Sänger's sketch of a stratosphere rocket plane.

assumed burning time of 20 min he arrived at a total flight time of a few minutes over one hour and an average speed of 1600 mi per hr. Of course he also tried to visualize the appearance of such a stratosphere speedster; Fig. 78 shows a copy of his sketch. When it was first published in 1933 it reminded one of Captain Ferber de Rue's prediction about the *projectile dirigeable*. Now it looks more like a first sketch of the much later American research plane X-1. It isn't, of course.

A few years later Dr. Sänger was hired by the Hermann Göring Institute, the research laboratories of the Luftwaffe. He informed me of this by writing me an innocuous private letter on office stationery, but he naturally could not tell me anything else. I now know that much of his time was spent not on rockets but on ram jets which the Germans call Lorin ducts. However, he did not abandon rockets completely and one piece of work which he did during the Second World War led, as we'll see later, to one of the most confused postwar episodes in rocket history.

Doctor Sänger did not have anything to do with the rocket-propelled airplanes built or planned by the Germans. One of those was produced and flown operationally, the Messerschmitt Me-163B Komet. One, a short-range reconnaissance plane for one rocket unit, the DFS 228, was under development when the Third Reich collapsed. Another, a scout bomber with two rocket units, the DFS 346, theoretically capable of reaching an altitude of 100,000 ft and a speed of 1700 mi per hr, was in the design stage.

All these were powered by rocket motors which had been developed in the factory of Hellmuth Walter of Kiel. As has been mentioned earlier, high-strength hydrogen peroxide became available for the first time in Germany in 1936. Some of Hellmuth Walter's motors utilized that as an oxidizer with a fuel of some kind; these motors came to be dubbed "hot motors." Others utilized T-Stoff as a power source without fuel, by way of catalytic decomposition; those were the "cold motors."

Walter's work on hydrogen-peroxide rocket motors was chiefly responsible for the larger advance gained by the German chemical industries in the production of this substance. The process itself had been developed first by Albert Pietsch, the owner of the Elektrochemische Werke in Höllriegelskreuth in Bavaria. Quite likely Walter did not even know about it when it was first done. Nor did he contribute directly to the technology of manufacture. But his steady demands for higher and higher purities, backed by orders for larger and larger quantities (with stiff specifications appended), did improve the quality and expand the industry. The first Walter motor for airplanes (as distinct from rocket units for assisted take-off) was the R. I. which was test-flown in 1937 by being mounted in a Heinkel airplane which also retained its conventional engine. That motor produced a thrust of around 770 lb with a fuel consumption of 7.3 lb per sec.

During the same year the German Air Ministry approached Alexander Lippisch, who had meanwhile obtained his doctor's degree at Heidelberg, with the request for a design for a high-speed fighter plane. He was handed just two things: a demand for high speed and a statement that a motor with so-and-so much thrust would be available. The design made by Dr. Lippisch was labeled DFS 194, the letters standing for Deutsche Forschungsanstalt für Segelflug (German Research Institute for Soaring Flight), an organization with which Dr. Lippisch had been connected for years and which was a kind of offspring of the Rhön-Rossitten Gesellschaft. Dr. Lippisch himself stated in a letter that he thought back to his old experiments in the Rhön Mountains when he began the design work. In fact the resemblance between the little Storch model and the finished product is great; both had swept-back wings, a very short fuselage, the landing "ski."

In 1938, when the design was nearly finished, design and designer were shunted over to the Messerschmitt company which had experience in fast airplanes. The production model of the new plane was then called the Me-163. Wind-tunnel tests were made, but the question of the motor was still wide open. Heinz Gartmann, who worked with the Bavarian Motor Works and is one of the founders of the postwar German Rocket Society, recalls that one Me-163 stood around in one of the construction halls of BMW for some time. The BMW engineers were desperately busy with turbojets. They also designed take-off units, one of which was supposed not to be used for take-off but to increase the maneuverability of the plane in the air;[2]

[2] This idea was expressed for the first time to my knowledge by Dr. Sänger in 1935, in an article published in the *Schweizerische Bauzeitung* (vol. 107, no. 2), the organ of the Swiss Society of Architects and Engineers. He assumed a thrust of 1 ton for 90 sec with a rocket motor of a "dry" weight of 205 lb and 745 lb of fuels.

they designed rocket motors of small sizes and probably hoped that they would get around to a larger motor of about 3000-lb. thrust.

But Professor Walter won out; the Me-163 was flown with his motor. It was a "cold" motor, operating on the decomposition of T-Stoff (H_2O_2) by Z-Stoff, calcium permanganate solution in normal speech. The two liquids were fed into the combustion chamber by pumps driven by a steam turbine operating on the same reaction, except that in the turbine's steam generator the catalyst was a "stone," a cement-like substance treated with permanganate against which the T-Stoff was sprayed.

The early flight tests were disappointing. "Not good enough for combat operations" was the verdict of the Air Ministry. Soon afterward Messerschmitt and Lippisch found themselves at odds with each other. Lippisch left. Messerschmitt's engineers made a number of changes, some based on Lippisch's suggestions, some on experience with the first model, some for the sake of better production. The new model was the Me-163B. Then there was a long wait for a new motor. At BMW the engineers started thinking again—one might still be able to get that contract. But Walter won out again, this time with a "hot" motor which was officially designated as the "109–509," but in Walter's factory called the R II-211. The table shows the differences:

	HWK, R II-203 FOR ME-163A		HWK, R II-211 (109-509) FOR ME-163B	
	U.S.	METRIC	U.S.	METRIC
Weight	**167 lb**	**75.5 kg**	**368.5 lb**	**153 kg**
Thrust	**440–1560 lb**	**200–700 kg**	**660–3300 lb**	**300–1500 kg**
Exhaust velocity	**3400 ft/sec**	**1035 m/sec**	**5800 ft/sec**	**1680 m/sec**
Fuel consumption (max.)	**15.4 lb/sec**	**7 kg/sec**	**17.5 lb/sec**	**8 kg/sec**

The fuel for the "hot" version was called C-Stoff. It consisted of 30 per cent hydrazine hydrate ($N_2H_4.H_2O$), 57 per cent methyl alcohol, and the rest water. Hydrazine hydrate can be used as a fuel by itself, and in combination with H_2O_2 as an oxidizer it has the advantage of being spontaneously combustible so that no ignition device is required. But as an energy source it is greatly inferior to alcohol, hence only enough of it was put into the mixture to make it self-igniting.

The motor 109–509 could provide power for 15 to 20 min at minimum fuel consumption; at full thrust the fuels lasted for 4 min 11 sec. In order to increase the time the plane could stay in the air, Walter developed a motor which was designated the 109–509-C. It differed from the other mainly in having an auxiliary "cruising chamber," a rocket motor of small size, positioned below the other, producing a thrust of 300 kg or 660 lb. This was just enough to keep the plane in the air. The main 109–509-C motor was also somewhat more powerful, about 10 per cent, than the 109–509, because the Me-163C plane for which it was developed was slightly larger than the Me-163B. The latter appeared in combat in 1944, with initial success.

Meanwhile the situation had been quite confused. The Allies had stepped up their bombing raids. Udet personally ordered production of the Komet (Me-163B).

But at that point the Messerschmitt organization had its hands more than just full. The fast Me-262 twin-jet fighter which they had built and of which they were proud had to be converted into a bomber. It was a very nearly impossible task, but Hitler had personally given the order. Consequently the responsibility for the Me-163 was shifted once more, to the firm of Focke-Achgelis, working in Messerschmitt's plant, without much valid agreement on who could order what. Later in 1944 one Junkers model, the Ju-287, was canceled and the Me-163 was shifted to Junkers. The engineers there redesigned it once more, in addition to producing the design they were given. The new version was called Ju-248, until an order came that it was to be known as the 8–263. The 8–263 never took to the air under its own power but did reach the glide test stage.

The Me-163B was a very small airplane; when I saw it for the first time I wondered whether anybody could get in, but the fact that I had just come out of a B-29 bomber may have contributed to that feeling. The span of its swept-back wings was only about 30 ft (for precise dimensions, see Appendix 3), the over-all length not quite 19 ft, the height 8 ft. Except for a vertical stabilizer with rudder it had no tail surfaces. Take-off was by means of a wheeled undercarriage which dropped off; landing by means of a retractable "ski" or "keel." Landing speed was around 100 mi per hr; top speed 515 mi per hr at sea level; 560 mi per hr at 40,000 ft.

During the latter part of the Second World War the development of rocket-propelled aircraft veered off in a still different direction. Both in Germany and in Japan it was realized that the growing superiority of the Allies in the air could not be checked by antiaircraft fire from the ground, since manned bombing squadrons were not robot bombs which had to stick to a given flight path and altitude. At least in Germany there were not enough defending fighter planes in the air (that was true in Japan too, for a later date), apparently more because of a lack of trained pilots than a lack of planes. Antiaircraft missiles of the type of Wasserfall and Rheintochter were promising, but it was evident that the development of a really reliable guiding method would take years. The one alternative that was still possible was what may be called the manned missile.

The concept had been voiced, quite tentatively, for the first time by Oberth in his *Wege zur Raumschiffahrt* (page 295). He wrote that "an idea related to that of the rocket airplane is the flying tank. This is a large heavy-walled combat rocket, with or without wings of armor plate, which is supposed to fly into the midst of enemy air fleets in order to destroy them either by gunfire or collision. . . ." In 1943 Dr. Alexander Lippisch made the same suggestion but because fifteen years had passed between Oberth's book and his suggestion and because those fifteen years had brought enormous practical advances, he could state the problem more specifically. The whole is known only through a released report (PB Report No. 54500, in its original German without translation) which consists in the main of an article by one Dr. P. Karlson advocating Dr. Lippisch's suggestion. The "ram rocket," as it was called, was to be a rocket with a heavily braced pointed steel nose. It was to have three sharply swept-back surfaces near the tail end, combining fins and control surfaces in their operation. The rocket was to have a liquid-fuel motor, with solid-fuel assisted take-off. Its ceiling was not stated specifically but would have had to be several times that of the bombing squadron to be attacked.

The "ram rocket" was to take off vertically or very nearly so, jettison its take-off

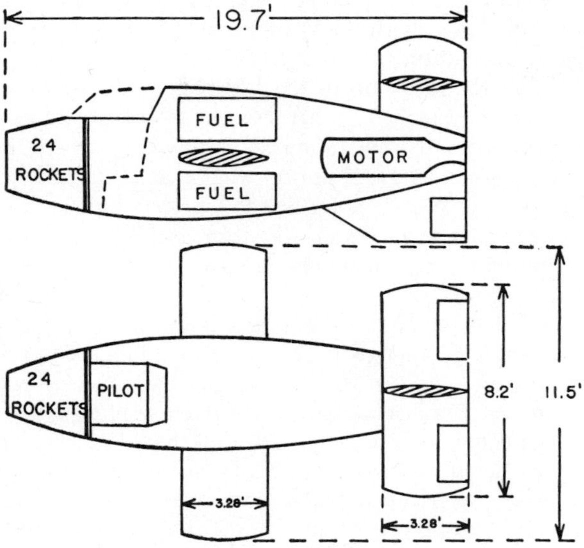

Fig. 79. German "manned missile" Natter (viper).

help, and then be aimed by its pilot for a plane, more specifically the wing or tail surfaces. When it broke through by sheer kinetic energy, the bomber would of course crash, while the rocket, relatively undamaged, would continue on its way. As soon as feasible the pilot would jump out of, or be ejected from, the rocket. Near the peak of its ascent, the latter would release its own parachute and float to the ground, to be reconditioned for service.

There can be little doubt that a "ram rocket" would perform as suggested and that it probably could be used over again because the thin sheet metal of airplane wings and control surfaces would offer little resistance to such an impact. The problem, as usual, would be to hit the target; even with a live pilot in control that would be difficult at the speeds involved. And the pilot would probably need as long a training time as a fighter pilot.

It may have been the obvious difficulty of pilot training on the one hand and the equally obvious need for some kind of piloted missile which led to the compromise solution that was called Natter (viper). As an official project it was initiated on August 1, 1944, by the Bachemwerke, a factory owned by a Mr. Bachem and the Luftwaffe Colonel Knemayer. The Natter (Fig. 79) was to be a small rocket airplane, designed to take off vertically from a short launching rack by means of its HWK–hydrogen-peroxide motor, assisted by several solid-fuel rockets manufactured by the factory of Schmidding. The design of the whole Natter–and this is important for an understanding of the proposed performance–was such that it could have been built by semiskilled labor in very small factories. Main construction material was wood. The missile consisted of three main sections. In front there was a tightly packed battery of twenty-four bombardment rockets, wired for ignition in such a way that they were all fired simultaneously. Then came the section for the pilot and, finally, the tail section with the Walter motor.

Upon reaching the altitude of the attacking bombers, the pilot was to level out his flight, fly in the direction of the bombers, and discharge his twenty-four rockets. Then, after unfastening his harness, the pilot was to push his stick all the way forward, thereby activating a mechanism which would break the Natter apart. First the nose section, now empty, was to be jettisoned. Then a parachute attached to the valuable Walter motor was to be ejected in the rear. Upon opening, it would stop the remainder of the missile, throwing the pilot, who had nothing in front of him any more, into the air where he would pull the rip cord of his own parachute.

It is easy to see that the Natter designer followed the reasoning which had led to Lippisch's "ram rocket," though avoiding the actual ramming collision. The Luftwaffe accepted the idea and in September 1944 models were tested in the high-speed wind tunnel at Braunschweig. Then fifteen specimens of the Natter were produced and subjected to launching and glide tests. The former usually meant the end of the specimen, for the Schmidding rockets were anything but reliable. They often exploded, ripping the wood construction apart in the process. If they did not explode they displayed an intolerable discrepancy of performance; their thrust varied as much as plus or minus 50 per cent of the figure specified. One glide test was made with a motorless Natter, the weight of the HWK motor being simulated by the proper amount of ballast. In that test the break-up mechanism did not operate, so that the pilot had to bail out in the manner he had been taught originally. But after landing safely he spoke enthusiastically of the flying characteristics of his strange craft.

Then a piloted take-off was tried, which ended badly. At about 500 ft altitude the cockpit cover fell off. Since the headrest for the pilot was attached to it, the pilot probably died at that instant from a broken neck. But the Natter continued climbing, flying upside down, at an angle of about 15 degrees. At 5000 ft the fuel flow was either choked off or the supply was exhausted, for the Natter, without propulsion, turned over and dived into the ground.

When the war ended there were a total of 200 "vipers" on order with Bachem, 50 by the Luftwaffe and 150 by the Waffen SS. None had been delivered. The German government is also known to have promised the plans of this weapon to the Japanese. But nobody could be found who knew whether this had actually been done. Both design and performance would have been exceedingly well suited to Japanese methods. But the Japanese never built one.

They did build a manned missile of another type, the suicide version, as part of their Kamikaze attacks. These suicide attacks against American fleet units could, and did, use any kind of plane which could carry a load of high explosives and be dived into a target. But one of them, the *Baka,* was designed specifically for such attacks. It was a small plane, about 20 ft long, with a wingspread of 16 ft 6 in. The nose consisted of a 1200-lb warhead; propulsion was furnished by several large solid-fuel rocket units. The Baka was carried by a Betty bomber and the two pilots were connected by telephone until the pilot of the carrier plane decided that the moment had come to cast the Baka off.

At the time the Germans drew their final plans for manufacture of the Natter, rocket airplanes were under discussion in the United States too. In December 1944 the Air Force (then still the Army Air Force) approached the engineers of Bell Aircraft Corporation about the design of a man-carrying airplane to be used for research, not for combat. It was a postwar project that was initiated while the war was still going on. The purpose was to gather information about flight at transsonic and supersonic speeds.

But here some background is necessary. As airplanes became faster and faster during the earlier stages of the war, pilots on landing occasionally told weird stories. Their planes had been misbehaving; they did not "feel" right. Occasionally, it was claimed, the movement of a control surface produced an effect the opposite of what it should have been. People began to talk about "compressibility effects," and exaggeration followed hard on the heels of discovery. The term "Mach number" was mentioned and since that Mach number has something to do with the velocity of sound some writers began to sound "authoritative" warnings in popular aviation magazines. Newspaper stories followed, and a legend was rapidly built up about a "sonic barrier" which had to be penetrated by sheer force provided that it could be penetrated at all. Dr. Theodore von Kármán later spoke jokingly about a "brick wall in the sky." The whole was a structure of misunderstood facts, carelessly accepted rumors, and half-digested theoretical concepts, cemented by superstition.

At one time during the war students (most of them in uniform) might be informed that one could not approach the speed of sound, or Mach number 1, or, simplified, $M = 1$. The students would carefully note this down, sitting in their lecture hall in the northeast corner of the town. At the same time, some 5 miles southwest of the same town, draftees would be practicing antiaircraft fire, watching the mechanical loaders ram the long cartridges into the breechblocks of their long-barreled A. A. pieces. The shells emerged from the gun muzzles at a speed of $M = 3$ and nobody even thought about it.

The Mach number, named after the Austrian physicist Dr. Ernst Mach, who first investigated such problems, is in itself a very simple idea. The velocity of sound, at average temperature, is around 760 miles per hour. But one cannot simply say that an airplane, flying at 760 miles per hour, is flying with sonic velocity. At an altitude of 1 mile the velocity of sound is a different figure, and at one of 3 miles a different figure still. And because the air has a different density in these altitudes the speed of the plane is likely to be different too. To deal with this complication the Mach number was introduced, defined very simply as

$$M = v/s$$

where M is the Mach number, v the speed of the plane (true air speed), and s the velocity of sound at that altitude. Obviously a plane which makes $M = 0.7$ near sea level may make $M = 1.0$ or $M = 1.3$ at a different height, depending entirely on conditions, since both v and s change. This is rather common knowledge by now among people interested in aviation. But most of them think that this changeability of the Mach number is caused by the different density of the air at various heights. In a way it is, because with different density airplane performance changes. But the velocity of sound has nothing at all to do with the density of the air; it is completely a function of the *temperature,* as the tables on pages 511–12 show.[3]

According to the Mach number involved, aeronautical engineers distinguish four speed ranges. The first is subsonic, from $M = 0$ to $M = 0.8$; the second is transsonic, from $M = 0.85$ to $M = 1.3$; and the third supersonic, anything above $M = 1.3$ to about $M = 3.0$. Above that the term "hypersonic" is used. Offhand it might seem that subsonic and supersonic would be sufficient, but there is a good reason for the intermediate transsonic. On a body moving with, say, $M = 0.9$ the air flow may be supersonic at some points while it is still subsonic at others. This could be called the speed range of confused flow patterns, but "transsonic" is not only a shorter, but indubitably a more dignified-sounding term.

In the subsonic speed range one is confronted by the fact that the air in the path of the plane does not behave like air. With this paradoxical statement I refer to an early lesson in elementary physics in which the young student is taught the fundamental difference between gases and liquids. He is told that a gas, for example air, can be easily compressed to a small fraction of its original volume, while a liquid, for example water, cannot be compressed. The instructor usually adds that with enormous pressures a liquid, too, will be compressed a little, but that this amounts to only fractions of 1 per cent of its original volume and is, for this reason, negligible for practical purposes. What he should add is that he is referring to a gas in an enclosed space, like a cylinder with a piston. In the open atmosphere, air, although it *can be* compressed, normally *is not.* It just tries to get out of the way, which it can do as long as the speed of the moving object is subsonic. Only when the speed becomes supersonic and the air cannot get out of the way is it actually compressed even in the open.

Doctor von Kármán phrased this by saying that at supersonic speed the area ahead of the plane, or missile, is the "area of no signal." This "no-signal area" is the area outside the "Mach cone" in Fig. 80. We have there a source of disturbance—let's say we have a pulsating point, although a pulsating point doesn't make any sense mathematically. As long as that pulsating point stays in one place the shock waves it produces spread out in concentric spheres. The spheres will gradually peter out because their surface is increasing steadily and they have only a certain amount of energy. But we are not interested in what will happen at a long distance from the source (although that is theoretically important); we are concerned only with the region where these shock waves are noticeable. When the source of disturbance begins to move, the spheres are no longer concentric; the "signal" does not carry as far in one direction as in the other. And when the speed of the disturbance gets to be faster than that of sound the disturbance overtakes its own signal. All the expanding spheres fit into a cone which has been named the Mach cone. Figure

[3] See Herbert Schaefer, "Machmeters for High-Speed Flight Research" in *Journal of the Aeronautical Sciences,* vol. 15, no. 6, June 1948. The proof boils down to the following:

$$s^2 = kRT$$

where R is the gas constant, T the absolute temperature, and k the ratio of the specific heats and constant pressure and constant volume (1.4 for atmospheric air). Therefore

$$M^2 = v^2/s^2 = v^2/kRT.$$

Since neither k nor R changes, the changes in Mach number must be due to the changes of T. (Schaefer, in his paper, followed the standard practice of aeronautical engineers in designating the velocity of sound as c. I have used s in my explanation because throughout this book the letter c is used to designate exhaust velocity.)

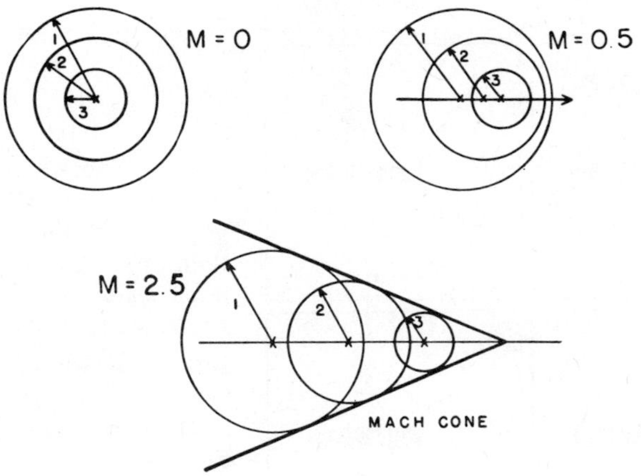

Fig. 80. The origin of the "Mach cone." A source of disturbance which does not move ($M = 0$) produces concentric (spherical) waves. When the source starts moving the waves are no longer concentric; their appearance at half the speed of sound ($M = 0.5$) is shown. When the velocity of movement is several times the speed of sound ($M = 2.5$) the source of disturbance overtakes its own "signal"; all the expanding spheres fit into a cone which is called the "Mach cone." The area outside the Mach cone is called the "zone of no signal."

81 shows how this actually looks on photographs of artillery projectiles taken with the so-called *schlieren* method. This is a German term, the plural of the word *Schliere,* which means an area of unequal density and was originally used mostly with reference to sections of poor quality in optical glass. There is no English equivalent that covers all the meanings, but for photographs of shock waves one might call it the "striation" method.

But the existence of a "no-signal" area is not the only difference between subsonic and supersonic flow. If air has to flow around a corner, either a protruding or an "indented" corner, it can just do so when it is subsonic. When it is supersonic it still can get around the protruding corner without much trouble, but if it is caught in an "indented" corner it would have to double up into its own flow. That obviously can't be done and the result is a shock wave, one of the shock waves that started the whole story. And when air is flowing through a tube there is another difference (Fig. 82). As long as the tube is straight, the speed of flow makes no difference. However, if the diameter of the tube shrinks gradually, forming a converging nozzle, subsonic flow gets faster, ending with $M = 1$ if the nozzle is long enough. A supersonic flow, in the same nozzle, slows down. In a widening (diverging) nozzle the subsonic flow slows down while the supersonic flow speeds up. That is the reason the exhaust nozzle of a rocket motor first converges, to bring the subsonic flow up to sonic speed, then, at just the right instant, it starts diverging, so that the flow is speeded up some more, ending with the highest possible exhaust velocity, or rather the highest practicable—the nozzle can't

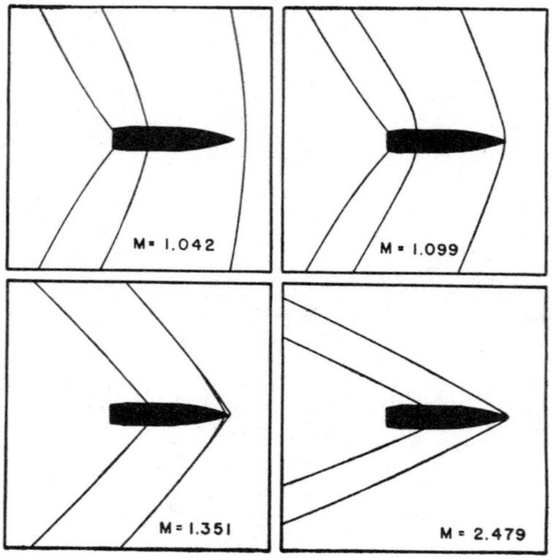

Fig. 81. Shock waves on 155-millimeter projectile. Tracings of photographs released by the Aberdeen Ballistics Laboratory, showing shock waves caused by an American 155-mm "Long Tom" projectile at various Mach numbers. Note the close similarity of the figure at the lower right to the Mach cone in Fig. 80.

very well be 200 ft long, which may be the theoretical requirement in a given case.

While all these things were known more or less clearly as theoretical concepts, it was very often uncertain how important the one or the other effect would be, especially with regard to airplane design, and there especially with reference to the speed range of confused flow patterns. The engineers, of course, wanted numerical values. The experimental airplane under discussion in December 1944 was to provide them. It was designated as the XS-1, the letters standing for "Experimental Supersonic." (Now it is just called the X-1.) It was to be able to climb to 35,000 ft and attain a speed of at least 800 mi per hr at that altitude, which would be $M = 1.21$. It was to be able to maintain that speed for not less than 2 min, preferably 5.

As the X-1 finally emerged it might be described as a V-2 rocket laid flat, with wings, tail assembly, and tricycle landing gear and with a pilot in the instrument compartment, replacing the 975 lb of equipment in the German version by 150 to 180 lb of much more versatile organic matter. But this description, while correct as to shape and appearance, would do injustice to the designers of the X-1, for the final similarity came about from necessity, not from choice.

To begin with, the X-1 was *not* planned as a rocket airplane from the outset. The designers had free choice of any power plant they could get, provided the performance of the plane lived up to preliminary specifications. On the other hand, they wanted the plane to look as much as possible like a normal airplane because they did not want to introduce too many new factors all at the same time. At first the designers thought, quite naturally, of a turbojet plane. But the turbojet engines then available would not produce the necessary speed. They might be able to approach $M = 1$ but only near sea level in dense air.

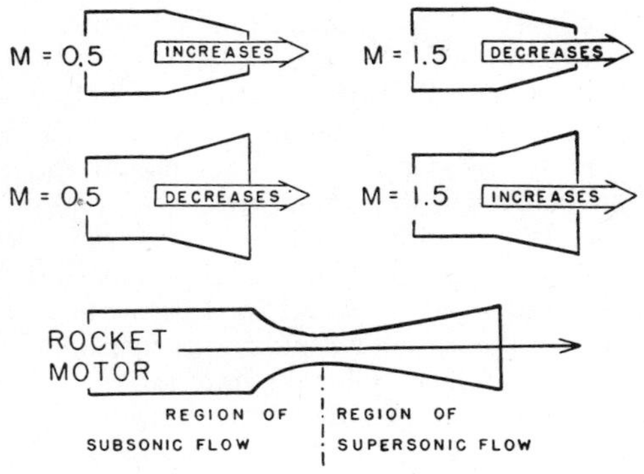

Fig. 82. Supersonic flow through nozzles. For a given nozzle shape, either converging or diverging, subsonic and supersonic flows produce opposite results. Application of this fact produces the typical shape of the rocket-motor exhaust nozzle.

I'll now let the designers speak for themselves:[4]

> Consideration was then given to the combination of a turbojet and rocket power plant. The primary purpose of the turbojet was for take-off, climb to altitude, and return to the home base while the rocket was to accelerate the airplane to, and maintain the desired speed after, the operational altitude had been reached. This combination resulted in an excessively large airplane. The turbojet performance fell off at altitude, resulting in a poor rate of climb, which in turn called for a large amount of fuel. The speed at which the airplane was flying when the operational altitude was reached was also low, requiring a considerable amount of rocket fuel for acceleration purposes. The use of two such widely different power plants also increased the installation and operational problems. Although the fuel consumption of an all-rocket powered airplane was high, the rate of climb was also high, averaging better than 20,000 ft per min between sea level and 35,000 feet, with a climbing speed of 500 m.p.h. Thus, the fuel required for climb was relatively low and that required to accelerate from climbing speed to the desired test speed was less than with a combination turbojet and rocket. At higher altitudes the potential climbing speeds and rate of climb were even higher, reaching maximum values of 120,000 ft per min at an altitude of 120,000 feet, flying at nearly 1400 m.p.h. After a comparative analysis of the design studies made around the various power plants, it was decided to proceed with the all-rocket-powered airplane.

(All the figures mentioned were, of course, the results of mathematical analysis and not test results.)

> Many propellants were considered before liquid oxygen and ethyl alcohol were selected. Hydrogen peroxide was discarded because the motors in opera-

[4] The quotations are from a lecture delivered by R. M. Stanley, chief engineer, and R. J. Sandström, chief preliminary design engineer, both of Bell Aircraft Corporation, at the Annual Meeting of the Institute of the Aeronautical Sciences, January 28, 1947, and published in *Aeronautical Engineering Review,* August 1947.

> tion at that time exhibited low specific impulses. It was available in this country only in limited quantities, and it must be handled with a great deal of care. Acid and aniline were not considered suitable for a man-carrying aircraft. These propellants are spontaneously combustible, and special precautions must be taken to keep them separated, since any dual leak in the air might initiate a disastrous fire. Acid is also injurious to personnel, which complicates the handling problem. Nitro-methane, being a mono-propellant, would simplify the fuel system considerably, but it is subject to detonation under certain conditions which at that time were not too well understood. Although it was believed that Nitro-methane had good possibilities, it was not felt that it had been developed to the point of reliability desired. Gasoline and liquid oxygen were studied, but with a regeneratively cooled motor it would have been necessary to carry a third tank containing water aboard for this purpose, as gasoline alone is not a suitable cooling medium. The propellants selected, liquid oxygen and alcohol, are readily available, have a good specific impulse, are relatively safe and easy to work with, are not spontaneously combustible, and are non-injurious to personnel (small spatters of liquid oxygen are not harmful). The motor is regeneratively cooled by circulating the fuel through a cooling jacket before it is injected into the combustion chamber. To aid in cooling, one part of water is mixed with three parts of ethyl alcohol. It was found that the addition of this amount of water had a small effect on the overall impulse but aided appreciably in the cooling problem.
>
> (Stanley and Sandström)

The job of building the rocket motor fell to Reaction Motors, Inc., in New Jersey, the same firm which later produced the motor for the rocket Viking. The motor was designated the 6000 C4, a designation which makes sense for a change, for the 6000 refers to the thrust of the motor in pounds, and it has four chambers which can be operated separately. The total dry weight of this motor is 210 lb and the dimensions are: length, 56 in., diameter, about 19 in. Actually it is a motor consisting of four rockets.

> The rockets may be run separately or in combination as selected by the pilot. The motors are started by a small igniter located in the forward end of the cylinder, which ignites a stream of fuel and gaseous oxygen by means of a spark plug. When the chamber pressure reaches a value of approximately 50 lb per sq in. the propellant valves open admitting the fuel and liquid oxygen. No provision has been made to throttle the individual rockets, and hence the pilot has at his command only 25, 50, 75, and 100 per cent of full thrust.
>
> To supply the propellants to the rocket motor, a turbine driven pump was specified. The development of this item was recognized quite early as being one of the items most likely to interfere with the early flight of the airplane since it involved the development of a highly reliable piece of apparatus of a type hitherto unknown. To forestall probable flight delays, a temporary installation of pressurized tanks has been employed. This has reduced the airplane's endurance by about 1.5 minutes, has increased its landing weight by approximately 2,000 lbs, and has run the landing wing loading up to approximately 55 lbs per sq ft.
>
> (Stanley and Sandström)

When this compromise had to be made the engineers probably felt pretty bad about it. As it turned out it did not matter at all, because the war ended before the X-1 was finished and B-29 bombers were available to carry the research plane close to operating altitude (Plate XXVII). But they had another problem. One can buy compressed nitrogen, at pressures of either 2000 or 2200 lb per sq in. But unless they wanted to install very large pressure tanks in the X-1, something they decidedly did *not* want to do, that was not high enough. However, one could buy liquid nitrogen and could get higher pressures by evaporating it. An evaporator was built, a steel sphere 36 in. in diameter with walls 3 in. thick. Made in two halves and welded together, it was tested to stand as high as 9000 lb per sq in. That sphere was partly filled with liquid nitrogen which then emerged as a gas under the desired pressure.

> During the testing of the evaporator we had a rather interesting experience. While operating the evaporator, it was noticed that droplets of liquid dripped from the frosted lines in which flowed liquid nitrogen. This could not be condensed moisture in view of the low temperatures of the lines, and was quite baffling in view of the many pressure tests that had been made. It was finally discovered that liquid oxygen was being condensed out of the air on coming in contact with the cold nitrogen lines. (Stanley and Sandström)

With the X-1 ready for flight tests and a B-29 available as a carrier plane it remained to work out a plan for the tests. Preliminary investigation proved that the airflow would provide a separating force. To guard against backward drift some wooden guideposts were attached to the B-29 and smeared with fresh red paint just prior to take-off. If red paint were afterward found on the X-1 it would show that it had touched these guideposts. (None ever was.)

Normally all these flight tests should have taken place in California at Muroc Air Base, some 75 miles east of Los Angeles. The Air Force Base, which soon afterward was rechristened Edwards AFB, is situated at the "shore" of Rogers Dry Lake (more often, but incorrectly, called Muroc Dry Lake), an old lake bed shaped like a poorly written figure 8, perfectly flat and so hard that rain water does not sink in but has to evaporate. The days there are nice and hot and at night it often freezes, up to an inch of ice if there is enough moisture around for an inch of ice. But when everything was ready for testing (except that the X-1 was still minus rocket motor, the weight of which was simulated), it was February 1946, and at that season the Dry Lake could be expected to be flooded. Therefore the glide tests, ten of them, had to be made at Pinecastle, Fla. Then the rocket motor arrived and everything was shipped to Muroc where the complete plane was thoroughly tested.

> After four additional glide flights, during one of which a complete fuel load was jettisoned, the first powered flight was made. On December 9, 1946, in perfect weather, the X-1 was launched at 27,000 feet over Muroc Dry Lake. Approximately 10 seconds later, No. 1 rocket chamber was ignited, followed shortly thereafter by No. 2. The speed of the aircraft picked up so rapidly, however, that thrust was cut back to 25 per cent for reasons of safety and a slow climb was made to 35,000 feet. The cabin noise level was reported very low. At 35,000 feet, 50 per cent thrust was again applied and a Mach No. of .79 quickly attained. All power was then shut down and the airplane glided down to 15,000 feet and with all four rocket motors running 100 per cent thrust was applied during a brief climb. The pilot experienced very high accelerations which he compared to that of a fighter during take-off employing water injection. On its first powered flight, the X-1 reached a Mach No. of .795 during brief application of 50 per cent thrust, demonstrated satisfactory motor control and general handling characteristics, and adhered closely to the flight plan. The eleven subsequent flights have been equally satisfactory. (Stanley and Sandström)

The early flights were made by a test pilot, Chalmers ("Slick") Goodlin, who then quit with an unusual amount of publicity. His place was taken by an Air Force test pilot, Captain (later Major) Charles E. ("Chuck") Yeager, who was then 25 years old and had a most distinguished war record as a fighter pilot with the Eighth Air Force in England. (On one occasion he shot down five Messerschmitt Me-109's in succession.) He made his, and the world's, first supersonic flight on October 14, 1947. For no known, or even guessable, reason the Air Force refused to release the results for several years and rumor exaggerated the speed. Actually it was 760 miles per hour, which, at the altitude at which it took place, was clearly supersonic. When at the approximate anniversary of the flight, I asked Captain Yeager how it had felt, he had no ready answer. Well, he just flew. That's all. His instruments told him that he was faster than sound. Otherwise it had felt like any other flight in the same plane.

There were two specimens of the X-1, to prevent the test program from being interrupted because of some minor damage—"they are forever busting up the nose-wheel," a Bell engineer said complainingly to me—and Captain Yeager made about thirty flights in the X-1. The design speed of the X-1, the best possible performance that it could give, was Mach 1.5 at 60,000 feet, about 980 miles per hour. This was almost reached in 1948 when the plane accelerated to 967 miles per hour. On one occasion in 1949 peak altitude was reached with 70,140 feet. And in January 1949 Captain Yeager did with the X-1 what it had been designed for originally; he took off from Dry Lake without a carrier plane. With all four rocket chambers going full blast the X-1 became airborne after a run of 2300 feet and reached 23,000 feet 100 seconds later. In August 1950, its work done, the X-1 was retired from service and turned over to the National Air Museum in Washington, D.C.

By then another supersonic airplane had arrived at Edwards AFB, the Douglas D-558-II, better known as the *Skyrocket,* designed by Ed Heinemann. Although the Skyrocket was a Navy project it was always flown by Douglas test pilots, mainly Gene May and William Bridgeman. The research scope assigned to the Skyrocket was much wider than that assigned to the X-1. The main purpose of the X-1 had been to become supersonic and to fly at supersonic speed for at least 2 minutes to find out how a plane would handle at supersonic speeds. The Skyrocket was to investigate handling characteristics near the speed of sound, both above and below; it was considered somewhat as a prototype for fighter planes to come.

The Skyrocket was completed near the end of 1947 and made its first flight in February 1948. The plane was built mostly of 75-ST aluminum alloy. Unlike the X-1 both its wings and tail surfaces were swept back, the wings at an angle of 33 degrees, the tail surface at an angle of 40 degrees. Wing span was 25 feet; over-all length 45 feet, 3 inches; height 11 feet, 6 inches; take-off weight about 15,000 pounds. Heinemann did just what the Bell engineers had rejected doing—he built the Skyrocket for both turbojet and rocket propulsion. The turbojet was a Westinghouse 24C with 3000-pounds thrust; the rocket motor, by Reaction Motors, was very similar to that of the X-1, with a total thrust of 6000 pounds when all four chambers were burning and a fuel consumption of almost 1 ton per minute.

When the Skyrocket was fully fueled the jet engine alone could not get it off the ground; two of the rocket chambers had to help. Or else, as was often done, the rocket fuel tanks were left empty and two take-off (JATO) rocket units helped the plane to take off. Many of the flights made just to find subsonic flying characteristics were performed as pure jet flights. For the first fifty-nine flights Gene May was pilot; with the sixtieth William Bridgeman became the pilot. By that time it had become customary for the experimental flights to be accompanied by another pilot "flying chase" in a production jet. The chase pilot for the Skyrocket tests was often Charles E. Yeager, or else Lieutenant Colonel Frank K. ("Pete") Everest. Later, when the testing program became more difficult there were two chase planes in the air.

By November 1949 it became clear that the Skyrocket could not be expected to take off from the ground, become supersonic, and *stay* supersonic for a reasonable length of time. Hence the Skyrocket also was to be air-dropped from a B-29, but the existing Skyrocket lacked the necessary attachments for this purpose. Douglas built two more, one precisely like the first except that it did have the attachments for air drop, the other as a pure rocket plane.

The story of the Skyrocket bears some resemblance to the story of the Viking. There were times when everything went wrong. Six times the B-29 took off with William Bridgeman aboard and the Skyrocket dangling from its cut-away bomb-bay and six times the air drop had to be called off. Usually something went wrong with the pressurization of the rocket system. Once one of the two chase planes had to make an emergency landing and they had to wait for another one to take its place. That one also developed an emergency and by that time so much of the oxygen had evaporated that there would have been little sense in continuing. And the first air drop that did take place took place by mistake. Bridgeman was in his cockpit, tightly encased in a most uncomfortable pressure suit. The pilot of the B-29 was giving the final countdown when Bridgeman saw one of the pressure gauges dropping below the critical mark. He shouted into his microphone "Stop! don't drop!" but the communications system only operated one way at a time. The pilot of the B-29 held his button down so that Bridgeman could hear *him* counting down. Bridgeman, in great hurry and some desperation, flipped "on" again switches which he had already switched to "off." He was dropped and made, considering circumstances, a fine flight. His Skyrocket, when supersonic, also developed an unsuspected trouble which received the learned designation "lateral oscillations"; in the language of the pilot the ship simply went out of control.

One occasion that could have ended with a crash was not the fault of the plane but the fault of the suit. The face plate tended to frost over with the moisture from Bridgeman's breath so that he simply could not see. Yeager shepherded him down into warmer air until the frosting disappeared just before the actual landing. Yeager had been prepared to "talk him down" until the wheels touched Dry Lake in the manner in which an airport traffic controller will talk a plane down in foul weather. But in spite of all his troubles Bridgeman with the Skyrocket was for some time the "fastest man in the air."[5] On August 15, 1951 he reached a maximum speed of 1238 miles per hour at a radar-measured height of 79,000 feet.

After the Navy's testing program had been finished, the Skyrocket was turned over to the NACA which, according to Bridgeman, had been waiting for it with ill-concealed impatience. Well, maybe they had been impatient, but they accomplished results. On August 21, 1953, Lieutenant Colonel Marion Carl (of the

[5] None of these flights is accepted by the *Fédération Aéronautique Internationale* as making either speed or altitude records because they do not start from sea level.

Marine Corps) pushed the Skyrocket to 83,235 feet. On October 14, 1953, NACA research pilot Scott Crossfield got the Skyrocket to fly not only faster than sound but even faster than twice the speed of sound (Mach 2.01 to be precise) by flying at the rate of 1327 miles per hour at a height of about 60,000 feet.

Meanwhile, however, the successor to the X-1, the Bell X-1A, and Major Yeager had done still better. The X-1A had a design speed of 1700 miles per hour. In external appearance there was little difference; the plane was slightly larger than the X-1 and the pilot's cockpit had been moved forward for better visibility. Internally, the fuel was now pumped to the rocket motor by means of a peroxide-driven fuel pump (as had already been done in the all-rocket version of the Skyrocket) and the arrangement of the tanks was different. On December 12, 1953, the X-1A reached 1650 miles per hour at 70,000 feet. This time Yeager experienced the same thing that had happened to Bridgeman in the Skyrocket. The plane went out of control and he fell a long distance—there is, after all, safety in altitude—until the X-1A would "fly" again. By the time he was really and fully in control he was down to 20,000 feet.

In the summer of 1954 Major Arthur Murray took the X-1A up to 90,000 feet.

The end of the X-1A came on August 8, 1955. It had been taken up as usual by a B-29, piloted by Stanley P. Butchart. Research pilot Joseph A. Walker had already left the B-29 and was sitting in the X-1A's cockpit. The B-29 was at 31,000 feet, nearing the altitude and the point above the ground that had been picked for the air drop. The oxygen in the X-1A had been "topped off" once more from a supply carried in the B-29 for the purpose. It was not quite 2 P.M. Pacific Standard Time. There was an explosion in the X-1A. Walker reported it to Butchart as a "minor explosion," but the pilot of one of the chase planes, who was quite close, said that it nearly blew his canopy off. Then there seemed to be a fire, though nobody was quite sure. Of course, there could be no air drop of the X-1A, or rather there might have to be an air drop very fast, but without the pilot, for the fuel might explode at any moment. Two men in the B-29, John W. Moise and Charles W. Littleton, each placing one foot on the X-1A and the other foot on the curving inside of the B-29's bomb-bay, pulled pilot Walker out of his cockpit. His oxygen line broke in the process and his face turned blueish but before he blacked out somebody else provided him with a spare bottle of oxygen. Meanwhile pilot Butchart jettisoned the oxygen in the tank of the X-1A; it shot out without much prompting. Then he jettisoned the alcohol, but it was obvious that he could not get all of it out of the tank. And the peroxide for the fuel pumps could not be jettisoned at all.

While the question of whether it would be safe to land was being debated—some men checked to see whether the X-1A had been partly dislodged by the explosion and reported that it projected below the extended landing gear—Butchart had brought his B-29 down to 6500 feet. Conversation with the base established where it would be safe to drop the X-1A. A point was picked and the damaged research plane was released. It fell tail first, hit the ground at 2:17 P.M., exploded on impact, and started a brush fire. As an editorial writer summed it up, the rocket failed, but the men did not. They were all decorated later.

By that time there was an X-2.

The X-2 had a different purpose. Built of stainless steel and the nickel alloy known as Monel K, the X-2 was to investigate a new problem. It had been calculated that a plane flying at Mach 1 at sea level would heat up by 100 degrees Fahrenheit merely by way of skin friction—100 degrees above the temperature it already has, that is. Sunlight will produce additional heat, and the plane's engines some more. Ten miles up you are about 100 degrees better off as far as air temperature is concerned. And naturally there is less skin friction because there is less air. On the other hand, the sun shines more strongly. The X-2 was designed to find out how fast one could go at what altitude with what degree of heating up.

In addition to being built of materials that could heat up without losing strength, the X-2 had a much more powerful rocket motor (a Curtiss Wright of 12,000 pounds total thrust), a heavily insulated cabin, and a landing skid instead of wheels, like the Me-163B. As a safety measure the front end of the plane could be detached by the pilot in an emergency; the Skyrocket had had that feature too. In the case of the X-2 the device was to work as follows: the front "capsule" would be detached from the remainder of the plane, it would be slowed down by a ribbon parachute, and when it was slow enough and low enough the pilot would jump from it with his own parachute.

Luck was not with the X-2. The first one suffered an explosion while still hanging from the bomber. It was an accident much like the one which befell the X-1A but the explosion was much more powerful and the pilot of the X-2, "Skip" Ziegler, and a crew member in the carrier plane were killed. This took place over the Niagara Falls area and the damaged X-2 was jettisoned into Lake Ontario. The second X-2 made its first glide flight on August 5, 1954, and then quickly proceeded to snatch a few records. Captain Iven C. Kincheloe took it to a height of 126,000 feet and late in July Lieutenant Colonel Frank K. Everest reached a speed of 1900 miles per hour (Mach 2.9 for the altitude where it was done).

The last flight of the X-2 took place on September 27, 1956. The X-2 was taken up by a B-50 bomber which had replaced the old B-29. The pilot of the X-2 was Captain Milburn G. Apt, who had much jet experience but was sitting at the controls of a rocket-propelled research plane for the first time. Everything went in a normal-seeming manner; in fact everything went smoothly. The X-2 was dropped. Captain Apt fired the rocket motor and reached a speed of 2200 miles per hour (Mach 3.3). Naturally the chase plane, piloted by Captain Kincheloe, "lost" the rocket plane while the latter raced away, but this is normal—they always find each other later with the help of radio. And there was radio contact with the X-2 even after the fuel supply was exhausted.

Then there was sudden radio silence.

It was established later that the X-2 had fallen from 75,000 feet to 45,000 feet. At that height Captain Apt disengaged the "capsule" from the plane. He was later found dead in the desert with the capsule. What probably happened, the Air Force said, was that, during the fall of the plane Captain Apt was shaken violently in the cockpit and that after separating the capsule he was either unconscious or too dazed to jump from the falling capsule. Anybody who has read the accounts by Bridgeman and Yeager knows that what happened to Apt was not novel; it was just novel to him. Of course something might have gone wrong which could not be established from the wreckage (Apt's last remark over the radio was unintelligible and even the FBI's experts could not make it out from the recording), but it is also possible that Captain Apt, to whom this was a new experience, became panicky too early, and that the tragedy might have been avoided if he had waited a few more minutes before disengaging the capsule.

The next (and eminently successful) rocket-propelled research aircraft of the United States was the X-15, but before discussing its performance we have to look at an interlude consisting of attempts to make a rocket-propelled airplane do some work other than research.

It is not surprising that Dr. Eugen Sänger was the author of one of these attempts. As has been told (p. 218), the Peenemünde group had a winged rocket (the A-9) as one of its advanced projects. Such a winged rocket would have to re-enter the denser layers of the atmosphere at a very shallow angle. Sänger began to think about what would happen if the angle were *not* shallow. There should be a novel aerodynamic effect and that hitherto unconsidered effect should lead to a novel manner of long-range flight. Both Dr. Sänger and his assistant, the mathematician Dr. Irene Bredt,[6] must have worked day and night, because they produced, in a surprisingly short time, a report of 400 typewritten pages, which was ready for distribution in August 1944.

Here the story of the "Sänger-Bredt Report"—often called for brevity "the Sänger"—begins. Actually there are two stories, one about the contents of the report, one about its fate. The latter story began in a routine fashion. After Sänger and Bredt had finished their work, a hundred copies of the report were made. Somebody who was duly authorized to do so slapped a big rubber stamp on the front page, reading (in German, of course):

> This is a State Secret . . . must be passed personally from hand to hand, or, if with personal address, in double envelope with return receipt. To be transmitted by courier if at all possible, in exceptional cases by mail insured with more than 100 marks. Any copying, photographing, etc., verboten. To be kept in locked steel safe in rooms which are guarded 24 hours a day. Any neglect of these regulations will be most severely punished.

A list of recipients, eighty of them, was drawn up and appended to the report itself. The State Secret was on its way, by courier in double envelopes with return receipt, in exceptional cases by mail. Professor Heisenberg, the German atomic energy expert, got one. So did Dr. von Braun in Peenemünde. So did General Dornberger, and Professor Willy Messerschmitt. Also Professor Tank of Focke-Wulf aircraft, Professor Dornier of Dornier aircraft, Professor Heinkel of Heinkel aircraft, Professor Mader of Junkers aircraft, Professor Prandtl of the Aerodynamical Research Center in Vienna, and Professor Pröll of the Engineering College in Hannover. All very important people, but also very busy people, especially in view of the military situation at that time, for D-Day was not far away. Most of them probably found time to read through the Report, but nobody had time to do anything about it.

Then came V-E Day. All of those just mentioned, including Dr. Sänger himself, were interrogated by the Allies. English and American Intelligence Teams found

[6] Soon after the end of the war Dr. Sänger and Dr. Bredt were married. For some time Dr. Sänger was head of the Institute of Jet Propulsion Physics in Stuttgart (see Appendix 2, section 6); he had to quit this post because it was alleged that he had helped to develop military rockets for Egypt. He denied these allegations and later became professor of space flight technology at West Berlin's Technological University. He died of a heart attack on February 10, 1964, while lecturing to his students. (For obituary, see *The New York Times*, February 11, 1964, much of this obituary is based on an interview with General Walter Dornberger who happened to be in West Berlin at the time.)

or were handed copies of the Report, which by that time was mentioned in such a way that you could hear the capital R. In the east the same thing took place. The Russians collected a copy of the Report in the library of a scientist who could not get away in time. Later on they found two more copies. The Report was read by an expert who wrote a digest in Russian which was passed on to higher levels. And to higher levels. Only weeks after the capture of the Report, the digest landed on the desk of Yosif Vissarionovitch Djugashvili, known to history as Stalin. He ordered a full translation made and then summoned a conference about "the Sänger." Present were Vassili Stalin, aviator son of the *voshd* (boss), Molotov, Malenkov (Politburo), Beriya (Secret Police), Voznesenski (Politburo), Voroshilov—also two experts, Colonel Serov and Lieutenant Colonel Grigory A. Tokayeff (pronounced: *toe-KAH-yeff*), both of whom had read the Report. A preliminary discussion about V-2 rockets was cut short by Malenkov with the statement that their range was too limited—"Do you think we are going to fight Poland?"

Tokayeff, at Stalin's order, delivered a forty-minute lecture on the contents of the Sänger report. Stalin wanted to know the whereabouts of Sänger. Tokayeff didn't know—"Possibly Paris?" The outcome of the meeting was that Serov and Tokayeff were ordered to find Sänger and bring him to Russia in "a voluntary-compulsory manner." Vassili Stalin was to accompany them. They took off, first for Berlin. In strictest incognito. With fighter escort. They "looked for Sänger everywhere for months." They also looked for Dr. Bredt, "but no trace of her was ever found."

Of course I had no idea about any of these events then, but a year or so later Tokayeff found himself an official reason to go to London, where he remained, writing his memoirs for the *Daily Express.* In the installment of these memoirs which was devoted to rockets (published Sunday, January 23, 1949) Tokayeff mentioned that the meeting had taken place in April 1947. At just about that time Dr. Sänger had written to me from Paris. No secrecy was involved; his letter was uncensored. Irene Bredt, incidentally, was working with him there. I leave unanswered the question of how young Stalin, Serov, and Tokayeff spent their "months of looking," and turn to the report itself.

As has been mentioned, Sänger wondered about what would happen if a *winged* rocket entered the denser layers of the atmosphere—say the 25-mile layer—too fast and too steeply. The answer was that it would ricochet, like a flat stone hitting the surface of a lake. The ricochet would throw it into layers far too attenuated to support it. Some distance away the rocket would strike the denser layers again and bounce off once more. The path, seen as a whole, would be a wavy line, the "undulations" slowly diminishing in magnitude, both vertically and horizontally. One might say that it would resemble a roller coaster. This type of path increased the possible range of a winged rocket to a surprising extent and in an entirely unforeseen fashion. And on that Sänger built the conception of an antipodal bomber.

Figure 83 shows how the rocket plane itself might look. It would be about 92 ft long, with a wingspread of nearly 50 ft. Its "dry weight" would be 20 metric tons, the fuel and bomb load 80 metric tons, resulting in a take-off weight of 100 metric tons. But with such a weight an unassisted take-off would use up the greater part of the fuel supply. Nor would it be easy to provide a plane of such take-off weight with enough take-off assistance in the usual manner. The way around this

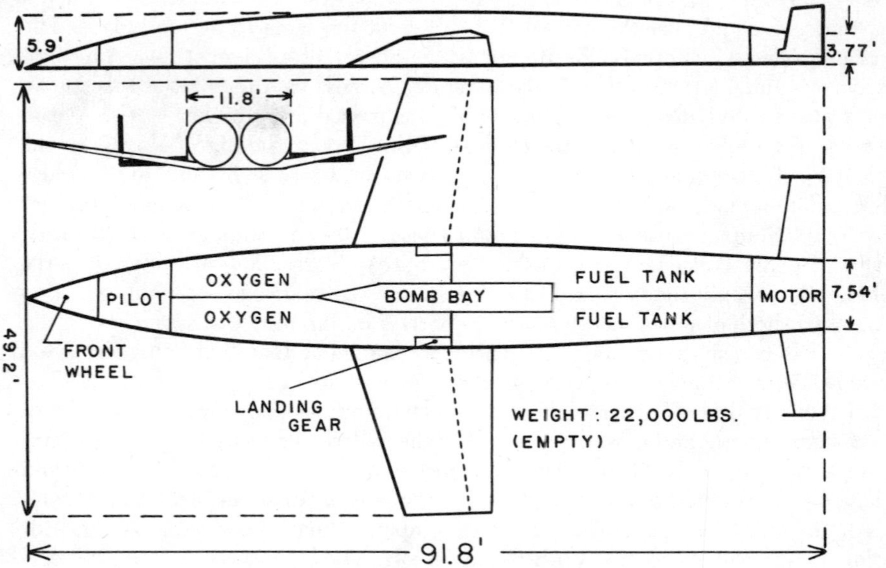

Fig. 83. Dr. Sänger's antipodal bomber.

difficulty, as designed by Dr. Sänger, was a long straight take-off track, 3 km or roughly 2 mi long. The plane would sit on a kind of sled to which any required number of rocket units could be attached. They would not even need to be high-grade units since they were not supposed to leave the ground. The rocket sled would work for about 11 sec, at the end of which the plane would be at the end of the runway, moving at 500 meters per sec or 1640 ft per sec, fast enough to make it lift. Under its own power it would then climb.

Assuming an exhaust velocity of 3000 meters per second the rocket plane would reach a maximum velocity of 6000 meters per sec or 3.73 mi per sec. That would carry it to a peak altitude of 261 km or 162 mi. But what would occur is best shown in the following table:

	ALTITUDE		DISTANCE FROM TAKE-OFF POINT	
	MI	KM	MI	KM
First peak	162	261	1,553	2,500
First low	25	40	2,796	4,500
Second peak	78	125	3,573	5,750
Second low	25	40	4,350	7,000
Third peak	75	120	5,033	8,100
Third low	25	40	5,810	9,350
Fourth peak	56	90	6,214	10,000
Fourth low	25	40	6,711	10,800
Fifth peak	51	82	7,208	11,600
Fifth low	25	40	7,643	12,300

After that there would be four more peaks, each at 60 km altitude, with corresponding low points at 40 km, the horizontal distance from low point to low point being of the order of 1000 km or 600 mi, growing slightly shorter. The ninth low point would be 16,800 km (9818 mi) from take-off point, measured along the ground. Then the plane would remain at 40 km altitude for some time. Some 23,000 km (14,290 mi) from take-off point it would finally begin to lose altitude, and 500 km (310 mi) later, halfway around the earth, it would land. The landing speed, surprisingly, would be only 90 mi per hr, which any pilot can handle. And any large airport could receive the plane. But for this trip it could carry only 300 kg (660 lb) of payload, in addition to its single pilot.

All this requires an exhaust velocity of 3000 meters per sec, which is not yet standard. Before progressing to this example Dr. Sänger had calculated shorter ranges. These, however, involved the necessity of turning around, if the plane was to be used for military missions and was not supposed to fall into the hands of the enemy the first time it was used. Turning the plane around at a rate of speed close to 1 mi per sec was found to be exceedingly tough. In all probability many things would go wrong, and enormous quantities of fuel would be needed. Turning around was not completely impossible, but going straight ahead to land at an antipodal base was much easier. That was simpler in other respects too. If the planes started from a base in Germany, say Berlin, for bombing missions, the problem was just one of proper direction. If you are finally to land at an antipodal point it is unimportant whether you go east or west. By the same token the return flight could be utilized as a bombing run too, either on the same target or an another one.

The scheme is, in principle, easy to understand. It did involve a terrifying amount of mathematical work. But it calculated neatly—except for a few geographical facts which could not be changed. To begin with, the antipodal point for points in Germany happens to be in the Australia and New Zealand area. A deviation of a few hundred or even a thousand miles from the mathematical point would not have mattered much, but if the areas where the planes would have to land were not already occupied by New Zealand, Australian, or even American forces, they soon would be. Furthermore the target cities were not located just where the "flight plan" needed them. Any bomb-dropping would have to be done in a low point of the trajectory. Even then the bombs would cover a horizontal range of some 200 to 400 mi, depending on the speed at the moment of release. That not only made aiming hard, it also required a target city within such a range of a low point.

It is unlikely that the people who founded the larger cities belonging to the Allies had such a possibility in mind. But, for flights from German soil, the large cities are in locations which would have been chosen if the prospect had been foreseen. Everything important is located under a peak. With one exception—and for that reason Allied officers found maps on which New York was labeled "*Ziel Eins*" (Target no. 1). New York could have been hit from a low point, with the bomber proceeding to Japan or at least to a part of the Pacific then still under Japanese domination.

There was one more possibility. Why stop at the antipodal point? Why not go all the way around to take-off base again? Calculation showed that this would need an exhaust velocity of 4000 meters per second, which would result in a maximum velocity of 7000 meters per second (4.4 mi per sec), with the first peak at 280 km (174 mi) altitude, 3500 km (2175 mi) from take-off point, and the first low at the usual altitude of 40 km, 6750 km (4194 mi) from take-off point. There would also be nine peaks, with the ninth low 27,500 km (17,088 mi) from take-off point. Landing, on the take-off point, would take place 13,060 sec after take-off, or, in round figures, 3 hr 40 min.

The Report concluded by recommending the one-base concept as the more useful one, and then listed the research projects which would have to be carried out to make it possible.

It is easy to understand why none of the important Germans who read this report did anything about it; it was clearly too late to begin the development of an entirely new weapon. Besides, from the military point of view, the concept itself was of doubtful value. The weapon did not look very versatile, and a bomb load of 660 lb was hardly intriguing. A year later, after the United States had demonstrated that atomic bombs were possible, a small bomb weight no longer looked so useless, which was probably the reason the Russian top command became interested.

The later American missile project Dynasoar (for "dynamic ascent and soaring flight") was ultimately derived from the Sänger report, but after some time this project was canceled.

But the concept of a winged rocket, returning into the denser atmosphere for a prolonged supersonic glide without ricochet maneuvers, looked as if it might lead to a supersonic passenger transport. One of the first to think about this was Dr. Hsue-shen Tsien, at the time (1949) Goddard Professor at the California Institute of Technology. (He later returned to China.) His idea is shown in Fig. 84. Counting on a take-off weight of 50 tons, he had in mind a nearly vertical take-off. At an altitude of somewhat over 100 miles the fuel supply would be exhausted, but the ship would coast to a peak altitude of over 300 miles, about 600 miles from the take-off site. Re-entry would take place about 1200 miles from the take-off site. Dr. Tsien assumed that aerodynamic equilibrium (meaning the altitude at which the air would support a winged rocket with a given wing loading and the assumed speed) would take place at 27 miles. Time elapsed since take-off would be about 15 minutes; the supersonic glide would last for about 45–50 minutes and would

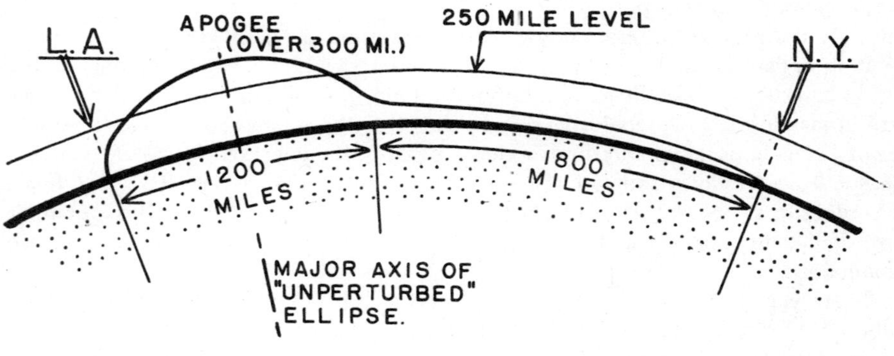

Fig. 84. Flight path of a winged passenger-carrying rocket, as conceived by Dr. Tsien. Flying time would be on the order of 1 hr.

carry the craft over another 1800 miles. The landing speed would be on the order of 150 mph.[7]

Meanwhile Dr. Walter Dornberger, in collaboration with Krafft A. Ehricke, had also been thinking about a rocket-propelled passenger ship for long-distance travel via open space. The outcome (Fig. 85) was a two-stage ship. Both stages would be delta-winged piloted rockets with the passengers in the second stage but, as the drawing shows, the two stages would not be one on top of the other. They should rather be called the ventral stage and the dorsal stage, since one cannot even make the distinction that stage no. 2 will be the second to be ignited.

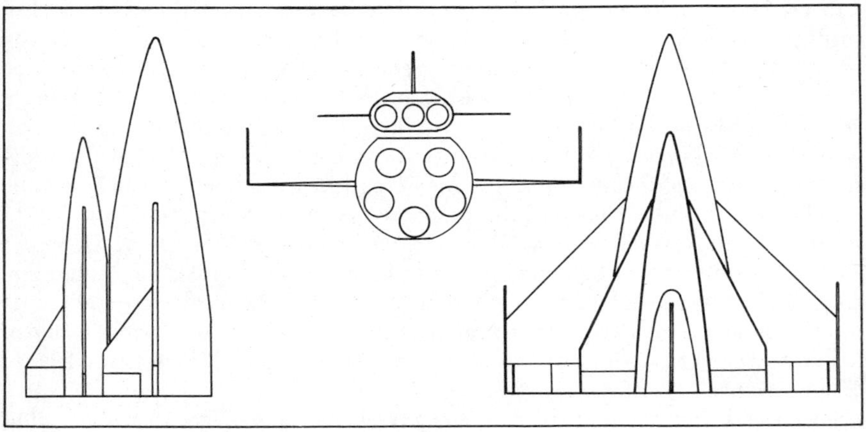

Fig. 85. Dornberger's passenger-carrying rocket ship.

There would be vertical take-off, with all rocket motors firing, five in the ventral stage and three in the dorsal stage. But the three dorsal-stage motors would get their fuel from the tanks of the ventral stage as long as the two were connected. By that method the thrust of three extra motors is obtained without adding to the take-off weight. Separation of the two stages would take place 130 seconds after take-off; the ventral stage would be landed by its pilot while the dorsal stage continued. Highest acceleration would take place when the maximum velocity of 8450 miles per hour was reached; it is then 3.5 *g*, which any healthy person can stand even without previous experience or training. For the sake of the passengers one motor

[7] In a paper read at the Eleventh Annual Meeting of the American Rocket Society (New York, November 26–29, 1956) G. Harry Stine re-examined Dr. Tsien's suggestion from a different point of view. Such a craft might be developed both for high-altitude research and for the training of space pilots. Assuming the pilot's compartment, with refrigeration system, life-support system, etc. (and the pilot himself), to weigh 1450 lb, the structure of the rocket would weigh 20,550 lb and the fuels (30,000 lb of liquid oxygen, 13,000 lb of gasoline, plus 500 lb of pressurizing gas) 43,500 lb, so that the take-off weight would be 65,500 lb. If the rocket motor had a thrust of 120,000 lb the fuel would last for 90 sec. With a vertical climb a peak altitude of 55 mi would be reached.

would be kept burning at a reduced rate to provide ¼ *g* where zero-g would normally prevail. The time required for a transcontinental flight would be 75 minutes and the peak of the flight would be 27.5 miles up.

The foregoing was written ten years ago, in 1957.

Now, in 1967, supersonic airliners with turbojet propulsion are underway, the joint British-French project Concorde and Boeing's American counterpart. Whether the supersonic jet transport has shelved the idea of a passenger-carrying winged rocket permanently is impossible to say. Such winged rockets will be needed eventually as shuttle rockets to and from the manned space station. Once they exist they might be adapted to general use.

And now for the X-15.

It was even more *un projectile dirigeable* to use Captain Ferber de Rue's words, than the X-1 and the X-2. After some preliminary studies NACA, in 1954, came to the conclusion that a research aircraft capable of reaching an altitude of 250,000 ft and a speed of $M = 6$ was desirable. It was assumed that the aircraft would be air-launched like the X-1. Later in 1954 a formal memorandum signed by NACA, Air Force, and Navy completed the initial organization of the project and in December of the same year the Air Force sent out requests for bids to twelve manufacturers. After a study of the proposals received, North American Aviation was selected in November 1955. Three identical craft were to be built. In September 1956 a contract for a suitable rocket engine was given to Reaction Motors, Inc., which had meanwhile become a division of Thiokol Chemical Corporation. The rocket engine was to have a thrust range from 17,000 to 57,000 pounds, and the fuel tanks were to be dimensioned in such a manner that full thrust could be had for 90 seconds.

Since the development of the rocket engine would take longer than the building of the plane, the first version of the X-15 utilized an existing engine with a thrust of 8000 lb. Two such engines were installed so that their combined thrust about equaled the minimum thrust of the engine under development.

The first X-15 was completed October 15, 1958.

The over-all length of the X-15 was 50 ft, its wingspan an inadequate-looking 22 ft 6 in. Its wing area was 200 sq ft. It had a retractable nosewheel and two retractable landing skids under the tail assembly. The X-15 was the first plane to have a dorsal as well as a ventral vertical tail fin. The horizontal tail surfaces were built with a slight slope downward. There were hydrogen-peroxide thrusters for attitude control around the nose and also on the wings.

The fuselage had a well for the nosewheel ahead of the pilot compartment; behind the pilot there was a tank with liquid nitrogen (for cabin pressurization) and a tank for hydrogen peroxide (for the attitude thrusters). The main section, as in a true rocket, was the tank section, with an oxygen tank capable of holding 1034 gallons of liquid oxygen, followed by a fuel tank holding 1445 gallons of anhydrous ammonia. Cylindrical oxygen and fuel tanks surrounded an elongated tank for compressed helium which formed the center line of the fuel-tank section. The helium was for pressurization of the fuel tanks to aid the fuel pumps. Behind the tank section there was another hydrogen-peroxide tank (powering the fuel pumps) and then the rocket motor—motors in the first version. The fuel flow for full thrust for the final rocket engine (designated the XLR-99) was 210 lb per sec. The launch weight of the X-15 was 31,275 lb, the dry weight 12,971 lb.

The ideal flight path of the X-15—not actually achieved until Spring 1962—was as follows: The X-15 would be dropped from a B-52 bomber at an altitude of 45,000 feet. The pilot would permit his craft to fall for about 1000 feet, start the rocket engine and climb at an angle of 35°. At an altitude of 160,000 ft the fuel would be used up, and the plane, moving at the rate of 4000 mph, would begin a weightless "parabolic flight" (quite similar to Fig. 75, except that there would be a climb instead of a dive preceding the "parabola") with its peak at 250,000 ft, where 99.997 per cent of the atmosphere would be below the craft. Then the descending leg would begin, with re-entry at about 130,000 ft, after a little more than 3 min of zero *g*. Pull-out of the descending leg would begin at about 80,000 ft and be completed at 65,000 ft, at a distance of 200 miles, measured horizontally, from the point where the X-15 was dropped from its carrier plane.

But the first flight of the X-15 in March 1959 was a so-called "captive flight," which means that the X-15 remained attached to the bomber from take-off to landing. On June 8, 1959, there was a glide flight without power, using plane no. 1 with Scott Crossfield at the controls. On September 17, 1959, Crossfield (using plane no. 2) made the first powered flight, leading to an altitude of 52,341 ft, and with a maximum speed of 1393 mph ($M = 2.11$). On the third powered flight on November 5, 1959 (with no. 2), there was an engine fire and the craft was badly damaged in landing. However, it was repaired, and no. 2 made its next flight on February 11, 1960, to an altitude of 88,116 ft.

The first flight with the XLR-99 rocket engine, again with craft no. 2 and again with Crossfield at the controls, was made on November 15, 1960. It went to 81,200 ft with a maximum speed of 1960 mph. After Crossfield had made all the preliminary flights, a selected group of X-15 pilots took over. They were Major Robert A. Rushworth (USAF), Major Robert M. White (USAF), Milton O. Thompson (NASA), Joseph A. Walker (NASA), and John B. McKay (NASA). Slowly and carefully the X-15 was taken to higher and higher altitudes and higher and higher speeds. On November 9, 1961, Major White flew faster than 4000 mph for the first time—the precise figure was 4093 mph ($M = 6.04$). The first flight of craft no. 3 took place on December 20, 1961; on this flight, the craft was taken only to 81,000 ft. On April 30, 1962, pilot Walker took craft no. 1 to 246,700 ft, and on June 21, 1962, Major White took craft no. 3 to precisely the same altitude. Only a week later, on June 27, pilot Walker pushed craft no. 1 to 4105 mph during a relatively shallow flight. On July 17, 1962, he took craft no. 3 to 314,750 ft, thereby achieving the rank of astronaut, which is given to any pilot who makes a flight above 50 miles.

After the long succession of perfect flights with the X-15 nobody was too perturbed when craft no. 2 was badly damaged in an emergency landing on November 9, 1962. Instead of merely being repaired, it was modified for higher speeds (5000 mph) and modification of the others is to follow. The X-15, in a new version, is going to embark on a new program.

Aviation and rocket research came together in one other application: to get airplanes off the ground, off the water, and later off the flightdecks of aircraft carriers.

This became known as JATO, the letters standing for jet-assisted take-off. Of course the rockets could not be efficient, but the choice was to get the planes into the air with poor rocket efficiency or not at all.

The idea of using rocket thrust to assist in the take-off of airplanes originated soon after the First World War. At first its application was considered only where it was needed most: for pontoon-equipped seaplanes. In about 1921 somebody suggested jetting a stream of water from the pontoons to help in the take-off. Then aviation engineers began to wonder about a more widespread application of the principle, on the theory that any airplane needs more power for take-off than for flight. If extra take-off power could be provided, the engines of the airplane need be only powerful enough to sustain it in flight.

Professor Hugo Junkers (then still alive) rejected this reasoning as unsafe, but he was interested in take-off help for special cases. He took one of his single-engined pontoon-equipped airplanes—the same type that made one of the first transatlantic flights—and overloaded it to such an extent that it was incapable of taking off. Then he had a battery of Sander rockets installed. Apparently he either feared ridicule or else thought that a great commercial advantage might be gained soon—in any case he was tight-lipped about the details. The normal take-off weight of these planes was around 3 tons; Junkers did not say how much the overloaded specimen weighed. Nor did he state how many pounds of thrust were supplied by the rockets. In any event, the take-off of the overloaded plane with the aid of the battery of rockets was successful. The experiment had been made in great secrecy in August 1929 but in spite of all secrecy the fact leaked out, and Professor Junkers finally permitted Winkler to publish the information. But nothing else was published for many years to come.

The first stories about practical use of rocket take-off boosters came from Holland at the time of the Battle of Britain. The Germans, it was reported, launched overloaded bombers from small field airdromes in that manner. The units were described as metal gratings fitting the bottom of the fuselage of the planes and holding large numbers of solid-fuel rockets which were ignited electrically by the pilot. Once the plane was airborne, either the pilot jettisoned the empty grating or it dropped off automatically.

This may have been a quick improvisation, utilizing solid-fuel rockets (possibly Schmidding rockets) which were on hand. Hellmuth Walter's hydrogen-peroxide take-off units had performed well as early as 1938, when take-offs of a Heinkel He-112 airplane with Walter units were demonstrated and filmed. The British later captured the films, which incidentally were of perfect photographic quality and must have been made by professional motion-picture cameramen. The take-off units were usually suspended from the wings near the engine nacelles and in shape were as nearly spherical as possible. They were to be jettisoned immediately after use so that they would not cause drag afterward. That was done without parachutes, with the hope that their spherical shape would keep them from being damaged when they hit the ground, so that they could be used again.

Over water this might have worked, but when jettisoned over land, even over the testing area, they smashed into tree trunks and fence posts, hit rocks or roadways, and were salvageable afterward only in the most exceptional cases. Walter's engineers then put a second container around the unit and filled the space between the containers with a ring-shaped parachute. That worked. All the early units had motors of the "cold" type, but as the weight of the planes increased, "hot" motors were introduced.

Later on, the field of liquid-fuel take-off units seems to have been divided about equally between Walter in Kiel and Bavarian Motor Works. Two models, produced near the end of the war, can be considered typical; their specifications are given in the following table.

	HWK RI-209/109-502 FOR T-STOFF AND FUEL		BMW 109-718 FOR SALBEI AND FUEL	
	U.S.	METRIC	U.S.	METRIC
Thrust	3300 lb	1500 kg	1320–2750 lb	600–1250 kg
Time	30 sec	30 sec	112–180 sec	112–180 sec
Oxidizer	486 lb	220 kg	1350 lb	612 kg
Fuel	61 lb	27.7 kg.	410 lb	185 kg
Rate, maximum	18 lb/sec	8.15 kg/sec	15.6 lb/sec	7 kg/sec
Exhaust velocity	5750 ft/sec	1750 m/sec	5750 ft/sec	1750 m/sec
Motor weight	500 lb	228 kg	176 lb	80 kg

The HWK RI-209 take-off unit was designed and used for two airplanes, the Heinkel He-III and the Junkers Ju-88. The BMW 109–718 unit was designed specifically for the turbojet fighter Messerschmitt Me-262; the turbine of the jet engine also drove the fuel pumps of the take-off unit. The BMW unit's force-feed mechanism, therefore, was powered by the airplane's jet engine, which explains why it was so much lighter than the independent HWK unit.

The Japanese had been made acquainted at an early date with the German methods of manufacturing high-strength hydrogen peroxide. Pietsch's Elektrochemische Werke had sold the whole process to Mitsubishi Seishi Kaisha. But the Japanese Navy showed no interest in hydrogen-peroxide rockets (and very little in rockets of any kind) until June 1944. Then research work was started but it was not aimed at the development of take-off units; it was to produce a rocket power plant for the Shusui airplane then in the design stage. The Kaiten torpedo also was to profit from this work.

But there were three Japanese solid-fuel take-off units, operating on the double-base powder described in Chapter 7 in connection with Japanese bombardment rockets. The smallest of these units was 46.6 inches long with an outside diameter of only 7.6 in. It developed a maximum thrust of 2310 lb (1050 kg), the mean was 1254 lb (570 kg), the duration 4.03 sec. The other two had a burning time of approximately 10 sec (one a fraction short of 10 sec, the other a fraction above). The mean thrust of one was 814 lb (370 kg), its maximum thrust 1927 lb (876 kg); the mean thrust of the other was 1430 lb (650 kg) and its maximum thrust 3747 lb (1703 kg). The latter was also the propulsion source of the Baka.

In America development of take-off units was carried on in several places, but the only group to produce a finished job, ready for manufacture, was Dr. von Kármán's creation, the Jet Propulsion Laboratory of the Guggenheim Aeronautical Laboratory of the California Institute of Technology, JPL of GALCIT for short. In December 1938, General H. H. Arnold requested a research and development program for several important devices, including take-off units. The National Academy of Science told JPL to go ahead.

Experimental work began in 1939, the first test unit being a 2-ft length of steel pipe, with a wall thickness of 1 in. One end was plugged, the other equipped with a flange to which an exhaust nozzle could be bolted. Because the thrust of a take-off booster must be relatively low but of comparatively long duration (otherwise the booster might just tear itself loose, damaging the plane) the researchers decided that their charge should be permitted to burn on the cross section only. That meant a wall-fitting charge, but with a wall-fitting charge heat transfer along the wall might cause explosions. Hence they fitted the charge into a liner which would not conduct heat well and then fitted the liner into the pipe. The final result naturally looked very much like one of Sander's branders—there are only so many different shapes which a rocket can assume.

After two years and a few hundred tests the Caltech men had a small but reliable booster unit weighing 12.7 lb, about 2 lb of which was propellant. The unit was capable of delivering 28 lb of thrust for 12 sec. The propellant was called GALCIT 27; its loading had to proceed along the lines used by the fireworks makers of old. The charge had to be fed into the chamber in twenty-two separate increments, each subjected to a pressure of 18 tons. Using less pressure in loading or larger increments resulted in unsafe charges because there were then sections of unequal density. Since everybody was eager to find out how a take-off booster influenced the take-off characteristics of a plane, a number of flight tests were made in August 1941. The plane was a small 750-lb Ercoupe. Six of the small units were used in each test. The pilot was Air Force Captain Homer Boushey.

It might have looked at that point as if it were merely necessary to scale this unit up to a useful size. But the problem ahead of the experimenters was more complex by far.

The powder stick which had been produced with so much trouble and labor shrank in storage. That, of course, produced a space between charge and liner and transformed the restricted burning rocket into an unrestricted burning rocket, with results which do not need to be described. GALCIT 27 had to be discarded.

In May 1942 an improved propellant was ready, GALCIT 46. The number alone is indicative of the large amount of hard work that had been done in the meantime. GALCIT 46 did not shrink in storage. But it was sensitive to temperature in another way; one could not count on satisfactory performance in Africa or Greenland. The same was true for ballistite which had been investigated in the meantime. A 1000-lb thrust unit with a ballistite charge would deliver the 1000 lb when the temperature was 90° F. But at 40° F it would deliver only 600 lb. In the latter case the duration was longer, so that if the unit had been used for a bombardment rocket the difference in range would probably have been tolerable. But that additional duration was of little use if the problem consisted in getting a plane off the short flight deck of a small carrier.

Since the available smokeless powders, single-base and double-base and in all conceivable combinations, had failed to meet the specialized requirements of a take-off booster, the research group tried something entirely new. Its number was GALCIT 53 and early tests were so promising that it was soon given priority over all other developments still running. GALCIT 53, like all solid propellants, was a mixture of fuel and oxidizer. The oxidizer was potassium perchlorate. The fuel was asphalt, with a small percentage of oil with an asphalt base.

> The mixture was prepared by heating the asphalt and oil in a mixing kettle to a temperature of 350° F., then stirring in the perchlorate. Before the combustion chambers were loaded with the finished propellant, they were lined with a hot mixture of asphalt and oil. When the propellant had cooled sufficiently, it was scooped into the combustion chambers, which were bounced a few times to assure uniform settling, then set aside for the propellant to harden. In its finished form, GALCIT 53 is a black plastic, at ordinary temperatures resembling stiff paving tar. It can be detonated with difficulty, if at all. Only with patience can it be ignited with a match flame; but once ignited it burns fiercely, emitting a white light and dense white smoke. Burning in a combustion chamber under pressure of 1800 lbs. per squ. inch, the propellant gives an average exhaust velocity of 5300 feet per sec. at an average burning rate of 1.25 inches per second.
>
> Units loaded with the new propellant were recommended to be fired at temperatures between 40° F. and 100° F. Much above the recommended temperature it became viscous and flowed. Therefore it was imperative that invariably the JATO units be stored right side up. The rocket motor designed for use with GALCIT 53 . . . was approximately 13 inches long and 5½ inches in diameter.
>
> (Galcit Report)

The U. S. Navy then contracted with the Aerojet Engineering Corporation (then of Pasadena, now of Azusa, Calif.) for a number of rocket units loaded with GALCIT 53 for experimental purposes. In due course of time they were ordered in quantities. As the orders continued, the thrusts called for rose from 200-lb and 500-lb to 1000-lb units. The development of these units was done partly by the Galcit Jet Propulsion Laboratory, partly by the firm of Aerojet. "Since 1942," the Galcit Report states, "most of the work done by the Laboratory to improve solid propellants has been directed toward widening the temperature limits for safe operation. In 1943, the Laboratory developed GALCIT 61-C, which the Navy continued to use until the war ended."

The choice of these solid-fuel units was probably due to logistic considerations; the Navy's own development of take-off units at Annapolis had been one of liquid-fuel units. The Truax group produced a unit of 1500-pound thrust, two of which were needed to help a PBY flying boat take off. They were attached to the wing struts. Goddard had worked on a unit with higher thrust which was built into the tail of the flying boat.

Naturally the Laboratory also paid attention to the problem of a liquid-fuel take-off unit. The first finished model delivered a thrust of 1000 lb for 25 sec, operating on nitric acid and aniline. It was flight-tested with a Douglas A-20-A bomber at Muroc during the period between April 7 and 24, 1942. Since then larger units of the same type have been made. In 1945 two types called XCALT-6000 and X40ALD-3000 were demonstrated at Wright Field. Their thrusts were what the designations indicated: 3000 and 6000 lb respectively.

The life history of the JATO units is a case of one wartime development's being killed off by another one. A few years after the war, say from 1950 on, the armed services as well as the commercial airlines could have ordered take-off units of almost any desired performance. They could even have specified whether they wanted liquid or solid fuel, and, if the latter, whether they wanted one large unit or a cluster of small ones. One or two airlines actually bought solid-fuel clusters for experimental use at airports at high altitudes, such as that of Mexico City, but they did not pursue the idea. It was probably too expensive.

Now such units are no longer needed, unless the circumstances are exceptional. That other wartime development, the jet plane, has made them obsolete, since jets don't need them.

APPENDIX 2

The International Situation

When we try to pick out anything by itself, we find it hitched to everything else in the universe.

—John Muir

1. Space Travel and International Law

That space travel would pose legal problems probably did not occur to its pioneers, but lawyers, especially those engaged in the practice of international law, realized this at an early date. Some time in 1931 a member of the VfR, Dr. Vladimir Mandl, who was a practicing lawyer in Pilsen (Czechoslovakia) sent me a manuscript on the legal aspects of space travel, requesting my help in getting it published. It was published in 1932 under the title *Das Weltraum-Recht.* Dr. Mandl's study tried chiefly to show that existing legislation—he used the very strictly codified German law as the means of presenting his case—covered most of the foreseeable legal problems. But Dr. Mandl failed to established a unified legal point of view. This was done (as far as I know for the first time) during a lecture delivered at the first Hayden Planetarium Symposium on Space Travel (October 12, 1951) by Oscar Schachter of the legal department of the United Nations (published in *JBIS,* January 1952, somewhat expanded as a chapter in *Across the Space Frontier*). Schachter took the position that open space should be regarded like the open sea, as not belonging to any power and available to the use of all.

Most legal experts who have published on the theme agree with this fundamental concept.[1] The main difficulty arises from the natural fact that it is impossible to reach open space without passing through air space which is considered sovereign territory by all nations, largely as a by-product of the First World War.

At a time when there were only free balloons and a very few airships and the air-

[1] To mention a few: Andrew J. Haley's lecture, "Basic Concepts of Space Law," at the IAF Congress in Rome, November 1955, published in *Jet Propulsion,* November 1956; "International Law and Activities in Space" by C. Wilfred Jenks, published in *The International and Comparative Law Quarterly,* January 1956; and "Projecting the Law of the Sea into the Law of Space" by Rear Admiral Chester C. Ward, a lecture at the ARS Annual Meeting, November 1956, ARS Preprint 377–56.

plane still had to be invented, in 1902, a legal expert named Paul Fauchille presented a draft of a "proposed convention on the regulation of aerial navigation" at a meeting of the Institute of International Law. In this proposal the air was to be as free as the ocean. It was adopted in somewhat modified form in 1906. But then the First World War came, and it was one of the neutrals who defended the idea of sovereignty over the air space most vigorously. This was Holland, which had the bad luck to lie directly under the shortest air route between England and Germany. It is true that its air space was violated; disabled seaplanes had to land in Dutch waters and a couple of zeppelins drifted off course in foggy weather. The zeppelins, as a matter of principle, hid in drifting clouds to avoid being spotted, which might have been a useful protective measure but did not improve navigation.

Just after the First World War, in October 1919, the Paris Convention for the Regulation of Air Navigation was signed and "freedom of the air" was completely ruined in the very first article of this convention: *Les Hautes Parties Contractantes reconnaissent que chaque Puissance a la souveraineté complète et exclusive sur l'espace atmosphérique au-dessus de son territoire*—"The High Contracting Parties recognize that every Power has complete and exclusive sovereignty over the air space above its territory." The "territory" included, of course, the 3 miles of sea offshore which by longstanding international agreement are considered to belong to the country bordering the sea. The important legal point is that the sovereignty of a country over these 3 miles of sea water is somewhat limited. If a vessel, in order to pass from one tract of open sea to another one, has to navigate through sovereign waters it can do so without asking anybody's special permission; it has the right of "innocent passage." This right applies to passenger liners, freighters, expeditions of any kind, and rescue missions; whether a naval vessel, in time of peace, has the right of innocent passage is disputed. (But in time of peace this is no problem because such fleet movements take place with prior agreement.)

The Paris Convention was followed by a welter of additional conferences dealing with all kinds of side issues—such as cases of infectious disease discovered on an international flight—and by separate international agreements, especially among the nations of the Western Hemisphere. Then came the Second World War and near its end another important conference on civil aviation took place, this time in Chicago. The conference ended in December 1944, and again the first article of the convention adopted read the same, nearly word for word, as the first article of the Paris Convention. But something new had been added: "No aircraft capable of being flown without a pilot shall be flown without a pilot over the territory of a contracting state without special authorization by that state." This was the first legal recognition of the existence of guided missiles.

At first glance this convention seems to rule out any artificial satellites for whatever purpose unless there is a prior agreement of all countries. (In the case of the International Geophysical Year such an agreement was established by the fact that virtually all nations participated in it.) But this is not really the case, as Admiral Ward explained:

"Professor Cooper, and other distinguished authorities, have pointed out that our development of a law of outer space is not restricted by our present agreements affirming each nation's sovereignty over the air, as the 'air space,' above it. These agreements relate only to the 'air space.' They apply only as far as the upper limits of the region in which air is sufficiently dense to support the flight of conventional aircraft—that is, those aircraft supported through reaction with the air. We are therefore free to develop a law of outer space, to apply in areas above this region of relatively dense air. . . . The lawmakers wait only for the physical facts of space to be supplied by the explorers, the scientists, the mathematicians, and the physicists. With the physical facts in hand, we can attempt to set the upper limits of national sovereignty."

The subsequent technological developments with artificial satellites and manned spacecraft led to the general acceptance of the lower limit at which such flight can be carried out—100 miles, in round figures—as the limit of space beyond which everything is in space where there is no national sovereignty. The International Aeronautical Federation accepts an altitude of 100 kilometers (ca. 62 miles) as the limit of space; on the other hand, the United States bestows astronaut's wings to any pilot who has taken a vehicle of any type beyond 50 miles. It may be added here that there cannot be any national sovereignty over the area covered by the oceans; hence the freedom of the high seas and the freedom of space merge at the top of the wave crests for three-quarters of the earth's surface.

An important legal development was initiated in May 1966 by President Lyndon B. Johnson when he proposed an international treaty to limit military activities in outer space. The treaty was negotiated in the Outer Space Committee of the United Nations and adopted by acclamation by the General Assembly of the United Nations on December 19, 1966. The treaty forbids the placing of nuclear weapons or other weapons of mass destruction in orbit or on the moon or other planets; it forbids military installations on the moon or other planets and bans military maneuvers originating from the moon or another planet. (It does not forbid the orbiting of military spacecraft without heavy weapons, or the use of artificial satellites for military reconnaissance or communciations.) Treaty-signing ceremonies were held simultaneously in London, Moscow, and Washington on January 27, 1967, when representatives of sixty-two nations affixed their signatures.

2. International Cooperation with NASA

NASA realized very soon after it was created in 1958 that its activities would involve a great deal of international cooperation, at first because of the need for tracking stations in other parts of the world. By the end of 1962 sixty-one foreign nations had agreed to cooperate with NASA, and it was only natural that this cooperation was expanded into fields other than tracking and reporting.

NASA agreed to furnish rockets and facilities for the launching of foreign-built satellites—so far three British, two Canadian, one French, and two Italian satellites have been launched with NASA rockets (for details see the sections under the names of these countries in this Appendix).

Another program in which financially weaker nations could participate was the exploration of the upper atmosphere with sounding rockets. The country in question furnishes the launching site, facilities, and personnel; NASA furnishes the rockets and, if necessary, instruction. No exchange of funds takes place. Among the countries participating in such joint experiments, two, France and Japan, have space programs of their own. The others do not, with Argentina constituting an in-between case.

Argentina has an unofficial organization called the Inter-American Committee for Space Research and also an official National Committee for Space Research. Professor Teofilo Tabañera is president of both. The space center is located at Chamical, 150 miles from Cordoba, in an area consisting mainly of large salt flats.

An Argentine design for a sounding rocket is called Gamma-Centauro; it is a solid-fuel rocket with a solid-fuel booster. Take-off weight is 70 lb, over-all height slightly more than 7½ ft. The rocket is capable of carrying an 11-lb payload to about 15 mi. First launches of American solid-fuel sounding rockets from Chamical took place in December 1964.

But in November 1966 Argentine rockets were launched from Wallops Island. The name of the rockets was Orion II. They were about 12 ft long, with a diameter of 8 in and a take-off weight of 286 lb. They could carry payloads weighing 30 to 40 lb to altitudes of 55 to 70 mi. Of the three launches the one of November 4 was successful, the one of November 7 exploded at T + 6 sec, and the one of November 9 was successful again. One payload, coming down by parachute, was retrieved in midair by a helicopter.

Cooperation with India was first planned with a view of having a launch site close to the equator. The site is at Thumba, near the southern tip of India. At first, American-built rockets were launched there, but on March 16, 1967, an Indian-built instrument package was launched from Wallops Island by a Nike-Apache rocket with great success.

Japanese-American launchings were also made from Wallops Island in March and April, 1966. John F. Spurling of Wallops Station was in charge of the American rockets, Maosuke Arizumi of the Japanese Weather Bureau of the Japanese rockets. The rockets carried meteorological instruments and had the following characteristics:

	MT-135 (JAPANESE)	ARCAS (U.S.)
Total length with payload	**10½ ft**	**7½ ft**
Diameter	**5½ in.**	**4½ in.**
Take-off weight	**150 lb**	**77 lb**
Payload weight	**22 lb**	**12 lb**
Burning time	**10.5 sec**	**29 sec**
Average thrust	**1830 lb**	**336 lb**
Peak altitude	**38 mi**	**32 mi**

The rockets were launched in pairs, the Japanese rocket first, with the American rocket following 15 minutes later. The purpose was meteorological research.

The agreement with The Netherlands calls for firings from Surinam with release of sodium vapor at altitude. The launchings from the Andoya range in Norway, in which Sweden and Denmark were also involved, had the purpose of investigating noctilucent clouds. Experiments of this kind have also been made from New Zealand.

As regards national programs, Argentina has offered its Chamical range for use by any other Latin American country, while India's Thumba range is open to any Asian country. So far nobody has taken advantage of these offers.

3. Canada

Canada became involved in upper-atmosphere and space research during the International Geophysical Year by furnishing a firing site in high altitudes. The site was Fort Churchill, which was then the northernmost point in Canada that could be reached by rail. During the International Geophysical Year several dozen high-altitude rockets (Aerobees, and later solid-fuel two-stage rockets of the DAN type) were fired from Fort Churchill. The firing site is still used on occasion.

But then Canada entered the field of high-altitude-research rockets with rockets of its own, known as the Black Brant series. The development work was done under the auspices of CARDE (Canadian Armament Research and Development Establishment). The first rocket was Black Brant I, which used a body supplied by the Bristol Aeroplane Company in Great Britain. The subsequent Black Brant rockets were all Canadian. The first Black Brant II was test-flown in October 1960, after a dozen rounds with Black Brant I rockets.

The dimensions and characteristics of the various types can be seen from the following table. All Black Brant rockets use solid fuel and all are single-stage rockets except Black Brant IV, which is a two-stage rocket.

MARK	I	II	IIA	III	IV	V
DIMENSIONS (IN.)						
Over-all length	291.6	332.4	320.6	216.6	435.3	320.6
Diameter	10.2	17.3	17.3	10.2	1st Stage 17.2 2nd Stage 10.2	17.3
Nose-cone length	62.4	86.2	95.8	64.7	64.7	95.8
WEIGHTS (LB)						
Total without payload	1609	1945	2240	630	2990	2640
Nominal payload	150	150	150	40	40	150
PERFORMANCE						
Nominal altitude in miles for firing angle of 85°	140	170	145	110	617	240

While the Black Brant rockets are exclusively Canadian, the other Canadian space-research programs are all joint Canada/U.S. ventures. The most unusual is the 16-inch smoothbore "space gun," conceived by Professor Gerald Vincent Bull of McGill University and located at Barbados. The projectiles launched from this gun weigh 190 lb and have reached altitudes exceeding 350,000 ft. At the peak of their flight the projectiles release tri-methyl-aluminum which produces fluorescent trails that make upper air winds visible. The "space gun" was financed by the Canadian government and the U.S. Army Ballistic Research Laboratories.

The Canadian satellite projects have the names Alouette (French for "skylark") and ISIS (for International Satellite for Ionospheric Studies). Alouette I, with a diameter of 42 in., a height of 34 in., and a weight of 319 lb, was launched by a Thor/Agena B rocket from the Pacific Missile Range on September 29, 1962. The Pacific Missile Range had to be chosen because the orbit was to be nearly polar (inclination to the equator 80.5°) and circular at an altitude of 600 mi. The actual orbit is, of course, slightly elliptical, but very slightly so, for the difference between perigee and apogee is only 22 mi. Orbital period is 190.5 min.

Alouette II, weighing 320 lb, was launched November 28, 1965, by the same rocket that launched the American satellite Explorer XXXI. Perigee is at 315 mi, apogee at 1853 mi. Orbital period is 121.4 min.

The ISIS series of three satellites is still in preparation.

4. Egypt

The fact that Egypt had a rocket development program was made known to all the world on July 21, 1962, which was the tenth anniversary of the removal of King Farouk and also the day on which four large rockets were fired in the presence of representatives of several governments and a few newspapermen. The first two of these rockets, whose appearance clearly proclaims that they are descendants of the V-2, were named El Qahir (Conqueror) and were said to have a range of 370 miles. The other two were named El Zafir (Victor) with a range reported as 200 miles. Measurements of photographs indicate an over-all length for El Zafir of about 18 feet and of El Qahir of about 36 feet. At a later date the Egyptian government announced that it had a two-stage rocket named El Ared (Pioneer) with a range of about 600 miles. It was also said that an improved El Ared would be used to launch a small satellite.

These facts would not have taken up as much newspaper space as they did if they had not been accompanied by a political story, resulting from the fact that the scientists in charge of the Egyptian rocket establishment are mostly German. Three of them were mentioned by name—Doctors Wolfgang Pilz, Hans Görcke, and Hans Kleinwächter. The government of Israel not only denounced them as an "evil crew" and as Nazis but also requested the West German government to make them quit Egypt. Of course the West German government could do no such thing; the West German constitution does not place restrictions on the place of residence or type of work of its citizens. But the West German government does seem to have offered them better jobs, for at least one, Dr. Wolfgang Pilz, returned to West Germany in June 1965. Ironically, his place is said to have been taken by two American citizens identified only by their last names (Smith and Hicks).

None of the Egyptian rockets went into action during the "six-day war" with Israel in June 1967. The Israelis did capture an Egyptian missile site in the vicinity of Bitter Lake, but the rockets they captured were Russian SAM-3 antiaircraft missiles which were known to have been acquired by the Egyptians. What actually happened to the Egyptian rocket-development program is as mysterious at the moment as were the hieroglyphs before they were deciphered.

5. France

Around the year 1950, when it became clear that the United States and the Soviet Union were about to start a long-lasting missile race, the Direction des Études et Fabrications d'Armement (DEFA—the French ministry of national defense) decided that France, too, should have some experience in rocket-building.

The ministry ordered the Laboratoire de Recherches Balistiques et Aérodynamiques at Vernon to begin a suitable design. The result was the rocket known as Véronique.

The designers did not utilize the rocket built by Robert Esnault-Pelterie during the last years of his active life but began with the design of the V-2, which they scaled down. The fuels chosen were stable fuels—Diesel oil and fuming nitric acid. Ignition was by a hypergolic igniter using furfuryl alcohol. When the rocket was fired an interesting arrangement of four tethers of steel cable was used to keep it on a vertical path. By the time the rocket was stable the tethers were disconnected. Test firings took place in North Africa, and a total of fifteen Véroniques was fired in 1953 and 1954.

The characteristics of the rocket were as follows:

	U.S.	METRIC
Over-all length	**24 ft**	**7.3 m**
Body diameter	**21.5 in.**	**55 cm**
Take-off weight	**2204.6 lb**	**1000 kg**
Fuel weight	**1542.2 lb**	**700 kg**
Payload (max.)	**132 lb**	**60 kg**
Altitude (average)	**71.5 mi**	**115 km**
Altitude (max.)	**84 mi**	**135 km**
Time to peak	**175 sec**	

The maximum altitude given in the table was reached with a shot on February 21, 1954. After this promising beginning there came a pause for financial reasons, but later a slightly heavier Véronique flew again from the Hammaguir firing range in the Algerian Sahara not far from Colomb-Béchar near the Moroccan border. The new Véronique was a foot shorter than the first model, but its diameter had increased to 27 in. and the total weight (without payload) to 3086 lb, of which 2204 lb were fuels. The oxidizer for the new version was still nitric acid, but the fuel proper was a turpentine derivate, known as terebenthine. Payload weight had been increased to 132 lb. The new Véronique was ready in 1960 and a number of test flights were made in 1960 and 1961. One of these flights carried a rat that was recovered after a flight to 66 mi.

By that time many rocket and missile designs were on French drawing boards and a number of government organizations as well as industrial firms were preparing France for her entry into the space age. Compared to the German attitude (confused and uncertain) and the British attitude (timid and parsimonious), the French attitude was refreshingly clear. France wanted to be an independent missile power and France wanted to go into space, *coûte que coûte!*

Since the agencies involved are mentioned by their customary abbreviations, these and the full names should be listed first. They are:

SEREB, Société pour l'Étude et la Réalisation d'Engins Balistiques
S.E.P.R., Société d'Étude de la Propulsion par Réaction
CNES, Centre National d'Études Spatiales
CNET, Centre National d'Études des Télécommunications
CERMA, Centre d'Études et de Recherches de Médicine Aéronautique
CNRS, Centre National de la Recherche Scientifique

The controlling agency is CNES, established in 1962 and under the direction of

General (retired) Robert Aubinière. Soon the "activity reports" from CNES began to read like NASA's reports. In condensed form, the following table shows what CNES did between July 1, 1964, and July 1, 1965.

TYPE OF ROCKET AND NUMBER	LAUNCH DATE	LAUNCH SITE	ALTITUDE REACHED MI	ALTITUDE REACHED KM	RESEARCH ESTABLISHMENT
Dragon 10	Aug. 1, 1964	Vik (Iceland)	280	450	CNRS
Dragon 11	Aug. 7, 1964	Vik (Iceland)	280	450	CNRS
Centaure 62	Oct. 5, 1964	Hammaguir	92	148	CNET
Centaure 63	Oct. 7, 1964	Hammaguir	72	118	CNET
Véronique 53	Nov. 4, 1964	Hammaguir	95	153	CRNS and Utrecht Univ.
Centaure 47	Nov. 5, 1964	Hammaguir	114	185	CNRS
Centaure 69	Nov. 5, 1964	Hammaguir	108	175	CNRS
Centaure 64	Nov. 5, 1964	Ile du Levant	93	150	CNRS
Centaure 67	Nov. 5, 1964	Reggane	unknown*		CNRS
Centaure 72	Nov. 5, 1964	Reggane	unknown		CNRS
Centaure 71	Nov. 6, 1964	Ile du Levant	81	130	CNRS
Centaure 66	Nov. 6, 1964	Reggane	unknown		CNRS
Véronique 52	Nov. 8, 1964	Hammaguir	60	98	CNRS
Bélier	Nov. 10, 1964	Ile du Levant	42	68	Laboratoire de Physique de l'Atmosphère
Centaure 75	Nov. 27, 1964	Hammaguir	118	191	Max Planck Institut (Munich)
Centaure 76	Nov. 30, 1964	Hammaguir	128	206	Max Planck Institut
Dragon 16	Dec. 3, 1964	Hammaguir	242	390	Max Planck Institut
Véronique 56	Feb. 12, 1965	Hammaguir	59	95	CNRS
Centaure 85	Mar. 4, 1965	Hammaguir	85	136	CNRS
Véronique VA77	May 27, 1965	Hammaguir	43	70	CNRS
Rubis 1	June 5, 1965	Hammaguir	1243	2000	CNES, Satellite Div.

* There is no tracking equipment at Reggane.

The French program is carried forward by two "families" of rockets, the Bélier family of solid-fuel sounding rockets, built by Sud-Aviation in collaboration with CNET, and the SEREB research and space rockets, all of which bear the names of gems.

To begin with the Bélier family (the word means "ram" but is used more in the military sense of battering ram), the Bélier has the following characteristics:

	U.S.	METRIC
Length	13 ft 1.9 in.	4 m 1.1 cm
Diameter	12.0 in.	30.5 cm
Fin span	30.8 in.	78.2 cm
Launch weight	694 lb	315 kg
Altitude (max.)	50 mi	80 km
Burning time	21 sec	
Time to peak	2 min 10 sec	

The next rocket in this family is the Centaure, which carries the Bélier as its second stage. Figures for the Centaure are:

	U.S.	METRIC
Length (with Bélier)	19 ft 9 in.	6 m 2 cm
Diameter	11 in.	27.9 cm
Fin span	28.5 in.	72.5 cm
Take-off weight (with Bélier)	1029 lb	457 kg
Altitude (for upper stage)	90 mi	140 km
Burning time	4.5 sec	
Time to peak (upper stage)	3 min 12 sec	

The next member of the family is the Dragon, powered by a solid-fuel motor known as the Stromboli. It also carries a Bélier as its second stage. Figures for the Dragon are:

	U.S.	METRIC
Length (with Bélier)	23 ft 3.2 in.	7.1 m
Diameter	21.9 in.	55.6 cm
Fin span	48.2 in.	122.5 cm
Take-off weight (with Bélier)	2550 lb	1157 kg
Altitude (for upper stage)	250 mi	400 km
Burning time	19 sec	
Time to peak (upper stage)	5 min	

The next bigger one is the Pegase, which is the booster for a Dragon carrying a Bélier. Its specifications:

	U.S.	METRIC
Length (all three stages)	33 ft 11.2 in.	10 m 43.3 cm
Diameter	22 in.	55.8 cm
Fin span	61.1 in.	155.3 cm
Take-off weight (with upper stages)	4512 lb	2047 kg
Altitude (for top stage)	590–630 mi	950–1000 km
Burning time	19 sec	
Time to peak (for top stage)	9 min 10 sec	

This completes the Bélier family, but Sud-Aviation also builds the Eridan, a two-stage sounding rocket for payloads varying from 130 to 700 lb. Both stages are powered by the Stromboli. Characteristics of the Eridan are:

	FIRST STAGE		SECOND STAGE	
	U.S.	METRIC	U.S.	METRIC
Length	10 ft 2.7 in.	311.7 cm	14 ft 6.4 in.	443 cm
Diameter	22 in.	55.8 cm	21.9 in.	55.6 cm
Fin span	61.1 in.	155.3 cm	48.2 in.	122.5 cm
Take-off weight, total	4422 lb	2006 kg		
Burning time	19 sec		19 sec	

The maximum altitude, between 105 and 250 miles, depends on the payload. The upper stage of the Eridan can be flown by itself as a single-stage sounding rocket. When so used it is called the Aigle (eagle).

A few French rockets belong neither to the Bélier family nor to the family bearing the names of gems. Nord-Aviation's Véga research vehicle—which has possibilities as an antiaircraft missile—consists of a solid-fuel booster 12 ft 5 in. long with a diameter of 22 in. The booster carries a ramjet missile capable of operating at altitudes from 65,000 to 115,000 feet at Mach 2 and 3. Total length of the Véga that was actually test-flown at Hammaguir early in 1961 was 32½ ft.

The four-stage solid-fuel Bérénice with an over-all length of 43½ ft was designed by SEPR jointly with ONERA (Office National d'Études et de Recherches Aerospatiales) for the purpose of testing re-entry nose cones. It is quite similar in function to the American X-17. The two lower stages (diameter near 22 in.) fire on the way up, while the two upper stages (diameters 13.23 in. and 8.66 in., respectively) fire on the downward leg of the trajectory that has a peak of 170 mi and a re-entry point 70 mi from the take-off area (Ile du Levant).

The SEPR-25 engine is a three-chamber engine for hypergolic fuels, with a total thrust of 9930 lb. This engine is built into the tail of the Trident fighter airplane, which is then sustained in flight by two French-built Armstrong Siddeley Viper turbojets with 1650 lb of thrust each, located at the wingtips of the aircraft.[1]

And now for the "gems": Agate, Diamant, Emeraude (emerald), Rubis (ruby), Saphir, and Topaze. But these are not five different rockets, for some of them are compounds of the others, as in the Bélier family. Saphir is an Emeraude with a Topaze as the upper stage; Diamant, the satellite launch vehicle, is Saphir plus a third stage, while Rubis is an Agate with the third stage of the Diamant as its second stage. Of these rockets only the Emeraude is designed for liquid fuels (the same fuels used in the second version of Véronique); all the others are solid-fuel rockets.

The Rubis made a successful flight from Hammaguir on April 22, 1966, carrying a 79.3-lb payload to 1264.5 mi. The characteristics of the Rubis on that flight were:

	U.S.	METRIC
Total weight	**7650 lb**	**3470 kg**
Propellant weight	**5600 lb**	**2540 kg**
Over-all length	**31.8 ft**	**9 m 61 cm**
Diameter, first stage	**31.7 in.**	**80.5 cm**
second stage	**25.8 in.**	**65.6 cm**
Acceleration (max.)	**35 g**	**35 g**
Ignition of second stage	**28 mi**	**45 km**

The Emeraude, launched with dummy stages, at first gave trouble. There were three failures in a row, in June, August, and November 1964, but the fourth launch,

[1] Operational French military missiles are:

AS-20, air-to-surface, length 8.5 ft, diameter 9.84 in., fin span 2.5 ft, launch weight 293 lb; radio-controlled, solid-fuel.

AS-30, a larger version of AS-20; length 12.5 ft, diameter 13.5 in., fin span 3.3 ft, weight 1100 lb.

Entac, an antitank missile with a range of 6500 ft, adopted by the U.S. Army in Europe.

Masurca, a ship-to-air missile, all solid fuel, two-stage with an over-all length of 27.9 ft and a launch weight of 3828 lb. Beam-rider type.

Malafon, an antisubmarine weapon; two solid-fuel boosters launch an airplane-shaped missile that drops a homing torpedo in the target area.

R.511 and R.530, air-to-air missiles by S. A. Engins Matra—R.511 weighs 397 lb and has a length of 10.14 ft and a diameter of 0.85 ft; R.530 is slightly heavier and a foot longer.

early in 1965, was a success. Saphir (Emeraude plus Topaze) was successfully launched on July 5, 1965.

The dimensions of two of the basic rockets, Agate and Topaze, are quite similar, as the following table shows:

	AGATE	TOPAZE
Length	28.1 ft	25.76 ft
Diameter	31.5 in.	31.5 in.
Weight, take-off	7496 lb	7506 lb
propulsion unit	5688 lb	6206 lb
instrument capsule	915 lb	584 lb
nose cone	893 lb	716 lb
propellant	4189 lb	5015 lb
Number of exhaust nozzles	1	4
Burning time	18 sec	39 sec
Thrust at launch	190 kN*	120 kN

* The abbreviation kN means kilo-Newtons; a Newton is the force that imparts a speed of 1 m/sec^2 to a mass of 1 kg.

The Emeraude could not be launched without carrying dummy stages, but the dimensions, as used for the first stage of the Diamant, were: length 32.0 ft, diameter 4.4 ft, weight empty 4290 lb, weight of acid 21,340 lb, weight of fuel 6750 lb, take-off thrust 59,700 lb. The third stage of the Diamant vehicle was 6.5 ft long, diameter 26.0 in., weight empty 150 lb, propellant weight 1410 lb. The over-all length of the Diamant (with satellite) at take-off was 62 ft.

But the first French satellite, designated FR A-1, was launched by a Scout rocket from the Pacific Missile Range. It weighs about 135 lb, plus about 23 lb for the separation equipment. FR A-1 was followed into space on February 17, 1966, by D.1A, launched from Hammaguir by a Diamant three-stage rocket. Two other satellites, named Diadème, followed in February 1967. The weight of the D.1A satellite was only 44 lb, but the Diadèmes weigh around 200 lb. The orbits of the four satellites are:

NAME OR DESIGNATION	LAUNCH DATE	PERIGEE		APOGEE		ORBITAL PERIOD (MIN)
		MI	KM	MI	KM	
FR A–1	Nov. 26, 1965	327.5	527	1118	1799	108.7
D.1A	Feb. 17, 1966	311	502	1701	2738	118.7
Diadème–1	Feb. 8, 1967	354	570	775	1247	104.3
Diadème–2	Feb. 15, 1967	367	591	1168	1880	110.2

By the end of 1967 Hammaguir will be past history. Under the terms of an international agreement, the French will have to leave the area. Two new French test centers are under construction. One is also in Africa but in Guiana; this is where the ELDO[2] satellite carriers are going to take off. The other is in France itself, on the west coast about halfway between Boulogne and Biarritz. It will be known as CEL (Centre d'Essais des Landes) and will be able to accommodate missiles with ranges up to 1200 miles.

[2] See section 12 of this Appendix.

6. Germany (West) (*Deutsche Bundesrepublik*)

West Germany gained the right to conduct aeronautical and space research in 1955. However, as Professor A. W. Quick of the Technical University of Aachen pointed out during a lecture in Washington, D.C., in December 1960, for some time both the government and the public were apathetic. The general feeling was that aeronautical research was a warlike activity and the Germans had had enough of war. As far as the peaceful aspects, such as commercial air transportation, were concerned, the Germans felt that during the ten years since the end of the Second World War the other nations had made such strides that the best solution would be just to buy French and American airliners. Judging by the complaints in German scientific and engineering journals, this attitude is to a great extent still present; criticisms of the German Federal Government for being uninterested in space research are frequent.

In the organization of the West German Federal Republic, space research is partly under the jurisdiction of the Ministry of Transportation, partly under that of the Ministry of Research; but since telegraphy and radio are sections of the postal administration, the postal authorities are in charge of everything that involves radio. Interestingly enough, the first item of German space research after the war was initiated by the post office, which wanted to know more about transmission through the ionosphere. The activity consisted in designing and building instruments that were carried aloft by the French Véronique rocket (see section 5 of this Appendix). The establishment of a joint German-French research institute was one of the results of this venture. But German space research from 1955 to about 1960 dealt with instrumentation and space medicine, not with propulsion.

The only attempt to create new means of propulsion was made by Dr. Eugen Sänger and his wife Dr. Irene Bredt in a small research institute located at the Stuttgart airport. They investigated the possibilities of a steam rocket as a booster for ramjet-propelled airplanes, keeping in mind that such a steam rocket might also be a booster (heavy, but cheap) for high-altitude sounding rockets. While this work did not produce a practical steam rocket, a new generation of German rocket engineers became acquainted with the principles of rocketry by working with Sänger and Bredt.

By 1963 German space plans had crystallized to the extent that an industrial firm, Boelkow Entwicklungen KG, in collaboration with a government research station and several universities, began work on two projects. One was called Project 622; it is the third stage of the Europa-1 rocket (see section 12 of this Appendix). The other, Project 625, involved a multipurpose satellite weighing an estimated 3300 lb. To launch this satellite would require an Atlas-Agena rocket and the Germans hoped that NASA would furnish it. But NASA refused on two grounds: it had never furnished a heavier rocket than a Thor for a foreign government, and its feeling was that, as one official put it, the Germans were trying "to skip the junior semester." The Germans then approached the U.S. Air Force about an Atlas-Agena vehicle. The Air Force was willing to sell them an Atlas (with launch from U.S. territory) for six million dollars. This ended the discussions, since Boelkow did not have a government contract for its satellite but had paid for the design studies out of company funds.

Meanwhile another aircraft company, Dornier, went ahead with Project 621, a sounding rocket with paraglider wings. Since the rocket (liquid fuel) is to be recovered, each rocket is to make at least half a dozen flights. Because of the recovery, the data obtained by the rocket will be stored on board, making telemetry systems unnecessary. Another advantage of recovery is that instruments can be recalibrated after the flight, which makes it possible to judge the accuracy of the data obtained. Preliminary weight estimates were: take-off weight 530 lb, return weight 165 lb, plus payload weight.

In 1965 a ground station was constructed on Sardinia with the necessary equipment to receive data and transmit guidance commands to the paraglider rockets. The rockets were full-scale models with functioning recovery systems, but they did not take off under their own power. The tests were drop-tests from aircraft for the check-out of the recovery system. The rockets landed within an area 660 ft in diameter; usually they were undamaged and could be used again.

Dornier's paraglider rocket is about the only German project where progress has been steady, even if slow. All other announcements from West Germany between 1960 and 1964 were tentative, sometimes incomplete, and occasionally subject to reversal: at one point the international aviation journal *Interavia,* which is published in Switzerland in several languages, asked whether the German space plans were reality or fantasy. But by 1965 the Germans had achieved a few firm plans.

The construction of the third stage of the Europa-1 rocket is going ahead, though the organizational chart for this one undertaking looks about as complicated as the organizational chart for all of NASA. And the overly ambitious Boelkow satellite has been trimmed down to a new project called the 625 A-1 Small Research Satellite, which is to operate in a polar orbit and can be launched by a Scout rocket. The eight-sided satellite will have a length of 3.8 ft, a diameter of 2.5 ft, and a total weight of 132 lb, of which 26 lb will be scientific instrumentation.

7. Israel

The most secret missile program in the world is not that of either Communist China or the Soviet Union, but that of Israel. In the summer of 1961 Shimon Perez, Deputy Defense Minister of Israel, admitted that the country had a rocket program but said that, for security reasons, the names of the scientists involved could not be released. Early in July 1961 (precise date not stated) a high-altitude solid-fuel rocket was fired from an undisclosed spot at the shore of the Mediterranean Sea. It was named Shavit II (*shavit* is Hebrew for "comet") and its weight was stated to be "above 250 kilograms" (or above 550 lb). The rocket carried no instruments but released a cloud of sodium vapor. Its peak altitude was kept secret, and questions as to whether there had been a Shavit I were not answered. Western observers estimated that Shavit II, if fired for range, would have covered a distance of about 70 miles.

But whatever the goal of the program it either was not attained or was canceled for budgetary reasons, for in 1966 it became known that the government of Israel conducted negotiations with the government of France for the purchase of a military version of the Diamant rocket with a range of about 500 miles. The outcome of these discussions was never made public.

During the "six-day war" against the Arab states the Israeli Air Force employed rocket-accelerated bombs with delayed fuzes which did much damage to the runways of enemy airports. Whether these bombs were manufactured in Israel or elsewhere has not been disclosed.

8. Italy

Italy was still without a space program at the time that the Soviet Union had already achieved manned orbital flight and the United States was on the verge of doing so. Hence Italy wisely decided not to embark on an expensive program of rocket development but to enter space research via collaboration, both with the United States and with ELDO (European Launcher Development Organization). The responsible government agency in Rome bears the name Centro Ricerche Aerospaziali.

The collaboration with the United States was called the San Marco Project and had as its goal the placing of an Italian-designed and Italian-built satellite in an orbit which would remain over the geographical equator. There were to be two suborbital test launches (Operation Shotput) and two attempts at orbit. The latter would use American Scout rockets, the former a two-stage rocket, also called Shotput, built by NASA for this purpose. The agreement was that there would be no exchange of funds between Italy and the United States. Italy would supply the satellites and the launch crews; NASA would supply the rockets, the necessary instruction, and tracking facilities.

The first Shotput rocket was ready early in April 1963. The main stage of the rocket was a Pollux E-6 booster by Thiokol, with two Thiokol Recruit rockets as take-off help. This stage developed a take-off thrust of 120,000 lb. The upper stage was an Altair rocket (by Hercules Powder Co.). The whole vehicle had a take-off weight of 11,000 lb, a diameter of 33 in. and a length of 32 ft. The San Marco satellite weighed 177.5 lb. The first Shotput flight took place on April 20, 1963; the second on August 2 of the same year. During the second flight the San Marco satellite was carried to a peak altitude of 183 mi and a distance of 606 mi. No recovery was planned; both shots were made from Wallops Island, as was the first of the two satellite shots.

The latter was launched by an Italian crew at 3:20 P.M. (EST) on December 15, 1964, with a somewhat heavier satellite weighing 254 lb. The Scout rocket worked perfectly and so did the satellite, which went into an orbit ranging from 128 to 510 mi with an orbital period of 95 min. The first estimate of the useful lifetime in space was 3 months and this proved to be nearly correct, but the satellite stayed in space half a year longer—it re-entered on September 13, 1965.

Having been launched from Wallops Island, San Marco I could not go into an equatorial orbit without large and fairly complicated orbit corrections for which no provisions had been made. Its inclination to the equator was 38°. In order to have an equatorial satellite, the launch pad itself should be in an equatorial location. The Italians achieved this for San Marco II by launching it from a Texas tower type of structure located in the ocean under the equator east of Africa. San Marco

II was launched, again using a Scout rocket, on April 26, 1967, and attained an orbit inclined to the equator by only 3°, with a perigee at 134.8 mi, apogee at 497.7 mi, and an orbital period of 94.9 min.

9. Japan

After the Japanese rocket ordnance of the Second World War had been seized and presumably destroyed, a number of Japanese scientists at Tokyo University, under the leadership of Professor Hideo Itogawa, decided that they would really study rockets from a scientific point of view. They decided in favor of solid fuels and began with a truly modest rocket which they called the Pencil. It was not quite 12 in. long with a diameter of 0.7 in. The little rocket could be fired safely near Tokyo, testing a variety of fuels, nozzle shapes, etc. In 1955 about 150 Pencil rockets were built. Then a lower stage of the same diameter was added, and then a somewhat larger two-stage rocket, called Baby, was designed. It had a diameter of 3 in. and a total propellant charge of 2.2 lb. Then came a rocket which Professor Itogawa called "nearly man-sized"—it stood 4 ft 5 in. tall.

The success of these tiny rockets caused the Japanese Science Council to participate in the atmospheric research programs of the International Geophysical Year. Two Japanese companies were charged with the development of new fuels and the result was a series of two-stage solid-fuel rockets called the Kappa-6. In 1958 these rockets were successfully fired to altitudes up to 37 miles. A launch center had to be chosen for these shots; the one selected was located on the west coast of the island of Nippon at Michikawa. Later another test center was added, the Niijima Center, based on an island off the east coast of Nippon. By 1960 "space" was an item in the Japanese budget, though a very modest item. Expenditures for 1960 were the equivalent of $750,000; for 1961 this figure was doubled; in 1962 the figure reached $2¼ million, in 1963, $3¼ million. It jumped to $7¼ million in 1964 and nearly reached $10 million mark in 1965. In 1966 it was about $11 million.

By adding a third stage to the Kappa-6 a rocket called Kappa-9L was produced; in 1961 it carried small payloads to altitudes of over 200 miles. The next rocket was a new design, called Kappa-9M, a two-stage rocket with an over-all length of 37.7 ft, which was capable of lifting a payload of 110 lb to an altitude of 220 mi. During the same year, 1963, the two-stage Lambda-2 made its appearance, with a payload capacity of about 380 lb and a maximum altitude of 310 mi.

At this point the Japanese rocket researchers had to consider a geographical fact. They fired westward from Michikawa Launch Center, usually at an angle of 80° since they were shooting for altitude. But suppose that one of the Lambda rockets malfunctioned and went off at a lesser angle. A rocket with an altitude capability of 310 miles can have a horizontal range of a little over 600 miles. This is 200 miles more than the average width of the Japan Sea, on the other side of which lies Russian territory.

A new launch center was established at Kagoshima, the southern tip of the southernmost Japanese island. From that point, in a southerly direction, there are thousands of miles of open sea and the remote limit of the open sea is Antarctica. From Kagoshima the new Lambda-3, a three-stage rocket, took Japanese instrument packages to altitudes of over 600 miles.

THE ROCKETS OF THE KAPPA AND LAMBDA SERIES

	K-6	K-6H	K-8L	K-8	K-9L	K-9M	L-2	L-3
Number of stages	2	2	2	2	3	2	2	3
First flight	1958	1960	1962	1960	1961	1963	1963	1964
Take-off weight (lb)	595	727	771	3306	3417	3306	13,890	15,400
Peak altitude (mi) (80° launch)	37	50	100	124	217	217	310	620
Payload (lb)	44	44	55	110	33	110	396	375
Over-all length (ft)	18.4	22.6	24.3	35.8	31.0	37.7	54.1	62.6
Diameter (ft)								
First stage	0.82	0.82	0.82	1.38	1.38	1.38	2.41	2.41
Second stage	0.49	0.49	0.52	0.82	0.54	0.82	1.38	1.38
Third stage					0.54			1.38

The next larger rocket in the Lambda series was the Lambda-4S with an over-all length of 55 ft, a base diameter of 29 in., and a take-off weight of 18,700 lb. The take-off thrust was 82,500 lb. The Lambda-4S was used for a few "dress rehearsals" for a satellite shot, but for the actual satellite shot the Japanese had a new booster, called M-14, with a thrust of 220,000 lb, and a diameter of 64 in. The four-stage rocket which it lifted off the ground was called Mu-2. Meanwhile a test satellite with a weight of 57.2 lb had been built. It was spherical in shape and battery-powered. The batteries were expected to last for only ten or eleven orbits.

The launch date was September 26, 1966, but while the take-off looked fine the result was disappointing. Apparently the top stage had failed to ignite and the satellite failed to go into orbit.

10. Union of Soviet Socialist Republics

The first name in Russia's space activities is, of course, that of Ziolkovsky (see pp. 93 ff.) and for a long time his seemed to be the only one. Actually Professor Ivan Meshchersky of the St. Petersburg Polytechnic Institute had dealt with the mathematics of rocket propulsion just as competently as Ziolkovsky himself and had written a textbook (*Collection of Problems in Theoretical Mechanics,* first published in 1904) which was in the library of many Russian students of engineering. But Meshchersky did not become well known and many Russians of later generations may have wondered who he was when *Komsomolskaya Pravda,* in 1959, proclaimed that "it is to his labors that Sputnik owes its existence." The reason for the sudden acclaim was that 1959 was the hundredth anniversary of Meshchersky's birth.

Ziolkovsky acquired company in 1929, when two societies for the study of rockets were formed (see p. 108) and when Yuri V. Kondratyuk of Novosibirsk published a book on space travel. Kondratyuk suggested ozone instead of oxygen as an oxidizer; theoretically a fine suggestion, marred by the fact that liquid ozone is unreliable and often reverts, with explosive violence, to normal oxygen. He also suggested re-entry by means of a glider and advocated a manned space station orbiting the moon. The Moscow group at that time made large solid-fuel rockets for assisted take-off of aircraft. But the first experimenter in the Soviet Union in the modern sense seems to have been a Lett, Friedrich Arturovitch Zander (or Tsander) who began designing and testing rocket motors in 1928.

The early period of Russian experimentation, from 1928 to 1933, is clouded by the fact that Russian historians disagree among themselves. Zander's OR-1 was built in 1930, but there are grave doubts whether it was ever tested, even though some say that Zander did so. OR-2 was tested by the Moscow group in 1933, only 10 days before Zander's death in March 1933. He died in a sanatorium in the Caucasus Mountains near the town of Kislovodsk where his monument now stands. (Ziolkovsky's monument is, of course, in Kaluga, his birthplace.)

Zander's rockets are described as "semiliquid" by Russian witers; they probably were the first hybrids. On August 17, 1933, the rocket ORM-50, begun by Zander, was fired. It stood 8 ft tall, weighed 22 lb at take-off, and had a diameter of 4 in. The official statement said it worked well, but the altitude was not revealed. The official credit for the shot went to Mihail Konstantinovitch Tikhonravov. Tikhonravov then went on to build "an all-liquid version" of the same rocket which was launched on November 25 of the same year. (One Russian source says that Tikhonravov's first rocket made its first flight in 1932, but this may have been a simple typographical error.) Russian rocket research seems to have gone on until the outbreak of the Second World War, but the emphasis was on battlefield rockets.

After the war Stalin, as was mentioned in Appendix 1, was much impressed with the Sänger-Bredt report, and was greatly annoyed that the Peenemünde staff had gone over to the Americans and that U.S. troops had even cleaned out the Mittelwerk factory near Niedersachswerfen. But two complete V-2s, intended for training purposes, and the V-2 manuals, had fallen into Russian hands, and the Russians promptly set up a factory for building V-2 rockets. To save what could be saved—or, in my opinion, to impress Stalin—General Ivan Serov of the Russian Secret Police rounded up "several thousand" German scientists and technicians and brought them to Russia. Among them were about 200 men from Mittelwerk, under the leadership of Hans Gröttrupp, and their families. This action took place during the second half of October 1946.

Fantastic stories about this episode appeared in magazines in many countries (but not in the Soviet Union). All of these had in common that their writers substituted hearsay and their own imaginations for the facts they did not know. In reality the Germans did not build anything for the Russians, did not "supervise" the firings, and did not "introduce innovations." Instead they were kept in ignorance and the work they did was paper work. The external circumstances of the life of these Germans were well described by Mrs. Gröttrupp, who kept a diary and had it published after her return to Germany.

The Germans were housed—well housed considering conditions in the Soviet Union at the time—on the island of Gorodomliya in Seeliger Lake, which can be found on a map halfway along a straight line connecting Leningrad and Moscow. They were also paid, in fact highly paid; the drawback was that there was no place to spend the money. Their main occupation was to answer written questions in writing. As long as the Russians fired Russian-built V-2s, a few of the engineers were taken to the proving ground at Kapustin Yar, 50 miles east of Stalingrad (now Volgograd), as consultants. As soon as the Russians had progressed to the rocket they called Pobyeda ("victory"), no German ever set foot into the proving ground again. The Pobyeda rocket seems to have been an elongated V-2 with a longer range.

Those Russians who had actual contact with the Germans (with the exception of a few political commissars, they were all officers of the Red Army) tried to be friendly and were disappointed that all the Germans really wanted was to go home. By 1952 most of the Germans had been released and at the time that Sputnik went into space there was not a single German rocket man left in the Soviet Union.

The whole episode has a flavor of meaninglessness, which is my reason for thinking that General Serov was mainly interested in impressing Stalin. There is no doubt that the Russians themselves are pleased that their rocket development was essentially done by Russians. Beginning around the year 1950 Russian rocket work began to fall into a pattern. The first objective was military, culminating in intercontinental ballistic missiles. These ICBMs then could be used for space shots, and, as regards space, the development went straight in the direction of manned flight. The Russians did not make any particular attempts at diversification until after manned flight had been accomplished.

The first set of Russian-developed missiles was the so-called "T-series," beginning with the T-1, which can be called an enlarged and improved V-2. The T-series consisted of the following weapons:

DESIGNATION	NUMBER OF STAGES	LENGTH (FT)	DIAMETER (FT)	ESTIMATED LAUNCH WEIGHT (TONS)	RANGE (MI)	REMARKS
T-1 (M-101)	1	50	5–6	20	600–750	liquid
T-2 (M-103)	2	80	7–10	35	1800	liquid
T-3 (M-104)	3	90–120	16	85	5000	liquid
T-4 (M-102)	2	55	6.5	35	1000	liquid
T-5B	2 ?	30	ca. 2		25	tank-launched
T-5C	1	25	ca. 2		15–20	tank-launched
T-6	2	10	5 in.		ca. 14	AA missile, solid
T-7	1				60	AA missile, liquid
T-8	2				15	booster solid, rocket liquid

The T-3A had a range of 6000 mi, the T-3B a range of 6500 mi. The designation T-4A was reserved for an antipodal bomber of the Sänger type; apparently the development was stopped early. T-7 was originally a research rocket, adopted as an antiaircraft weapon. The three missiles called J-1, J-2, and J-3 are winged target drones with turbojet propulsion, but could be used in the same way as the German V-1. At the same time rockets named Golem were developed by the Soviet Navy. Golem-1 was a liquid-fuel, 54-ft rocket with a range of 400 mi. Golem-2 was a two-stage liquid-fuel rocket a few feet longer than Golem-1, with a 1200-mi range, while Golem-3 was a single-stage solid-fuel rocket 17 ft in length, with a range of 8 mi.

The newer Soviet missiles (1965) are known by their NATO designations, with the exception of two older rockets that are still in use. One of these is the M-100A, an 8-ft solid-fuel unguided rocket either for air-to-air or air-to-ground use; the other is the antitank rocket ME, also called Igor, 2 ft long, 3⅓ in. in diameter, with a range of about 1 mi. The unguided solid-fuel battlefield rockets of the Soviet Union have diameters of 140 mm, 200 mm, 240 mm, and 280 mm and are fired from multiple launchers carried on armored trucks, or sometimes on tanks.

More recent antitank missiles, carried in triple and quadruple launchers by armored vehicles are called Snapper and Swatter by NATO. They are solid-fuel missiles.

DESIGNATION	LENGTH (IN.)	DIAMETER (IN.)	FIN SPAN (IN.)	WEIGHT (LB)	RANGE (FT)
Snapper	48	5	33	44	6000 +
Swatter	48	5½	26	55	6000 +

Heavier battlefield missiles are called Frog and Scud, The Frogs are solid-fuel rockets while the Scuds are propelled by liquid fuels. The Frogs are launched by tanks, carrying one missile apiece; the Scuds need special carrier vehicles.

DESIGNATION	LENGTH (FT)	DIAMETER (FT)	LAUNCH WEIGHT (TONS)	RANGE (MI)
Frog				
1	31.5	1.8	4	40
2	29.5	1.0	2.3	15–17
3	34.5	1.3	2.1	22–28
4	33.5	1.3	2.0	30 +
Scud				
A	34.8	2.8	ca. 5	50 min.; 102 max.
B	37.8	3.0	ca. 6	60 min.; 136 max.

Of the big missiles, the types Shyster and Sandal are of intermediate range and apparently single stage; the intercontinental missiles Savage (solid) and Scrag (liquid) have several stages. Scrag is the one that U.S. newspapers call Big Brother.

DESIGNATION	LENGTH (FT)	DIAMETER (FT)	LAUNCH WEIGHT (TONS)	RANGE (MI)
Shyster	69	5¼	26	500–600
Sandal	72	5¼	28	900–1000
Savage	64	5¾	40	5000 +
Scrag	100 +	8½	200	6000 +

The submarine-launched Golem missiles have been replaced by new types. One is called Snark by NATO; the second does not have a name yet; the third is launched from surface vessels (patrol boats).

	LENGTH (FT)	DIAMETER (FT)	LAUNCH WEIGHT (TONS)	RANGE (MI)
Snark	46	5.4	19.5	1200
New type	35	5.4	15.0	1200
Patrol-boat launched	24.6	2.7	3.9	40

The dimensions of the known newer antiaircraft missiles are:

	LENGTH (FT)	DIAMETER (FT)	FIN SPAN (FT)	RANGE (MI)	LAUNCH WEIGHT (TONS)
Long-range two-stage	48.5	2.5	13	150	10
Medium ("Guideline")	23	1.5	6¼	20–25	1.4
Medium (new type)	23	1.8	8¼	25	ca. 3
Short-range	17¼	1.0		18	0.7
Tank-launched	24.6	2.8	7.9	45	2.2

The tank-launched antiaircraft missile is a ramjet, with solid-fuel boosters.

The Russian Meteo rocket was not only used for high-altitude research during the International Geophysical Year, it also initiated aeromedical research using dogs and other animals. The first dog, carried to a height of about 70 miles by a Meteo rocket in 1955, was named Malyshka ("little one"; it was a very small white spitz). The rocket was a two-stage liquid-fuel rocket, weighing 1500 lb fueled. The first stage was surrounded by a ring of six solid-fuel boosters weighing together 520 lb (of which 180 lb was fuel). Total launch weight, then, was a little over 2000 lb. The height of the rocket was 29 ft. At the peak of the flight the instrument section (or animal capsule) was separated from the rocket. Both were equipped with parachutes.

Larger (but unspecified) rockets carried heavier payloads with animal capsules to higher altitudes in the years to follow. On August 27, 1958, one such flight was made, in which a payload of 3726.5 lb was carried to an altitude of 279.6 mi. There were two dogs in the payload, Pestraya ("Checky") and Belyanka ("Whitey"), and they were successfully recovered. On July 2, 1959, a payload of 4400 lb was carried to over 200 mi. The payload included two dogs and a hare, also recovered unharmed.

After a number of such flights, orbital experiments followed, the first of them being Sputnik-II with the dog Laika. Sputnik-III did not carry an animal capsule. Whether Sputnik-IV did is not known, but because an attempt was made to de-orbit this satellite, it seems likely that it did. Sputnik-IV went into orbit on May 15, 1960, with perigee at 189 mi and apogee at 222 mi. On May 19 a radio command was sent out to fire the retro-rockets, but the capsule was in the wrong position at the moment and the retro-rockets accelerated it into a larger orbit. The satellite broke into eight pieces, all of which re-entered in time. Sputnik-V was put into a precisely circular orbit at a distance of 198.8 mi from the ground on August 19, 1960, and de-orbited the following day. The animal capsule held two dogs, several smaller animals, microscopic life forms and plants and seeds. This flight was the first in which living animals were successfully recovered from orbit. Sputnik-VI, orbited December 1, 1960, did not fare so well. Its orbit was 116.4 mi at perigee and 164.6 mi at apogee. De-orbited on December 3, it made a bad re-entry and the animal capsule with two dogs in it burned up.

Sputnik-VII, fired February 4, 1961 (orbit 139/204 mi, with a 90-min orbital period), served no known function; it re-entered February 26. Sputnik-VIII was orbited one day after Sputnik-VII and fired the first Russian Venus probe. Sputnik-IX was orbited on March 9 (orbit 110/155 mi, orbital period 88.2 min) and de-orbited the same day with successful recovery of the dogs. Sputnik-X was put into a very similar orbit (110/153 mi, orbital period 88 min) on March 25 and also de-orbited the same day with successful recovery. By then the Russian scientists and designers must have felt certain that they had mastered the techniques of de-orbiting and re-entry, because the next flight was that of Yuri Gagarin.

The other manned flights followed as listed in the following table:

MANNED FLIGHTS BY SOVIET COSMONAUTS

SPACECRAFT	COSMONAUT	CODE CALL IN FLIGHT	LAUNCH DATE	ORBITS COMPLETED	FLIGHT DURATION
Vostok 1	Gagarin	*Swallow*	April 12, 1961	1	1 hr 48 min
Vostok 2	Titov	*Eagle*	Aug. 6, 1961	17	25 hr 18 min
Vostok 3	Nikolayeff	*Falcon*	Aug. 11, 1962	64	94 hr 22 min
Vostok 4	Popovitch	*Golden Eagle*	Aug. 12, 1962	48	70 hr 57 min
Vostok 5	Bykovsky	*Hawk*	June 14, 1963	81	119 hr 06 min
Vostok 6	Tereshkova	*Seagull*	June 16, 1963	48	70 hr 50 min
Voskhod 1	Komarov, Feoktistov, Yegorov		Oct. 12, 1964	16	24 hr 17 min
Voskhod 2	Belyayev, Leonov		Mar. 18, 1965	17	26 hr 02 min
Soyuz 1	Komarov		Apr. 23, 1967	17	ca. 24 hr.

The full names of the Soviet cosmonauts and their military ranks at the time of flight are: Gagarin, Yuri Alexeyevitch (Major); Titov, Gherman Stepanovitch (Major); Nikolayeff, Andrian Grigoryevitch (Colonel); Popovitch, Pavel Romanovitch (Colonel); Bykovsky, Valery Fedorovitch (Lieutenant Colonel); Tereshkova, Valentina Vladimirovna (Junior Lieutenant); Komarov, Vladimir Mihailovitch (Colonel); Feoktistov, Konstantin Petrovitch (scientist, civilian); Yegorov, Boris Borisovitch (M.D., civilian); Belyayev, Pavel Ivanovitch (Colonel); Leonov, Alexei Arkhipovitch (Lieutenant Colonel). Miss Tereshkova, criticized by *Pravda* for wearing stiletto heels on the launch pad, married Colonel Nikolayeff after her flight, taking the name Nikolayeva-Tereshkova. On June 8, 1964, she became the mother of a baby girl, whom she named Yelena after her own mother.

SOVIET SATELLITES

The first set of Russian research satellites, called Sputnik, was soon followed by more specialized satellites, bearing the names Polyot ("flight"), Elektron, Proton, Molniya ("lightning"), and Kosmos, in approximate order of their appearance. The Polyot (accent is on the second "o") satellites could change their orbit; only two of them were fired and the orbital changes accomplished by the second one were not announced.

	LAUNCH DATE	PERIGEE (MI)	APOGEE (MI)	ORBITAL PERIOD	REMARKS
Polyot 1					
first orbit	Nov. 1, 1963	210	368	94 min	
final orbit		213	892	102 min	orbiting
Polyot 2	Apr. 12, 1964	191	291	92.3 min	orbiting

The Elektron satellites were launched in pairs. Their purpose was stated to be radiation research and from the orbits it is clear that they investigated the Van Allen Belts. All four are still orbiting.

	LAUNCH DATE	PERIGEE (MI)	APOGEE (MI)	ORBITAL PERIOD (MIN)
Elektron 1	Jan. 31, 1964	250	4,423.5	169.3
Elektron 2	Jan. 31, 1964	363	42,152.5	1356.3
Elektron 3	July 11, 1964	251	4,350	168
Elektron 4	July 11, 1964	283	40,886	1314

The Proton satellites were enormously heavy—one of the few cases in which the Soviet Union announced the weights—and their purpose was (and remains) mysterious. From the great weight one would guess that they were prototypes of manned spacecraft, but then one would expect that they would have been de-orbited by means of retro-rockets; instead they were left in orbit to undergo natural orbital decay.

	LAUNCH DATE	PERIGEE (MI)	APOGEE (MI)	ORBITAL PERIOD (MIN)	WEIGHT (LB)	DATE OF RE-ENTRY
Proton 1	July 10, 1965	108.7	279.5	90.9	26,896	Oct. 11, 1965
Proton 2	Nov. 2, 1965	118.5	396	92.6	26,840	Feb. 6, 1966
Proton 3	July 6, 1966	114.6	361	92.3	ca. 26,000	Sept. 16, 1966

If the purpose of the Proton satellites remained unexplained, that of the Molniya satellites was announced: they are communication satellites, capable of transmitting radio and television (including color television) broadcasts.

	LAUNCH DATE	PERIGEE (MI)	APOGEE (MI)	ORBITAL PERIOD	REMARKS
Molniya I	April 23, 1965	375.5	24,780	12 hr 0.3 min	orbiting
2nd Molniya I	Oct. 13, 1965	348	24,695	11 hr 59 min	orbiting
3rd Molniya I	Apr. 25, 1965	310	24,509	11 hr 50.5 min	orbiting
4th Molniya I	Oct. 20, 1966	308.8	24,863	12 hr 9 min	orbiting

The designations 2nd, 3rd, and 4th Molniya I are probably meant to indicate that there was no design change but that the new satellites were put into orbit when the preceding satellite ceased to transmit.

The Kosmos series of satellites is very evidently a multipurpose series. Kosmos-1 was launched March 16, 1962, Kosmos-159 and Kosmos-160 on May 17, 1967. Western trackers have found that the Kosmos satellites, in irregular sequence, went into orbits of four different inclinations to the earth's equator—namely, 49°, 51°, 56°, and 65°. Some of them are cerainly surveillance satellites, while Kosmos-154 and Kosmos-155 are believed to have been prototypes of the Soyuz spacecraft in which Colonel Komarov lost his life. Kosmos-154, launched April 8, 1967, spent two days in an 115/144-mi orbit and, according to British sources, had a length of 30 ft, with a diameter of 10 ft and a weight of 13,230 lb. Kosmos-155, launched April 12, 1967, stayed for 8 days in an 126/178-mi orbit; it was smaller than Kosmos-154 but weighed enough for a manned spacecraft—that is, 8820 lb.

SOVIET SPACE PROBES

The designations of Soviet space probes are: Cosmic Rocket, Zond, Venera (the Russian name of the planet Venus), and Luna (the Russian name of the moon).

The Venera probes were obviously bound for Venus, while those named Luna were lunar probes, either for soft landing or for going into orbit around the moon.

See Appendix 3 for a comparison of Russian and American probes, listed by launch dates.

11. United Kingdom

The thinking about rockets in London and in British research centers after 1945 began, naturally, with the V-2. A farsighted suggestion then made was to develop two V-2-type rockets, one with a range of 300 to 350 miles and one with a range of 3000 miles. One can only wonder where Great Britain would now be in space if that suggestion had been heeded. The only tangible result of that early suggestion is the Woomera firing range in Australia, a result of an agreement reached by the governments of Great Britain and Australia in 1946.

The Woomera firing range is located near the Australian south coast, the impact area in the western part of the northern shore, providing an overland missile range about 1250 miles in length. The range can be extended over water to Christmas Island in the Indian Ocean, which would add another 1000 miles. Practically all British missiles have been test-fired at Woomera, since no missile with a range longer than 20 miles could be test-fired in Great Britain itself—sea-going missiles being the obvious exception.

The Australians' main contribution was the testing facility for Great Britain—at one time, before the Atlantic Missile Range was created, there was some talk about making the facility available to the United States, too—but two Australian-designed sounding rockets have also been fired at Woomera. Both are two-stage solid-fuel rockets, using the same booster. The smaller of the two, Aeolus, stands 21 ft tall and can carry a 50-lb instrumentation cone to 45 mi. Its take-off weight is about ½ ton. The larger one, Long Tom, weighs 1 ton at take-off, stands 27 ft tall, and can carry the same instrumentation load to 100 mi.

The big missiles for the sake of which the Woomera range was established were never built. The years from 1946 to 1960 saw activity in three different areas: (1) small rocket weapons for battlefield use and air defense; (2) aircraft rocket engines for take-off assistance and in-flight use; and (3) preparations for a large missile of so-called "intermediate" range.

Quite a number of small missiles of various types were designed and tested, with the following reaching operational status:

GROUND-TO-AIR

Seaslug (Navy). Two-stage, solid-fuel, mounted on warships. Length 19.5 ft, diameter 1.5 ft, wing span 4.5 ft. Beam-rider type guidance. The range is given only as "long."

Seacat (Navy). Two-stage, solid-fuel, mounted on warships, four missiles to one battery. Length 4 ft 10.3 in., diameter 7.5 in., wing span 25.6 in. Radio command guidance. Short-range weapon against low-flying aircraft. (Has been purchased by the Australian, Dutch, New Zealand, Swedish, and West German governments for their navies.)

Bloodhound I (RAF). Air-breathing missile with a range of about 35 mi. Length 24 ft, launch weight with four solid-fuel boosters 4500 lb. Radar homing guidance. Sustainer engine takes over after boosters have been jettisoned. (Has been purchased by the Swedish government.)

AIR-TO-AIR

Fireflash (RAF). Length 7 ft 5 in., diameter 5.5 in., wing span 28 in., launch weight 300 lb. Beam-riding missile with a range of 6–8 miles.
Firestreak (RAF). About the same dimensions and range as Fireflash. Infrared homing guidance. Also used by the Royal Navy.

ANTI-TANK WEAPONS

Pye P.V. Solid-fuel wire-guided missile with a launch weight of about 80 lb, a length of 5 ft, and a fin span of 2 ft. Performance detail is classified.
Vigilant. Solid-fuel wire-guided missile with a launch weight of 26 lb and a length of 3 ft. Range is 1700 yd.

(The largest of the solid-fuel missiles was to be Blue Water, a 25-ft missile with a range of 70 mi., a tactical battlefield missile designed to be fired from a mobile carrier. In August 1962, to the dismay of interested circles in Great Britain, the development of the Blue Water was stopped.)

AIR-TO-SURFACE

Blue Steel (RAF). Air-launched missile with a range of about 400 mi. Liquid-fuel, inertial guidance. Length 36 ft, wing span about 13 ft, launch weight about 15,000 lb.

The first of Britain's take-off units was the de Havilland Sprite, a "cold" rocket based on the decomposition of hydrogen peroxide; A. V. Cleaver was the man in charge of the development. The weight of the Sprite empty was 350 lb, weight charged was 925 lb, over-all length was 84¼ in., and it could deliver 5000 lb of thrust for 9 sec. A later development, the Super Sprite, also a "cold" rocket, developed 4200 lb of thrust for 40 sec. It was designed to be jettisoned with a parachute after the aircraft was airborne.

The first British take-off unit using actual combustion of the fuel (methanol and hydrogen peroxide) was Armstrong Siddeley's Snarler. Its thrust was 2000 lb at sea level for a duration of about 3 min. Empty weight was 215 lb. (The current Bristol Siddeley BS 605 rocket engine weighs 240 lb and produces a thrust of 4000 lb. These engines are normally used in pairs; fuel is kerosene; oxidizer is hydrogen peroxide.)

The most successful rocket engine of this type is de Havilland's Spectre, also burning kerosene with hydrogen peroxide, with an over-all length of 56.5 in. and a maximum diameter of 32 in. The Spectre which has variable thrust (there is also one which does not) can deliver a maximum thrust of 8000 lb and a minimum of

800 lb. The Saunders-Roe SR 53 intercepter prototype (1957) had a mixed propulsion system, a Spectre rocket, and an Armstrong Siddeley Viper jet engine.

As in the United States, after Sputnik some people began to clamor for a space program; one can only say that their government has disappointed them. In the summer of 1959 a government spokesman said that no real program had been decided on, but that it had been decided what Great Britain would *not* do. It would not build planetary probes, it would not put animals into orbit, and it would not form a program for manned space flight. Nor would it spend more than "a few hundred thousand pounds" for space.

But at that time Englishmen still had a few space-pointing projects to console them. A solid-fuel motor named Raven was under development; it would produce 11,500 lb of thrust for 30 sec and it would power the Skylark upper-atmosphere-research rocket. The Skylark was three-finned, unguided, and 25 ft long with a diameter of 17½ in. Then there was the Black Knight test missile, which was to fly nose cones for the long-range missile Blue Streak. The Black Knight was powered by four "gamma engines," each gimbal-mounted, with a total thrust of 16,400 lb at sea level and 19,100 lb at altitude. The engines burned kerosene with hydrogen peroxide. The rocket was 35 ft tall, with a diameter of 3 ft. Reports from Woomera stated that the Black Knight had carried nose cones to altitudes of more than 300 miles and had performed well.

The Blue Streak, for which the Black Knight did the preliminary research, was a 61.5-ft rocket with a diameter of 10 ft. It was thin-skinned like the Atlas, and the empty weight was 6.1 tons, while the take-off weight was 92.2 tons. Each of its two RZ-2 engines, burning the same fuels as the Black Knight, developed a sea-level thrust of 137,000 lb. The range of the Blue Streak was to be 2800 mi. On April 13, 1960, the Blue Streak was canceled amid uproar in the House of Commons.[1]

The news was not quite as bad as it first sounded. The cancellation only concerned the Blue Streak *as a missile;* a restricted amount of work on its development continued, and a new version of the Blue Streak is now going to be the first stage of the ELDO rocket Europa-1 (see section 12 of this Appendix).

However, the amount of money that the British government was willing to spend on space research was enough to design and build several scientific research satellites of high efficacy and great ingenuity. The ingenuity of the designers was especially strained, because they did not wish to duplicate any American or Russian work that had already been done. The result of their efforts was a satellite that is known both as UK-1 and as Ariel-I. It was followed by UK-2 and UK-3, all three put into orbit by American rockets. The third one was fired from the Pacific Missile Range with an inclination of 80° to the equator.

DESIGNATION	LAUNCH DATE	PERIGEE (MI)	APOGEE (MI)	ORBITAL PERIOD (MIN)	SATELLITE WEIGHT (LB)
UK-1	Apr. 26, 1962	242	754	95	132
UK-2	Mar. 27, 1964	186.4	825	101.2	150
UK-3	May 5, 1967	306	373	95.6	198

[1] Development of the Raven is described in *Spaceflight,* vol. IX, no. 10 (Oct. 1967).

12. ELDO and ESRO

Since the European countries are smaller and less financially powerful than the USA and the USSR, they realized that individually they could not have space programs comparable to those of the two main competitors. But they could hope to achieve a useful space-research program by uniting for the purpose. Consequently two international organizations were formed. ELDO (European Launcher Development Organization), first proposed in 1960 by the United Kingdom and France, was formally organized in 1964, with headquarters in Paris and Count Renzo de Carrobio as Secretary General. Its purpose was to develop a large satellite launch vehicle and a satellite. The member nations, listed by the size of their contributions to the ELDO budget, are: United Kingdom (38.79%), France (23.93%), West Germany (22.01%), Italy (9.78%), Belgium (2.85%), and The Netherlands (2.64%). There is also a seventh member which is not European and does not contribute money; this is Australia, which makes the Woomera firing range and its facilities and personnel available to ELDO.

ESRO (European Space Research Organization) was started on March 20, 1964, in order "to provide for and to promote collaboration among European States in space research and technology exclusively for peaceful purposes." The membership is larger than that of ELDO. Again listed by the size (percentage) of contributions it consists of: United Kingdom (25%), West Germany (22.56%), France (19.14%), Italy (11.17%), Sweden (5.17%), Belgium (4.41%), The Netherlands (4.24%), Switzerland (3.43%), Spain (2.66%), and Denmark (2.21%). Austria and Norway have "observer" status. Headquarters is also in Paris, under the leadership of Professor Pierre Auger. Since ESRO's activities are more diversified than those of ELDO—these include fundamental research, high-altitude probes, and satellite design—ESRO has the following subdivisions:

ESDAC: European Space Data Center, in Darmstadt, West Germany

ESLAB: European Space Research Laboratory, in Noordwijk, The Netherlands

ESRANGE: European Space Range (for sounding rockets), near Kiruna, Sweden

ESRIN: European Space Research Institute, in Frascati, Italy

ESTEC: European Space Technology Center, in Noordwijk and Delft, The Netherlands

ESTRACK: European Space Tracking network, with stations in Fairbanks, Alaska; Norway; Spitsbergen; Port Stanley (Falkland Islands); and Redu, Belgium.

ESRO also plans a number of satellites to be launched by American rockets (Scout and Delta), the first three of which are:

	LAUNCH VEHICLE	SATELLITE WEIGHT (LB)	MAIN MISSION	PRIME CONTRACTOR
ESRO-1	Scout	176	data over North Pole	L.T.C., France
ESRO-2	Scout	190	Van Allen belt	Hawker-Siddeley, U.K.
Heos A	TAD (Delta)	226	Earth's magnetosphere	Junkers, W. Germany

The ESRO-2 satellite, holding seven experiments designed by British and Dutch universities and the French atomic energy agency, was ready for firing on May

29,1967. It was to go into a polar orbit ranging from 215 to 680 miles, and because of the polar orbit the firing site was Vandenberg AFB in California. The rocket, a Scout, suffered one of the rare malfunctions of this rocket. The malfunction was in the third stage and as a result that ESRO-2 splashed somewhere in the southern part of the Pacific Ocean.

The first goal of ELDO is the rocket called Europa-1, with a British first stage, a French second stage, a German third stage, and an Italian satellite protected by an Italian nose cone.

The French stage, when forming part of the Europa-1, is called Coralie, but when it is part of the assembly it has neither tail fins nor a nose cone. Coralie with fins, a dummy upper stage, a dummy satellite, and a nose cone is called Cora. The first test flight of a Cora rocket took place from Hammaguir November 27, 1966, but a circuit failure at T + 62 sec cut the flight short. The second flight took

SPECIFICATIONS OF SATELLITE LAUNCH VEHICLE ELDO-A (EUROPA-1)

	METRIC	U.S.
Total height	31.68 m	103.8 ft
Max. diameter	3.69 m	12.1 ft
Launch weight (with payload)	104,670 kg	230,340 lb
Payload for 200 km (124 mi) orbit	1,000 kg	2,205 lb
Payload for 500 km (311 mi) orbit	850 kg	1,874 lb
FIRST STAGE (Blue Streak, *British*)		
Height	18.75 m	61.5 ft
Diameter, tank section	3.05 m	10.0 ft
Max. weight	89,406 kg	197,107 lb
Dry weight	6,997 kg	15,425 lb
Propellant weight (kerosene and liquid O_2)	82,409 kg	181,681 lb
Take-off thrust (2 engines) total:	135,800 kp	289,386 lb
Fuel consumption	547 kg/sec	1,206 lb/sec
Burning time	156 sec	
SECOND STAGE (Coralie, *French*)		
Height	5.7 m.	18.7 ft
Diameter	2.0 m	6.6 ft
Max. weight	11,894 kg	26,222 lb
Dry weight	2,099 kg	4,627 lb
Propellant weight (UDMH and N_2O_4)	9,750 kg	21,495 lb
Thrust (4 engines), total:	28,000 kp	61,729 lb
Fuel consumption	101 kg/sec	223 lb/sec
Burning time	96.5 sec	
THIRD STAGE (*West German*)		
Height	3.36 m	10.8 ft
Diameter	2.0 m	6.6 ft
Max. weight	3,370 kg	7,430 lb
Dry weight (with guidance)	610 kg	1,345 lb
Propellant weight (hydrazine/UDMH and N_2O_4)	2,760 kg	5,070 lb
Thrust		
main engine	2,300 kp	5070 lb
two vernier engines	twice 40 kp	176 lb

place on December 18, 1966. This time the two engines burned for their full duration of 100 sec and Cora reached the peak of the flight—34 mi—176 sec after lift-off.

The first Europa-1 rocket, taking off at Woomera on May 24, 1966 (first stage only, with dummy upper stages), was not quite successful; at T + 130 sec there was an unexplained malfunction. The second Europa-1 rocket was fired (also from Woomera) on August 4, 1967, and also suffered a malfunction after ten postponements because of bad weather and other reasons. This time the nature of the malfunction was known: the Coralie had failed to ignite. The next shot of a Europa-1 will be from the new French range in West Africa.

APPENDIX 3

Notes and Tables

Composition of Blackpowder Rocket Charges (Addendum to Chapter 3)

The composition of the propulsive charge of blackpowder rockets has remained virtually unchanged, as a comparison of various recipes shows. All figures refer to percentages by weight.

	CHINESE	GREEK	ARABIC	ENGLISH
		(ALL 13TH CENT.)		
Saltpeter	61.0	69.2	69.5	41.2
Charcoal	18.3	23.0	15.7	29.4
Sulfur	18.3	7.8	14.8	29.4
Mi-to-sing	2.4	—	—	—
	100.0	100.0	100.0	100.0

The Chinese recipe is that given by Reynaud and Favé in their work on early incendiaries; the "Greek" recipe is Recipe No. 33 in the *Liber Ignium* of Marchus Graecus; the Arabic is that of Hassam er-Rammah; and the English that of Roger Bacon as deciphered by Lieutenant Colonel Hime. Reynaud and Favé placed a question mark after "Mi-to-sing," presumably indicating that they had failed to identify this substance. I was informed by Dr. Ven Te Chow of the University of Illinois that this is the classic Chinese trade name for litharge, or fused lead monoxide. More modern recipes seem to show less saltpeter, which merely means that they had saltpeter of higher purity.

	ENGLISH		GERMAN	
	17TH CENT.	20TH CENT.	17TH CENT.	20TH CENT.
Saltpeter	60.0	61.6	56.3	60.0
Charcoal	25.5	23.0	18.7	25.0
Sulfur	14.5	15.4	25.0	15.0
	100.0	100.0	100.0	100.0

The English seventeenth-century recipe is that of Nathanael Nye; the twentieth-century one is taken from the *Treatise on Ammunition.* The older German recipe is that of Colonel von Geissler; the twentieth-century German is what was used in 1910 for large-sized signal rockets and lifesaving rockets.

Hail Rockets (Addendum to Chapter 5)

The controversy about the dispersion of hail clouds by small explosions that was started by R. Baur's report in 1906 (see pp. 97–98) is still going on. After the First World War it was pointed out that no hail had fallen during the months-long violent battle of Verdun, and the continuous artillery barrage was cited as the probable reason. Later, the same was said about the battles of the Marne. The implication was that maybe Baur had been right after all. New experiments with hail rockets were made during the twenty years between the two world wars. In Emmishofen in Switzerland a fireworks manufacturer named Müller produced such rockets as an "off-the-shelf" item. His anti-hail rockets were 1½ inches in diameter, had a length of 14 inches and a vertical range of 900 yards, and carried a charge weighing slightly less than 2 ounces. He claimed that one explosion inside the hail cloud changed the hail to snow and that additional rockets changed the snow to rain.

In 1933 experiments were carried out both in Austria and in Southwestern Germany. The Austrian tests were witnessed by the Viennese editor Professor Erich Doležal who wrote me an account of them. The rockets used had a diameter of 27 millimeters and there were two types, one somewhat longer than the other. The shorter one had a vertical range of 900 yards, the longer a range of 1100 yards. Both set off their explosive charge near the peak of the flight. "The explosion," Professor Dolezal continued, "tears a hole into the hail cloud and the horizontal wind below the cloud then enters this hole and tears the cloud apart. The tests have, so far, yielded good results. Instead of hail raindrops fall. . . . It is not a case of just moving the clouds so that one village may escape and the next one get it twice as strongly; the hail is dissolved into water. . . ."

The tests in South Germany were carried out at the instigation and with financial aid of the Department of Agriculture of the state of Württemberg. In two districts (Blaubeuren and Ebingen) fire was opened on hail clouds (once in each of the two districts) and was continued until the cloud had been dispersed, without regard for the number of rockets expended. In nineteen other districts fire was opened on hail clouds, in each case with six rockets or fewer. In three districts hail fell anyway, but sixteen other districts claimed success.

After the Second World War the Department of Agriculture of West Germany announced that large-scale tests would be repeated, but no report was issued. The Russians, however, use anti-hail rockets officially over the wine-growing regions of Georgia. Their rockets have nose cones that not only explode but also scatter silver-iodide crystals in the process; the newspaper *Red Star* (the newspaper of the Red Army) claimed, late in 1961, that "thousands of acres of vineyards" were saved from possible damage. The newspaper's description of what has been observed is quite similar to what Dolezal wrote—the explosions break up the clouds and the break-up process is then completed by natural events. It is interesting that the Russians bought their first set of anti-hail rockets from Angelo Patti in Somma Campagna,

Italy. Patti manufactures two types of rockets: one type carried an explosive charge and the other silver iodide with a small bursting charge. Most of Patti's customers are, of course, Italian vintners and farmers.

Another group of people that swears to the efficacy of shooting rockets into hail clouds is the fruit growers of Queensland, Australia. Their rockets can climb to 4000 feet and carry both explosive and silver iodide.

Rocket Mail (Addendum to Chapter 5)

One of the hopes of the early rocket experimenters in Europe had been to utilize long-range rockets for the rapid transportation of mail. The idea of the mail rocket cannot be found in Ziolkovsky's works, or in Goddard's first Smithsonian report, or even in the first edition of Oberth's book. The earliest reference I could find is in the March 1928 issue of *Die Rakete,* in a summary of a lecture delivered in Vienna February 9, 1928, by Dr. Franz von Hoefft who was then the president of an Austrian (actually only Viennese) Society for High-Altitude Exploration. With reference to a proposed rocket type he called RH IV, Dr. von Hoefft said, "The upper stage will carry a mailbag; it should be capable of reaching any point on the globe along a Keplerian ellipse in about one hour." The June 1928 issue of *Die Rakete* contains the text of the lecture delivered by Professor Oberth during the same month at the annual meeting of the WGL in Zoppot (see Chapter 5, pp. 109–110). Near the end of the lecture he said: "I would suggest, therefore, as a beginning the building of small rockets with automatic guidance that can bridge distances of 1000 to 2000 km (600 to 1200 mi) and carry a payload of 10 to 20 kg (22 to 44 lb). Several factors that I cannot discuss now because of lack of time happen to facilitate the automatic guidance of a rocket so that, in my opinion, it should be possible to predetermine the place at which the rocket returns into the atmosphere with an uncertainty of only a few miles. This rocket, therefore, seems suitable for transporting urgent mail over long distances in a very short time. The rocket would have to land by means of a parachute; some other means of transportation would then carry the apparatus to its precise destination. At a later date, I would equip such a rocket with a powerful booster that would result in transoceanic ranges. The booster rocket would drop off after one minute and should be easy to locate."

Of course all these figures were rather on the optimistic side, but they planted the idea of the mail rocket. In 1929, when Oberth was working on what was hoped would be the first high-altitude sounding rocket, the Ufa Company explored a few of the legal aspects. Its representative asked for an interview with the American Ambassador in Berlin, Dr. Shurman. Ambassador Shurman wisely refrained from having, or at least stating, opinions about technological problems. "Just two years ago," he told the Ufa representative, "the chair you are sitting in was occupied by Dr. Eckener of the Zeppelin works who talked to me about transatlantic mail by dirigible." But the Ambassador did voice an opinion about the procedure to be followed. *If* it had been demonstrated that the rockets were reliable and would not endanger the life or the property of American citizens, the proper thing to do would be to ask for permission at the United States Embassy. The Embassy would forward the request by cable to the Secretary of State who would pass it on to the President who would then call a cabinet meeting which would make the final decision. Of course nothing happened after this conversation.

In 1931 it became known that an Austrian engineer named Friedrich Schmiedl was operating a rocket mail route in Styria, mostly between Schöckel and Radegund, which had some kind of semiofficial status. Schmiedl used 6-ft powder rockets of his own manufacture which ejected the mailbag (equipped with parachute) shortly before crashing, roughly half-way down the descent. Extensive correspondence convinced me that Schmiedl was on his way to making rocket mail practical for certain purposes and under certain conditions. He was mainly thinking of mountainous territory where two villages might be separated by only a few miles, but still be a day's *walk* apart. Such mail shooting might also be practical for islands fringed by a heavy surf. I knew at that time that in the past this had been done somewhere by means of modified Congreve rockets. Since then I have found out that the place was the island of Niuafo'ou, one of the Tonga group south of Samoa in the Pacific. But the method had not proved satisfactory; either the rockets fell short and sank, or else the letter container burst upon impact and the mail was ruined by sea water. Beginning in 1902 the rocket mail was gradually abandoned and replaced by so-called tin-can mail, a soldered can (buoyant) being thrown overboard and retrieved by native canoemen who could brave the roughest seas. That delivery by rocket had failed to work in a specific case, and with rockets made for giving distress signals, did not invalidate the idea.

A list of Schmiedl's mail shots, compiled by the experimenter himself in 1933, follows. (*V* stands for *Versuchsrakete,* meaning "experimental rocket"; beginning with R-1 the mail was received officially by the local post offices.)

			LETTERS TRANSPORTED	
DATE	ROCKET	FLIGHT	TOTAL	REGISTERED.
Feb. 2, 1931	V-7	Schöckel to Radegund	102	—
Apr. 21, 1931	—	"Kalte Rinne"	79	—
Oct. 28, 1931	V-8	Grazer Feld	84	(night test)
Sept. 9, 1931	R-1	Hochtrötsch to Semriach	333	36
May 28, 1932	V-9	Schöckel to Radegund	228	125
same	V-10	Schöckel to Radegund	192	113
June 11, 1932	V-11	Grazer Feld	28	(storm test)
July 23, 1932	V-12	Schöckel to Kumberg	231	187
same	V-13	Schöckel to Kumberg	200	100
March 16, 1933	V-14	Garrach to Arzberg	283	—

After the highly successful tests of 1932 Schmiedl was trying to establish a regular schedule, but the Schuschnigg government suddenly forced him to destroy all his work, including the launching ramp at Schöckel (Styria). No reason or explanation was given, but dealing with Schuschnigg's government was hardly less dangerous than taunting the Gestapo, so Schmiedl had to consent to destroy his own work.

Schmiedl's work not only showed that there was a place for short-range mail rockets under special circumstances, it also proved that rocket "stamps" found a market. On April 15, 1931, Tiling put a bundle of 188 numbered and autographed postcards into one of his rockets. A year or so later a German experimenter named Gerhard Zucker started building mail rockets. To get himself into the good graces of the Hitler government he made his flights for the Nazi Winter Relief (*Winterhilfe*). Then he went to England, and then returned to Germany, since his shots in England had not been at all successful. Zucker's rockets were exceedingly simple in design: a centrally located steel tube served as container for the blackpowder rocket; the mail was put into the compartments between steel tube and outer jacket. Similar experiments (in approximate chronological order) were made in Holland, the U.S.A., Australia, India, Cuba, and Mexico. The bill was usually footed by stamp collectors who hoped for an official rocket mail one day and wanted to accumulate "forerunners."

The experiment in Cuba resulted in the issue of the only existing official rocket mail stamp. When the shot was being planned, the Departamento de las Comunicaciones printed 200,000 copies of the then current 10-centavo stamp in a bright emerald-green color (the stamp was normally black), overprinting them in black "*Experimento de Cohete Postal, Año de 1939*" to commemorate the event. The experiment was a failure, but the stamp is official; it is no. C31 in Scott's catalogue. In the wake of the launching of artificial satellites and especially following manned orbital flights, a great many countries issued stamps honoring rocket pioneers, astronauts, and specific spacecraft. But the Cuban stamp is still the only official stamp for rocket mail.

It now seems doubtful that long-range rocket mail as visualized in 1929 will come into existence. At the time that the Ufa Company approached Ambassador Shurman the argument ran as follows: the rocket would need about 45 minutes for the trip; if one allowed for 1 hour's delay between shots and 1½ hours for retrieval and bringing the mail ashore at the receiving end, the time would add up to about 3½ hours, sometimes to 4 hours. This could be compared favorably to steamer mail that needed a minimum of one week. But now jetliners cross the Atlantic in 6½ to 7 hours and the supersonic passenger liners under development will cut this time to at most 4 hours, the time once calculated for rocket mail.

Rocket Societies (Addendum to Chapter 6)

In the wake of the founding of the VfR an increasing number of other rocket societies were formed, the first of which was the Oesterreichische Gesellschaft für Höhenforschung (Austrian Society for High-Altitude Exploration), founded by Dr. Franz von Hoefft. It was later superseded by the Oesterreichische Gesellschaft für Raketentechnik (Austrian Society for Rocket Technology) founded by von Pirquet and Rudolf Zwerina.

Oldest of the American societies is the American Rocket Society in New York, which was founded in March 1930 by G. Edward Pendray and David Lasser under the name of the American Interplanetary Society. This society was the first in this country actually to build a rocket; it built several. The first was a two-stick repulsor patterned frankly after the VfR model. It was damaged during its first test run in autumn 1932. Most of its parts, especially the tanks and the motor, could be salvaged, and the new rocket, called "No. 2," was tested at Marine Park, Great Kills, Staten Island, in May 1933. The society's journal, *Astronautics* (no. 26), reported on the test as follows:

> Experimental Rocket No. 2 of the American Interplanetary Society was shot at twenty minutes past eleven o'clock, Sunday, May 14. It reached an altitude of approximately 250 feet after firing a trifle more than two seconds. At that

height the flight was brought to an abrupt end by the bursting of the oxygen tank.

After this three more rockets, called Nos. 3, 4, and 5, were planned. No. 5 turned out to be hopeless nonsense; No. 3 was built but could not be fueled and was therefore useless. No. 4, a one-stick repulsor, was tested September 9, 1934. It took off with high acceleration and rose almost vertically to about 300 ft. At that point one of the four nozzles of the motor burned out and the rocket assumed a strange weaving motion. The peak altitude is estimated to have been 382 ft. It covered a horizontal distance of 1300–1400 ft. The rocket struck the water with great violence and was so badly smashed that it could not be repaired. The newly devised parachute release had no chance to function.

Since that experiment only ground tests have been performed. The record is a 12-sec run on June 22, 1941, when a ceramic-lined motor designed by Alfred Africano produced a recoil of more than 260 lb for about 2 sec. Lack of a permanent workshop and proving ground, combined with an almost frantic desire to avoid "conventional methods," resulted in a rather erratic program and may explain the comparatively small amount of work done between 1935 and the outbreak of the Second World War.

The Experimental Committee of the Society, or at least several of its members, then became the nucleus of the firm of Reaction Motors, Inc., in New Jersey.

After the Second World War the American Rocket Society officially abandoned an experimental program of its own and devoted itself to organizational activities and the publication of technical periodicals. Its annual meetings became important and newsworthy events. But there was an undercurrent of slight misgivings. Its directors were accused of emphasizing rocket and missile technology, and disregarding space travel, in order to demonstrate the "respectability" of the society. These misgivings led to the founding of the American Astronautical Society, which is now a flourishing society, too. While the American Astronautical Society has remained independent, in 1963 the American Rocket Society merged with the older Institute of the Aeronautical Sciences founded soon after the First World War by Major Lester B. Gardener. The new organization adopted the name American Institute for Aeronautics and Astronautics, or AIAA.

The second rocket society in the United States was the Cleveland Rocket Society, founded by a German named Ernest Loebell who resided in Cleveland. Loebell, a professional engineer, built a test stand and carried out a few experiments with liquid-fuel rocket motors.[1] The Cleveland Rocket Society existed from about 1933 to 1937 when Loebell decided to return to Germany. Since then nothing has been heard of him. Two active rocket societies, both in California, were founded after the war. One was the Glendale Rocket Society (later renamed Reaction Research Society), the other the Pacific Rocket Society. Both were located in the Greater Los Angeles area and both experimented with solid-fuel rockets in the Mohave desert. A California state law expressly forbidding rocket experiments spelled the end of both societies.

During the five years following the launching of Sputnik I the desire of many

[1] A comprehensive article about the Cleveland Rocket Society by John Tascher appeared in *Technology and Culture,* vol. VII, no. 1 (Winter 1966.)

young people to experiment with rockets created both a legal and a safety problem. After a few young people had been killed or injured, bills designed to outlaw such experiments were introduced in many state legislatures and voted into law in a few of them. Attempts were to allow amateurs to experiment under carefully controlled conditions. The best solution was found by G. Harry Stine, formerly an engineer with the Martin Company, who almost single-handed founded the National Association of Rocketry (NAR). The solution consisted in having a pyrotechnic company manufacture an absolutely reliable solid-fuel propulsion unit. Members of the NAR build their rockets around these manufactured units; any tampering with the unit itself disqualifies both the design and its designer. There are annual competitions, called NARAMs that are great fun for members and visitors alike.

In Great Britain the British Interplanetary Society (B.I.S.) was founded in 1933 by Phil E. Cleator in Liverpool. Its headquarters was later transferred to London. Other similar societies were founded elsewhere in Great Britain, all of them prevented by the Guy Fawkes Act from carrying out experimental work. They all suspended activities during the Second World War; after the war, due mainly to the efforts of L. J. Carter, secretary of the London B.I.S., they were amalgamated. The first meeting of the "new" B.I.S. was held December 8, 1945; the official day of incorporation was December 31, 1945. Membership at the time was about 280 and included George Bernard Shaw.

In Germany two societies were founded immediately after the war. Hans K. Kaiser who, after the demise of the VfR, had tried to gather its members into a society called Gesellschaft für Weltraumforschung (Society for Space Research), founded a new one with the same name preceded by "North-West German." Another society, founded in Stuttgart, took the name of Kaiser's original unsuccessful society. The two merged after a year or two of minor squabbles; later the name was changed to Deutsche Gesellschaft für Raketentechnik und Raumfahrt (German Society for Rocket Technology and Space Travel). The society is now a respected engineering organization and attempts are being made to merge it with the older VDI, the engineering society that originally so harshly condemned space-travel efforts.

An account of the international organization now comprising all these societies follows.

The International Astronautical Federation

In June 1949 the Gesellschaft für Weltraumforschung, successor to the VfR, passed a resolution requesting its directors "to explore the possibilities of forming an international association" of societies with goals similar to its own. The three leading members of the German society—Heinz Gartmann, H. H. Koelle, and Dr. Günter Loeser—got in touch with the British Interplanetary Society and the Groupement Astronautique Française. The director of the latter, Alexandre Ananoff, organized the First International Congress in Paris. It took place September 30, 1950, and was attended by representatives of astronautical societies in Argentina, Austria, Denmark, France, Germany, Spain, Sweden, and the United Kingdom. This first congress, actually a preliminary meeting, agreed on three points: (1) that such an association was desirable; (2) that it should be established one year later at another meeting in London, to be organized by the B.I.S.; and (3) that an *ad hoc* committee should be formed for study and coordination of proposals made, or to be made, by the member societies. Dr. Eugen Sänger was selected to head this committee, partly because of his international reputation, partly because he was not at the moment a member of any of the societies.

The Second International Congress was opened at Caxton Hall, London, on September 3, 1951. This time American representatives were present—Fred C. Durant III and Andrew G. Haley of the American Rocket Society, and E. V. Sawyer and David Elliot representing a group of amateur rocket societies—as were representatives of societies in two other countries that had not participated in the First Congress—Italy and Switzerland. The International Astronautical Federation was definitely established. Dr. Sänger became its president, Dr. Loeser and A. G. Haley vice-presidents, and Josef Stemmer of Switzerland secretary.

The Third International Congress took place in 1952 in Stuttgart, West Germany. During this congress the constitution of the I.A.F. was formally adopted and its headquarters were established at Baden, Switzerland. Registration, therefore, was under Swiss law. Logically enough, the next congress, the fourth (1953), took place in Zürich. During that congress Fred C. Durant, then president of the ARS, succeeded Dr. Sänger as president of the I.A.F. A. G. Haley remained a vice-president; the other vice-president chosen was Professor Dr. Friedrich Hecht, president of the Austrian society. Josef Stemmer remained secretary.

The fifth congress (1954) took place in Innsbruck, Austria, and the I.A.F. grew by the addition of societies in Egypt, Japan, and Brazil, and of another American organization, the American Astronautical Society. An official journal, *Astronautica Acta,* was created. Durant remained president, and Professor Teofilo M. Tabañera of Argentina and Erlin Buch Anderson of Denmark became the vice-presidents. The sixth congress (1955) took place in Copenhagen, where for the first time the Soviet Union was represented with two delegates—Academician Leonid I. Sedov and Professor K. F. Ogorodnikov of Leningrad University. Again Durant remained president, Professor Tabañera was re-elected a vice-president, and the other vice-president elected was General Gaetano A. Crocco of Italy.

The seventh congress took place in Rome in 1956, where Pope Pius XII bestowed the Apostolic Blessing upon the delegates and gave Papal approval to astronautical activities. The former chairman of the B.I.S., Dr. Leslie R. Shepherd, became the new president. The number of vice-presidents increased to five: General J. P. Bergeron (France), Fred C. Durant (USA), J. Marial (Spain), Professor L. I. Sedov (USSR), and Professor Tabañera (Argentina). Josef Stemmer of Switzerland became secretary again after a lapse of one year. The eighth congress took place in Barcelona in 1957. The dates were October 6–12, just the time Sputnik-I went into orbit; the eight-man Russian delegation looked understandably pleased. At the end of the congress Andrew G. Haley was elected president; Bergeron, Sedov, and Tabañera were re-elected vice-presidents, and the new vice-presidents were: Dr. J. M. J. Kooy (The Netherlands), Dr. L. R. Shepherd (U.K.) and Professor Kazimierz Zarankiewicz (Poland).

The ninth congress (1958) was held in Amsterdam. Societies in Bulgaria, Greece, and Israel became members, and societies in India, Belgium, Czechoslovakia, Iran, and Canada were granted observer status. The president and all the vice-presidents were re-elected for another year. The tenth congress (1959) was again organized by the B.I.S. and took place during the first week of September at Church

House, Westminster, London. It was marred by the sudden death of Professor Zarankiewicz, while he was presiding over a plenary session.[1]

The eleventh congress took place in Stockholm in 1960 and since then congresses have continued to meet annually in different countries. The one in Athens attracted world-wide attention because both American astronauts and Russian cosmonauts were present, exchanging hugs and pins.

By the end of 1966 the countries represented in the I.A.F. totaled thirty-three. The number of societies represented was fifty, but only one society per country is a voting member. For example, the American Institute of Aeronautics and Astronautics in New York is the American voting member, while the American Astronautical Society (Washington, D.C.), the Aerospace Medical Association (Washington, D.C.), the National Space Club (Washington, D.C.), and the Section on Space Medicine of the Medical Society of the State of New York (New York City) are members without a vote.

Maxim's Rocket Charges (Addendum to Chapter 7)

The first large solid-fuel rockets were *almost* invented in 1910 by Hudson Maxim, the younger brother of Sir Hiram Stevens Maxim who acquired fame and fortune for the gun named after him. Hudson Maxim was interested in the question of whether the type of explosives now called double-base powders could be used as fuel in internal combustion engines. He wrote an article[2] in which he first pointed out that double-base powders were far more expensive than normal automotive fuels. He emphasized that he was speaking from experience: "I have conducted more experiments in the use of explosive materials for driving motors than any man living or dead. I have spent more than sixty thousand dollars in these experiments." But he did advocate such materials for driving naval torpedos because in that case the price did not matter much.

"In my system of driving torpedoes, I employ a material called motorite, consisting of seventy per cent nitroglycerine and thirty per cent guncotton. In the form in which it is used it is no longer explosive, but burns continuously without explosion. Bars of the material seven inches in diameter and five feet long are forced into steel tubes. These tubes are closed at one end and the other end screws into a combustion chamber. Water is forced into the combustion chamber instantly upon ignition of the exposed end of the motorite in the combustion chamber. Under three hundred pounds pressure to the square inch, motorite burns at the rate of a foot per minute. . . . The products of combustion and steam together are utilized to drive a turbine. . . . A pound of motorite will evaporate about two pounds of water, so that each pound of motorite consumed represents three pounds of steam and products of combustion."

What Hudson Maxim had there was a solid-fuel rocket charge five ft long and seven in. in diameter; it is remarkable that he never had an accident which opened his eyes to the fact that he had made a large rocket.

[1] Two of the three Germans who had started the idea of the I.A.F. had also died since the first congress, Heinz Gartmann of a heart attack and Dr. Günter Loeser in a helicopter crash in the United States.

[2] I am quoting from the signed typescript of this article; penciled notations about type size to be used indicate that it was published, but I don't know when and where.

Characteristics of Nebelwerfer 41, 15-Centimeter Smoke Projectile (Addendum to Chapter 7)

	U.S.	METRIC (COMPUTED)
DIMENSIONS		
Over-all length	40 5/16 in.	103 cm
Length, propellant section	18 3/4 in.	47.6 cm
Length, smoke composition container	14 in.	35.5 cm
Length, cap	7 13/16 lb	20 cm
Diameter, (OD) propellant section	5 1/2 in.	14 cm
Diameter, cap (max)	6 3/16 in.	15.5 cm
WEIGHTS		
Over-all	79 1/4 lb	36 kg
Cap	3 1/2 lb	1.6 kg
Smoke container, empty	22 1/2 lb	10.2 kg
Smoke container, filled	32 1/2 lb	14.7 kg
Bursting charge	2 lb 14 oz	1.315 kg
Booster charge	9/10 oz	25 g
Propellant container	24 1/2 lb	11.1 kg
Propellant	14 lb	6.3 kg
Spacers, adapters, etc.	1 3/4 lb	0.8 kg
RANGE (max)	20,000 ft	6 km

Principal Source: PB Report No. 4543 (Library of Congress)

Characteristics of German Naval Rocket-Propelled Flare 21-Centimeter R-LG (Addendum to Chapter 7)

	U.S. (COMPUTED)	METRIC (FROM GERMAN DRAWING)
Over-all length (5.1 cal.)	43 in.	109 cm
Outside diameter (max)	8.4 in.	21.4 cm
Over-all weight	132 lb	60 kg
Propellant weight (diglycol)	33.5 lb	15.2 kg
Flare weight	16.5 lb	7.5 kg
Propellant, length	20.1 in.	51 cm
Propellant, OD/ID	7.5/4.2 in.	19.2/10.8 cm
Ejecting charge (blackpowder)	1 1/4 oz	35 g
Burning time, rocket	3.3 sec	3.3 sec
Burning time, flare	120 sec	120 sec
Nozzles (6), length	3 in.	7.5 cm
Nozzles, expansion angle	12°	12°
Nozzles, cant angle	9°	9°
Velocity (max)	1840 ft/sec	560 m/sec
Altitude (max)	17,000 ft	5200 m

Source: Published report of Combined Intelligence Objectives Sub-Committee (H. M. Stationery Office, London, File No. XXXII-114)

United States Rocket Ordnance, Second World War (Addendum to Chapter 7)

PORTABLE ROCKET WEAPONS

2.36 in. Bazooka. Anti-tank rocket with shaped charge. First production model had an effective range of about 600 ft; next model, often called "Superbazooka," had the same dimensions but a heavier propelling charge extending effective range to 2000 ft. Max diameter 2.36 in., length 18 in., weight 3.3 lb.

3.5-in. Bazooka. Enlarged version of preceding, developed for use against heavier tanks. Weight 8.5 lb. First operational use, Korea.

AIRBORNE ROCKETS

Retro-rocket. Anti-submarine weapon fired from planes backward with velocity matching plane speed, so that the weapon fell vertically. There were three different rocket bodies to match the flying speeds of three airplane types. The head of the rocket carried 35 lb of high explosive. Retro-rockets sank the last "probable" German submarine in the war, on April 30, 1945.

Retro-float light. Same as preceding, but fitted with flare.

3.5-in. anti-submarine rocket. Rocket with fixed fins. Length 54½ in., total weight 54 lb. Propelled by one 8.5-lb "grain," fitted with 20-lb solid steel heads.

3.5-in. anti-submarine rocket. Same as preceding, but fitted with 50-lb 5-in. high-explosive warhead.

4.5-in. Aircraft Rocket M-8. Total weight 38 lb; fitted with warhead containing 5 lb of high explosive. At first with fixed fins, replaced later by folding fins. Was fired from three tube clusters carried under the wings.

115-mm Aircraft Rocket. Also called the "super 4.5-in." Length 6 ft, total weight 103 lb, weight of warhead 40 lb, with four fixed fins. Had longer range and was more accurate than M-8. Two types of heads: one with 8.5 lb of high explosive in fragmentation casing, the other a semi-armor-piercing type.

5-in. HVAR (High Velocity Aircraft Rocket). Nicknamed "Holy Moses." Over-all length 6 ft, over-all weight 140 lb. Introduced in combat July 1944; became the most used air-to-ground rocket of the U. S. armed forces.

6.5-in. Aircraft Rocket "Ram." Developed from 5-in. HVAR in 1950 (in less than one month) for use in Korea. First airborne rocket with shaped charge.

"Tiny Tim." Total length 10 ft 3 in.; diameter 11.75 in., over-all weight 1284 lb. Warhead weight 590 lb, containing 150 lb TNT. Propelling charge four "grains," over-all weight 146 lb. First operational use against Okinawa; later in Korea.

GROUND-TO-GROUND AND BARRAGE ROCKETS

3.5-in. smoke rocket. Development model of aircraft rocket, utilized for smoke screens.

4.5-in. M-8. Same as Aircraft Rocket M-8. 4000-yd range when used as a ground-to-ground weapon. Used with fixed fins by Navy as a barrage rocket and with folding fins by Army. Multiple launchers developed for this weapon were Xylophone (8 tubes) and Calliope (60 tubes).

4.5-in. smoke rocket. Adaptation of M-8 for laying smoke screens. Range 1100 yd. Smoke-producing agent first "FS liquid smoke" (mixture of sulfurtrioxide and chlorosulfonic acid), later white phosphorus. The 3.5-in. aircraft rocket was also adopted as a smoke rocket.

7.2-in. demolition rocket. Warhead carried 32 lb of high explosive. The Army launcher devised for this rocket was nicknamed Whiz Bang, while the Navy 120-round launcher for the same rocket was called Woofus.

7.2-in. chemical rocket. Range up to 3000 yd. Was used for smoke composition and could have been used for poison gas. Each rocket carried 20 lb of chemicals. Had a special 24-round launcher, nicknamed Grand Slam.

4.5-in. spinner M-16. Spin-stabilized rocket of same over-all weight as M-8, but with a range of 5250 yd and slightly heavier warhead. Launchers for this weapon: Honeycomb (24 tubes) and Hornet's Nest (60 tubes). A proximity fuze for these rockets was developed but was not used in action.

3.5-in. spinner rocket. Experimental only, not introduced.

5-in. spinner rocket. Range 700–1000 yd, for penetrating flat angle fire. Over-all weight 50 lb, over-all length 30 in. Like the 115-mm Aircraft Rocket it was equipped with semi-armor-piercing head and general-purpose head.

5-in. spinner barrage rocket. Over-all weight 50 lb, over-all length 32 in., range 5000 yd. Two other spinner barrage rockets of the same caliber, but with ranges of 2500 and 1250 yd, respectively, were under development, but only the 5000-yd-range type saw action.

SPECIAL APPLICATIONS

"Minnie Mouse." The rocket fired from the "mousetrap" submarine killers. The warheads were fitted with impact fuzes which would work only under water. A weapon of the same purpose as the old depth-charge thrower.

"Window Rocket." Adaptation of 3.5-in. aircraft rocket carrying "radar window," length of aluminum foil which causes spurious readings in enemy radars. Each rocket carried over 75,000 pieces of foil, 3/16 in. wide and from 1½ to 16 in. long.

AA target rocket. 3.5-in. rocket with large plywood fins, range of 1 mi, speed of 360 mph, used for training in antiaircraft gunnery.

Harpoon rocket. Line-carrying rocket with a pointed head with four folding spades and two 4.5-in. rockets in tandem. The first rocket did the actual carrying; the second, ignited by delay fuze, forced the pointed head deep into the ground. Pull on the line expanded the spades.

"Snake." A ski-like structure 100 ft long, assembled from metal plates 5 ft long and 5 in. wide, with TNT clamped to plates; the whole propelled by a 4.5-in. rocket mounted at the head of the device. Used to detonate a path through enemy mine fields.

Japanese Rocket Ordnance, Second World War (Addendum to Chapter 7)

TYPE	WEIGHT KG (LB)	CALIBER CM (IN.)	BURNING TIME (SEC)	ROCKET CHARGE KG (LB)	MAXIMUM RANGE M (FT)
Combat rocket	90 (198)	20 (7.9)	2.01	8.1 (17.8)	1800 (5900)
same, modified	80 (176)	20 (7.9)	2.14	11.7 (24.9)	4000 (13,100)
Anti-tank	47.4 (104)	20 (7.9)	0.96	4.9 (10.8)	500 (1640)
Depth charge	34.6 (76)	15 (5.9)	1.73	3.9 (8.6)	2500 (8200)
Incendiary anti-aircraft	23.9 (52.6)	12? (4.73)	0.92	3.4 (7.5)	1500 (4920)
Combat high explosive	23.9 (52.6)	12? (4.73)	0.92	3.4 (7.5)	4800 (15,700)
Anti-tank	10.35 (22.8)	10 (3.94)	0.25	0.8 (1.8)	100 (328)
Anti-tank	5.4 (11.9)	8 (3.15)	0.35	0.4 (0.9)	100 (328)
Antiaircraft	6 (13.2)	?	0.37	1.0 (2.2)	?
Antiaircraft	7 (15.4)	?	0.37	1.8 (4)	?
250-kg bomb	370 (815)	—	3.37	49.6 (109)	5000 (16,400)
60-kg bomb	102 (224)	—	4.03	10.6 (23.3)	?

The types listed were actually used in the Second World War; in addition there were numerous experimental types of which a few were occasionally tested in action. Source: PB Report 16,800.

Characteristics of the Long-Range Rocket A-4 (V-2) (Addendum to Chapter 8)

The following figures are based on Naval Research Laboratory Report R-2955:

DIMENSIONS

	U.S.	METERS (COMPUTED)
Over-all length, German	46 ft 11 in.	14.300
Over-all length, White Sands	47 ft 0 in.	14.325
Warhead, German	7 ft 6 in.	2.285
Warhead, White Sands (NRL)	7 ft 7 in.	2.310
Control compartment	4 ft 7 in.	1.400
Fuel-tank section	20 ft 5 in.	6.225
Tail section	14 ft 5 in.	4.395
Length of fins	12 ft 11 in.	3.935
Diameter of body, center	5 ft 5 in.	1.650
Diameter across fins	11 ft 8 in.	3.555

WEIGHTS

		LB	KG (COMPUTED)
Warhead, German			
Shell	551 lb (250 kg)		
Amatol	1654 lb (750 kg)		
Total		2205	1000
Warhead, NRL			
Shell	1055 lb (478.5 kg)		
Equipment	1055 lb (478.5 kg)		
Counterweight	200 lb (91.0 kg)		
Total		2310	1048
Control compartment		1058	480
Fuel-tank section (empty)		1636	742
Rocket motor		2053	931
Tail section with fins		1885	855
Fuels, total		19,392	8796
"Dry" weight, German		8837	4008
"Dry" weight, White Sands		8942	4056
Take-off weight, German		28,229	12,805
Take-off weight, White Sands		28,334	12,853

PERFORMANCE

Turbine
- Diameter of blades 18.5 in.
- Working pressure about 310 lb/sq in. (21 atm)
- 5000 rpm, 675 hp
- Steam consumption 1.68 kg/sec; 3.7 lb/sec

Oxygen pump
- Impeller diameter 10.55 in.
- 5000 rpm, 320 hp
- Delivery 75 kg/sec; 165 lb/sec
- Delivery pressure about 360 lb/sq in. (24 atm)

Alcohol pump

Impeller diameter	13.45 in.
5000 rpm	355 hp
Delivery	50 kg/sec; 110 lb/sec

Delivery pressure about 370 lb/sq in. (25 atm)

MOTOR	U.S.	METRIC
Over-all length	5 ft 8 in.	1.725 m
Diameter, combustion chamber	3 ft 1 in.	0.940 m
Diameter, throat	1 ft 4 in.	0.405 m
Diameter, nozzle exit	2 ft 5 in.	0.735 m
Fuel consumption	275 lb/sec	125 kg/sec
Duration	68 sec	68 sec
Exhaust velocity (sea level)	6560 ft/sec	2000 m/sec
Temperature in chamber	3630 F	2000 C
Pressure in chamber	215 lb/sq in.	14.5 atm
Thrust, sea level	59,500 lb	27 tons
Thrust, 25 miles up	70,000 lb	31.8 tons
Acceleration (effective g)	1 → 6	1 → 6
Velocity, max	5575 ft/sec	1700 m/sec
Operational range	180–190 mi	290–306 km
Maximum observed range	220 mi	354 km
Operational peak altitude	60 mi	97 km
Vertical peak altitude	116 mi	186.7 km

The fuel of the V-2 was ordinary ethyl alcohol—in this case made from potatoes—to which enough water had been added to bring its strength down to 75 per cent by volume. The reason for the addition of water was the following: the exhaust velocity of a rocket motor is determined, in the main, by combustion pressure and temperature. But the nature of the exhaust is important too; the exhaust velocity will be higher, if the other factors remain unchanged, for an exhaust consisting of lighter gas molecules. The combustion products of burning ethyl alcohol are CO_2 and H_2O, and of course the CO_2 molecule is by far the heavier. By adding water to the alcohol the proportion of water molecules in the exhaust is increased and the average molecular weight depressed. This addition also decreases combustion temperature, but at a lesser rate. The mixture ratio which was actually used is the optimal mixture for the purpose.

Record of Early V-2 Shots from Peenemünde in 1942 and 1943 (Addendum to Chapter 8)

NO.	DATE		BURNING TIME	RANGE KM	RANGE MI	REMARKS
	1942					
2	June	13	36	1.3	0.8	rolled, unstable
3	Aug.	16	45	8.7	5.4	nose broke off
4	Oct.	3	58	190	118	too steep; success
5	Oct.	21	84	147	91.3	steam generator misbehaved
6	Nov.	9	54	14	8.7	vertical, height 67 km
7	Nov.	28	37	8.6	5.3	tumbled, lost vanes
9	Dec.	12	4	0.1	0.06	H_2O_2 explosion
	1943					
10	Jan.	7	—	0	0	explosion on ignition
11	Jan.	25	64.5	105	65.2	too steep, rolled
12	Feb.	17	61	196	121.8	too shallow
13	Feb.	19	18	4.8	3.0	fire in tail
16	Mar.	3	33	1.0	0.6	vertical; explosion
18	Mar.	18	60	133	82.4	too steep; rotated
19	Mar.	25	28	1.2	0.7	tumbled, exploded
20	April	14	66	287	178.3	fell on land
21	April	22	59	252	156.6	fell on land
22	May	14	62	250	155.3	cut-off switch failed
26	May	26	66.5	265	164.6	
25	May	26	40	27	16.8	Brennschluss at 40 sec
24	May	27	55	138	85.7	
23	June	1	62	235	146.0	premature Brennschluss
29	June	11	63.5	238	147.9	
31	June	16	60.5	221	137.3	premature Brennschluss
28	June	22	62.5	75	46.6	exploded after 70 sec
30	June	24	65.1	287	178.3	cut-off switch failed
36	June	26	64.9	235	146.0	
38	June	29	15	3	1.8	fell on airport
40	June	29	63.6	236	146.6	impact not observed
33	July	1	—	—	—	Brennschluss at take-off; exploded
41	July	9	—	—	—	fell on pump building
34	July	9	—	—	—	Brennschluss at take-off

The numbers in the first column are the production numbers of the rockets; the dates give the sequence of firing. No. 1 was used only as a "mock-up"; nos. 1 to 6 weighed about 1 ton more than the later A-4. No. 8 was used for crew training; nos. 14, 15, and 17 were stored in the open for more than a year with tarpaulin covers only to test for storability; they were fired after 1943. Nos. 27, 32, 35, 37, and 39 had various experimental equipment for development purposes; they were fired after 1943.

Source: *V-2—Der Schuss ins Weltall* by Walter Dornberger (German edition), and personal correspondence

Characteristics of the German Project A-9 + A-10, Peenemünde (Addendum to Chapter 8)

EARLIER VERSION

	A-9		A-10	
	U.S.	METRIC (COMPUTED)	U.S.	METRIC (COMPUTED)
Length (each step)	46 ft	14.0 m	?	?
Diameter	5.5 ft	1.7 m	11.5 ft	3.5 m
Diameter over fins	undetermined		undetermined	
Take-off weight (each step)	28,660 lb	13,000 kg	191,800 lb	86,960 kg
Weight empty	8820 lb	3814 kg	55,100 lb	25,020 kg
Propellants	17,640 lb	8000 kg	136,700 lb	61,490 kg
H_2O_2 + Permang. (Ca.)	410 lb	186 kg	nitrogen pressure	
Payload	2200 lb	1000 kg	The A-9	
Fuel consumption	260 lb	118 kg	2728 lb	1237 kg
Time	68 sec	68 sec	50 sec	50 sec
Thrust at sea level	55,130 lb	25,000 kg	441,000 lb	200,000 kg
Velocity at Brennschluss	9200 ft/sec	2800 m/sec	3937 ft/sec	1200 m/sec
Altitude at Brennschluss	100 mi	160 km	80,000 ft	24 km
Mass-ratio	2.7	2.7	2.63	2.63
Range	3105 mi	5000 km	——	——

Total take-off weight A-9 + A-10 = 220,460 lb or 99,960 kg
Effective mass-ratio A-9 + A-10 = 7.1:1.

LATER VERSION

	A-9		A-10	
	U.S.	METRIC (COMPUTED)	U.S.	METRIC (COMPUTED)
Length (each step)	46.5 ft	14.2 m	65.5 ft	20 m
Diameter	5.4 ft	1.65 m	13.5 ft	4.15 m
Diameter over fins	undetermined		29.5 ft	9 m
Take-off weight (each step)	35,850 lb	16,260 kg	152,240 lb	69,060 kg
Weight empty	6615 lb	3000 kg	37,470 lb	17,000 kg
Propellants	26,260 lb	11,910 kg	111,470 lb	50,560 kg
H_2O_2 + Permang. (Ca.)	772 lb	350 kg	3300 lb	1500 kg
Payload	2200 lb	1000 kg	The A-9	
Fuel consumption	275 lb/sec	125 kg/sec	2231 lb/sec	1012 kg/sec
Time	95 sec	95 sec	50 sec	50 sec
Thrust at sea level	56,000 lb	25,400 kg	441,000 lb	200,000 kg
Velocity at Brennschluss	9200 ft/sec	2800 m/sec	3940 ft/sec	1200 m/sec
Altitude at Brennschluss	100 mi	160 km	80,000 ft	24 km
Mass-ratio	4.07	4.07	2.56	2.56
Range	3105 mi	5000 km	——	——

Total take-off weight A-9 + A-10 = 188,090 lb or 85,320 kg
Effective mass-ratio A-9 + A-10 = 10.42:1.

Adapted from *JBIS*, xxxv, article by Gatland, Dixon, and Kunesch.

Characteristics of Some German Missiles, Second World War (Addendum to Chapter 8)

ANTIAIRCRAFT MISSILE WASSERFALL (PEENEMÜNDE) (See Fig. 42.)

	U.S.	METRIC (COMPUTED)
Over-all length	25.6 ft	780 cm
Diameter	2.9 ft	88.5 cm
Diameter over fins	8.2 ft	250 cm
Take-off weight	8400 lb	3810 kg
Dry weight	3872 lb	1756 kg
Fuel (Visol)	770 lb	350 kg
Oxidizer (Salbei)	3300 lb	1500 kg
High explosive in warhead	330 lb	150 kg
Fuel consumption	96.8 lb /sec	31.6 kg/sec
Thrust, designed	17,500 lb	7950 kg
Thrust, actual	17,160 lb	7780 kg
Time, designed	45 sec	45 sec
Time, actual	40–42 sec	40–42 sec
Take-off acceleration (approx.)	$1\ g_{eff}$	$1\ g_{eff}$
Max. acceleration (approx.)	$3.4\ g_{eff}$	$3.4\ g_{eff}$
Max. velocity	2500 ft/sec	760 m/sec
Exhaust velocity	5800 ft/ sec	1768 m/sec
Max. altitude	60,000 ft	18.3 km
Max. range	16.5 mi	26.4 km

ANTIAIRCRAFT ROCKET TAIFUN (PEENEMÜNDE) (See Fig. 41.)

	U.S.	METRIC (COMPUTED)
Over-all length	75.6 in.	192 cm
Diameter	3.9 in.	10 cm
Diameter over fins	8.6 in.	22 cm
Take-off weight	65 lb	29.5 kg
Dry weight	42.6 lb	19.3 kg
Fuel (Visol or Tonka)	5.7 lb	2.6 kg
Oxidizer (Salbei)	15.9 lb	6.75 kg
Fuel consumption	9.6 lb/sec	4.35 kg/sec
Time	2–3 sec	2–3 sec
Thrust	1320–2200 lb	600–1000 kg
Take-off acceleration	3.1 g	3.1 g
Max. acceleration	4.5 g	4.5 g
Exhaust velocity	3200 ft/sec	975 m/sec
Max. altitude	46,000–52,000 ft	14–15.8 km

ANTIAIRCRAFT MISSILE ENZIAN (HWK, MESSERSCHMITT, HOLZBAU KISSING K-G) (See Fig. 42.)

	U.S.	METRIC (COMPUTED)
Over-all length	11.5 ft	350 cm
Diameter	3 ft	91.5 cm
Span	13.1 ft	400 cm
Take-off weight	4350 lb	1973 kg
Remaining weight with warhead	2090 lb	950 kg
High explosive in warhead	660 lb	300 kg
Take-off rockets, together	705 lb	320 kg
Motor	230 lb	105 kg
Fuel (gasoline)	242 lb	110 kg
Oxidizer	990 lb	450 kg
T-Stoff (H_2O_2)	66 lb	30 kg
Solid fuel in take-off rockets	330 lb	150 kg
Thrust at take-off	12,100 lb	5490 kg
Thrust in flight (mean)	3300 lb	1500 kg
Thrust, take-off rockets, time	6 sec	6 sec
Thrust, missile, time	65–70 sec	65–70 sec
Take-off angle	30°	30°
Velocity at booster-drop	885 ft/sec	270 m/sec
Exhaust velocity	5000 ft/sec	1500 m/sec
Max. altitude	53,000 ft	16,155 m
Range	16 mi	25½ km

ANTIAIRCRAFT MISSILE SCHMETTERLING (BMW AND HENSCHEL) (See Fig. 42.)

	U.S.	METRIC (COMPUTED)
Over-all length	11.8 ft	360 cm
Over-all length, fitted with V-T fuze	14.1 ft	430 cm
Diameter	1.1 ft	33.5 cm
Span	6.6 ft	200 cm
Take-off weight	970 lb	440 kg
Remaining weight with warhead	395 lb	180 kg
High explosive in warhead	51 lb	23 kg
Take-off rockets, together	420 lb	285 kg
Motor	330 lb	150 kg
Fuel (Tonka)	31 lb	14 kg
Oxidizer (Salbei)	122 lb	55 kg
Solid fuel in take-off rockets	165 lb	75 kg
Thrust, take-off	7280 lb	3300 kg
Thrust, in flight	815 lb	370 kg
Velocity at Brennschluss	690 lb	210 m/sec
Time	33–57 sec	33–57 sec
Fuel consumption (highest)	4.8 lb/sec	2.3 kg/sec
Exhaust velocity	5600 ft/ sec	1700 m/sec

ANTIAIRCRAFT MISSILE RHEINTOCHTER (RHEINMETALL-BORSIG) (See Fig. 42.)

	RHEINTOCHTER-I		RHEINTOCHTER-III	
	U.S.	METRIC (COMPUTED)	U.S.	METRIC (COMPUTED)
Length, over-all	20.7 ft	630 cm	20.6 ft	628 cm
Length, without booster	13.1 ft	400 cm	16.4 ft	500 cm
Diameter	1.8 ft	55 cm	1.8 ft	55 cm
Diameter over fins	9 ft	275 cm	10 ft	305 cm
Take-off weight	3860 lb	1750 kg	3300–3400 lb	1500-1540 kg
Dry weight	1650 lb	750 kg	1540 lb	700 kg
Fuel (Visol)	not reported		not reported	
Oxidizer (Salbei)	not reported		not reported	
Booster fuel (solid)	not reported		not reported	
High explosive in warhead	220–330 lb	100–150 kg	220 lb	100 kg
Time	10 sec	10 sec	45 sec	45 sec
Time, booster	0.6 sec	0.6 sec	0.9 sec	0.9 sec
Thrust at take-off	165,000 lb	74,850 kg	62,000 lb	28,125 kg
Thrust in flight	8800 lb	4000 kg	3900 lb	1770 kg
Fuel consumption	not reported		not reported	
Exhaust velocity	5800 ft/sec	1768 m/sec	5800 ft/sec	1768 m/sec
Max. velocity	1180 ft/sec	360 m/sec	980 ft/sec	300 m/sec
Max. altitude	23,000 ft	7 km	not reported	

AIR-TO-AIR MISSILE X-4 (BMW AND RUHRSTAHL A.-G.) See Fig. 43.)

	U.S.	METRIC (COMPUTED)
Over-all length with warhead	6.5 ft	123.5 cm
Length without warhead	4.0 ft	200.0 cm
Diameter	8.6 in.	22.0 cm
Span	2.8 ft	85.0 cm
Weight, over-all	132.0 ft	60.0 kg
Warhead	44.0 ft	20.0 kg
Salbei	14.1 ft	6.4 kg
Tonka-250	3.5 ft	1.6 kg
Max. fuel consumption	1.95 lb/sec	0.88 kg/sec
Thrust (diminishing)	308 → 66 lb	140 → 30 kg
Max. velocity	760 ft/sec	230 m/sec
Time	17 sec	17 sec
Exhaust velocity	5000 ft/sec	1500 m/sec
Range	2 mi	3.2 km

This wire-guided air-to-air missile X-4 must not be confused with the missile Fritz-X of Rheinmetall-Borsig which was a radio-guided, gravity-propelled, air-to-surface (air-to-ship) bomb, similar in principle to the American Razon bomb.

Principal source: PB Reports 23742 and 27795

Characteristics of the Rocket Rheinbote (Addendum to Chapter 8)

DIMENSIONS	U.S.	METRIC (COMPUTED)
Booster, length	6 ft 3 in.	190.0 cm
Booster, diameter	1 ft 9 in.	53.5 cm
First Step, length	11 ft 6 in.	350.0 cm
First Step, diameter	10.6 in.	26.8 cm
Second Step, length	11 ft 6 in.	350.0 cm
Second Step, diameter	10.6 in.	26.8 cm
Third Step, length	13 ft 1 in.	400.0 cm
Third Step, diameter	7.5 in.	19.0 cm
Total length after assembly	37 ft 5 in.	1140.0 cm
WEIGHTS	**LB**	**KG**
Booster, over-all	1529	695
Booster, propellant	539	245
First Step, over-all	935	425
First Step, propellant	308	140
Second Step, over-all	869	395
Second Step, propellant	308	140
Third Step, over-all	352	160
Third Step, propellant	132	60
Warhead, high explosive	88	40
Total, propellant	1287	585
Total take-off weight	3773	1715
Max. range	124¼ mi	220 km

Data from *JBIS*, No. XXIII

Record of V-2 Firings from White Sands Proving Ground and Florida Missile Testing Range (Addendum to Chapter 9)

NO.	DATE	TIME	WEIGHT (LB) EMPTY WITH WARHEAD	WEIGHT (LB) TAKE-OFF	BURNING TIME (SEC)	ALTITUDE (MI)	RANGE (MI)
	1946						
1	March 15	—	—	—	57 *B*	static firing	
2	April 16	2:47 P.M.	8,530	?	19 *R*	5	0
3	May 10	—	8,190	27,850	59 *I*	70	31.0
4	May 29	2:12 P.M.	8,696	28,400	60.2 *I*	70	37.6
5	June 13	4:40 P.M.	9,286	28,396	58.5 *I*	73	40.0
6	June 28	12:25 P.M.	9,807	29,146	66.8 *B*	67	41.0
7	July 9	12:25 P.M.	8,977	27,850	60.6 *T*	83	61.0
8	July 19	12:11 P.M.	9,167	28,840	28.5 *E*	3	0.5
9	July 30	12:36 P.M.	8,562	28,000	68.6 *B*	104	68.0
10	Aug. 15	11:00 A.M.	9,012	28,694	18.5 *R*	2	0.7
11	Aug. 22	10:15 A.M.	9,152	27,220	6.5 *R*	—	0.1
12	Oct. 10	11:02 A.M.	9,164	28,760	67.7 *B*	102	12.0
13	Oct. 24	12:15 P.M.	9,070	29,000	59.8 *B*	65	17.0
14	Nov. 7	1:31 P.M.	8,684	28,000	31.0 *R*	—	5.0
15	Nov. 21	9:55 A.M.	8,885	28,670	62.5 *B*	62	12.6
16	Dec. 5	1:08 P.M.	9,050	28,782	69.0 *B*	104	111.1
17	Dec. 17	10:12 P.M.	8,797	28,952	69.6 *B*	116	21.0
	1947						
18	Jan. 10	2:13 P.M.	9,434	29,100	60.0 *B*	72	25.0
19	Jan. 23	5:22 P.M.	9,140	28,355	59.0 *B*	31	10.3
20	Feb. 20	11:16 A.M.	9,390	28,455	58.0 *B*	68	13.9
21	March 7	11:23 A.M.	9,182	28,847	63.0 *B*	100	35.0
22	April 1	1:10 P.M.	8,800	28,095	57.0 *T*	80	24.0
23	April 8	5:10 P.M.	8,840	27,460	57.0 *T*	64	19.0
24	April 17	4:22 P.M.	9,061	28,726	66.0 *B*	87	45.0
26	May 15	4:08 P.M.	9,827	29,492	63.5 *B*	76	35.0
29	July 10	12:18 P.M.	9,522	29,187	32.0 *R*	10	1.4
30	July 29	5:55 A.M.	8,533	28,198	62.5 *B*	99	1.0
27	Oct. 9	12:15 P.M.	9,107	28,772	62.5 *T*	97	28.0
Spec.	Nov. 20	4:47 P.M.	9,249	28,914	39.0 *M*	13	1.5
28	Dec. 8	2:42 P.M.	9,493	29,158	61.5 *B*	65	28.0
	1948						
34	Jan. 22	1:12 P.M.	9,548	29,000	67.0 *B*	99	48.0
36	Feb. 6	10:17 A.M.	8,789	28,814	65.8 *B*	70	1.4
39	March 19	4:10 P.M.	9,659	29,224	25.0 *B*	3	1.0
25	April 2	6:47 A.M.	9,742	29,100	69.5 *B*	89	48.0
38	April 19	12:54 P.M.	9,169	28,834	57.0 *R*	35	32.0
B-1	May 13	6:43 A.M.	8,573	27,938	64.5 *I*	70	32.0
35	May 27	7:15 A.M.	10,411	30,076	62.4 *T*	87	41.0
37	June 11	3:22 A.M.	10,161	29,826	57.3 *M*	39	17.0
40	July 26	11:03 A.M.	9,942	29,607	60.8 *B*	60	23.0
43	Aug. 5	5:07 A.M.	8,993	28,658	65.5 *B*	104	53.0
B-2	Aug. 19	7:45 A.M.	8,082	27,447	33.8 *M*	8	0.9

NO.	DATE	TIME	WEIGHT (LB) EMPTY WITH WARHEAD	TAKE-OFF	BURNING TIME (SEC)	ALTITUDE (MI)	RANGE (MI)
33	Sept. 2	6:00 P.M.	8,422	28,037	63.0 *T*	94	40.0
B-3	Sept. 30	8:30 A.M.	8,052	27,417	56.5 *R*	93	22.0
B-4	Nov. 1	7:24 A.M.	8,105	27,470	28.5 *E*	3	1.0
44	Nov. 18	3:35 P.M.	8,858	28,523	65.5 *B*	90	29.0
42	Dec. 9	9:08 A.M.	8,234	28,106	60.6 *T*	67	25.0
	1949						
45	Jan. 28	10:20 A.M.	9,537	29,450	56.5 *R*	37	10.5
48	Feb. 17	10:00 A.M.	9,652	29,217	63.5 *B*	79	37.0
B-5	Feb. 24	3:14 P.M.	8,732	28,297	61.0 *I*	63	21.5
41	March 21	11:43 P.M.	9,971	29,636	65.5 *B*	80	32.4
50	April 11	3:05 P.M.	9,530	28,973	62.5 *T*	53	20.0
B-6	April 21	5:17 P.M.	8,651	27,936	48.0 *M*	31	0.4
46	May 5	8:15 A.M.	9,200	28,940	25.6 *M*	5	1.4
47	June 14	3:35 P.M.	10,575	30,171	67.3 *B*	83	37.0
32	Sept. 16	4:19 P.M.	9,996	28,839	24.7 *E*	3	0.5
49	Sept. 29	9:58 A.M.	9,276	28,911	65.5 *B*	94	43.5
56	Nov. 18	9:03 A.M.	10,414	29,315	66.4 *T*	77	32.0
31	Dec. 8	12:15 P.M.	10,211	29,695	65.0 *B*	79	37.8
	1950						
53	Feb. 17	11:00 A.M.	9,900	29,433	65.0 *B*	92	40.5
B-8	July 24	9:29 A.M.	8,575	27,937	— *I*	not fired for altitude	
B-7	July 29	6:25 A.M.	8,665	28,245	— *I*		
51	Aug. 31	10:09 A.M.	10,683	30,183	64.9 *B*	85	36.1
61	Oct. 26	4:02 P.M.	8,807	28,180	49.7 *E*	—	—
	1951						
54	Jan. 18	1:14 P.M.	9,296	29,227	44.0 *B*	1	0.5
57	March 8	8:16 P.M.	10,407	30,067	18.5 *E*	2	0.2
55	June 14	6:48 A.M.	9,192	28,795	0.0 *E*	0	0.0
52	June 28	2:43 P.M.	9,781	29,431	22.0 *E*	4	0.4

The numbers are those of the rockets; they were not all fired in numerical sequence. Numbers preceded by "B-" refer to the Bumper Program. The heaviest rockets listed are those of "Operation Blossom," a separate upper-air research project using modified rockets, which stood about 65 inches taller than normal. Interestingly enough, three out of seven of the "Blossom" V-2's (nos. 32, 52, and 57) had tail explosions some 10 seconds after take-off, for which the modification must have been responsible in some unexplained manner; the only other tail explosion that ever occurred was in one of the Bumper rockets.

The time is the time of the White Sands Proving Ground which is Mountain Standard Time, except for Bumper no. 7 and no. 8, which were fired from Florida where the time is Eastern Standard. The letters following the figure for the burning time indicate how the burning was terminated. *B* means Brennschluss, burning until exhaustion of fuels (no. 39 ran out of H_2O_2 long before the propellants proper were used up); *R* means cut-off by radio (emergency cut-off); *T*, cut-off by time switch; *I*, cut-off by integrator; *E*, explosion. In the case of no. 55 it was not a portion of the propulsion system which exploded, but the high explosives carried for separating the warhead from the body of the rocket on the descending leg of the trajectory. *M* means a malfunction of some kind, for example, the premature closing of a valve.

Source: *Final Report. Project Hermes, V-2 Missile Program.* General Electric, Report No. R52A0510, September 1952.

Record of Viking Firings (Addendum to Chapter 9)

NO.	DATE	WEIGHT (LB) TAKE-OFF	WEIGHT (LB) PAYLOAD	BURNING TIME (SEC)	ALTITUDE (MI)	TIME TO PEAK (SEC)	MAX. VEL. (FT/SEC)
I	May 3, 1949	9,650	464	54.5	50	164	3450
II	Sept. 6, 1949	9,985	412	49.5	32	133	2675
III	Feb. 9, 1950	11,050	528	59.6	50	169	3440
IV	May 11, 1950	11,440	959	74	105	242	5160
V	Nov. 21, 1950	11,390	675	79	108	248	5150
VI	Dec. 11, 1950	10,890	373	70	40	145	4030
VII	Aug. 7, 1951	10,730	394	72	136	266	5865
VIII	June 6, 1952	12,810	none	61	4	50?	—
IX	Dec. 15, 1952	14,615	765	99	135	287	5795
X	May 7, 1954	14,750	830	100	136	290	5720
XI	May 24, 1954	15,005	825	103	158	309	6300
XII	Feb. 4, 1955	14,815	887	102	144	299	5900

All firings except IV from White Sands Proving Ground.

Summary:

I and II, premature cut-off, turbine leaks.
III, cut-off by radio because of excessive recurrent drift.
IV, shipboard firing.
V, thrust 18,800 lb instead of 20,450 lb.
VI, night firing failure of fins.
VII, altitude record for first type of Viking.
VIII, rocket destroyed by breaking loose on static firing.
IX, perfect.
X, motor explosion during static test; rebuit rocket made perfect flight.
XI, altitude record for single-step liquid-fuel rocket without booster.
XII, perfect.

Source: *The Viking Rocket Story* by Milton W. Rosen (New York: Harper, 1955).

Characteristics of the Research Rocket Aerobee (Model AJ 10-25) (Addendum to Chapter 9)

DIMENSIONS	U.S.	METRIC
Over-all length (without booster)	18 ft 11 in.	5.76 m
Length of booster	6 ft 7 in.	2 m
Diameter	15 in.	38 cm
Payload volume	6 ft^3	0.17 m^3

	LB	KG
WEIGHTS		
Payload	120–200	55–90
Empty	300	136
Oxidizer	497	225.4
Fuel	181	82.1
Helium	5	2.27
PERFORMANCE		
Thrust of rocket motor	4000	1814
Thrust of booster	18,000	8165
BURNING TIME		
rocket	34 sec	
booster	2.5 sec	
WITH 120-LB PAYLOAD	U.S.	METRIC
Brennschluss altitude	80,000 ft	24.4 km
Brennschluss velocity	4880 ft/sec	1487 m/sec
Peak altitude	80 mi	128.75 km
WITH 200-LB PAYLOAD		
Brennschluss altitude	74,000 ft	22.5 km
Brennschluss velocity	4300 ft/sec	1310 m/sec
Peak altitude	66 mi	106.2 km

Source: ARS Preprint 345 (1956), lecture on Aerobee-Hi Development by John W. Townsend, Jr., and Robert M. Slavin.

Weights and Performances of Nike-Deacon (DAN) Sounding Rockets (Addendum to Chapter 10)

WEIGHTS	LB	
Booster with charge	1170	
Fins	109	
Adapter	45	
	1324	1324
Deacon with charge	151.5	
Fins, shroud, etc.	25.5	
Nozzle extension	5.0	
Nose cone and instrumentation	34.0	
Total weight Second Stage	216.0	216
Total take-off weight		1540

Note: The weight of the nose cone and the instrumentation for no. 2 was 39 lb, hence the second stage of no. 2 weighed 221 lb and the total take-off weight of no. 2 was 1545 lb.

DAN (continued)

PERFORMANCES:	NO. 1	NO. 2
Angle of elevation	75°	75°
Burning time, first stage	3.5 sec	3.5 sec
Altitude at Brennschluss, first stage	4900 ft	5200 ft
Coasting period	13.7 sec	9.45 sec
V_{max} after firing of Second Stage	5150 ft/sec	5289 ft/sec
Altitude at max. velocity	47,060 ft	39,339 ft
Nose-cone ejection at X+	52 sec	52 sec
Peak altitude	356,000 ft	350,000 ft
Time to peak	161 sec	156 sec
Impact, horizontal distance	340,000 ft	320,000 ft

Note: Impacts were not observed; figures are based on tracking data. Theoretical maximum altitude for vertical shots: 385,000/487,000 ft (smaller figure for 60-lb, larger for 10-lb payload).

Source: NACA, TN 3739. (Report dated December 5, 1955; published, July 1956.)

The Atmosphere from Sea Level to 20 Kilometers, with Speed of Sound Added (Addendum to Chapter 10)

HEIGHT	TEMPERATURE		PRESSURE	WEIGHT	SPEED OF SOUND
KM	C	K	MM OF Hg	KG/M ^{3}	M/SEC
0	+15.0	288.0	760.00	1.2255	341
1	+ 8.5	280.5	674.09	1.1120	338
2	+ 2.0	275.0	596.23	1.0068	334
3	− 4.5	268.5	525.79	0.9094	329
4	−11.0	262.0	462.26	0.8193	326
5	−17.5	255.5	405.09	0.7363	321
6	−24.0	249.0	353.77	0.6598	317
7	−30.5	242.5	307.87	0.5896	313
8	−37.0	236.0	266.89	0.5252	308
9	−43.5	229.5	230.45	0.4664	305
10	−50.0	223.0	198.16	0.4127	301
15	−55.0	218.0	90.65	0.1931	296
20	−55.0	218.0	41.41	0.0883	296

Source: NACA Report No. 218

Some Values from the Grimminger Report (Addendum to Chapter 10)

	TEMPERATURE		PARTICLES	MEAN FREE	SPEED OF SOUND
KM	C	K	PER CU CM	PATH (M)	(M/SEC)
137.1	+171.1	444.1	$3.44 \cdot 10^{12}$	0.727	470
152.4	+232.5	505.5	$1.25 \cdot 10^{12}$	2.000	502
182.9	+355.3	628.3	$2.30 \cdot 10^{11}$	10.90	no value
228.6	+539.4	812.4	$3.19 \cdot 10^{10}$	78.60	no value
259.1	+662.2	935.2	$1.09 \cdot 10^{10}$	229.00	no value
274.3	+723.6	996.6	$6.75 \cdot 10^{9}$	371.00	no value
300.0	+827.0	1100.0	$3.21 \cdot 10^{9}$	779.00	no value

Source Grimminger Report, Table 14.

The Atmosphere from 20 to 120 Kilometers (Addendum to Chapter 10)

HEIGHT	TEMPERATURE		PRESSURE	WEIGHT	SPEED OF SOUND	MEAN FREE PATH (MM)	
KM	C	K	KG/M²	KG/M³	M/SEC	NACA	GRIMMINGER
20	− 55.0	218.0	568.4	$8851 \cdot 10^{-5}$	296.0	0.0011	0.001
25	− 55.0	218.0	261.0	$4059 \cdot 10^{-5}$	296.0	0.0022	0.003
30	− 55.0	218.0	120.1	$1864 \cdot 10^{-5}$	296.0	0.0047	0.006
35	− 33.0	240.0	56.48	$795 \cdot 10^{-5}$	310.6	0.011	0.014
40	+ 3.7	276.7	29.35	$358 \cdot 10^{-5}$	333.5	0.024	0.03
45	+ 40.3	313.3	16.56	$178 \cdot 10^{-5}$	354.9	0.05	0.07
50	+ 87.0	350.0	9.970	$96 \cdot 10^{-5}$	375.1	0.09	0.11
55	+ 87.0	350.0	6.167	$59 \cdot 10^{-5}$	375.1	0.15	0.18
60	+ 87.0	350.0	3.820	$36.6 \cdot 10^{-5}$	375.1	0.24	0.29
65	+ 46.4	319.4	2.315	$24.3 \cdot 10^{-5}$	358.3	0.36	0.46
70	+ 15.9	288.9	1.335	$15.5 \cdot 10^{-5}$	340.7	0.56	0.68
75	− 14.7	258.3	.725	$9.4 \cdot 10^{-5}$	322.2	0.92	1.06
80	− 33.0	240.0	.3675	$5.1 \cdot 10^{-5}$	310.6	1.68	2.2
85	− 25.7	247.3	.1877	$2.4 \cdot 10^{-5}$	325.2	3.39	4.2
90	− 7.5	265.5	.1029	$1.2 \cdot 10^{-5}$	347.1	6.64	8.0
95	+ 10.8	283.8	.0602	$.61 \cdot 10^{-5}$	369.2	12.1	14.5
100	+ 29.0	302.0	.0373	$.34 \cdot 10^{-5}$	391.5	20.7	29.6
105	+ 47.3	320.3	.02406	$.21 \cdot 10^{-5}$	403.2	34.1	48.4
110	+ 65.5	338.5	.01589	$.13 \cdot 10^{-5}$	414.5	54.4	76.6
115	+ 83.3	356.8	.01074	$.08 \cdot 10^{-5}$	425.5	84.8	117.5
120	+102.0	375.0	.00740	$.05 \cdot 10^{-5}$	436.3	129.0	192.6

Source: NACA Technical Note no. 1200

Some Values from the ARDC Model Atmosphere (1956) (Addendum to Chapter 10)

KM	TEMPERATURE (K)	M (GR/MOL)	P (MILLIBARS)
20.06	216.66	28.966	5.4748×10^{1}
25.10	216.66	same	2.4886×10^{1}
47.35	282.66	same	1.2044×10^{0}
75.89	196.86	same	2.4521×10^{-2}
91.29	196.86	same	1.8154×10^{-3}
128.55	(278.88)	(25.02)	1.4510×10^{-5}
179.95	(686.13)	(24.45)	6.1895×10^{-7}
314.86	(1024.67)	(19.30)	1.4473×10^{-8}

The temperature is the real kinetic temperature in degrees Kelvin; M is molecular weight in grams per mole; P is pressure. Figures in parenthesis are subject to minor revision.

Source: Report of the Geophysics Research Directorate, Air Force Cambridge Research Center, ARDC (September 1956).

Professor Oberth's Table of Mass-Ratios (Addendum to Chapter 10)

$M_0/M_1 = e^{v/c}$. Values for v and c are in meters per sec.

c \ v	500	1000	2000	3000	4000	5000
1000	1.64	2.72	7.39	20.0	54.5	148
2000	1.29	1.64	2.72	4.48	7.39	12.2
3000	1.18	1.39	1.94	2.72	3.78	5.29
4000	1.13	1.29	1.64	2.11	2.72	3.49
5000	1.10	1.22	1.49	1.82	2.22	2.72

c \ v	6000	7000	8000	9000	10,000	11,000
1000	405	1089	2987	8060	22,070	60,000
2000	20.0	33.0	54.5	89.6	148.7	243.5
3000	7.39	10.25	14.35	20.0	27.95	39.0
4000	4.48	5.76	7.39	9.5	12.20	15.75
5000	3.32	4.06	4.95	6.06	7.39	9.02

c \ v	12,000	13,000	14,000	15,000
1000	163,100	444,000	1,200,000	3,290,000
2000	402	662	1091	1805
3000	54.6	76.1	106.3	148.7
4000	20.0	25.8	33.2	42.7
5000	11.0	13.47	16.42	20.0

Source: *Die Möglichkeit der Weltraumfahrt.*

"The Problem of Synergy" (Addendum to Chapter 11)

At any one moment while a rocket is burning, part of the total energy of the fuels is transferred to the rocket and part to the exhaust gases. The problem of transferring as much energy as possible to the rocket and as little as possible to the exhaust was named "the problem of synergy" by Professor Oberth. This concept, therefore, refers to the most advantageous movement of the rocket, as distinct from problems of design and construction. One might say that the synergic problem is the equivalent of efficient driving, the problem of getting the most out of the vehicle on hand.

Oberth expressed this problem in the formula

$$\frac{dA}{dm} = c \cdot \cos a \cdot v$$

which gives the ratio between the increase of energy (of the rocket) dA and the mass expenditure dm. As usual c is the exhaust velocity and v the rocket velocity, while a is the angle between the direction of rocket movement and direction of the thrust P.

Reasoning from this formula, Oberth developed the following three points:

1. c must be as high as possible—but the formula shows that the maximum value of c is only part of what has to be done for optimum fuel economy.

2. since $\cos a$ becomes largest when $a = 0$, the direction of P and of the rocket movement should coincide if at all possible.

3. the factor v indicates that the faster the rocket is moving when the additional burning takes place, the more will dA increase.

Hence the following demands should be made.

A. The rocket should ascend with high acceleration (this, of course, is valid only *after* air resistance has been eliminated by altitude).

B. The rocket should not ascend vertically. Theoretically a horizontal "ascent" would be best (but is impossible because of air resistance).

C. The horizontal component of the ascent curve should point east in order to utilize the earth's rotation.

D. *Zusammenlegung der Stösse* (condensation of power application) is needed to comply with point 3 because it will result in generally higher rocket velocities while burning than would occur if the applications of power were separated by considerable intervals of time.

The compromise between this set of demands and the existence of air resistance is the synergic curve illustrated and discussed in Chapter 11.

Elliptical Orbits around Earth (Addendum to Chapter 11)

INITIAL VELOCITY AT PERIGEE PER SEC		DISTANCE OF APOGEE FROM SURFACE		VELOCITY AT APOGEE PER SEC		ORBITAL PERIOD		
KM	MI	KM	MI	KM	MI	D	H	M
8.00	4.971	287.0	178.3	7.656	4.757		1	27.2
8.20	5.095	962.5	598.0	7.076	4.397		1	34.6
8.40	5.219	1,849.7	1,149.3	6.512	4.046		1	43.4
8.60	5.344	2,819.2	1,745.0	5.965	3.706		1	53.8
8.80	5.468	3,948.2	2,453.3	5.435	3.377		2	6.5
9.00	5.592	5,294.1	3,289.6	4.919	3.056		2	22.0
9.20	5.717	6,914.2	4,296.2	4.416	2.744		2	41.6
9.40	5.841	8,891.4	5,524.6	3.926	2.440		3	6.8
9.60	5.965	11,372.7	7,066.6	3.449	2.143		3	40.7
9.80	6.089	14,581.0	9,060.2	2.987	1.856		4	24.8
10.00	6.214	18,816.2	11,691.8	2.532	1.573		5	28.6
10.20	6.338	22,330.8	13,875.7	2.078	1.291		7	7.0
10.40	6.462	34,386.0	21,366.4	1.627	1.001		9	59.4
10.50	6.524	40,465.6	25,144.1	1.430	.889		11	59
10.60	6.586	49,164.7	30,549.4	1.217	.756		15	2
10.70	6.649	61,411.4	38,159.2	1.007	.626		19	43
10.80	6.711	79,902.4	49,648.7	.798	.496	1	3	32
10.90	6.773	111,067.1	69,013.7	.592	.367	1	18	32
11.00	6.835	174,672.5	108,536.4	.388	.241	3	7	12
11.02	6.847	196,263.4	121,952.1	.347	.216	3	21	36
11.03	6.854	209,084.1	129,918.5	.327	.203	4	6	0
11.04	6.860	223,562.9	138,915.2	.306	.190	4	16	6
11.05	6.866	240,083.2	149,180.5	.286	.178	5	4	6
11.06	6.872	259,090.9	160,990.7	.266	.165	5	18	18
11.07	6.878	281,223.6	174,743.9	.246	.153	6	11	30
11.08	6.885	307,310.3	190,953.3	.225	.140	7	8	42
11.09	6.891	338,501.7	210,334.7	.205	.127	8	11	12
11.10	6.897	376,452.0	233,916.0	.185	.112	9	21	0
11.11	6.903	423,717.1	263,285.1	.165	.103	11	17	30
11.12	6.909	483,993.0	300,738.7	.145	.065	14	5	48

Figures used in the calculation: Earth's radius R = 6,378.4 km = 3,963.4 mi. Circular velocity v_c = 7.914 km per sec = 4.9734 mi per sec.

Parabolic velocity (escape velocity) v_p = 11.172 km per sec = 6.942 mi per sec.

The perigee for these orbits would lie at about 170 mi; the assumption is that the orbiting body is in a circular orbit at that altitude and then accelerated to the velocity given in the first column, taking on the elliptical orbits tabulated.

Adapted from a table in *Die Weltraumfahrt hat begonnen* by Professor Dr. Karl Schütte, published by Verlag Herder, Freiburg im Breisgau, 1958.

The Vanguard Program (Addendum to Chapter 11)

DESIGNATION	DATE	RESULTS AND REMARKS
TV-0	Dec. 8, 1956	Successful shot for testing telemetry, etc. This was actually Viking No. 13.
TV-1	May 1, 1957	Successful shot; Viking No. 14 carrying third stage (no second stage) for testing.
TV-2	Oct. 23, 1957	Successful shot; Vanguard first stage with dummies for second and third stages.
TV-3	Dec. 6, 1957	First test of complete vehicle; lost thrust after 2 seconds and burned on pad.
TV-3 back-up	Feb. 5, 1958	Rocket broke up in mid-air at 20,000-ft altitude at T + 57 seconds.
TV-4	March 17, 1958	Put "Vanguard I," the 3.25-lb test satellite, into orbit.
TV-5	Apr. 28, 1958	First to carry 21.5-lb satellite. Third stage failed to ignite. Rocket coasted to 358-mi altitude, then fell into Atlantic.
SLV-1	May 27, 1958	Premature cut-off of second stage. Rocket rose to 1850 mi before falling.
SLV-2	June 26, 1958	Premature cut-off of second stage. No orbit.
SLV-3	Sept. 26, 1958	Failure because second stage did not develop full thrust. Seems to have nearly completed one circuit around earth.
SLV-4	Feb. 17, 1959	Put "Vanguard II" (21.5-lb satellite) into orbit.
SLV-5	Apr. 13, 1959	Second stage faulty; vehicle tumbled. No orbit.
SLV-6	June 22, 1959	Malfunction in second stage. Rocket ruptured at T + 40 seconds.
SLV-7	Sept. 18, 1959	Put "Vanguard III" (50-lb satellite) into orbit. The rocket was the "back-up" for TV-4 with more powerful third stage.

TV stands for Test Vehicle; SLV stands for Satellite Launching Vehicle. The Vanguard rocket in the Smithsonian Institution is the shell of the "back-up" for TV-2.

Solid-Fuel "Off-the-Shelf" Hardware and Research Rockets (Addendum to Chapter 11)

NAME	LENGTH (IN.)	DIAMETER (IN.)	WEIGHT, LOADED (LB)	BURNING TIME (SEC)	THRUST, SEA LEVEL (LB)
Algol	358	40	22,460	36	104,500
Altair	58	18	514	38	3,060
Antares	115	30	2,300	37	14,500
Nike Booster	150	16	1,170	3.5	50,000
Castor	244	31	8,870	27	55,000
Recruit	107	9.5	350	1.5	35,000
Cajun I	108	6.75	172	2.8	8,100
Lance	185	15.1	1,682	ca. 7	38,800
Meteor	57.4	18	431	37	2,600
Asp	107	6.5	192	ca. 5	6,000

These solid-fuel units were used in various combinations, resulting in rockets with such names as Javelin, Strongarm, and Journeyman. Strongarm had an Honest John as the first stage, two Nike boosters as the second and third stages, a Recruit as the fourth stage and a T-55 JATO unit as the top stage. Journeyman's first stage consisted of a three-rocket cluster (one Sergeant and two Recruits); a Lance rocket was the second stage, another Lance the third stage, and an Altair the fourth stage. This was not the most efficient method for carrying an 85-lb instrument package into the Van Allen Belt, but it was the simplest, since these rockets were available. The Navy's Iris rocket has a length of 11 ft 4 in. to which a nose cone with instrumentation is added so that the total length is about 17–18 ft. Diameter of this solid-fuel rocket is 12 in. It can carry a 100-lb payload to 200 mi. Iris replaced the Aerobee-Hi for shipboard use. While Iris was under development, a solid-fuel rocket named Arcon was used. Its diameter was only 6 in., total length 11 ft 2 in. The take-off weight was 250 lb; the payload that could be carried weighed 40 lb.

Delta Launches from 1960 to the End of 1966 (Addendum to Chapter 12)

DELTA NO.	DATE	SATELLITE	DELTA NO.	DATE	SATELLITE
1	May 13, 1960	Echo-A (F)	22	Dec. 21, 1963	Tiros-VIII*
2	Aug. 12, 1960	Echo-I*	23	Jan. 21, 1964	Relay-II*
3	Nov. 23, 1960	Tiros-II*	24	Mar. 19, 1964	Beacon-Explorer (F)
4	Mar. 25, 1961	Explorer-X*	25	Aug. 19, 1964	Syncom-III*
5	July 12, 1961	Tiros-III*	26	Oct. 3, 1964	Explorer-XXI*
6	Aug. 16, 1961	Explorer-XII*	27	Dec. 21, 1964	Explorer-XXVI*
7	Feb. 8, 1962	Tiros-IV*	28	Jan. 22, 1965	Tiros-IX*
8	Mar. 7, 1962	OSO-I*	29	Feb. 3, 1965	OSO-II*
9	Apr. 26, 1962	Ariel-I*	30	Apr. 6, 1965	Early Bird*
10	June 19, 1962	Tiros-V*	31	May 29, 1965	Explorer-XXVIII*
11	July 10, 1962	Telstar-I*	32	July 1, 1965	Tiros-X*
12	Sept. 18, 1962	Tiros-VI*	33	Aug. 25, 1965	OSO-C (F)
13	Oct. 2, 1962	Explorer-XIV*	34	Nov. 6, 1965	Geos-I*
14	Oct. 27, 1962	Explorer-XV*	35	Dec. 15, 1965	Pioneer-VI**
15	Dec. 13, 1962	Relay-I*	36	Feb. 3, 1966	ESSA-I*
16	Feb. 14, 1963	Syncom-I*	37	Feb. 28, 1966	ESSA-II*
17	Apr. 2, 1963	Explorer-XVII*	38	May 25, 1966	Explorer-XXXII*
18	May 7, 1963	Telstar-II*	39	July 1, 1966	Explorer-XXXIII*
19	June 19, 1963	Tiros-VII*	40	Aug. 17, 1966	Pioneer-VII**
20	July 26, 1963	Syncom-II*	41	Oct. 2, 1966	ESSA-III*
21	Nov. 26, 1963	Explorer-XVIII*	42	Oct. 26, 1966	Intelsat-II*

Note: F indicates that the launch was a failure. One asterisk (*) means in orbit; two asterisks (**) mean in circumsolar orbit.

Number of Meteorites Striking the Earth in One 24-Hour Period (Addendum to Chapter 13)

VISUAL MAGNITUDE	NUMBER IN EACH MAGNITUDE	TOTAL NUMBER MAGNITUDE GIVEN AND LARGER	INDIVIDUAL MASS (MG)	DIAMETER FOR STONY METEORITES (MM)
−3	28,000	28,000	4000	13.0
−2	71,000	99,000	1600	9.66
−1	180,000	280,000	630	7.07
0	450,000	730,000	250	5.20
1	1,100,000	1,900,000	100	3.83
2	2,800,000	4,700,000	40	2.81
3	7,100,000	12,000,000	16	2.08
4	18,000,000	30,000,000	6.3	1.52
5	45,000,000	75,000,000	2.5	1.12
6	110,000,000	120,000,000	1.0	0.82
7	280,000,000	470,000,000	0.4	0.60
8	710,000,000	1,200,000,000	0.16	0.45
9	1,800,000,000	3,000,000,000	0.063	0.33
10	4,500,000,000	7,500,000,000	0.025	0.24
15	$45 \cdot 10^{10}$	$75 \cdot 10^{10}$	$2.5 \cdot 10^{-4}$	0.05
20	$45 \cdot 10^{12}$	$75 \cdot 10^{12}$	$2.5 \cdot 10^{-6}$	0.01
25	$45 \cdot 10^{14}$	$75 \cdot 10^{14}$	$2.5 \cdot 10^{-8}$	0.0024
30	$45 \cdot 10^{16}$	$75 \cdot 10^{16}$	$2.5 \cdot 10^{-11}$	0.0005

Excerpted from Grimminger, *Journal of Applied Physics,* October 1948, and Fletcher G. Watson, *Between the Planets,* Philadelphia: Blakiston, 1941.

Probable Number of Meteorite Hits That Would Be Sustained by a Rocket in the Vicinity of the Orbit of Earth (Addendum to Chapter 13)

Figures are for an exposed area of 1000 sq ft.

	PROBABLE NUMBER OF HITS OF GIVEN MAGNITUDE ONLY		PROBABLE NUMBER OF HITS OF GIVEN MAGNITUDE AND LARGER	
VISUAL MAGNITUDE	NUMBER PER HOUR	TIME IN HOURS BETWEEN HITS	NUMBER PER HOUR	TIME IN HOURS BETWEEN HITS
− 3	$1.84 \cdot 10^{-10}$	$5.44 \cdot 10^{9}$	$1.86 \cdot 10^{-10}$	$5.36 \cdot 10^{9}$
0	$2.95 \cdot 10^{-9}$	$3.38 \cdot 10^{8}$	$4.78 \cdot 10^{-9}$	$2.09 \cdot 10^{8}$
5	$2.95 \cdot 10^{-7}$	$3.38 \cdot 10^{6}$	$4.90 \cdot 10^{-7}$	$2.04 \cdot 10^{6}$
8	$4.66 \cdot 10^{-6}$	$2.14 \cdot 10^{5}$	$7.75 \cdot 10^{-6}$	$1.29 \cdot 10^{5}$
10	$2.95 \cdot 10^{-5}$	33,800	$4.90 \cdot 10^{-5}$	20,400
12	$1.84 \cdot 10^{-4}$	5440	$3.10 \cdot 10^{-4}$	3230
14	$1.18 \cdot 10^{-3}$	846	$1.96 \cdot 10^{-3}$	511
16	$7.22 \cdot 10^{-3}$	138	$1.23 \cdot 10^{-2}$	81
18	$4.66 \cdot 10^{-2}$	21.4	$7.75 \cdot 10^{-2}$	13
20	$2.95 \cdot 10^{-1}$	3.4	$4.90 \cdot 10^{-1}$	2
22	1.8	0.5	3.1	0.3
24	11.8	$8.46 \cdot 10^{-1}$	19.6	$5.11 \cdot 10^{-2}$
26	72.2	$1.38 \cdot 10^{-2}$	123.0	$8.10 \cdot 10^{-3}$
28	466	$2.14 \cdot 10^{-3}$	775	$1.29 \cdot 10^{-3}$
30	2950	$3.39 \cdot 10^{-4}$	4900	$2.04 \cdot 10^{-4}$

Excerpted from Grimminger, *loc. cit.*

Gravitational Characteristics of Planets (Addenum to Chapter 13)

	MASS	*g* AT SURFACE	ESCAPE VELOCITY		ORBITAL VELOCITY	
PLANET	EARTH = 1	EARTH = 1	MI/SEC	KM/SEC	MI/SEC	KM/SEC
Mercury	0.04	0.27	2.2	4.3	29.7	47.7
Venus	0.81	0.85	6.3	10.3	21.7	35.1
Earth	1.00	1.00	7.0	11.2	18.5	29.7
Mars	0.11	0.38	3.1	5.04	15.0	24.1
Jupiter	317.0	2.64	37.0	59.5	8.1	13.0
Saturn	95.0	1.17	22.0	35.4	6.0	9.6
Uranus	14.7	0.92	13.0	21.6	4.2	6.8
Neptune	17.2	1.12	14.0	22.8	3.4	5.4
Moon	0.012	0.16	1.47	2.37	0.64	1.03

United States Satellite Launch Vehicles (Addendum to Chapter 14)

NAME	TOTAL HEIGHT (FT)	TAKE-OFF WEIGHT (LB)	TAKE-OFF THRUST (LB)	CAPABILITY
Scout[1]	72	38,500	104,500	240 lb (345 mi orbit)
Thor-Agena D	76.3	(classified)	170,000	1,150 lb (115 mi orbit)
Delta[2]	90	114,200	170,000	550 lb (575 mi orbit)
Atlas-Agena D	104	(classified)	368,000	5,000+ lb (115 mi orbit)
Atlas-Centaur	113	300,000	368,000	8,500 lb (115 mi orbit)
Titan II	109	300,000	430,000	7,000 lb (115 mi orbit)
Titan III-C	125+	ca. 1,400,000	ca. 2,000,000[3]	20,000 lb (115 mi orbit)
Saturn I[4]	190	1,165,000	1,500,000	22,000 lb (115 mi orbit)
Saturn V	363	6,100,000	7,500,000	240,000 lb (115 mi orbit)

[1] The stages of the Scout are: first stage Algol IIB, second stage Castor, third stage Antares IIA, fourth stage Altair.
[2] The "Thrust-Augmented Delta" (or TAD) has three strap-on Castor rockets for take-off help. This adds 27,400 lb to the take-off weight but increases the take-off thrust to 330,000 lb, nearly doubling the satellite weight that can be put into orbit.
[3] This figure applies to the two solid-fuel boosters for take-off.
[4] Figures are for Saturn I with an S-IV as a second stage and payload nose cone.

The first version of the solid-fuel intercontinental missile Minuteman is being replaced by later versions which will make several dozen Minuteman missiles available for space launches.

The Fischer Hoax (Addendum to Appendix 1)

Discounting the mandarin Wan-Hoo, whose historicity cannot be established, the first man to fly with rocket power was Friedrich Stamer, on June 11, 1928; Fritz von Opel followed on September 30, 1929. The next was Ettore Cattaneo of Milan in 1931. Then there followed an absolutely outrageous hoax which originated in the London weekly *Sunday Referee,* with a report that a German named Fischer had made a rocket flight to an altitude of 6 miles on October 29, 1933. The story, as I know to my sorrow because of the mail I received, was reprinted in abridged versions all over the planet. The original, slightly condensed, read as follows:

> A sensational secret demonstration of the practicality of the rocket principle applied to flight was made here [Rügen] last Sunday, when Herr Otto Fischer was shot 6 miles into the air within a 24-foot steel rocket and returned to earth safe and sound, though shaken. The pilot who risked his life in this experiment is the brother of the designer and constructor of the rocket, Herr Bruno Fischer. Owing to the disastrous results of a similar experiment made on Rügen in the spring of last year, when the original inventor was killed, the demonstration was made under cover of absolute secrecy, under the auspices of the Reichswehr, the German War Ministry. The inhabitants of the island knew nothing of the proposed experiment, and no members of the press were called to witness it. For some months the two brothers [Otto and Bruno Fischer] had been working day and night in Barmbeck, a small village near Hamburg, to complete the rocket, Bruno Fischer having been an assistant in the building of the first projectile.

When the projectile was completed it was transported to Rügen with great secrecy. On Sunday morning, at six o'clock, Otto Fischer shook hands with his brother and the small group of Reichswehr officials present to witness the experiment, and crawled into the rocket through the small steel door.

Bruno Fischer and the three officials then retired to a small hollow in the ground about 200 yards away and Fischer closed the switch that sent the rocket on its journey. There was a blinding flash and a deafening explosion and the slim torpedo-shaped body was gone from the steel framework in which it had rested.

A few minutes later it came into sight again, floating nose upward from a large parachute that had automatically been released when it had begun to descend. As it drifted nearer, the steel fins on the outside of the body could be seen moving as its pilot manipulated the rocket so that it would land on the island. A few seconds later it came to rest on the sands a few yards away and Fischer crawled through the door of the rocket, white and shaken, but smiling triumphantly. The journey through space had lasted 10 minutes and 26 seconds.

"It was a tremendous sensation," he said to the men who had rushed forward to congratulate him. "When the rocket left the ground I was conscious of a deafening roar, and an unbearable weight seemed to be crushing me against the floor of the rocket. Then I lost consciousness for a moment, due to the tremendous acceleration which drained the blood from my head. When I came to my senses and looked at the altimeter before my face, it flickered at 32,000 feet—a fraction over 6 miles—and then began to drop rapidly. I had completed my climb and was descending. Peering through the little glass window in the side of the compartment I could just see the tip of the parachute billowing above me. The next thing that occupied my attention was the tremendous heat of the asbestos floor on which I was standing. The reason was that the rocket had merely been propelled about 200 yards by the initial explosion, and had been driven the remainder of the distance by the rockets in its tail. . . ."

Nowadays virtually everybody would know that this could not be a truthful report, since everybody has seen on television the take-off of large rockets. Of course, there is no "initial explosion," and large rockets do not "disappear in a flash." The crowning piece of ignorance was, of course, the description of the moving steel vanes for steering the rocket to a certain spot. The vanes could move as much as they wanted; that would not noticeably influence the drift of a load suspended from a parachute.

The editor of the *Sunday Referee,* when questioned by letter about these facts, replied that his policy of editorial secrecy prevented him from disclosing his source of information and passed lightly over the technological absurdities by saying that his reporter, not being a trained engineer, might have made a few minor mistakes. The point was, of course, that the mistakes were not minor but fundamental; it was as if he had written that the new Boeing passenger liner took off by flapping its huge wings.

Since so many people wrote asking for more information, I checked the membership list of the VfR; though Fischer is a fairly common German name, there were no Fischers among our members; the only thing that was true was that there actually is an island named Rügen.

The anonymous reporter of the *Sunday Referee* was bested by another writer —or was he the same one?—who signed himself as W. J. Makin and published an article entitled "Space Explorers" in the May 1935 issue of *Nash's Pall Mall Maga-*

zine in London. In that article "Mr. Makin" explained that he had heard that there were some doubts about the secret Rügen ascent and that he set out to investigate it:

> I decided to begin my investigations by flying to the first most elaborately equipped rocket aerodrome, the Raketenflugplatz. A young engineer led me to the records chamber, where not only the work of the Raketenflugplatz is carefully recorded, but scientific results from rocket aerodromes in other parts of the world are collected. [There were no others.—*W. L.*]
>
> But even as I was combing these records, a heavily built man with a charming smile and a curiously soft voice was introduced to me.
>
> "Herr Otto Fischer."
>
> Even as I shook hands with him I realized that I was meeting the one man who had traveled through space inside a rocket and lived to tell the tale—the first passenger to enclose himself in a steel rocket of some 24 feet, which was shot 6 miles into the air. The rocket was designed and constructed by his brother, Herr Bruno Fischer. In great secrecy it was transported to the island of Rügen. . . .

"Mr. Makin"—I understand that this was a pen name and that the English journalist hiding behind it was killed during the Second World War—was careless, to say the least. At the date he gave for his visit to the Raketenflugplatz, it was no longer active.

But this hoax, although I exposed it wherever I could, is apparently not quite dead even now; only two years or so ago I received an inquiry from England asking me whether I "still maintained that the story about the brothers Fischer was a hoax in view of what has been learned about the German experiments at Peenemünde which is quite close to Rügen."

Characteristics of Interceptors Me-163B and Me-163C (Komet) (Addendum to Appendix 1)

Me-163B

DIMENSIONS	U.S.	METRIC
Span	30.5 ft	930 cm
Length	18.7 ft	570 cm
Height	8.2 ft	250 cm
Wing, area	210 sq ft	19.6 m^2
Wing, loading, take-off	43 lb/sq ft	209 kg/m^2
Wing, loading, landing	22 lb/sq ft	107 kg/m^2

WEIGHTS	LB	KG
Fuselage	611.5	278
Jettisonable carriage	176	80
Wings	867	394
Tail	48.5	22
"Ski" and tail wheel	176	80
Controls	127.5	58
HWK R II–211 motor	365	166
Fuel tanks, etc.	447	203
Electrical and hydraulic systems	321	146
Radio, etc.	130	59
Armor	365	166
Guns (2)	275	125
Total weight, empty	3,909.5	1,777
Ammunition	121	55
T-Stoff	3,410	1,550
C-Stoff	1,029.5	468
Total take-off weight	8,470	3,850

PERFORMANCE	U.S.	METRIC
Speed, max.	560 mi/hr	900 km/hr
Speed, landing	99.4 mi/hr	160 km/hr

CLIMB TO			CLIMB TO		
KM	MI	MIN	KM	MI	MIN
2	1.24	1.48	8	4.97	2.84
4	2.48	2.02	10	6.21	3.19
6	3.72	2.27	12	7.45	3.45

Me-163C

Developed from the Me-163B. Generally enlarged version, with a span of 32.2 ft, a length of 23.1 ft, and a take-off weight of 11,280 lb. Max. speed 590 mi per hr, ceiling 52,500 ft. Mounted two MK-108 30-mm automatic cannon with 60 rounds per gun.

Characteristics of the "Piloted Missile" Natter (Addendum to Appendix 1)

Designation: Bachem 8–349.

	U.S. (COMPUTED)	METRIC
Total take-off weight with boosters	4840 lb	2200 kg
Take-off weight without boosters	3740 lb	1700 kg
Fuel weight	1430 lb	650 kg
Span, wings *	11.8 ft	360 cm
Span, tail (horizontal)	8.2 ft	250 cm
Area, wings	38.7 sq ft	3.6 m²
Area, tail (horizontal)	27 sq ft	2.5 m²
Design speed, max. (sea level)	497 mi/hr	800 km/hr
Radius of action (at 12 km)	25 mi	40 km
Armament: 24 73-mm rockets, each	5.7 lb	2.6 kg
Charge (explosive) per rocket	14 oz	0.4 kg

* Drawing in report gives span as 350 cm.

Source: Combined Intelligence Sub-Committee Report (H. M. Stationery Office, London, No. XXX-107)

Characteristics of Typical American Solid-Fuel JATO Unit (Addendum to Appendix 1)

Manufactured by Aerojet, Inc., Azusa, Calif. Model: 14AS-1000D-4 and D-5, C.A.A. Engine Type Certificate No. 249.

Ignition: electric, hot-wire type, requiring 12 v, 25 amp for 0.110 sec, or 24 v, 50 amp for 0.032 sec. Full thrust from 0.2 to 0.4 sec after closing of circuit.

May be jettisoned; can be reloaded and fired again after servicing.

	U.S.	METRIC
Length	35.4 in.	89.9 cm
Diameter	10.25 in.	26.1 cm
Frontal area	0.57 sq ft	0.05 m²
Weight, loaded	200 lb	91 kg
Weight, empty	120 lb	54 kg
Weight, propellant (ALT-161)	80 lb	37 kg
Exhaust velocity	5635 ft/sec	1717 m/sec

Rating (normal)	1000 lb/t (455 kg/t), any altitude, 14 sec	
Rating (equivalent hp)	350 hp/100 mi/hr (160 km/hr), 14 sec	
Storage temperature	0–130 F	−18–54 C

Source: Manufacturer's Information Sheet, dated February 2, 1949

U.S. Missile Arsenal, 1966

BATTLEFIELD WEAPONS

Honest John. Unguided solid-fuel rocket, two versions. Range of type M-31 is 12 mi, of type M-50 20 mi. Can carry nuclear warheads. Launch weight 5900 lb. (See Fig. 34.)
Little John. Unguided solid-fuel rocket, launch weight 800 lb, range 10 mi.
Lacrosse. Solid-fuel missile with cruciform wings. Warhead high-explosive or nuclear, command guidance, range up to 30 mi, launch weight 2300 lb, length 19 ft, diameter 20½ in., wing span 108 in. *Honest John, Little John,* and *Lacrosse* will be replaced by *Lance,* a missile using prepackaged liquid fuels, with a 42-mi range.
Davy Crockett. Solid-fuel missile carrying small atomic warhead, launched like Bazooka, for battlefield use; performance classified.
SS-10 and SS-11. Anti-tank weapons, developed by Nord Aviation of France. Wire-guided solid-fuel missiles with high-explosive warheads; four large fins.

	SS-10	SS-11
Launch weight	33 lb	63 lb
Length	34 in.	46 in.
Diameter	6 in.	6 in.
Fin span	30 in.	20 in.

Entac. Replacement of SS-10 and SS-11, developed by French army. Solid-fuel; warhead is shaped charge. Wire-guided, weight 40 lb in launch box.
MAW (Medium Anti-Tank/Assault Weapon). Solid-fuel rocket, wire-guided, shoulder-fired; launch weight 27 lb, range up to 1500 yd.

SURFACE TO UNDERWATER

Alfa. Solid-fuel rocket carrying depth charge; launch weight 500 lb, range up to 1000 yd.
Asroc. Solid-fuel. Launch weight 1000 lb, range up to 8 mi. Carries torpedo.

AIR-TO-AIR WEAPONS

Folding Fin Aircraft Rocket. Single-stage, solid-fuel, 48 in. long, diameter 2.75 in., unguided. Launch weight 18 lb. Has several types of warheads. Also used against ground targets.
Sidewinder. Solid-fuel rocket with high-explosive warhead and infrared homing device. Launch weight 155 lb, length, approx. 9 ft, diameter 5 in. *Chaparral,* a somewhat modified *Sidewinder,* is used from the ground as an antiaircraft rocket.
Genie. Solid-fuel, unguided rocket with proximity fuze. Nuclear warhead. Launch weight 800 lb, range approx. 7 mi.
Phoenix. Liquid-fuel long-range missile, range over 50 mi, launch weight 950 lb, all other information classified.
Sparrow-III. Solid-fuel missile with high-explosive warhead, electronically controlled homing. Launch weight 350 lb, length approx. 12 ft, diameter 8 in.
Falcon. Name of a missile family, all air-to-air, with various kinds of warheads and guidance systems. Launch weights from over 100 to over 200 lb, ranges around 6 mi.

AIR-TO-GROUND WEAPONS

Bullpup. Bullpup A is a solid-fuel rocket with a launch weight of 570 lb and a range of 3 to 6 mi. Length 11 ft, diameter 1 ft, warhead weight 250 lb. Bullpups B and C are prepackaged liquid-fuel rockets, general dimensions and performances the same as Bullpup A, but with warhead weighing 750 lb. Command guidance. (Some air-to-ground missiles, such as *Skybolt,* were canceled before development was finished; some others, such as *Sram,* which has a 110-mi range, are still under development.)

ANTIAIRCRAFT MISSILES

Redeye. Shoulder-launched solid-fuel missile against low-flying aircraft. Infrared homing, high-explosive warhead. Length 44 in, diameter 2.75 in. Launch weight 22 lb.

Hawk. Vehicle-mounted single-stage solid-fuel antiaircraft missile, mainly against low-flying aircraft. Launch weight 1275 lb, length 16.8 ft, diameter 48 in. Radar homing, high-explosive warhead. Range 22 mi, ceiling up to 45,000 ft. (Has been sold to Sweden, Saudi Arabia, Israel, and Japan.)

Terrier. Shipboard-mounted solid-fuel missile, consisting of one stage and booster. Beam rider, high-explosive warhead. Range ca. 11 mi, ceiling ca. 60,000 ft. Launch weight 3000 lb, length 15 ft (27 ft with booster), diameter 12 in.

Tartar. Shipboard-mounted single-stage solid-fuel missile. Homing guidance, high-explosive warhead. Launch weight 1500 lb, length 15 ft, diameter 12½ in. Performance like that of *Terrier.* (Both *Terrier* and *Tartar* will be replaced by a standard Navy antiaircraft missile.)

Talos. Shipboard-mounted ramjet missile with solid-fuel booster. Range 70 mi, ceiling above 75,000 ft, beam rider, can carry nuclear warhead. Launch weight 7000 lb (3000 lb without booster), length 30 ft, diameter 30 in.

Nike-Ajax. Two-stage. Booster solid-fuel, rocket liquid-fuel. Command guidance, high-explosive warhead. Range ca. 30 mi, ceiling 60,000 ft. Launch weight 2300 lb, length (with booster) 31 ft, diameter 12 in.

Nike-Hercules. Replacement of *Nike-Ajax,* both stages solid-fuel, command guidance with high-explosive or nuclear warhead. Range 90 mi, ceiling 150,000+ ft. Launch weight 10,000 lb, length (with booster) 39 ft, diameter 31.5 in.

Nike-Zeus. Two-stage solid fuel anti-missile missile with nuclear warhead. Launch weight 22,800 lb, length 48 ft, diameter of rocket 36 in., fin span of booster 10 ft. Performance classified. To be replaced by *Nike-X.*

Sprint. Conical two-stage solid-fuel anti-missile missile, base diameter 4.5 ft, overall length 27 ft. To be used in conjunction with *Nike-X.*

LONG-RANGE ROCKETS

TACTICAL RANGE

Corporal. Single-stage liquid-fuel (mono-ethyl aniline $C_6H_5NH_2$ and red fuming nitric acid) missile with 90-mile range. Launch weight 11,000 lb. Height 45 ft, diameter 30 in., fin span 7 ft. Warhead high explosive or nuclear. Control by vanes in exhaust, like V-2 and Redstone.

Sergeant. Single stage solid-fuel, replacement of *Corporal.* Range ca. 90 mi., inertial guidance. Height 34.5 ft, diameter 31 in. (First successful test launch at White Sands September 1961; first successful launch by a tactical unit of the U.S. Army in October 1962.)

INTERMEDIATE RANGE

Redstone. Single-stage liquid-fuel (kerosene and liquid oxygen) missile; after proper heading and velocity have been reached, tank section is separated from the forward section which consists of guidance equipment and warhead. Range 200 mi, launch weight 62,000 lb, height 70 ft, diameter 70 in. Inertial guidance, nuclear warhead. (Used for suborbital flights in Mercury Program.)

Pershing. Two-stage all-solid-fuel rocket with a range of 235 mi. Replacement of *Redstone.* Inertial guidance, nuclear warhead. Launch weight ca. 10,000 lb. Dimensions classified; can go from traveling condition to firing position and firing of missile within a few minutes.

Thor. Single-stage liquid-fuel (kerosene and liquid oxygen) missile with a range of about 1700 mi. Inertial guidance, nuclear warhead. Launch weight 100,000+ lb, height 65 ft, diameter 8 ft. Take-off thrust 150,000 lb for earlier version, 165,000 lb for later version. (A modified Thor is the first stage of NASA's Delta satellite launch vehicle.)

Jupiter. Single-stage liquid-fuel (kerosene and liquid oxygen) missile with a range of 1700 mi. A development of the Redstone. Take-off thrust 150,000 lb, launch weight 110,000 lb. Height 60 ft, diameter 105 in. Inertial guidance, nuclear warhead.

Polaris. Two-stage solid-fuel missile for underwater launch from submarines. Launch weight approx. 30,000 lb. Height 28 to 31 ft, depending on type; diameter ca. 4½ ft. Range: Polaris A-2 1350 mi, Polaris A-2 1750 mi, Polaris A-3 2800 mi. Inertial guidance, nuclear warhead. (*Poseidon,* under development, will have height of 34 ft and a diameter of about 6 ft, but will fit into *Polaris* launch tubes; its range will be over 3500 mi, slated to replace *Polaris.*)

INTERCONTINENTAL RANGE

Atlas. Liquid-fuel rocket of "1½ stages," counting the booster section as a half stage. Range ca. 7000 mi, launch weight 270,000 lb, take-off thrust 380,000 lb. Height 82 ft, diameter 10 ft. (*Atlas* was used for the orbital flights of the Mercury Program and as a launch vehicle for heavy satellites.)

Minuteman. Three-stage all solid-fuel rocket with a range of ca. 7000 mi. Inertial guidance, nuclear warhead. Launch weight ca. 69,000 lb, height 54 ft, other information classified. Minuteman-II, which is to replace Minuteman, will have higher acceleration. After replacement, the first version of Minuteman will become available as a satellite-launch vehicle.

Titan-II. Larger version of *Titan-I,* for different fuels (unsymmetrical dimethyl inertial guidance and carry nuclear warheads. Fuels for Titan-I are kerosene and liquid oxygen. Launch weight 220,000 lb. Height ca. 98 ft. First stage has a diameter of 10 ft and develops a thrust of 300,000 lb; second stage has a diameter of 8 ft and develops a thrust of 80,000 lb.

Titan-II. Larger version of *Titan-I,* for different fuels (unsymmetrical dimethyl hydrazine and nitrogen tetroxide, N_2O_4). Both stages have a diameter of 10 ft. Over-all height 103 ft. First stage develops 430,000 lb of thrust, second stage 100,000 lb. Launch weight 330,000 lb. (Titan-II was used for the Gemini Program.)

Titan-III. Modified *Titan-II;* not a missile but the core for the Titan III-C launch vehicle.

Comparison of U.S. and USSR Probes, Listed by Launch Dates

Aug. 17, 1958	(USA)	Thor-Able rocket with lunar probe. First stage exploded at T + 77 sec.
Oct. 11, 1958	(USA)	Thor-Able rocket carrying lunar probe Pioneer I. Upper stage weak, peak altitude 71,300 mi. Re-entered 43 hr 17.5 min after lift-off.
Nov. 8, 1958	(USA)	Thor-Able rocket with Pioneer II. Top stage did not ignite; peak altitude 963 mi; fell into Atlantic 42.4 min after lift-off.
Dec. 6, 1958	(USA)	Juno (Jupiter plus extra stages) with Pioneer III. Upper stages too weak, peak altitude 66,654 mi, reentry 38 hr 6 min after lift-off.
Jan. 2, 1959	(USSR)	Cosmic Rocket I (named Metchta). Missed moon by 4600 mi. In orbit around sun: perihelion 90,968,000 mi, aphelion 122,535,000 mi, orbital period 447 days.
Mar. 3, 1959	(USA)	Juno rocket with Pioneer IV intended to be a near miss, missed moon by 37,300 mi, in orbit around sun, orbit like that of Metchta.
Sept. 10, 1959	(USA)	Atlas-Able I lunar probe. Exploded on launch pad.
Sept. 12, 1959	(USSR)	Cosmic Rocket II. Struck moon in the area of Mare serenitatis 35 hr after lift-off.
Oct. 4, 1959	(USSR)	Cosmic Rocket III (satellite 1959 Theta). Looped behind moon to take pictures; re-entered atmosphere May 19, 1960.
Nov. 26, 1959	(USA)	Atlas-Able II lunar probe. Exploded in midair at T + 45 sec.
Mar. 11, 1960	(USA)	Pioneer V space probe. Put into circumsolar orbit between orbits of Venus and earth; orbital period 311 days.
Sept. 25, 1960	(USA)	Atlas-Able III. Upper stages too weak; rocket fell back.
Dec. 15, 1960	(USA)	Atlas-Able IV. Rocket exploded at 40,000 ft above ground.
Feb. 12, 1961	(USSR)	Venera I, fired from orbiting Sputnik-VIII. Must have passed Venus about May 20, 1961, but no signals received. In orbit around sun, perihelion 66.7 million mi, aphelion 94.6 million mi, orbital period 300 days.
Aug. 23, 1961	(USA)	Lunar probe Ranger I, carried by Atlas-Agena B. Upper stage failed to work, rocket coasted to 120 miles, then fell back.
Nov. 18, 1961	(USA)	Lunar probe Ranger II. Same fate as Ranger I.
Jan. 26, 1962	(USA)	Ranger III. Intended to impact moon; missed moon by 23,000 mi. Now in orbit around the sun. Perihelion 91,503,000 mi, aphelion 108,134,000 mi, orbital period 406.4 days.

Apr. 23, 1962	(USA)	Ranger IV. Spacecraft tumbled and missed moon's leading edge by 900 mi, impacted on far side of moon.
July 22, 1962	(USA)	Mariner I. Intended to go to Venus; failure on take-off.
Aug. 27, 1962	(USA)	Mariner II. Full success, passed Venus on Dec. 14, 1962, at a distance of 21,648 mi, transmitted data over a distance of more than 50 million mi. In orbit around sun, perihelion 62.5 million mi, aphelion 113.8 million mi, orbital period 346 days.
Oct. 18, 1962	(USA)	Ranger V. Fired for near miss of moon with picture-taking assignment. Power failure soon after take-off; nothing worked. Passed moon on Oct. 20 at a distance of 450 mi. In orbit around sun; orbit similar to that of Ranger III.
Nov 1, 1962	(USSR)	Mars-1. Fired from parking orbit. Intended to pass Mars at a distance of 600 mi in June 1963 but radio contact was lost Mar. 31, 1962, probably because spacecraft started tumbling. Must be in orbit around sun, orbital data unknown.
Jan. 4, 1963	(USSR)	Lunar probe in parking orbit. Ignition failed and probe continued to orbit earth until re-entry after natural decay.
Feb. 5, 1963	(USSR)	Lunar probe. Booster failed; spacecraft fell into Pacific near Midway.
Apr. 2, 1963	(USSR)	Luna-4 (apparently the three "cosmic rockets" were, in retrospect, considered as Luna-1, -2, and -3), fired from parking orbit. Missed moon by 5300 mi; in orbit around sun, orbital data unknown.
Jan. 30, 1964	(USA)	Ranger VI. Made perfect flight and impacted moon in the vicinity of the crater Arago. TV cameras failed to work; reason still unknown.
Apr. 2, 1964	(USSR)	Zond-I. Launched for Venus. Nothing worked, no data.
July 28, 1964	(USA)	Ranger VII. Perfect flight; impacted moon near crater Guericke; transmitted 4316 pictures during last 15 min before impact.
Nov. 5, 1964	(USA)	Mariner III. Put into orbit leading to Mars. Failed to emerge from shroud, therefore useless. In orbit around sun; orbit must be similar to that of Mariner IV.
Nov. 28, 1964	(USA)	Mariner IV. Passed Mars at a distance of about 6000 mi on July 14, 1965. Transmitted 19 useful pictures, measured density of Martian atmosphere, established absence of a magnetic field. In orbit around the sun, perihelion 103 million mi, aphelion 146 million mi, orbital period 567 days.

Nov. 30, 1964	(USSR)	Zond II. Trailed Mariner IV. Power supply failed completely after a few weeks; must have passed Mars about a week after Mariner IV and must be in similar orbit around the sun.
Feb. 17, 1965	(USA)	Ranger VIII. Impacted on moon in the area of Mare tranquillitatis, 2.59° from lunar equator after 64.9 hr of flight. Transmitted 7137 pictures.
Mar. 21, 1965	(USA)	Ranger IX. Impacted in crater Alphonsus 64.5 hr after take-off, transmitted 5814 pictures.
May 9, 1965	(USSR)	Luna-5. Intended for soft landing, destructive impact instead.
June 8, 1965	(USSR)	Luna-6. Midcourse correction malfunctioned; missed moon by over 100,000 mi; now in orbit around sun.
July 18, 1965	(USSR)	Zond III. Was maneuvered past the moon and transmitted a series of good pictures, covering 3000 sq mi. Now in orbit around sun.
Oct. 4, 1965	(USSR)	Luna-7. Intended for soft landing; crashed on moon Oct. 7.
Nov. 12, 1965	(USSR)	Venera II. Passed Venus at a distance of 14,912 mi on Feb. 27, 1966; no transmissions received. Must be in orbit around sun.
Nov. 16, 1965	(USSR)	Venera III. Entered planet's atmosphere Mar. 1, 1966, and crashed. No transmission received.
Dec. 3, 1965	(USSR)	Luna-8. Intended for soft landing; crashed Dec. 6.
Dec. 16, 1965	(USA)	Pioneer VI. Placed into orbit between Venus and earth; perihelion 75.6 million mi, aphelion 90.7 million mi, orbital period 311 days.
Jan. 31, 1966	(USSR)	Luna-9. Reached moon on Feb. 3 and made soft landing. Transmitted about 30 pictures.
Mar. 31, 1966	(USSR)	Luna-10. Fired for orbit around the moon; orbit achieved April 2. Periselenion 217 mi, aposelenion 632 mi.
May 30, 1966	(USA)	Surveyor I. Landed on moon Apr. 2; transmitted nearly 10,400 detailed pictures over a period of nearly a month.
Aug. 10, 1966	(USA)	Lunar Orbiter I. Orbit achieved Aug. 14, aposelenion 1159 mi, periselenion 117 mi (this was later lowered to 36 mi). High-resolution camera worked poorly, but 215 pictures were obtained. On Oct. 29 the spacecraft was ordered to fire all its retro rockets so that it would crash. Crashed on the far side of the moon.
Aug. 17, 1966	(USA)	Pioneer VII. Put into orbit around sun between orbits of earth and Mars. Perihelion 92 million mi, aphelion 102 million mi, orbital period 400 days.

Aug. 24, 1966	(USSR)	Luna-11. Orbit around moon achieved Aug. 27. Periselenion 99 mi, aposelenion 745 mi. Orbital period 2 hr 58 min.
Sept. 20, 1966	(USA)	Surveyor B. Tumbled in flight, crashed Sept. 23. If successful, designation would have been Surveyor II.
Oct. 22, 1966	(USSR)	Luna-12. Achieved orbit around moon Oct. 23, orbital period 3½ hr; had transmission difficulties.
Nov. 7, 1966	(USA)	Lunar Orbiter II. Orbit achieved Nov. 10. Aposelenion is 1143 mi, periselenion first 125 mi, later lowered to 31.3 mi. Orbital period 3 hr 28 min. Transmitted hundreds of excellent pictures.
Dec. 21, 1966	(USSR)	Luna-13. Soft-landed in Oceanus procellarum after 80-hr flight. Transmitted good pictures; drove spike into lunar surface to test its strength.
Feb. 4, 1967	(USA)	Lunar Orbiter III. Achieved orbit around moon Feb. 8; orbits the moon going over both its poles. Aposelenion was 1118 mi, periselenion 131 mi. Periselenion was lowered for picture-taking to 32 mi, changing aposelenion to 1147 mi. Orbital period 3 hr 35 min. Very good pictures obtained.
Apr. 17, 1967	(USA)	Surveyor III. Soft-landed on moon (in Oceanus procellarum) after 65-hr flight. Equipped with small scoop that dug trenches about 1 ft long and several inches deep, placed a scoopful of lunar soil on its own footpad for contrast.
May 4, 1967	(USA)	Lunar Orbiter IV. Achieved orbit around moon May 7, with periselenion at 1625 mi and aposelenion at 3850 mi. After it had photographed 99 per cent of the lunar surface, the orbit was tightened to a periselenion at 48 mi and aposelenion at 2450 mi. Orbital period 5 hr 44 min.
June 12, 1967	(USSR)	Planetary probe Venera-IV fired for planet Venus. Reached Venus Oct. 19, 1967; spacecraft entered planet's atmosphere and released parachute-equipped landing capsule. Atmosphere was found to consist mainly of carbon dioxide (80 per cent or higher), and highest temperature recorded was 536 degrees Fahrenheit. It seems that capsule stopped transmitting before reaching planet's surface.
June 14, 1967	(USA)	Mariner-V fired for planet Venus, made flyby at a distance of 2480 mi 18 hr after Venera-IV entered the atmosphere. Flyby confirmed the findings of both Mariner-II and Venera-IV: the planet is definitely uninhabitable.
July 14, 1967	(USA)	Lunar probe Surveyor IV fired for landing in Sinus medii, a dark area in center of moon's visible surface. Contact was lost at 9:06 P.M., Eastern Standard

		Time, on July 16, about 2 min before landing. Apparently retro-rocket exploded on ignition.
July 19, 1967	(USA)	Satellite Explorer-XXXV, weighing 230 pounds, was put into orbit around the moon by a Delta rocket. Aposelenion is 4600 mi, periselenion 500 mi. Purpose was the study of the earth's magnetic field at the distance of the moon.
Aug. 2, 1967	(USA)	Lunar Orbiter V achieved orbit around moon August 5, periselenion at 125 mi, aposelenion at 3760 mi.
Sept. 8, 1967	(USA)	Lunar probe Surveyor V fired for landing in Mare tranquillitatis. After midcourse correction a helium valve began leaking; in spite of this handicap a soft landing was accomplished in target area on Sept. 10, with all equipment in working order.
Nov 7, 1967	(USA)	Surveyor VI. Soft-landed on moon in the area of Sinus medii on Nov. 9, 1967 and is functioning normally.

CHARACTERISTICS OF SATURN V.

	FIRST STAGE	SECOND STAGE	THIRD STAGE
Length	**138 ft**	**82 ft**	**58 ft**
Diameter	**33 ft**	**33 ft**	**21.7 ft**
Weight, fueled	**5,028,000 lb**	**1,064,000 lb**	**265,000 lb**
Thrust	**7,500,000 lb (5 F-1 engines)**	**1,000,000 lb (5 J-2 engines)**	**200,000 lb (1 J-2 engine)**
Kerosene	**214,200 gal**	———	———
Liquid hydrogen	———	**267,700 gal**	**66,900 gal**
Liquid oxygen	**346,400 gal**	**87,400 gal**	**20,400 gal**

INSTRUMENT UNIT
(on top of third stage)

Diameter	**21.7 ft**
Thickness	**3 ft**
Weight	**4100 lb**

APOLLO SPACECRAFT

Command module	**13 ft in diameter; weight 11,000 lb**
Service module	**13 ft in diameter; weight 52,000 lb; length 22 ft; engine thrust 22,000 lb**
LEM	**32,000 lb total weight; engine thrust can be varied from 1050 to 10,500 lb**

Selected Bibliography

When the first edition of this book was written I provided a complete bibliography of books and pamphlets, disregarding only articles in magazines and professional journals. In later editions I kept enlarging the bibliography to maintain its reputation and, up to the time of the first Sputnik, this could still be done.

However, in 1967, it has become an impossibility. Sputnik, Project Vanguard, Project Mercury, and Project Gemini resulted in a "publication explosion." By now, a complete bibliography would have to be a separate book. So the problem was to decide what a selected bibliography should accomplish. Its obvious purpose should be to guide the reader to information which this book, for one reason or another, does not contain. One category (not labeled as such) in the old bibliography could be eliminated completely—namely, older popular accounts that are now obsolete. Another expendable category would be the highly specialized engineering or scientific treatises. If anyone is designing a new rocket engine or computing the orbit of a new type of satellite he also knows where to find specialized literature he needs without referring to the bibliography of this book, and those who don't design rocket engines and compute orbits do not need such references.

The logical arrangement of the selected bibliography was to correspond to the chapters of the book. This is what has been done, and I hope that it will be found convenient and useful.

Chapter 1

Günther, Ludwig. *Kepler's Traum vom Mond.* Leipzig, 1898. Translated from the original Latin and with an introduction by Dr. Günther. The introduction contains a survey of similar literature in concise and condensed form. The original Latin work was first printed in Frankfort in 1634 as *Joh. Keppleri Mathematici olim imperatorii SOMNIUM seu Opus posthumum de astronomia lunari.* It was republished in Frankfort in 1870 as vol. VIII of *Joannis*

Kepleri Astronomia Opera Omnia. Dr. Günther's translation into German is the first translation of the work into a modern language and remained the only translation for half a century. An English translation made by Everett F. Bleiler was published in *Beyond Time and Space,* an anthology edited by August Derleth (New York: Pellagrini & Cudahy, 1950), which also contains Lukian's *True History* and Bishop Godwin's *Man in the Moone.*

Lear, John. *Kepler's Dream.* Berkeley: University of California Press, 1965. A good translation by Patricia Frueh Kirkwood—the first in English—of Kepler's full text and notes of *Somnium,* accompanied by an appraisal mainly from the literary point of view.

Ley, Willy. *Watchers of the Skies. An Informal History of Astronomy from Babylon to the Space Age.* New York: Viking, 1963. 528 p., with many illustrations. Second printing, 1964.

Lucian of Samosata. "*True History.*" The most recent separate edition in English seems to be the one by A. H. Bullen in 1902. The first English edition appeared in 1634 in Oxford under the title: *Certaine Select Dialogues of Lucian, Together with his True Historie,* translated from the Greeke into English by Mr. Francis Hicks. *Whereunto is added the Life of Lucian gathered out of his owne Writings. . . .* (*See* Günther, Ludwig.)

McColley, Grant. "Introduction to Bishop Francis Godwin's *The Man in the Moone* and *Nuncius Inanimatus.*" *Smith College Studies in Modern Languages,* XIX, no. 1, Oct. 1937. The reprint of *The Man in the Moone* is from the first London edition of 1629 and that of the *Nuncius Inanimatus* from the first London edition of 1638; both first editions were represented by unique copies. The *Nuncius* is accompanied, on opposite pages, by the English translation of Dr. Thomas Smith of Magdalen College, Oxford, which accompanied the Latin text of the second London edition of 1657.

Nicolson, Marjorie Hope. "Introduction to *Navis Aeria* by Bernardo Zamagna, published in Rome in 1786." *Smith College Classical Studies,* no. 12, March 1939. Translation from the Latin is by Mary B. McElwain.

———. "Cosmic Voyages." *ELH,* VII, no. 2, June 1940.

Nicolson, M. H., and Mohler, Nora M. "Swift's 'Flying Island' in the Voyage to Laputa." *Annals of Science,* II, no. 4, October 1937.

———. "The First 'Electrical' Flying Machine." *Smith College Studies in Modern Languages,* October 1939.

Preysing, Count Konrad (transl.). *Des Heiligen Hyppolitus von Rom Widerlegung aller Häresien* (*Philosophumena*). Munich: Kösel & Pustet, 1922. 291 p. This seems to be the only complete translation into a modern language. The work is prefaced by a biography of Hyppolitus. (There exist translations into Latin by P. Cruice and Duncker-Schneidewein.)

Rosen, Edward. *Kepler's Somnium.* Madison, Wis.: University of Wisconsin Press, 1967. 255 p. This is the most recent and most complete edition of *Somnium* in English. The translation of the book and all of Kepler's "Notes" was made by Professor Rosen himself. Numerous notes by the translator and several appendices enhance the value of this edition.

Chapter 2

Allott, Kenneth. *Jules Verne.* New York: Macmillan, 1941. 282 p. The first biography of Jules Verne in English, with a good bibliography of his numerous works: fact, fiction, and drama.

Bailey, J. O. *Pilgrims through Space and Time.* New York: Argus Books, 1947.

342 p. This is an enlarged version of the author's thesis on the "Scientific Romances of H. G. Wells." The added material is rather haphazard and highly incomplete, so that the book is mostly a treatise on the early Wellsian era.

Bleiler, E. F. (ed.). *The Checklist of Fantastic Literature.* Chicago: Shasta, 1948. 475 p. A list of more than 5000 book titles, all English, covering the whole field of fantastic literature, with introductory articles and index by title and by author.

Bretnor, Reginald (ed.). *Modern Science Fiction, Its Meaning and Its Future.* New York: Coward-McCann, 1953. 294 p. A symposium comprising chapters by eleven writers and editors of science fiction.

de Camp, L. Sprague. *Science-Fiction Handbook.* New York: Hermitage House, 1953. 328 p., well indexed. Ostensibly a handbook for beginning science-fiction writers, this is actually a fine history of the field, complete with personality sketches of a number of authors, living and dead.

Claretin, Jules. *Jules Verne.* (*Célébrités Contemporaines No.* 33.) 1883. 32 p.

Davenport, Basil. *Inquiry Into Science Fiction.* New York: Longmans, Green and Co., 1955. 87 p. A short, literate introduction to the field.

Flammarion, Camille. *Les mondes imaginaires et les monde réels.* Paris, 1865. Consists of two sections: the first dealing with the conditions on the surfaces of the planets of our solar system as they were then assumed to be; the second, almost three times as voluminous as the first, a survey of everything ever written about other planets (except strictly astronomical works of then recent date), philosophical as well as theological, including novelistic attempts, like Godwin's *Man in the Moone.* I own a copy of the 1892 edition 21st ed., 598 p.), which contains an added section about books published between 1865 and 1892.

de la Fuÿe, Marguerite Allotte. *Jules Verne.* Translated by Erik de Mauny. New York: Coward-McCann, 1956. 222 p., no illus., no index, no bibliography. Marguerite Allotte de la Fuÿe was the wife of Jules Verne's nephew, Louis Allotte de la Fuÿe; the marriage was severed by her husband's death after only a few months. The original of this book appeared under the title *Jules Verne—Sa vie, son œuvre,* in Kra in 1928.

Kretzmann, Edwin M. J. "German Technological Utopias of the Pre-War Period." *Annals of Science,* III, no. 4, October 1938. Much of the essay is devoted to Kurd Lasswitz.

Lafleur, Laurence J. "Marvelous Voyages." *Popular Astronomy,* 1942–43. A series of articles of scientific criticism of classic science-fiction novels.

Marcucci, Edmondo. *Giulio Verne e la sua opera.* 1930. 129 p. Contains a good bibliography of Jules Verne's works and a section on his Italian imitators.

Popp, Max. *Jules Verne und sein Werk.* Hartleben, 1909. 213 p. Devotes a great deal of space to Verne's imitators, especially those either writing in or translated into German.

Waltz, George H. *Jules Verne: The Biography of an Imagination.* New York: Holt, 1943. 223 p. Shorter than Allott's book; contains notes on the influence of Verne's work.

Chapter 3

Anderson, Robert. *The Making of Rockets. In Two Parts. The First containing the Making of Rockets for the meanest Capacity. The other: To make Rockets by a Duplicate Proposition, to 1000 pound Weight or higher.* London, Printed for

Robert Morden, at the *Atlas* in *Cornhil*. 1696. This small book of only xii + 48 p., dedicated to the Right Honourable Henry, Earl of Romney, is extremely rare. It contains the usual fireworks instructions with such helpful hints as "Rocket Moulds are made by Mr. *Cuggley* just without *Cripplegate*. Taper Bits for Rockets are made by Mr. *Goode* just within *Cripplegate*. Rods for Rockets are made by Mr. *Stateham* in *Token-House-Yard, Lothbury*. All three right good Workmen." The "estimation" of rockets is interesting: they can be "estimated" either by measure—i.e., inches and feet—or by weight—"if by Weight, then they are estimated by Cast Iron, that is, by the Gunners Rule. An Inch Rocket weigheth two Ounces and a quater: an Inch and a half Rocket weigheth half a Pound: a two Inch Rocket weigheth a little above a Pound: a three Inch Rocket weigheth almost four Pounds: a four Inch rocket weigheth nine Pounds: an eigth Inch Rocket weigheth seventy two Pounds, which may be called a Rocket Royal." On June 26, 1696, a Mr. Ayres Jr. fired a 4-inch and a 6-inch rocket, as witnessed by Mr. Dandrige, Mr. Rodway, the author of the book, and others. The 6-inch rocket rose at an angle of 78° "and the Horizontal Range was just 15 Chains, that is 330 Yards."

d'Arcet. *Notice sur les fusées incendiaires de Congrêve*. Paris, 1814.

Barber, F. M. *Lecture on Drifting and Automatic Movable Torpedoes, Submarine Guns and Rockets*. Newport, R.I.: U. S. Torpedo Station, December 1874.

Bem, Joseph. *Erfahrungen über die Congrev'schen Brand-Raketen, bis zum Jahre 1819 in der Königl. Polnischen Artillerie gesammelt und an seine Kaiserliche Hoheit, den Grossfürst Constantin, Général-en-Chef aller Königl. Polnischen Truppen, berichtet von Joseph Bem, . . . neben dem Französischen Original-Text in Teutscher Uebersetzung . . . herausgegeben von M. Schuh*. Weimar, 1820. In German and French.

Brock, Alan St. H. *Pyrotechnics*. London, 1922.

———. *A History of Fireworks*. London, George G. Harrap & Co. Ltd., 1949. 280 p., text illus., 40 plates, 8 in color. A very complete scholarly work about fireworks in Europe. Good bibliography.

Busch and Hoffmann. *Die Kriegsfeuerwerkerei der Königlich Preussischen Artillerie*. Berlin, 1851.

Chertier, François-Marie. *Nouvelles recherches sur les feux d'artifice*. Paris, 1854. 638 p. This is the most comprehensive book about fireworks rockets known to me.

Congreve, William. *A Concise Account on the Origin and Progress of the Rocket System*. Dublin, 1817. Congreve's name does not appear on the title page.

———. *The details of the Rocket System*. London, 1814.

———. *A Treatise on the Rocket System. London*, 1827. This book appeared in a German translation (Weimar, 1829); a French translation formed Part I of the book edited by J. Corréard (q.v.). An earlier translation by Ch. Puillon de Boblaye, Captain of the French artillery, with the title *Traité sur les fusées-Congrève* had been issued as a military manual only, not for general distribution. As distinct from the original it was richly illustrated; Fig. 12 of this book is taken from that manual.

Corréard, J. (ed.) *Histoire des fusées de guerre, ou recueil de tout ce qui a été publié ou écrit sur ce projectile*. Paris, 1841. In two parts: the first a translation of Congreve's *A Treatise on the Congreve Rocket System* (q.v.); the second a work by Montgéry, entitled *Recherches sur les fusées de guerre*.

Faber, Henry B. *Military Pyrotechnics*. "The History and Development of Military Pyrotechnics" with an Historical Introduction by Marvin Dana. Washington, D.C.: U.S. Government Printing Office, 1919. 3 vols. 229 + 228 + 230 p.

Well illustrated, contains several chapters on old war rockets and detailed descriptions of United States signal rockets.

von Geissler, Christoph Friedrich. *Neue, Curieuse und vollkommene Artillerie, Worinnen Dasjenige, so in 40 Jahren bey 25 Belagerungen, 24 Eroberungen und 3 Bataillen ausgeübet worden, in Vier nachfolgenden Wissenschaften als Büchsenmeisterey, Ernst-Feuerwerckerey, Petarden und Miniren angewiesen wird.* Dresden, 1718. Colonel von Geissler was Commander of the field artillery of the King of Poland and Prince of Saxony and was the first to make powder rockets surpassing 100 lb in weight. The report on these rockets is given on page 173 of the folio volume.

Goddard, Lt.-Col. Calvin. *Rockets.* Three articles on war rockets in the May/June, July/August, and Sept./Oct. issues of *Army Ordnance* (Washington, D.C.).

Guttman, Dr. Oscar. *Monumenta Pulveris Pyrii.* London, 1906. An enormous volume relating the early history of gunpowder and its applications, in largest folio, with many full-page engravings, bound in hardwood covers with brass clasps. The text is in English, German, and French in parallel columns. A masterpiece of research, typography, printing, and book binding, unfortunately published in a limited edition only.

Hale, William. *A Treatise on the Comparative Merits of a Rifle Gun and a Rotary Rocket.* London, 1863.

Hime, Henry W. L. *Gunpowder and Ammunition.* London: Longmans Green, 1904.

———. *The Origin of Artillery.* London: Longmans Green, 1915. 2nd ed. of *Gunpowder and Ammunition.*

von Hoyer, J. G. *System der Brandraketen nach Congreve und Anderen.* Leipzig, 1827.

Hume, J. *Remarks on Military Rockets.* London, 1811.

Jähns, O. *Geschichte der Explosivstoffe.* Vol. I. Berlin, 1897. Gives the history of war rockets in general.

Jones, Capt. Robert. *Artificial Fireworks.* London, 1776.

Kentish, Thomas. *The Complete Art of Firework-Making.* London: Chatto and Windus, 1905. 280 p., 24 plates, frontis.

Konstantinoff, General M. *Mémoire sur les fusées de guerre* and *Lectures sur les fusées de guerre.* Paris, 1858–1861. 2 vols. The two treatises form the first volume, the illustrations the second.

Latham, H. B. "The Rocket Service and the Award of the Swedish Decorations for Leipzig." *Journal of the Royal Artillery,* LVI, no. 4. January 1930.

Ley, Willy. *Shells and Shooting.* New York: Modern Age, 1942.

von Malinowsky and von Bonin. *Geschichte der brandenburgisch-preussischen Artillerie.* Vol. II. Berlin, 1841.

Montgéry. *Traité des fusées de guerre, nommées autrefois rockettes et maintenant fusées à la Congréve.* Paris, 1825.

Moore, W. *On the Motion of Rockets.* London, 1810.

Munroe, Innes, Esq. *A Narrative of the Military Operations on the Coromandel Coast against the Combined Forces of the French, Dutch and Hyder Ally Cawn from the Year 1780 to the Peace in 1784.* London, 1789. German edition in 1791.

Nye, Nathanael. *A Treatise of Artificiall Fire-Works for Warre and Recreation; Conteining a description to make sundry kinds of Fire-Works, both for use and pleasure, with lesse labour and cost than any hath heretofore been published.* London 1647.

Nye, T. *Thoughts on Aerial Traveling and on the Best Means of Propelling Balloons.* London, 1852.

Pictet, Adolph. *Essai sur les propriétés et la tactique des fusées de guerre.* Turin, 1848.

Pralon, A. *Une page de l'Histoire de l'Artillerie: Les Fusées de guerre en France.* Paris, 1883.

Rogier, Charles. *A Word for My King and Country: A treatise on the utility of a rocket armament, assisted by balloons where ships of war cannot be accessible; both defensive and offensive, to the annoying of the enemy's harbour.* Macclesfield, 1818.

von Romocki, S. J. *Geschichte der Explosivstoffe.* 2 vols. Berlin, 1898. Still the best book on the history of gunpowder, explosives, firearms, and military rockets.

Ruggieri, Claude-F'né. *Pyrotechnie Militaire, ou Traité Complet des Feux de Guerre et des Bouches à Feu,* . . . Paris, 1812.

Schmidlap, Johann. *Künstliche vnd rechtschaffene Feuerwerck zum schimpff, vormals im Truck nie aussgangen. Darinnen an tag geben vnd beschrieben . . . durch Johannem Schmidlap, von Schorndorf.* Nuremberg, 1591. To my knowledge the first printed book on fireworks only; all earlier books combined fireworks and gunnery.

Scoffern, J. *Projectile Weapons of War.* 1859. One of a large number of books on weapons and military matters published during the nineteenth century which contain descriptions of war rockets. Similar articles can be found in many others, for example, in J. Frost's *Book of the Army of the United States* (1845), in E. S. Farrow's *Military Encyclopedia* (1885), and in the British *Treatise on Ammunition* (His Majesty's Stationery Office), in several editions. The articles on war rockets are not the same in the various editions; generally speaking the more recent the edition the shorter the article. The last one is the edition of 1905. The articles on war rockets in these books and manuals deal with types contemporary or recent at the time the books were printed; they pay no attention to the earlier history.

Simienowicz, Kasimir. *Vollkommene Geschütz/Feuerwerk und Büchsenmeysterey Kunst.* Frankfort Main, 1676. Simienowicz was commander of the Polish Army; his book was written in Latin and published in that language, but the Latin text is only a part of the whole plan. The German edition, listed here, is a larger work, finished according to Simienowicz's original plan by Daniel Elrich, Ordnance Master of the city of Frankford-on-the-Main. The Latin section was translated into German by Leonhard Beer.

Susane, M. *Les Fusées de Guerre.* Metz, 1863.

Chapter 4

Kosmodemyansky, A. A. *Konstantin Tsiolkovsky, His Life and Work.* Moscow: Foreign Languages Publishing House, 1956. 102 p., photographs. This book began as a lecture about Ziolkovsky to the Society for the Dissemination of Political and Scientific Knowledge in Moscow in 1948. It was then expanded and printed by the Pravda Publishing House during the same year. The completely reworked Russian edition of 1955 is the basis of this translation.

Perelman, Yakov Isidorovitch. *Interplanetary Travels.* 1st ed. St. Petersburg, 1915. A popularization of Ziolkovsky's writings; the 8th ed. appeared in 1933. Dr. Perelman also wrote four books for young people: *Flight to the Moon* (1924), *Rocket to the Moon* (1st ed., 1930; 3rd ed. 1933), *Ziolkovsky, His Life, Inventions, and Scientific Works* (1932), *To the Stars in a Rocket* (1933). All in Russian, untranslated.

Ziolkovsky (Tsiolkovsky), Konstantin Eduardovitch. List of his writings in chronological order of publication:
On the Moon (science fiction), 1893, reprinted 1935.
Exploration of planetary space with reactive equipment, in *Na-ootchnoye Obosreniye,* 1903. The same title was used for several articles (1911–1914) in the aviation magazine *Vestnik vosdukhoplavaniya.*
Outside the Earth (science fiction), parts of it were written in 1896, appeared first as a serial in *Priroda i Ludi* and as a book in 1920.
The Rocket into Cosmic Space, 1924. Reprint in pamphlet form of the article of 1903 with additions. Introduction in German by A. L. Tchiyevsky.
The Exploration of Interplanetary Space, 1926. Essentially an enlarged and revised version of the articles that appeared in *Vestnik vosdukhoplavaniya.*
Cosmic Rocket, 1927.
Cosmic Reactive Trains, 1929.
The Aims of Astronautics, 1929.
To the Astronauts, 1930.
Reaction Airplane, 1930.
Semi-reactive Stratoplane, 1930.
In 1947 the Publishing House of the Defense Industry in Moscow published Ziolkovsky's *Collected Works* in several volumes, and they were later translated by NASA. The volume containing the *Works on Rocket Technology* is NASA's Technical Translation TT F-243, November 1965. A selection of the writings on space and on rockets, including his science-fiction stories, was published in English under the title *The Call of the Cosmos,* by the Foreign Languages Publishing House in Moscow. (No date, ca. 1963, 480 p.) *Outside the Earth,* translated by Kenneth Syers (with an Introduction by B. N. Vorobyev) was published under the title *Beyond the Planet Earth,* New York: Pergamon Press, 1960. 190 p.

Chapters 5 and 6

The books listed in this section represent the "pioneer period," all the authors who published prior to 1935. Though they appeared later, biographies of the pioneers are included here.

Ananoff, Alexandre. *Le Problème des voyages interplanétaires.* Paris: Société Astronomique de France, 1934. 24 p.
———. *La Fusée de guerre.* Paris: Société Astronomique de France, 1934. 12 p.
———. *La Navigation interplanétaire.* Paris: Société Astronomique de France, 1935. 32 p.
———. *Navigation interplanétaire.* Paris: Editions Elzévir, 1946. 64 p.
———. *Des Premières fusées au V-2.* Paris: Editions Elzévir, 1947, 64 p.
———. *L'Astronautique.* Paris: Librairie Arthème Fayerd, 1949. 498 p. A large work which relegates all of Ananoff's earlier writings to the rank of preliminary studies. Encyclopedic in scope.

Brügel, Werner. *Männer der Rakete.* Leipzig: Hachmeister & Thal, 1933. 144 p. Essays, autobiographical and pertaining to their rocket research, by the following: Hanns-Wolf von Dickhuth-Harrach, Robert Esnault-Pelterie, Robert H. Goddard, Franz von Hoefft, Willy Ley (report on the *Verein für Raumschiffahrt e. V.*), C. P. Mason (report on the American Rocket Society), Hermann Oberth, Guido von Pirquet, Nikolai A. Rynin, Friedrich Schmiedl, Johannes Winkler, and Konstantin E. Ziolkovsky.

Cleator, Phil E. *Rockets Through Space.* London: Allen and Unwin, 1936; New York: Simon and Schuster, 1936. 246 p., illus. The American edition contains a number of plates not in the British edition. Mr. Cleator is the founder of the British Interplanetary Society.

———. *Into Space.* London: Allen and Unwin, 1953; distributed in the United States under the imprint of Thomas Y. Crowell. 159 p., 26 text illus. 6 cartoons, 15 plates.

Esnault-Pelterie, Robert. *L'Exploration par fusées de la trés haute atmosphère et la possibilité des voyages interplanétaires.* Paris: Société Astronomique de France, 1928. 96 p. Originally a lecture before the French Astronomical Society.

———. *L'Astronautique.* Paris: Société Astronomique de France, 1930. 248 p. A much revised and greatly enlarged second edition of the preceding title.

———. *L'Astronautique, Complément.* Paris: Société Astronomique de France, 1935. 102 p.

Gussalli, Luigi. *Si può già tentare un viaggio dalle terra alla luna.* Milan: Società Editrice Libraria, 1923. Booklet devoted to several inventions and ideas of the author; the section on rocket propulsion forms only one chapter.

———. *Propulsori a reazione per l'astronautica, Seconda Communicazione.* Brescia: Giulio Vannini, 1941. 118 p. A greatly expanded version of the relevant section of the preceding.

———. *I Viaggi interplanetari per mezzo della radiazioni solari.* Brescia: Giulio Vannini, 1946. 92 p. With subtitle: *The Interplanetary Travels by the Solar Radiations; translated from the original text by Jeanne Wiest.* The first 60 pages are in Italian, the remaining 32 pages in a kind of English, as indicated by the subtitle. The main thesis is the elimination of large mass-ratios by utilization of solar energy.

Hartl, Hans. *Hermann Oberth.* Hannover: Theodor Oppermann Verlag, 1958. 240 p., with numerous photographs. A biography (in German) of the "Father of Astronautics."

Kondratyuk, Yuri Vassilyevitch. *The Conquest of Interplanetary Space.* In Russian, Novosibirsk, 1929. (Untranslated.)

Lasser, David. *The Conquest of Space.* New York; privately printed, 1931. 271 p.

Lehman, Milton. *The High Man; the Life of Robert H. Goddard.* With a Preface by Charles A. Lindbergh. New York: Farrar, Straus & Co., 1963. xvi + 430 p. 40 p. of photographs.

Ley, Willy. *Die Fahrt ins Weltall.* Leipzig: Hachmeister & Thal, 1926. 2nd ed., somewhat enlarged, Leipzig, 1929. 83 p.

———. *Die Möglichkeit der Weltraumfahrt.* Leipzig: Hachmeister & Thal, 1928. 344 p. Chapters by Karl Debus, Franz von Hoefft, Walter Hohmann, Willy Ley, Hermann Oberth, Guido von Pirquet, and Friedrich Wilhelm Sander.

———. *Grundriss einer Geschichte der Rakete.* Leipzig: Hachmeister & Thal, 1931. 16 p.

Marquis, Raoul. *Irons-nous dans la lune?* Paris: *Je sais tout,* 1932. 188 p.

Nebel, Rudolf. *Raketenflug.* Privately printed, 1932. 47 p.

Noordung, Hermann. *Das Problem der Befahrung des Weltraums.* Berlin: R. C. Schmidt, 1928. 188 p. Hermann Noordung is a pseudonym for Captain Potočnik of the old Austrian Imperial Army.

Oberth, Hermann. *Die Rakete zu den Planetenräumen.* Munich: R. Oldenbourg, 1923; 2nd ed., Munich: 1925. 92 p. Facsimile reprint, 1960.

———. *Wege zur Raumschiffahrt.* Munich: R. Oldenbourg, 1929. 431 p. Greatly enlarged third edition of *Die Rakete. . . .*

———. *Menschen im Weltraum.* Düsseldorf: Econ Verlag, 1954. 253 p., illus. A popular work with a mathematical appendix.

———. *Man into Space.* New York: Harper, 1957. 232 p. with diagrams. Translation of the original German *Menschen im Weltraum* (Düsseldorf: Econ Verlag, 1954) by G. P. H. de Freville. A collection of eight essays with a mathematical appendix.

———. *The Moon Car.* New York, Harper, 1959. 98 p., eng. diagrams. Translation of the original German *Das Mondauto* (Düsseldorf: Econ Verlag, 1959) by Willy Ley. Discussion of the engineering principles of a vehicle which, if necessary, can jump to a different level.

"Osoaviakhim." *Motion by Reaction,* 2 vols. Moscow and Leningrad, 1925. Report by the Section on Reaction of the Stratosphere Committee of the Osoaviakhim, a Russian semi-official flying society. In Russian, untranslated.

Philp, Charles G. *Stratosphere and Rocket Flight.* London: Pitman, 1935. 118 p. Two slightly different editions appeared during the same year. Unreliable in detail.

Roy, Maurice. *Propulsion by Reaction.* NACA Tech. Mem. no. 571, June 1930. Translated from *La Technique aéronautique,* Jan. 15, 1930.

———. *Recherches théoretiques sur le rendement et les conditions de réalisation des systèmes motorpropulseurs à réaction.* Paris: Dunod, 1930. 214 p.

———. *Tuyères, trompes, fusées et projectiles.* Paris: Dunod, 1948. (Ministère de l'Air Publication no. 203). 194 p.

Rynin, Nikolai Alekseyevitch. *Interplanetary Communications.* The so-called rocket encyclopedia, published in Leningrad, 1928–1932. The titles of the volumes are *I. Dreams, Legends and First Fantasies* (1928); *II. Cosmic Ships in Fantastic Literature* (1928); *III. Radiant Energy in Fantastic Literature and in Scientific Projects* (1931); *IV. Rockets* (1929); *V. Theory of Motion by Reaction* (1929); *VI. Super-Aviation and Super-Artillery* (1929); *VII. K. E. Ziolkovsky: His Life, His works and His Rockets* (1931); *VIII. Theory of Cosmic Flight* (1932); *IX. Astronavigation. History, Bibliography, and Index* (1932).

Scherschewsky, Alexander B. *Die Rakete für Fahrt und Flug.* Berlin: C. J. E. Volckmann, 1928. 134 p.

Schrenck, Martin, and Schiller, Melitta. *Die Rakete als Kraftmaschine.* xii + 13 p. This is *Tech. Mem. No. Vf.24/3,* published October 13, 1928, by the DVL, or Deutsche Versuchsanstalt für Luftfahrt, in a small and restricted edition.

Valier, Max. *Der Vorstoss in den Weltenraum.* Munich: R. Oldenbourg, 1924. 134 p. From 1925 to 1929 this book went through five printings without important changes; in 1930 a revised and greatly enlarged edition was published under the title *Raketenfahrt.* 240 p.

Zander (Tsander), Friedrich Arturovitch. *The Problem of Flight with the Aid of Reactive Devices.* Moscow, 1932. In Russian, untranslated.

NOTE: for the early book by Dr. Eugen Sänger which should be listed here as far as chronology goes, see listing of literature pertaining to the appendices.

Chapter 7

Boyce, Joseph C. (ed.). *New Weapons for Air Warfare.* Boston: Little, Brown, 1947. 292 p. Contains several chapters on rocket fuzes, bombardment rockets, and missiles.

Burchard, John E. (ed.). *Rockets, Guns, and Targets.* Boston: Little, Brown, 1948. 482 p. In the series Science in World War II. Consists of three parts, the first of which, about half the book, is devoted to American rocket ordnance; the other two parts to ballistic problems.

Huggett, Clayton, *et al. Solid Propellant Rockets.* Princeton, N.J.: Princeton University Press, 1960. 167 p., diagrams (paperback). Technical work; the co-authors are C. E. Bartley and Mark M. Mills.

Kit, Boris, and Evered, Douglas S. *Rocket Propellant Handbook.* New York: Macmillan, 1960. 354 p., numerous tables.

U.S. Navy Bureau of Ordnance in World War II, by Buford Rowland and William B. Boyd. Bureau of Ordance, Department of the Navy. (Not dated, about 1955, obtainable from Superintendent of Documents, Washington, D.C. 20025.) 539 p., illus. Chapter XIV, pp. 291–329, deals with Navy rockets.

Wimpress, R. N. *Internal Ballistics of Solid-Fuel Rockets;* subtitled: "Military Rockets Using Dry-Processed Double-Base Propellant as Fuel." New York: McGraw-Hill, 1950. 214 p. Technical work on artillery rockets, based on research work done at the California Institute of Technology during the Second World War.

Chapters 8 and 9

Bergaust, Erik. *Reaching for the Stars.* New York: Doubleday, 1960. 407 p., photographs. An "unauthorized biography of Wernher von Braun"—but W.v.B. did read the proofs of the book to insure accuracy of facts.

Buseman, Adolf. *Rockets Using Liquid Oxygen.* NACA Tech. Mem. no. 1144, April 1947. Translation of *R-Gerät mit Sauerstoff* in *Schriften der Deutschen Akademie der Luftfahrtforschung,* 1943.

Dornberger, Walter. *V-2; Der Schuss ins Weltall. Geschichte einer grossen Erfindung.* Esslingen: Bechtle Verlag, 1952. 295 p., 23 photographs. The story of the Peenemünde research institute and the development of the V-2.

———. *V-2.* American edition, New York: Viking, 1954, 281 p. with photographs. British Empire edition; London: Hurst & Blackett, 1954.

Gröttrup, Helmet. *Über Raketen.* Berlin: Verlag Ullstein, 1959. 243 p., 29 photographs and 151 diagrams. An introduction to rocket engineering for nonengineers. Since Helmut Gröttrup spent about six years in Russia as leader of the group of deported German engineers this book is rather disappointing, as it does not say a word about the author's experiences or even his own work.

Gröttrup, Irmgard. *Die Besessenen und die Mächtigen im Reiche der roten Rakete.* ("The Fanatics and the Mighty Men in the Realm of the Red Rockets.") Stuttgart: Steingrüben Verlag, 1958. Excerpts from the diary kept by Mrs. Gröttrup while in Soviet Russia. The English edition (London: A. Deutsch, 1959), has the title *Rocket Wife.*

Irving, David. *The Mare's Nest; The German Secret Weapons Campaign and Allied Countermeasures.* Boston: Little, Brown, 1965. 320 p. with photographs and diagrams. Well-researched story of the Allied (mainly British) reaction to reports about German secret weapons. First published in England (1964) by Wm. Kimber & Co.

Klee, Ernst, and Mark, Otto. *Damals in Peenemünde.* Hamburg: Gerhard Stalling Verlag, 1963. 120 p. A large-size book with many original photographs from Peenemünde and reproductions of original military and engineering documents. The American edition (New York: Dutton, 1965) has the title *The Birth of the Missile* and is a precise reproduction of the German original. The documents are accompanied by English translations.

Kooy, J. M. J., and Uytenbogaart, J. W. H. *Ballistics of the Future.* Haarlem, Holland: A. Stam, 1946; New York: McGraw-Hill, 1948. 472 p., folding plates. This book by two Dutch scientists who witnessed the V-2 firings from Dutch soil ranks with the major works of Oberth and Esnault-Pelterie. It is a rather complete exposition of vector calculus, exterior ballistics. The chapters on V-1 and V-2 are complete to small design detail, such as wiring diagrams of electric equipment.

Lusar, Rudolf. *Die deutschen Waffen und Geheimwaffen des 2. Weltkrieges.* Munich: J. F. Lehmann, 1959. 268 p. with 86 illustrations. Though in places uncritical, this is a useful compilation of German weapon developments, including those that remained unfinished. American edition, translated by R. P. Heller and M. Schindler (and slightly shortened) has the title *German Secret Weapons of the Second World War.* New York: Philosophical Library, 1959. 265 p. with almost all the illustrations of the original.

Miller, Col. Henry W. *The Paris Gun.* New York: Cape & Smith, 1930. 288 p. with plates and map sketches.

Newell, Homer E., Jr. *Sounding Rockets.* New York: McGraw-Hill, 1959. 334 p., photographs and drawings. A technical work on all the types of sounding rockets launched up to 1959. The chapters on particular sounding rockets are written by their designers. A total of twenty scientists contributed to this book.

Simon, Col. Leslie E. *German Research in World War II.* New York: Wiley, 1947. 218 p. This book deals with the organization of German research work rather than with the research problems, goals, and results.

Walter, Hellmuth. *Report on Rocket Power Plants Based on T-Substance.* NACA Tech. Mem. no. 1170, July 1947. Translation of *Bericht über die R-Triebwerke auf Grundlage des T-Stoffes,* in *Schriften der Deutschen Akademie der Luftfahrtforschung,* 1943.

Weber and Kehl. *Windtunnel Measurements on the Henschel Missile "Zitterrochen" in Subsonic and Supersonic Velocities.* NACA Tech. Mem. no. 1159, April 1948. Translation of *Windkanalmessungen am Henschel-Gerät "Zitterrochen" bei Unter- und Ueberschallgeschwindigkeiten; Deutsch Luftfahrtforschung, Untersuchungen und Mitteilungen Nr. 3122.* The missile *Zitterrochen* was apparently never completed and existed only as a test model. It looked like a very long artillery projectile with very short wings of very wide chord. The translators of the report translated *"Zitterrochen"* as "trembling ray" and seem to be puzzled by the word, as well they might be. It is a zoological term meaning "electric ray," just as *Zitteraal* means "electric eel."

Weyl, A. R. *Guided Missiles.* London: Temple, 1949. 140 p. An enlarged version of a long series of articles published originally in *The Aeroplane;* most of the material deals with German missiles of the Second World War.

Zborowski, Helmut. *Rocket Power Plants Based on Nitric Acid and Their Specific Propulsive Weights.* NACA Tech. Mem. no. 1145, May 1947. Translation of *Raketentriebwerke auf der Salpetersäurebasis und ihre spezifischen Antriebsgewichte,* in *Schriften der Deutschen Akademie der Luftfahrtforschung,* Heft 1071, no. 82, 1943.

PUBLICATIONS OF GOVERNMENT AGENCIES

The following reports cannot be purchased, but they were distributed to many libraries where they are available; some can also be obtained from the Library of Congress as photostats or on microfilm.

The Aerodynamic Development of the A.A. Rocket "Wasserfall." Translation of the original of Dr. Kurzweg. Headquarters Air Materiel Command, Wright Field, Dayton, Ohio. Release date August 1946, No. F-TS-615-RE; 35 p.

Control of the Surface-to-Air Missile "Rheintochter R. III." Headquarters Air Materiel Command, Wright Field, Dayton, Ohio. Release date July 1948, ATI No. 1957; 2 p.

German Developments in the Field of Guided Missiles. Headquarters Air Materiel Command, Wright Field, Dayton, Ohio. Summary Report by Col. D. L. Putt. No. F-SU-1122-ND. Release date July 12, 1946; 112 p.

High Altitude Research Using the V-2 Rocket. Johns Hopkins University Applied Physics Laboratory. July 1948. 90 p., 75 illus. By L. W. Fraser and E. H. Siegler. Covers the period March 1946–April 1947. Known as "Bumblebee Report No. 81."

A New Propellant Combination for Rockets, Using a Solid Oxygen Carrier and Ammonia as Fuels. By G. Damkoehler and W. Eggersgluess. Headquarters Air Materiel Command, Wright Field, Dayton, Ohio. Release date February 1947. No. F-TS-1041-RE. 49 p. Translation.

A Summary of German Missile Control Methods. Summary Report by Capt. Landon P. Hill. Headquarters Air Materiel Command, Wright Field, Dayton, Ohio. Release date June 7, 1946, No. F-SU-1115-ND. 50 p.

U.S. Rocket Ordnance; Development and Use in World War II. 57 p. Released March 31, 1946, by the Office of Scientific Research and Development (OSRD), War Department and Navy Department. Authoritative account of the development, characteristics, and tactical employment of American rocket ordnance.

Upper Atmosphere Research Report No. I. Naval Research Laboratory Report R-2955. Oct. 1, 1946. 200 p., of which 92 are full-page illus. "The report consists of a series of articles by various authors on research connected with the first cycle of V-2 firings." Covers firings Nos. 1-11.

Upper Atmosphere Research Report No. II. Naval Research Laboratory Report R-3030. Dec. 30, 1946. 164 p., 72 illus. Covers second cycle of V-2 firings, Nos. 12–15.

Upper Atmosphere Research Report No. III. Naval Research Laboratory Report R-3120. April 1947. 56 p., 14 illus.

Upper Atmosphere Research Report No. IV. Naval Research Laboratory Report R-3171. Oct. 1, 1947. 170 p., 130 illus. Covers third and fourth cycles of V-2 firings, Nos. 16–24.

Chapters 10 and 11

Bester, Alfred. *The Life and Death of a Satellite.* Boston: Little, Brown, 1966. 239 p., ill. with photographs. This is the story of a specific satellite, OSO-1 (for Orbiting Solar Observatory) from conception through test to orbiting.

Caidin, Martin. *Spaceport U.S.A. The Story of Cape Canaveral and the Air Force Missile Test Center.* New York: Dutton, 1959. 380 p., many illustrations.

Gantz, Kenneth F. (ed.). *The United States Air Force Report on the Ballistic Missile.* New York: Doubleday, 1958. Preface by General Thomas D. White, Chief of Staff, USAF, introduction by Major General Bernard A. Schriever. 338 p., photographs and diagrams. The twelve chapters and six appendices were written by sixteen authors, all Air Force officers.

Medaris, John B. *Countdown for Decision.* New York: Putnam, 1960. 303 p., photographs. General Medaris's story of the missile effort of the U.S. Army and the frustrations imposed upon it by Washington.

Pearson, Drew, and Anderson, Jack. *USA—Second-Class Power?* New York: Simon & Schuster, 1958. 334 p. An examination of the defense and scientific policies of the Eisenhower administration.

Shelton, William Roy. *Countdown: The Story of Cape Canaveral.* Boston: Little, Brown, 1960. Foreword by Maj. Gen. Donald N. Yates. 185 p., illustrated.

Chapter 12

Baar, James, and Howard, William. *Polaris!* New York: Harcourt, Brace, 1960. 245 p., photographs. The history of the trials and tribulations of the Polaris project.

Chapman, John L. *Atlas: The Story of a Missile.* New York: Harper, 1960. 190 p., photographs and diagrams.

Ley, Willy. *Ranger to the Moon.* New York: New American Library, 1965. Signet No. P2668. 127 p. with diagrams and 24 p. of photographs. Story of the flight of *Ranger-VII,* with background material.

———. *Mariner to Mars.* New York: New American Library, 1966. Signet No. P2932. 157 p. with diagrams and 24 p. of photographs. Story of the flight of *Mariner-IV* with background material.

Wheelock, Harold J. (ed.) *Mariner, Mission to Venus.* New York: McGraw-Hill, 1963. 118 p., many illustrations. This is a report of the staff of the Jet Propulsion Laboratory for NASA, concerning the successful flight of *Mariner-II* to Venus. Foreword by W. H. Pickering, the director of the laboratory.

Chapters 13 and 14

Alexander, Tom. *Project Apollo.* New York, Harper & Row, 1964. 234 p., no illustrations. A popular explanation of Project Apollo and why it took the shape it did.

Apollo Project. An over-all description was published by *The New York Times* in a series of long articles, the first running in the issue of July 30, 1962, the last in the issue of August 4, 1962. These articles, written by several staff members, were also published as a pamphlet.

Benson, General Otis O., Jr., and Strughold, Hubertus (eds.). *Physics and Medicine of the Atmosphere and Space.* New York: Wiley, 1960. 645 p., photographs and diagrams. The proceedings of an international symposium, sponsored by the School of Aviation Medicine, Brooks AFB, Texas.

Caidin, Martin. *The Astronauts.* New York: Dutton, 1960. 192 p., many illustrations. Written before Project Mercury had progressed to actual flights, telling the story of the preliminary research.

Carpenter, M. Scott, *et al. We Seven.* New York: Simon & Schuster, 1962. 352 p., many photographs. A series of essays, mostly personal, by the original seven astronauts.

Gantz, Kenneth, F. (ed.). *Man in Space: The United States Air Force Program for Developing the Spacecraft Crew.* New York: Duell, Sloan & Pearce, 1959. Preface by General Thomas D. White. 300 p., photographs and diagrams. By fifteen Air Force officers and scientists.

Gerathewohl, Siegfried J. *Principles of Bioastronautics.* Englewood Cliffs, N. J.: Prentice-Hall, 1963. 557 p., many illustrations. A comprehensive book on space medicine, including both theoretical studies and practical experiments up to the time the book was published.

Hanrahan, James S., and Bushnell, David. *Space Biology; The Human Factors in Space Flight.* New York: Basic Books, 1960. 263 p. plus 12 plates. Both authors are associated with the Air Force Missile Development Center. A carefully documented work with a good bibliography.

Simons, David G., and Schanche, Don A. *Man High.* Garden City, N.Y.: Doubleday, 1960. 262 p., no illustrations, no index. The story of the Air Force's high-altitude balloon flights with pure oxygen atmosphere.

Swenson, Loyd S., *et al. This New Ocean, A History of Project Mercury.* Washington, D.C.: NASA SP-4201, 1966. 681 p. with many photographs.

Appendices

Bolster, Rear Admiral Calvin M. *The Assisted Take-off of Aircraft.* Northfield, Vt.: Norwich University (James Jackson Cabot Fund Publication No. 9). Originally a lecture, covers the whole field of assisted take-off, including catapult take-off, but is chiefly devoted to JATO. Naturally the author is especially interested in naval applications, but others are not neglected.

Bridgeman, William, and Hazard, Jacqueline. *The Lonely Sky.* New York: Henry Holt, 1955. 316 p., 8 plates. The story of test-pilot Bridgeman and the Skyrocket research plane.

Eastman, Jacobs, and Shoemaker. *Tests on Thrust Augmentors for Jet Propulsion.* NACA Tech. Note no. 431. September 1932.

Gagarin, Yuri A. *Road to the Stars.* Moscow: Foreign Languages Publishing House, 1961. 197 p. with many illustrations. This is the English-language version of a 344-page book published soon after Gagarin's flight by the Pravda Publishing House. Most of it was written by two staff members of *Pravda,* N. Denisov and S. Borzenko.

Japanese Fuels and Lubricants. Research on Rocket Fuels of the Hydrogen Peroxide–Hydrazine Type. London: His Majesty's Stationery Office, 1946. 112 p. British Intelligence Objective Sub-Committee; Originating Agency: U.S. Naval Technical Mission to Japan. S.O. Code No. 51-8278-26. February 1946.

Krieger, F. J. *Behind the Sputniks.* Washington, D.C.: Public Affairs Press, 1958. 380 p., diagrams. This is the printed version of the RAND Research Memoranda RM-1760 (*Case Book on Soviet Astronautics,* mentioned on p. 536) and RM-1922. It consists of translations of writings by many Russian scientists and concludes with a detailed bibliography of Russian works.

Kurnosova, L. V. *Artificial Earth Satellites.* New York: Plenum Press; London: Chapman & Hall, 1960. Two vols. in one, 120 and 107 p., photographs and diagrams. English edition of the first two volumes of the Russian journal *Isskustvenniye sputniki zemli.*

Natter Interceptor Project. London: His Majesty's Stationery Office, 1945. 14 p. Combined Intelligence Objective Sub-Committee. Item no. 5. File no. xxx–107. July 1945.

Reitsch, Hanna. *Fliegen Mein Leben.* Stuttgart: Deutsche Verlags-Anstalt, 1951. 311 p., numerous photographs. Hanna Reitsch is best known as the German pilot who flew into and out of Berlin during the last week of the Second World

War, but before that she flew the rocket-propelled Messerschmitt Me-163 and the piloted version of the V-1 flying bomb. The American edition has the title *Flying Is My Life,* New York: Putnam, 1954, and was translated by Lawrence Wilson. 246 p., 13 plates.

Samson, D. R. (ed.) *Development of the Blue Streak Satellite Launcher.* New York: Macmillan, 1963. 128 p., diagrams and photographs. Subtitled: "Proceedings of the Second Space Engineering Symposium Held at Hatfield College of Technology, Hertfordshire, February 22, 1963"; contains six papers on various aspects of the Blue Streak rocket.

Sänger, Eugen. *Raketenflugtechnik.* Munich: R. Oldenbourg, 1933. 222 p.

———. *Neuere Ergebnisse der Raketenflugtechnik.* 22 p. This is the December 1934 "special issue" of the magazine *Flug,* published in Vienna. Dr. Sänger's article comprises the entire issue.

———. *Recent Results of Rocket Flight Technique.* NACA Tech. Mem. no. 1012, April 1942. Translation of the special issue of the magazine *Flug.*

Sänger, Eugen, and Bredt, Irene. *A Rocket Drive for Long Range Bombers.* Santa Barbara, Calif.: Dr. Robert Cornog, 1952. 175 p., 110 illus. Translated by M. Hamermesh. The famous Sänger-Bredt report on the round-the-world rocket bomber. This translation contains a number of photographs from Dr. Sänger's private archives, which were not part of the original report.

———. *Raumfahrt; Heute, Morgen, Uebermorgen.* Düsseldorf: Econ Verlag, 1963. 424 p., diagrams. This is Dr. Sänger's last work, intended to give a comprehensive picture of space travel activities. Much of the book is devoted to a description and discussion of American work, but it does not contain much that is new for an American reader.

Shternfeld, Ari. *Soviet Space Science.* New York: Basic Books, 1959, 361 p., photographs and diagrams. Foreword and epilogue by Willy Ley. Translation by Technical Documents Liaison Office, Wright Patterson AFB. Shternfeld was one of the recipients of the REP-Hirsch Prize while he lived in Paris. He later returned to Russia, where a Russian version of his prize-winning French manuscript was printed (in 1937). The current careful and valuable volume may be considered an up-dated version of the first work.

Soviet Writings on Earth Satellites and Space Travel. New York: The Citadel Press, 1958. 253 p., diagrams. Part I of the book is by Ari Shternfeld; Part II consists of translations of articles by Russian scientists, from Russian journals and newpapers.

Titov, Gherman, and Caidin, Martin. *I Am Eagle!* New York: Bobbs-Merrill, 1962. 212 p., photographs. Based in the main on interviews with Titov by Wilfred Burchett and Anthony Purdy, two British newspapermen.

Tregaskis, Richard. *X–15 Diary.* New York: Dutton, 1961. 317 p. plus 16 p. of photographs. The story of the X–15 rocket-propelled aircraft from its beginnings to August 1960.

A Few Recent General Books

von Braun, Wernher, and Ordway, Frederick I. *History of Rockets and Space Travel.* New York: Thomas Y. Crowell, 1966. 242 p., with many illustrations, some in color. A comprehensive work containing many photographs and other illustrations never before published in a popular book.

Clarke, Arthur C. (ed.) *The Coming of the Space Age.* New York: Meredith Press, 1967. 301 p., no illustrations. The book was planned by its author, in his phrase, as "an archival project"; all the selections had been printed before but in places difficult of access. Good casual reading.

Glasstone, Samuel. *Sourcebook on the Space Sciences.* Princeton, N.J.: D. Van Nostrand, 1965. 937 p., numerous illustrations. This book was written under the sponsorship of NASA with the assistance of many NASA scientists. It is a comprehensive survey of all the problems involved.

Stemmer, Josef. *Die Entwicklung des Raketenantriebes in allgemeinverständlicher Darstellung.* Zurich, 1944. 3 vols., *Hofmann Bibliothek* nos. 106, 107, and 108. Small pocketbook size; total of 368 p.

———. *Raketenantriebe.* Zurich: Schweizer Druck- und Verlagshaus A.-G., 1952. 523 p., 206 illus., 9 plates, some in color. A comprehensive work with contributions by Heinz Gartmann and Dr. Eugen Sänger. Extensive bibliography but no index.

Sutton, George P. *Rocket Propulsion Elements.* New York: Wiley, 1956. 483 p., photographs, many diagrams. Subtitled "An Introduction to the Engineering of Rockets," the book is precisely what the subtitle says. Especially valuable to engineering students. The first and much shorter edition of 1949 (294 p.) was translated into Russian.

Side Issues and Curiosities

von Baumgarten-Crusius, Artur. *Die Rakete als Weltfriedenstaube.* Leipzig, 1931. 174 p. This book by a retired German general is devoted to the proposition that long-range war rockets would be the decisive weapon upon first application and render wars impossible afterward. The author claimed that an otherwise unknown engineer Belz had completed such a weapon; he even published a photograph which was recognizable at first glance as a wood and canvas model patterned after the spaceship in *Frau im Mond.* About a year prior to this publication, General von Baumgarten-Crusius had written a novel, *Vengeance 193?*, anonymously published, around the same theme.

Crassus, Ing. *Der gefahrlose Menschenflug.* Hamburg, 1912. 37 p. The author's real name was Wilhelm Gaedicke; the book deals with the project of a helicopter with jet-driven rotor. Of historical interest only.

Piccard, Auguste. *Between Earth and Sky.* Translated from the French by Claude Apcher. London: The Falcon Press, 1950. 157 p., 60 plates. The author is the man who first reached the stratosphere by balloon, on May 27, 1931. The book is mostly an account of stratosphere exploration with balloons; the last few chapters are devoted to space travel. They are interesting in their own special way, for they read as if Professor Piccard based them on his knowledge as a physicist, without any acquaintance with rocket literature.

Pseudoman, Akkas. *Zero to Eighty.* Princeton, N.J.: Scientific Publishing Co., 1937. 283 p. The book has the subtitle: "Being my lifetime doings, reflections and inventions, also my journey around the moon." Akkas Pseudoman is the pseudonym of Dr. E. F. Northrup, inventor of the Ajax-Northrup high-frequency induction furnace, who died May 1, 1940. (For obituary see *The New York Times,* May 2, 1940.) The book, a semi-fictional autobiography, contains a great deal about the author's experiments with solenoid guns, including photographs, and a résumé of the theory of space travel which Dr. Northrup wanted to accomplish by means of rockets initially accelerated in large solenoid guns.

Index

Figures followed by an n refer to footnotes. Names occurring only in the Bibliography have not been indexed.